An Introduction to
CYTOLOGY

LESTER W. SHARP

Cornell University

MAVEN BOOKS

Chennai   Trichy   Tirunelveli   New Delhi

MAVEN BOOKS

An Imprint of **MJP Publishers**

ISBN 978-93-87867-78-9 **MAVEN Books**

No. 44, Nallathambi Street,
Triplicane,
Chennai 600 005

All rights reserved
Printed and bound in India

MJP   751   © Publishers, 2019

Publisher : **C. Janarthanan**

Publisher's Note

The legacy of a country is in its varied cultural heritage, historical literature, developments in the field of economy and science. The top nations in the world are competing in the field of science, economy and literature. This vast legacy has to be conserved and documented so that it can be bestowed to the future generation. The knowledge of this legacy is slowly getting perished in the present generation due to lack of documentation.

Keeping this in mind, the concern with retrospective acquiring of rare books has been accented recently by the burgeoning reprint industry. MAVEN Books is gratified to retrieve the rare collections with a view to bring back those books that were landmarks in their time.

In this effort, a series of rare books would be republished under the banner, "MAVEN Books". The books in the reprint series have been carefully selected for their contemporary usefulness as well as their historical importance within the intellectual. We reconstruct the book with slight enhancements made for better presentation, without affecting the contents of the original edition.

Most of the works selected for republishing covers a huge range of subjects, from history to anthropology. We believe this reprint edition will be a service to the numerous researchers and practitioners active in this fascinating field. We allow readers to experience the wonder of peering into a scholarly work of the highest order and seminal significance.

MAVEN Books

To

My Father

PREFACE

This book has been prepared for students of the biological sciences who desire a means of becoming more readily acquainted with the literature and problems of cytology. It does not pretend to be an exhaustive treatise for the use of experienced cytologists, though it is hoped that to them also some of its features may be of service.

For a number of years students of biology, especially those working along botanical lines, have been faced with the task of searching through a widely scattered literature for information on various cytological subjects. It is the purpose of this book not to render the consultation of that literature unnecessary, but only to make it easier; the student can scarcely be too strongly urged to derive his information from original sources wherever possible. The author does not presume to replace, but rather aims to supplement, Professor Wilson's well known book, *The Cell in Development and Inheritance*, which, though written twenty years ago and with the emphasis primarily on the zoological side, will remain invaluable to all workers for many years to come. The more recent works of Gurwitsch (*Morphologie und Biologie der Zelle*), Heidenhain (*Plasma und Zelle*), and Buchner (*Prakticum der Zellenlehre*) are of importance especially to the zoologist.

The living cell, or protoplast, which represents the organized protoplasmic unit of structure and function, obviously cannot receive complete description in structural terms. Until a comparatively recent period cytological researches dealt primarily with cell structure, including particulary the conspicuous changes undergone by this structure in connection with the reproduction of the cell (cell-division) and of the multicellular organism (maturation and fertilization). A gradual shifting of emphasis has since led to the opening of fruitful fields in other directions, and the important results already achieved have shown with increasing clearness the need for a closer acquaintance with the physiological aspects of cell activity, not only in metabolism and growth, but also in the reproductive phases of the life cycle. The present work, though dealing mainly with the structural aspects of the subject, may aid indirectly in fulfilling the above need by making the prerequisite data of cell morphology more readily available.

Throughout the book, which in many of its chapters treats chiefly of the plant cell, attention is focussed upon the protoplast: the cell wall is given only brief consideration, since it plays a relatively minor rôle in the processes of particular interest to the cytologist at the present time.

Because of their fundamental importance in connection with the problems confronting the geneticist, the phenomena of nuclear division, chromosome reduction, and fertilization are described with considerable fullness, and their relation to the problems of heredity is taken up in five special chapters. With regard to many of the subjects treated, it has not been found possible to formulate final conclusions, since in many cases nothing more than tentative general statements are warranted by the facts in our possession. In some chapters little more than catalogs of conflicting opinions can be given, but in such a form the state of certain questions is not inaccurately represented. The student entering upon the field of cytology will be impressed by the large number of special points which remain undetermined and general questions which await adequate answers. If he can look upon cytology as a developing science, and if he has reached the stage at which he no longer demands categorical answers to all his questions, this book will be of interest to him as much for the problems it raises as for those it helps to solve. Not the least of its functions is to indicate lines of research along which he can hope to make contributions to the subject.

In compiling his materials the author has not hesitated to draw very freely upon the writings of others. In many cases where direct quotation is not made, the language of the originals has been closely followed in order to lessen the liklihood of misrepresentation. His great debt to Professor Wilson's book will be apparent to all those familiar with that admirable work. The majority of the diagrams and a number of the other figures are new. Most of the latter, however, have been redrawn from works cited in the text, not only that the value of the book may be enhanced by the presence of authoritative illustrations, but also that the student may be encouraged to become more familiar with the original papers. The general systematic positions of organisms indicated in the text by their scientific names only may be ascertained by referring to the generic names in the index.

The illustrations are largely the work of Miss Mildred Stratton, in whose skill and spirit of cooperation the author has had invaluable assistance. The criticisms of the text kindly given by Professor C. J. Chamberlain of the University of Chicago and Professor R. A. Emerson of Cornell University have been very highly appreciated. Acknowledgements are also made to the author's other colleagues for their advice and continued encouragement. Further criticisms looking toward the improvement of future editions will be welcomed.

L. W. S.

Ithaca, New York,
September 8, 1920.

Note

Less than two months after the completion of the text of this book the author has received copies of the two new English works on cytology: W. E. Agar's *Cytology, With Special Reference to the Metazoan Nucleus* and L. Doncaster's *An Introduction to the Study of Cytology*. Both deal almost exclusively with animal cytology, the first being valuable for its account of chromosome behavior in animals, and the second for its discussions of gametogenesis, fertilization, parthenogenesis, and sex-determination. These works, together with the botanical portions of the present volume, should make an acquaintance with the general field of cytology much more readily attainable.

CHAPTER VIII

CHAPTER IX

CHAPTER X

CHAPTER XI

CHAPTER XII

CHAPTER XIII

CHAPTER XIV

CHAPTER XV

Mendelism—A typical case of Mendelian inheritance—The cytological basis of Mendelism—Mutation—Mutations accompanied by changes in chromosome number—Bearing on the origin of species and varieties— Mutations accompanied by no change in chromosome number—Conclusion.

CHAPTER XVI

Experimental evidence for sex-determination—Sex-chromosomes—Sex-chromosomes and Mendelism—Experimental alteration of the sex ratio Metabolic theories of sex—General discussion.

CHAPTER XVII

A typical case of linkage—Sex-linkage—Non-disjunction—Linkage groups— The chiasmatype theory—Application of the chiasmatype theory to the problem of linkage—General discussion—Other theories of linkage—Value of chromosome theory of heredity.

CHAPTER XVIII

Darwin's hypothesis of pangenesis—DeVries's theory of intracellular pangenesis—Nägeli's idioplasm theory—Weismann's theory—Some modern aspects of Weismannism—Non-factorial theories—A chemical theory of heredity—Conclusion—*Bibliography* 14 (for Chapters XIV-XVIII).

INTRODUCTION TO CYTOLOGY

CHAPTER I

HISTORICAL SKETCH

The history of cytology falls naturally into three periods, of which the first begins with the discovery of the cell by Robert Hooke in 1665, the second with the foundation of the Cell Theory by Schleiden and Schwann in 1838–9, and the third with the important researches of Strasburger, Hertwig, Bütschli, and others between 1870 and 1880. In the present sketch attention will be confined almost entirely to the first two periods, the work of the third, or modern, period being dealt with in the other chapters of the book.

Prior to the seventeenth century attempts to analyse the structure of organisms were necessarily unsatisfactory. Aristotle (384–322 B.C.) in his *De Partibus Animalium* distinguished the "homogeneous parts" and the "heterogeneous parts," the former corresponding in general to what we classify as tissues (bone, fat, cartilage, flesh, blood, lymph, nerve, membrane, nails, hair, skin, vessels, tendon, etc.), and the latter being the larger members of the body (head, face, hands, feet, trunk, etc.). Theophrastus, the pupil and successor of Aristotle, taught in his *Historia Plantarum* that the plant body is composed of "sap," "veins," and "flesh." Aristotle's classification was developed further by Galen (131–201 A.D.) and by his followers. Although we no longer regard the above components as elementary parts, but rather as tissues and organs, the ancients may be pardoned for not carrying the analysis further, for they did not possess the necessary instruments. Something was then known about the refraction of light, but it was not until many centuries later that suitable lenses were available. The first compound microscope was brought out in 1590 by J. and Z. Janssen, spectacle makers of Middleburg, Holland; and during the first part of the seventeenth century other improved models were designed by other workers. These instruments in the hands of men possessed of scientific curiosity soon led to many significant discoveries. A new world was opened to the eye of science, and the compound microscope has since remained an instrument of extraordinary value in biological research.

The Discovery of the Cell.—Cytology may be said to have begun with the discovery of the cell by Robert Hooke (1635–1703) in 1665. Hooke, who lived in London and has been described as a man of eccentric appearance and habits, showed a remarkably varied scientific activity. For a time he was a professor of geometry, and later became an architect. He performed many original experiments in mechanics and for a number of years was curator of experiments to the Royal Society. His interest in optics led him to examine all sorts of objects with the compound microscope. In charcoal and later in cork and other plant tissues he found small honeycomb-like cavities which he called "cells." He had no distinct notion of the cell contents, but spoke of a "nourishing juice," which he inferred must pass through pores from one cell to another. His many observations were embodied in his *Micrographia* (1665), a large work illustrated with 83 plates. The chapter containing his remarks on cells is entitled "Of the schematisme or texture of cork and the cells and pores of some other such frothy bodies." Quaint and crude as it now appears to us, the *Micrographia* takes its place as the earliest cytological classic.

Three other names even more prominent in the early history of microscopy are those of Malpighi, Grew, and Leeuwenhoek. Marcello Malpighi (1628–1694), an Italian physiologist and professor of medicine at Bologna, Pisa and Messina, is best known for his important pioneer work in anatomy and embryology. Most of his observations on plants were included in his *Anatome Plantarum* (1675) and had to do largely with the various kinds of elements making up the body of the vascular plant. Malpighi made a distinct step in advance in studying tissues with the cell as a unit; a clear fore-shadowing of the Cell Theory is seen in his remarks concerning the importance of the "utriculi" in the structure of the body. At Pisa Malpighi was associated with G. A. Borelli, who was one of the first to use the microscope on the tissues of higher animals.

Nehemiah Grew (1641–1712) was an English physician and botanist. He began a careful study of plant structure in 1664, and in 1670 read his first important paper before the Royal Society. Further contributions followed at intervals until 1682, when all of them were published under the title *The Anatomy of Plants*. Like Malpighi, an abstract of whose first work on plants was presented to the Royal Society in 1671, Grew was interested in tissues, and gave particular attention to the combinations of these tissues in different plant organs. He was strongly impressed by the manner in which the cells, which he also called "vesicles" and "bladders," appeared to make up the bulk of certain tissues: ". . . the parenchyma of the Barque," he said, "is much the same thing, as to its conformation, which the froth of beer or eggs is, as a fluid, or a piece of fine Manchet, as a fixed body" (p. 64). He further believed the walls of the cells to be composed of numerous extremely fine fibrils: in the vessels or longitudinal elements these fibrils were wound in the form of a close spiral, while the vessels themselves were bound together by a transverse series of interwoven threads. He accordingly compared the structure of the plant with that of a basket, and with "fine bone-lace, when the women are working it upon the cushion" (p. 121).

Antony van Leeuwenhoek (1632–1723) of Delft is remembered for his pioneer researches in the field of microscopy. He constructed a number of simple lenses of high power, and with these he was able to see for the first time certain protozoa, bacteria, and other minute forms of life. In the course of his investigations he observed the cells ("globules") in the tissues of higher organisms. His work, in spite of the fact that it was carried on without any definite plan, brought to light a number of important facts, but in general his accomplishments do not bear favorable comparison with those of Grew and Malpighi.

Preformation and Epigenesis.—After the death of Leeuwenhoek there ensued a period during which the actual investigation of the cell and the structure of organisms remained practically at a standstill. At that time, however, certain speculations were indulged in which should be recorded here, not because they can be regarded as scientific cytology but because of the influence they exerted upon the formulation of many cytological problems in later years. These speculations resulted in the division of the biologists of the day into two schools, the main question at issue being the manner in which the embryo develops from the egg. . The two theories formulated in answer to this question have been called the Preformation Theory and the Theory of Epigenesis.

According to the Preformation Theory, the basis for which was laid in the seventeenth century works of Swammerdam, Malpighi, and Leeuwenhoek, the egg contains a fully formed miniature individual, which simply unfolds and enlarges as development proceeds. Because of this unfolding the theory was also known as the Theory of Evolution, a phrase which has a quite different connotation today. In the eighteenth century the preformation idea was carried to an absurd extreme by Bonnet (1720–1793) and others, who argued that if the egg contains the complete new individual, the latter must in turn contain the eggs and individuals of all future generations successively encased within it, like an infinite series of boxes one within another. This theory of encasement (*emboîtement*) was a logical deduction from the since abandoned premise that everything, including organisms for all time, had been formed by one original creation, and that nothing could therefore be formed anew. The preformationists soon became separated into two groups: the spermists or animalculists, and the ovists. By the former the new individual was supposed to be encased in the spermatozoön, and figures were actually published showing a small human figure, or "homunculus," within the sperm head. The ovists, on the contrary, held that the individual is encased in the egg.   A bitter strife was carried on over this question by the two groups of preformationists, and various interesting compromises were made.   But all extreme forms of preformationism were to disappear in the light of more critical investigations, which went far to support the opposing Theory of Epigenesis.

Two of the early champions of the Theory of Epigenesis were William Harvey (1578–1667; *Exercitationes de Generatione Animalium*, 1651), and Caspar Friedrich Wolff (1733–1794; *Theoria Generationis*, 1759). As the result of many careful observations on the embryogeny of the chick Wolff was able to show beyond question that development is epigenetic: neither egg nor spermatozoön contains a formed embryo; development consists not in a process of unfolding, but in "the continual formation of new parts previously non-existent as such" (Wilson). Here there was room for the principle of true generation, or "the production of heterogeneity out of homogeneity."   The *Theoria Generationis* is to be regarded as one of the really great contributions to biological science, for the Theory of Epigenesis, to which it furnished substantial support, later became established with modifications as a fundamental principle of embryology, particularly through the work of von Baer in the nineteenth century.

In commenting on preformation and epigenesis Whitman (1894) emphasizes the fact that the tendency of modern biology has not been to show the entire falsity of either or both of these views, but to seek out the germs of truth possessed by each, and to relate them to modern biological conceptions.   "The two views missed the mark by over-shots in contrary directions," says Whitman.   The one theory claimed too much preformation: everything was preformed at the start.   The other theory claimed too much postformation: everything was formed anew.   Our present position, although it excludes both views in their crude original form, involves in a new sense both conceptions.   When we say that the egg is organized, possessing an architecture or mechanism in its cytoplasm or nucleus which largely predetermines development, we are making a modernized statement of the preformation idea.   When we say that the parts of the individual are in no way delineated in the egg, but are mainly determined by external conditions during the course of development, we are speaking in terms of modern epigenesis.   "The question is no longer whether all is preformation or all is postformation; it is rather this: *How far is post-formation to be explained as the result of pre-formation, and how far as the result of external influences?*"   When it is borne in mind, therefore, that one of the outstanding problems of modern cytology is that of identifying the factors involved in the development of an organized and highly differentiated individual from an organized but relatively undifferentiated egg cell, it is at once evident that our sketch of cytological history would be incomplete without the above reference to the early Theories of Preformation and Epigenesis.

Early Theories of Cell-formation.—The researches of Hooke, Malpighi, and Grew in the seventeenth century had shown that "cells," or "globules," are important structural elements in organisms. When attention was again directed to such matters in the eighteenth century there was very soon felt a need for a theory which would account for the origin of cells. We may briefly review some of the suggestions which were offered.

One of the earliest theories of cell-formation was that put forward by Wolff in the *Theoria Generationis.* According to Wolff, every organ is at first a clear, viscous fluid with no definite structural organization. In this fluid cavities (Bläschen; Zellen) arise and become cells, or, by elongation, vessels. These may later be thickened by deposits from the "solidescible" nutritive fluid. The cavities, or cells, are not to be regarded as independent entities; organization is not effected by them, but they are rather the passive results of an organizing force (*vis essentialis*) inherent in the living mass. Three important points in Wolff's theory should be noted because of the relation they bear to subsequent conceptions of the rôle of cells: the spontaneous origin of the cell, the organization of parts by differentiation in a homogeneous living mass, and the passive rôle of the cell in this organizing process. This theory was adopted in 1801 by C. F. Mirbel (1776–1854), who further believed that the cells communicate through pores in their walls.

K. Sprengel (1766–1833) stated that cells originate in the contents of other cells as granules or vesicles which absorb water and enlarge. Sprengel's observations seem to have been very poorly made, for he evidently mistook starch grains for the "vesicles" which were supposed to grow into new cells. But Sprengel's theory was upheld by L. C. Treviranus (1779–1864) in a work appearing in 1806, and both men fought many years for its support. Kieser (1812) further developed the theory that granules in the latex are "cell germs" which later hatch in the intercellular spaces to form new cells.

With a much clearer understanding of the nature of the problems involved a number of excellent observations were made by J. J. Bernhardi in 1805, by H. F. Link and K. A. Rudolphi in 1807, and by J. J. P. Moldenhawer in 1812. It is to be regretted that the deserved attention was not given to their views, for they promised to lead in the right direction.

A number of years later Mirbel, in a work on *Marchantia* (1831–1833), distinguished three modes of cell-formation: (1) the formation of cells on the surface of other cells, (2) the formation of cells within older cells, and (3) the formation of cells between older cells. The first mode apparently represented the budding of the germ tube arising from the spore, while the second and third modes were formulated as the result of a misinterpretation of the process of cell-multiplication in growing gemmæ.

Hugo von Mohl (1805–1872), in spite of his many valuable observations on the growth of algæ, in 1835 agreed essentially with Mirbel. He made a step in advance, however, when he described carefully for the first time the division of a cell. We shall see further on that von Mohl's later researches contributed largely to the upbuilding of an adequate theory of the cell.

F. J. F. Meyen (1804–1840) held that there are three fundamental forms of elementary organs: cells, spiral tubes, and sap vessels. He noted the wide occurrence of cell-division but did not describe the process in detail. Meyen apparently made the first attempt to distinguish cell-division from the free cell-formation described by previous workers. It has been pointed out by Sachs that if this short step had been clearly taken earlier the peculiar theory of cell-formation later developed by Schleiden would have been impossible. Von Mohl also had made observations ruling out Schleiden's idea, but his excessive caution prevented him from making a decisive statement on the subject. H. J. Dutrochet (1776–1847) in 1837 described the body as being composed of solids and fluids, the former being aggregations of cells of a certain degree of firmness, and the latter, such as blood, being made up of cells freely floating. He believed that although the cell contents may be more or less solid, the highest degree of vitality is compatible only with the liquid condition. He further recognized muscle fibers as elongated cells.

To all the above workers the important elementary unit was the "globule." It was customary to refer to this conception as the Globular Theory, in contradistinction to the curious and fanciful Fiber Theory put forth by Haller (1708–1777) many years before (1757), according to which the organism is made up of slender fibers cemented together by "organized concrete." For some the term "globule" stood for the granules seen in the cell contents, whereas for others it meant the cell itself. As observations multiplied and ideas became more definite the Cell Theory of Schleiden and Schwann was more and more distinctly foreshadowed. Before turning to the Cell Theory, however, we must notice briefly a few observations which had been made on the cell contents.

Early Observations on the Cell Contents.—Although the true nature and significance of the contents of cells were not recognized until many years later, a number of early investigators had seen protoplasm and had been impressed by certain of its activities. As early as 1772 Corti, and a few years later Fontana (1781) saw the rotation of the "sap" in the Characeæ and other plants. After being long forgotten these facts were rediscovered by L. C. Treviranus (1811) and G. B. Amici (1819), whereupon Horkel, an uncle of Schleiden, called attention to the earlier work of Corti. Protoplasmic circulation of the more complex type was discovered in the stamen hairs of *Tradescantia* by Robert Brown in 1831, and other workers, especially Meyen, soon added other cases.

During the first third of the nineteenth century no name is of greater interest to cytologists than that of Robert Brown (1773–1858). Although he is famous chiefly for his great taxonomic monographs and his morphological work, he is known in cytology as the man who is usually given the credit for the discovery of the nucleus, which he announced in 1831. Although it was Brown who was impressed by the probable importance of the nucleus, and who concluded in 1833 that it is a normal cell element, certain other observers, notably Fontana, who described a nucleus in 1781, and Meyen, who saw it in *Spirogyra* in 1826, should share the honor for its discovery. The phenomenon which has since been known as "Brownian movement" was seen by Brown in 1827.

The first period in the development of our subject is seen to have been one in which there was a tendency to indulge in speculation to an extent quite unwarranted by the facts at hand. As we have already pointed out, however, this speculation was of considerable importance to us, in that it had to do with questions which later became central problems of cytology. Carefully made observations were meanwhile increasing in number and varieyt, and the time eventually became ripe for the formulation of a theory which would correlate these data and give a definite trend to cytological investigations. Such a theory was soon forthcoming.

The Foundation of the Cell Theory.—The year 1838 marks an epoch in the history of biology. In this and the following year Schleiden and Schwann founded the Cell Theory, which, in view of its enormous influence upon all branches of biological science, may be regarded as second in importance only to the Theory of Evolution. We have seen that cells had been observed by various workers during many years, and had been recognized as being constantly present in the bodies of living organisms, but it remained for Schleiden and especially Schwann to formulate a comprehensive theory embracing the known facts and affording a starting point for further researches.

The Cell Theory stated primarily that *the body is composed entirely of cells and their products, the cell being the unit of structure and function and the primary agent of organization.* Subsidiary to this was Schleiden's theory of cell-formation, which should not be confused with the main thesis just stated.

Matthias Jakob Schleiden (1804–1881) is one of the most prominent and interesting characters in botanical history. He studied law at Heidelberg, medicine at Göttingen, and botany at Berlin, where he met Schwann and Robert Brown. The association of these men undoubtedly meant much to the future of botany and zoölogy. Eventually Schleiden became Professor of Botany at Jena, where he remained for 23 years. Schleiden was famous not merely because of his own work, but chiefly as the result of the tremendous impetus which he gave to investigation.

He sought to place botany on a scientific footing equal to that of physics and chemistry, and insisted upon accurate observation and developmental studies as the basis of morphology. Sachs says: "Endowed with somewhat too great love of combat, and armed with a pen regardless of the wounds it inflicted, ready to strike at any moment, and very prone to exaggeration, Schleiden was just the man needed in the state in which botany then was."

Theodor Schwann (1810–1882) was associated as a student with Johannes Müller, the great physiologist, first at Würzburg and later at Berlin. It was in the latter place that he put forth his statement of the Cell Theory. Immediately afterward he went to Louvain, where he was a professor for nine years, and later transferred to Liège. In disposition he contrasted strongly with Schleiden, being described as "gentle and pacific."

It is said that Schleiden, while dining with Schwann, discussed with him some of his ideas regarding cells in plants, which he had been studying in his laboratory. Schwann had been making similar observations on animals, and after the meal the two went to Schwann's laboratory, where they came to the conclusion that cells are fundamentally alike in both kingdoms. Schleiden's treatise on the subject, *Beiträge zur Phytogenesis*, appeared in 1838 and dealt mainly with the origin of cells. Robert Brown had recently discovered the nucleus, and about it Schleiden built up his theory of "free cell-formation," which was essentially as follows: In the general cell contents or mother liquor ("cytoblastema") there are formed, by a process of condensation, certain small granules (later called "nucleoli" by Schwann). Around these many other granules accumulate, thus forming nuclei ("cytoblasts"). Then, "as soon as the cytoblasts have attained their full size, a delicate transparent vesicle appears upon their surface." This vesicle in each case enlarges and forms a new cell, and, since it arises upon the surface of the cytoblast (nucleus), "the cytoblast can never lie free in the interior of the cell, but is always enclosed [*i.e.*, imbedded] in the cell wall . . ." Schleiden thus regarded new cell-formation as endogenous ("cells within cells") rather than the result of cell-division. With respect to the main proposition of the Cell Theory he says in the opening paragraphs: ". . . every plant developed in any higher degree, is an aggregate of fully individualized, independent, separate beings, even the cells themselves. Each cell leads a double life: an independent one, pertaining to its own development alone; and another incidental, in so far as it has become an integral part of a plant. It is, however, easy to perceive that the vital process of the individual cells must form the first, absolutely indispensable fundamental basis, both as regards vegetable physiology and comparative physiology in general; . . ."

Schleiden shared the results of his observations, including his errors, with Schwann, who was the one to formulate the Cell Theory in a comprehensive manner. Schwann announced the theory in concise form in 1838, and in 1839 published a very full account under the title "*Mikroskopische Untersuchungen über die Uebereinstimmung in der Struktur und dem Wachsthum der Thiere und Pflanzen.*" He says: "The elementary parts of all tissues are formed of cells in an analogous, though very diversified manner, so that it may be asserted *that there is one universal principle of development for the elementary parts of organisms, however different, and that this principle is the formation of cells.*" And further: "The development of the proposition that there exists one general principle for the formation of all organic productions, and that this principle is the formation of cells, as well as the conclusions which may be drawn from this proposition, may be comprised under the term *Cell Theory* . . ." ". . . all organized bodies are composed of essentially similar parts, namely, of cells . . ."

Elaboration of the Cell Theory.—The Cell Theory at once became established as one of the main foundation stones of biological research, but it underwent considerable modification as investigations proceeded. The main thesis, that the body is composed of cells and their products, remained, but other ideas associated with this in the minds of Schleiden and Schwann, particularly that concerning free cell-formation, were superseded. Soon after the formulation of the Cell Theory its elaboration was begun by Unger, von Mohl, and Nägeli, who based their conclusions on observations of a very high order. Franz Unger (1800–1870), in two works appearing in 1844 on vegetable growing points and the growth of internodes, argued for the origin of cells by division. Von Mohl, in two treatises (1835, 1844), maintained that there are two methods of cell-formation: by division and by the formation of cells within cells. He thought the "primordial utricle" (protoplast) must be absorbed to make way for the two new ones, or, less probably, the old one must divide into two. Like Schleiden, he thought the nucleus must be incorporated in the cell wall, but later (1846) concluded that it lies in the primordial utricle. It was in his paper of 1846 that von Mohl introduced the term "protoplasm" in its present sense.

Carl von Näegli (1807–1891) in 1844 produced an exhaustive treatise on the nucleus, cell-formation, and growth. In algæ and the microsporocytes of angiosperms he clearly showed that cells multiply by division, and Schleiden was forced to admit that this might be "a second kind of cell-formation." The continuation of Näegli's researches in 1846 completely overthrew Schleiden's conception of free cell-formation, establishing the significant fact that all vegetative cell-formation is by cell-division. Many similar observations had been made by Unger and von Mohl, but Nägeli elaborated a broad theory which took into account all of the data at hand. He distinctly defined cell-division and free cell-formation, and showed that what had been taken for the latter was only a special case of the former. Nägeli's conclusions were supported by new evidence furnished by other investigators, who further demonstrated that not only vegetative cells but also those reproductive cells (in thallophytes) which Nägeli thought in some cases might be formed freely, originate by a modified process of cell-division. It was now clear that *cells arise only from preëxisting cells,* a conception which had been emphasized by Remak (1841) and which Virchow (1855) expressed in the dictum "*omnis cellula e cellula.*"

Opinions concerning the origin of the nucleus and its rôle in cell-division varied greatly among these workers, reliable observations being as yet insufficient to allow the formulation of any definite conclusion. In 1841 Henle believed with Schleiden and Schwann that the nucleus was formed by the aggregation of "elementary granules," and that it was not constantly present. Goodsir looked upon the nucleus as the reproductive organ of the cell. Von Kölliker in 1845 asserted that nuclear division precedes the division of the cell, and Remak, as a result of his observations on blood cells in the chick embryo, formulated a definite theory of cell-division (1841, 1858). He believed cell-division to be a "centrifugal" process: the nucleolus, nucleus, cytoplasm, and cell membrane were supposed to divide in turn by simple constriction. Just such a process, though evidently very exceptional, has been observed at a more recent date by Conklin (1903). In describing a case of nuclear division Wilhelm Hofmeister (1824–1877) stated that the membrane of the nucleus dissolved, the nuclear material then separating into two masses around which new membranes were formed (1848, 1849). It was generally believed, however, that the origin of nuclei by division was of rare occurrence, and that ordinarily the nucleus dissolved just before cell-division, two new ones forming *de novo* in the daughter cells. Von Mohl (1851), who in the main agreed with Hofmeister, wrote as follows: "The second mode of origin of a nucleus, by division of a nucleus already existing in the parent-cell, seems to be much rarer than the new production of them . . ." And again, ". . . it is possible that this process [nuclear division] prevails very widely, since . . . we know very little yet respecting the origin of nuclei. Näegli thinks that the process is quite similar to that in cell-division, the membrane of the nucleus forming a partition, and the two portions separating in the form of two distinct cells."

It was not until many years later, in connection with researches upon fertilization and embryogeny, that the behavior of the nucleus in cell-division became known in detail, and its probable significance pointed out. In 1879 Eduard Strasburger (1844–1912) announced definitely that *nuclei arise only from preëxisting nuclei.* W. Flemming was led to the same conclusion by his studies on animal cells, and expressed it in the dictum "*omnis nucleus e nucleo*" (1882). (See footnote, p. 143.)

The Protoplasm Doctrine.—The Cell Theory and all of its corollaries were placed in a new light with the development of a more adequate conception of the significance of protoplasm. To its discoverers the cell meant nothing more than the wall surrounding a cavity; they spoke only in the vaguest terms of the "juices" present in cellular structures. The founders of the Cell Theory held a position but little in advance of this; they observed the cell contents but regarded them as of relatively slight importance. Even those who had been impressed by the phenomenon of protoplasmic streaming were not aware of the significance of the substance before their eyes.

Felix Dujardin (1801–1860) in 1835 described the "sarcode" of the lower animals as a substance having the properties of life. Von Mohl had seen a similar substance in plant cells, and in 1846, as noted above, he called it "Schleim," or "Protoplasma," the latter term having been used shortly before by Purkinje in a somewhat different sense. Nägeli and A. Payen (1795–1871) in 1846 recognized the importance of protoplasm as the vehicle of the vital activity of the cell; and Alexander Braun (1805–1877) in 1850 pointed out that swarm spores, which are cells, consist of naked protoplasm. An important point was reached when Payen (1846) and Ferdinand Cohn (1850) concluded that the "sarcode" of the animal and the "protoplasm" of the plant are essentially similar substances. In the words of Cohn:

"The protoplasm of the botanist, and the contractile substance and sarcode of the zoölogist, must be, if not identical, yet in a high degree analogous substances. Hence, from this point of view, the difference between animals and plants consists in this; that, in the latter, the contractile substance, as a primordial utricle, is enclosed within an inert cellulose membrane, which permits it only to exhibit an internal motion, expressed by the phenomena of rotation and circulation, while, in the former, it is not so enclosed. The protoplasm in the form of the primordial utricle is, as it were, the animal element in the plant, but which is imprisoned, and only becomes free in the animal; or, to strip off the metaphor which obscures simple thought, the energy of organic vitality which is manifested in movement is especially exhibited by a nitrogenous contractile substance, which in plants is limited and fettered by an inert membrane. in animals not so."

Protoplasm was now studied more intensively than ever. H. A. de Bary (1831–1888), working on myxomycetes and other plant forms, and Max Schultze (1825–1874), investigating animal cells, demonstrated the correctness of Cohn's view. The work of Schultze was especially important in that it firmly established in 1861 the Protoplasm Doctrine, namely, that *the units of organization are masses of protoplasm*, and that *this substance is essentially similar in all living organisms*. The cell, according to Schultze, is "a mass of protoplasm containing a nucleus, both nucleus and protoplasm arising through the division of the corresponding elements of a preëxisting cell." The cell wall, upon which the early workers had focussed their attention, turned out to be of secondary importance. The cell was thus seen to be primarily the organized protoplasmic mass, to which Hanstein in 1880 applied the convenient term *protoplast.*

Extensive studies on the physical nature of protoplasm were soon undertaken by Kühne (1864), Cienkowski (1863), and de Bary (1859, 1864); and there later followed the well-known structural theories of Klein, Flemming, Altman, and Bütschli.   (See Chapter III.)

Von Mohl as early as 1837 held that the plastid is a protoplasmic body.   The classic researches of Nägeli (1858, 1863) on plastids and starch grains laid the foundation for our knowledge of these bodies, which was greatly extended in later years by Meyer (1881, 1883, etc.) and Schimper (1880, etc.).   (See Chapter VI.)

It would be difficult to overestimate the value, both practical and theoretical, of the Protoplasm Doctrine, for its establishment has not only led to knowledge by which the conditions of life have been materially improved, but has also been an important factor in assisting man to a modern, rational outlook on organic nature, in which he has learned to include himself. It is not too much to say that the identification of protoplasm as the material substratum of the life processes was one of the most significant events of the nineteenth century.   The doctrine was furnished with a popular expression by Huxley in his well-known essay, *The Physical Basis of Life* (1868).

The New Conception of the Cell.—The conception of the cell had now developed into something quite different from what it had been in the minds of the founders of the Cell Theory.   The cell was now recognized as a protoplasmic unit, and the ideas of these men concerning the origin and multiplication of cells had been overthrown.   Future researches were to show more clearly the importance of the cell in connection with development and inheritance, and certain limits were to be set to the conception of the cell as a unit of function and organization.   To Schleiden and Schwann the multicellular plant or animal appeared as little more than a cell aggregate, the cells being the primary individualities; the organism was looked upon as something completely dependent upon their varied activities for all its phenomena.   "The cause of nutrition and growth," said Schwann, "resides not in the organism as a whole, but in the separate elementary parts—the cells."   This elementalistic conception of the organism as an aggregate of independent vital units governing the activities of the whole dominated biology for many years, notwithstanding its severe criticism by Sachs, de Bary, and many other later writers who pointed out that, owing to the high degree of physiological differentiation among the various tissues and organs, the cell cannot be regarded merely as an independent unit, but as an integral part of a higher individual organization, and that as such the exercise of its functions must be governed to a considerable extent by the organism as a whole (Wager). Such divergence of opinion led to much discussion over the question of organic individuality, which remains as one of the important problems of modern biology.

But in spite of all these changes we should not forget the great service rendered by Schleiden and Schwann in the formulation of the Cell Theory. Huxley (1853) estimated the value of their contribution in the following lines:

"Doubtless the truer a theory is—the more appropriate the colligating conception—the better will it serve its mnemonic purpose, but its absolute truth is neither necessary to its usefulness, nor indeed in any way cognizable by the human faculties. Now it appears to us that Schwann and Schleiden have performed precisely this service to the biological sciences. At a time when the researches of innumerable guideless investigators, called into existence by the tempting facilities offered by the improvement of microscopes, threatened to swamp science in minutiæ, and to render the noble calling of the physiologist identical with that of the 'putter-up' of preparations, they stepped forward with the cell theory as a colligation of the facts. To the investigator, they afforded a clear basis and a starting point for his inquiries; for the student, they grouped immense masses of details in a clear and perspicuous manner. Let us not be ungrateful for what they brought. If not absolutely true, it was the truest thing that had been done in biology for half a century."

Fertilization and Embryogeny.—*In Plants.*—Although it was known to the ancients that there is in plants something analogous to the sexual reproduction seen in animals, ideas of the organs and processes involved were very vague. Like Grew and others in the seventeenth century, the botanists of antiquity were aware of the fact that the pollen in some way influences the development of the ovary into a fruit with seeds. Definite proof that the stamens are (to speak somewhat loosely) the male organs was furnished in the well-known experiments of R. J. Camerarius (1691). But in spite of the excellent work of J. G. Koelreuter (1761), C. K. Sprengel (1793), and K. F. Gaertner (1849), all of whom proved the correctness of this conclusion, the idea of sexuality in plants was vigorously combatted in certain quarters for many years.

An important step in advance was made when G. B. Amici (1830) followed the growth of the pollen tube from the pollen grain on the stigma down to the ovule. Schleiden (1837) and Schacht (1850,1858) took up the study and made a curious misinterpretation: they regarded the ovule as merely a place of incubation for the end of the pollen tube, which they supposed to enter the ovule and enlarge to form the embryo directly. The work of Amici (1842), Tulasne (1849), and others showed the falsity of this notion, but an acrimonious discussion raged about the subject for a number of years, Schleiden (1842. 1844) using the most vigorous language in support of his position. After Hofmeister (1849) had followed the process with his characteristic thoroughness there could remain no doubt concerning the error of Schleiden and Schacht. Hofmeister clearly demonstrated that the embryo arises, as Amici contended, not from the end of the pollen tube, but from an egg contained in the ovule, the egg being stimulated to development by the pollen tube. He was wrong, however, in supposing that the tube did not open, but that a fertilizing substance diffused through its wall.

It was in the algæ that the union of the sperm cell with the egg cell (the act of fertilization) was first seen in the case of plants. In 1853 Thuret saw spermatozoids attach themselves to the egg of *Fucus*, and in 1854 he showed that they are necessary to its development. The actual entrance of the spermatozoid into the egg was first observed in 1856 by Nathanael Pringsheim (1824–1894) in *Œdogonium*. The fusion of the parental nuclei was seen by Strasburger (1877) in *Spirogyra*, but he thought they thereupon dissolved. This error was corrected shortly afterward by Schmitz (1879), who was thus the first to show clearly that the central feature of the sexual process in plants is the union of two parental nuclei to form the primary nucleus of the new individual.

That the same process occurs in fertilization in the higher plants was demonstrated by Strasburger, who in 1884 described the union of the egg nucleus with a nucleus brought in by the pollen tube. In 1898 and 1899 S. Nawaschin and L. Guignard completed the story by describing the phenomenon of double fertilization, whereby the second male nucleus contributed by the pollen tube unites with the two polar nuclei to form the primary endosperm nucleus. The subsequent work of Strasburger and others on the gymnosperms and angiosperms greatly cleared up the whole matter of fertilization and embryogeny in these plants. This work belongs to the modern period of cytology.

In Animals.—It is probable that the spermatozoön was first seen in 1677 by Ludwig Hamm, a pupil of Leeuwenhoek. The credit for the discovery, however, is usually given to Leeuwenhoek, since it was he who brought the matter to the attention of the Royal Society and pursued such studies further. He asserted that the spermatozoa must penetrate into the egg, but it was thought at that time and for many years afterward that they were parasitic animalcules in the spermatic liquid; hence the name "spermatozoa."

Although L. Spallanzani (1786) is usually said to have shown by his filtration experiment that the spermatozoön is the fertilizing element, it is pointed out by Lillie (1916) that Spallanzani did not draw the correct conclusion: he even denied that the spermatozoön is the active element, holding rather that the fertilizing power lies in the spermatic liquid. It was Prevost and Dumas who corrected this mistake and demonstrated the true rôle of the spermatozoön (1824). The spermatozoön was later shown by Schweigger-Seidel (1865) and La Valette St. George (1865) to be a complete cell with its nucleus and cytoplasm, as von Kölliker had maintained. That Schwann (1839) had been right in considering the egg as a cell was shown by Gegenbaur in 1861. The polar bodies formed at the time the egg matures are said to have been first seen by Carus (1824). Bütschli (1875) showed them to be formed as the result of the division of the egg nucleus, and Giard (1877) and Mark (1881) interpreted them as abortive eggs.

The penetration of the spermatozoön into the egg was not actually seen until Newport (1854) observed it in the case of the frog. In 1875 O. Hertwig (b. 1849) announced the important discovery that the two nuclei seen fusing in the fertilized egg are furnished by the egg and the spermatozoön—by the two parents. The rôle of the nucleus in fertilization was thus demonstrated in animals only shortly before it was in plants, and it is interesting to note that the first complete description of the union of the germ cells in animals was given by H. Fol in the same year (1879) that Schmitz described clearly the process in plants. It was now evident that fertilization in both kingdoms consists in the union of two cells (gametes), one from each parent (in diœcious forms), and that *the central feature of the process is the union of the two gamete nuclei, the new individual therefore deriving half of its nuclear substance from each parent.*

Although the cleavage of the fertilized animal egg to form the embryo had been seen many years previously, it was first definitely described by Prevost and Dumas in 1824 for the frog. At that time neither the egg nor the products of its division were known to be cells. The true meaning of cleavage was elucidated by M. Barry, who held that the blastomeres are cells and that their division is preceded by the division of their nuclei, and by a number of later writers, including A. von Kölliker, who traced in detail the long series of changes by which the multiplying embryonic cells become differentiated into the various tissues and organs. Embryogeny was thus shown to be a process of cell-division and differentiation. the fertilized egg cell initiating a series of divisions giving rise to all the cells of the body, and to the germ cells. The life cycle was now recognized as a cell cycle; and since the egg is the direct descendant of the egg of the previous generation it became evident, as Virchow pointed out in 1858. that there has been an uninterrupted series of cell-divisions from the beginnings of life on the earth in the remote past down to the organisms in existence today. The statement of this conception is known as the Law of Genetic Continuity. In the words of Locy (1915):

"The conception that there is unbroken continuity of germinal substance between all living organisms, and that the egg and the sperm are endowed with an inherited organization of great complexity, has become the basis for all current theories of heredity and development. So much is involved in this conception that . . . it has been designated (Whitman) 'the central fact of modern biology.' The first clear expression of it is found in Virchow's *Cellular Pathology*, published in 1858. It was not, however, until the period of Balfour, and through the work of Fol, Van Beneden (chromosomes, 1883) Boveri, Hertwig, and others, that the great importance of this conception began to be appreciated, and came to be woven into the fundamental ideas of development.'

The Beginning of the Modern Period in Cytology.—As Wilson (1900, p. 6) points out, the great significance of the many facts brought to light in the early days of cytology lies in the relation which they bear to the Theory of Evolution and to the problems of heredity, though for many years this was only vaguely realized. Darwin, aside from his Hypothesis of Pangenesis, scarcely mentioned the theories of the cell; and not until many years later was the cell investigated with reference to these matters. Researches on the origin of the germ cells, nuclear division, and fertilization, which brought the Cell Theory and the Theory of Evolution into intimate association, began shortly after 1870 with the works of Schneider (1873), Auerbach (1874), Fol (1875, etc.), Bütschli (1875, etc.), O. Hertwig (1875, etc.), van Beneden (1875, etc.), Strasburger (1875, etc.), Flemming (1879, etc.), and Boveri (1887, etc.). These men laid the foundations for the work which has followed; and their researches, greatly aided by the development of new refinements in microtechnique, ushered in modern cytology. A powerful stimulus to investigation was given when the zoölogists Hertwig, von Kölliker and Weismann, and the botanist Strasburger, concluded independently and almost simultaneously (1884–1885) that the nucleus is the vehicle of heredity, an idea which Haeckel had put forward as a speculation in 1866. The announcement of this conception led to an even more intensive study of the nucleus and of its rôle in heredity, a study which is now in progress, and which, more than any other one thing, can be said to characterize the work of our modern period.

Bibliography 1

A. Works dealing wholly or in part with the history of cytology, and general works on the cell:

AGAR, W. E. 1920. Cytology, with Special Reference to the Metazoan Nucleus. London.

BOVERI, TH. 1891. Befruchtung. Ergeb. d. Anat. u. Entw. **1**: 386–485.

BUCHNER, P. 1915. Prakticum der Zellenlehre. 1.

BURNETT, W. J. 1853. The cell; its physiology, pathology, and philosophy, as deduced from original investigations. Trans. Am Med. Assn. **6**.

CHUBB, G. C. 1910. Article on Cytology in Encycl. Brit., 11th ed.

DELAGE, Y. 1895. La structure du protoplasme et les théories sur l'hérédité. Paris.

DONCASTER, L. 1920. An Introduction to the Study of Cytology. London.

FLEMMING, W. 1882. Zellsubstanz, Kern und Kerntheilung. Leipzig.

1981–1897. Referate über Zelle. Ergeb. d. Anat. u. Entw. **1–7**.

GURWITSCH, A. 1904. Morphologie und Biologie der Zelle. Jena.

HAECKER, V. 1899. Praxis und Theorie der Zellen- und Befruchtungslehre.

HEIDENHAIN, M. 1907. Plasma und Zelle. Jena.

HENNEGUY, L. F. 1896. Leçons sur la Cellule. Paris.

HERTWIG, O. 1893. Die Zelle und die Gewebe. Jena. (Engl. Transl. by H. J. Campbell.)

1900. Die Entwicklung der Biologie im 19. Jahrhundert. Jena.

HOFMEISTER, W. 1867. Die Lehre von der Pflanzenzelle.

HUXLEY, T. H. 1853. The Cell Theory. Brit. and For. Med.-Chir. Rev. 12. Also in "Scientific Memoirs" 1. 1898.

JOHNSON, D. S. 1914. The evolution of a botanical problem. The history of the discovery of sex in plants. Science 39: 299–319.

KELLOGG, V. L. 1907. Darwinism Today. New York.

LILLIE, F. R. 1916. The history of the fertilization problem. Science 43: 39–53.

KOERNICKE, M. 1903. Der heutige Stand der pflanzlichen Zellforschung. Ber. deu. Bot. Ges. 21: (66)–(134).

LOCY, W. A. 1901. Malpighi, Swammerdam, and Leeuwenhoek. Pop. Sci. Mo. 58: 561–584.

1905. Von Baer and the rise of embryology. Ibid. 67: 97–126.

1915. Biology and its Makers. 3d ed. New York.

MARK, E. L. 1881. Maturation, fecundation, and segmentation in *Limax campestris*. Bull. Mus. Comp. Zool. Harvard Coll. 6: 173–625. pls. 5. (Early literature of mitosis and fertilization.)

MEVES, FR. 1896, 1898. Referate über Zelltheilung. Ergeb. d. Anat. u. Entw. 6, 8.

VON MOHL, H. 1851. The Vegetable Cell. (Engl. transl. by Henfrey.)

OSBORN, H. F. 1894. From the Greeks to Darwin.

RÜCKERT, J. 1893. Die Chromatinreduktion bei der Reifung der Sexualzellen. . Ergeb. d. Anat. u. Entw. 3: 517–583.

VON SACHS, J. 1875. History of Botany. (Engl. transl. 1889.)

STRASBURGER, E. 1907. Die Ontogenie der Zelle siet 1875. Prog. Rei Bot. 1.

1910. The minute structure of cells in relation to inheritance. In "Darwin and Modern Science" (Seward, editor).

THOMSON, J. A. 1899. The Science of Life. Chapters 9 and 10.

TURNER, W. 1890. The cell theory, past and present. Nature 43: 10–15.

TYSON, J. 1878. The Cell Doctrine: its History and Present State. Phila.

WAGER, H. 1911. Article on Plants: Cytology, in Encycl. Brit., 11th ed.

WALDEYER, W. 1888. Ueber Karyokinese und ihre Beziehung zu den Befruchtungsvorgängen. Arch. Mikr. Anat. 32: 1–122. (Engl. transl. in Quar. Jour. Micr. Sci. 30: 159–281. 1889.) (Early cell literature.)

WHITMAN, C. O. 1878. The embryology of *Clepsine*. Quar. Jour. Micr. Sci. 18: 215–315. pls. 12–15. (Early literature of mitosis and fertilization.)

1894. (1) Evolution and Epigenesis. (2) Bonnet's theory of evolution. Woods Hole Biol. Lectures 1894.

WHEELER, W. M. 1898. Caspar Friedrich Wolff and the Theoria Generationis. Ibid. 1898.

WILSON, E. B. 1900. The Cell in Development and Inheritance. 2d ed.

ZIMMERMANN, A. 1893–1894. Sammel-Referate aus dem Gesammtgebiete der Zellenlehre. Beih. Bot. Centr. 3 and 4. (Reviews of early literature).

For other reviews of early botanical cell literature see Jahresbericht über die Fortschritte der Anatomie und Physiologie, 15-20. 1886–1892; and Neue Folge, Vols. 1–13. 1892–1907.

B. Special works referred to in historical sketch:

AMICI, G. B. 1830. Note sur le mode d'action du pollen sur le stigmate. (Extrait d'une lettre de M. Amici à M. Mirbel.) Ann. Sci. Nat. Bot. I. 21: 329–332. Account of Amici's work in Atti della quarta Riunione degli scientiati Italiani tenuta in Padova nel Settembre del 1842.   Padova 1843.   See FACCHINI 1845. See also Giorn. Bot. Ital. Anno 2.

AUERBACH, L.   1874.   Organologische Studien.   Breslau.

VON BAER, K. E.   1828, 1837.   Über Entwickelungsgeschichte der Thiere.

BARRY, M.   1838-1841.   Embryological memoirs in Phil. Trans. Roy. Soc. London **128-131**.   See also: On the first changes consequent on fecundation in the mammiferous ovum.   Rep. Brit. Assn. Adv. Sci.   1840.

DE BARY, H. A.   1862.   Über den Bau und das Wesen der Zelle.   Flora **20**. 1864.   Die Mycetozoen.   2d ed. Leipzig.

VAN BENEDEN, E.   1875.   La maturation de l'oeuf, la fécondation et les premières phases du développement embryonnaire des mamifères d'apres des recherches faites chez le lapin.   Bull. Acad. Roy. Belg. **40**.

1876.   Contribution à l'histoire de la vésicule germinative et du premier noyau embryonnaire.   Ibid. **41**.

1883.   Recherches sur la maturation de l'oeuf, la fécondation et la division cellulaire.   Arch. de Biol. **4**.

BERNHARDI, J. J.   1805.   Beobachtungen über Pflanzengefässe.

BOVERI, TH.   1887a.   Ueber die Befruchtung der Eier von *Ascaris megalocephala*. Sitzungsber. Gesell. Morph. Phys.   München **3**.

1887b.   Ueber Differenzierung der Zellkerne während der Furchung des Eies von *Ascaris megalocephala*.   Anat. Anz. **2**: 688-693.

1887-1890.   Zellenstudien I, II, III.   Jenaische Zeitsch. **21-24**.

BRAUN, ALEX.   1850.   Betrachtungen über die Erscheinung der Verjüngung in der Natur, inbesondere in der Lebens- und Bildungsgeschichte der Pflanze.   (English transl., Ray Society 1853.)

BROWN, R.   1833.   Observations on the organs and mode of fecundation in Orchideæ and Asclepiadeæ.   Trans. Linn. Soc.   (Paper read and privately printed in 1831.) Also in: Misc. Bot. Works, Ray Society, 1866.

1866.   A brief account of microscopical observations on the particles contained in the pollen of plants; and on the general existence of active molecules in organic and inorganic bodies.   Misc. Bot. Works, Ray Society.   (Observations made in 1827.)

BÜTSCHLI, O.   1875a.   Vorl. Mitteilung über Untersuchungen betreffend die ersten Entwicklungsvorgänge im befruchteten Ei von Nematoden und Schnecken. Zeit. Wiss. Zool. **25**: 201.

1875b.   Vorl. Mitteilung einiger Resultate von Studien über Conjugation der Infusorien und die Zelltheilung.   Zeit. Wiss. Zool. **25**: 426.

1876.   Studien über die ersten Entwicklungsvorgänge der Eizelle, die Zelltheilung, und die Konjugation der Infusorien.   Abhandl. Senckenb. Naturforsch. Gesell. **10**.

CAMERARIUS, R. J.   1691.   De Sexu Plantarum.   1694.

CARUS, C.   1824.   Von den äusseren Lebensbedingungen der Weiss- und Kaltblütigen Thiere, etc.   Leipzig.

CIENKOWSKI, L.   1863.   Zur Entwicklungsgeschichte der Myxomyceten.   Jahrb. Wiss. Bot. **3**: 325-337.

COHN, F.   1850.   On the natural history of *Protococcus pluvialis*.   Nova Acta Acad. Caes. Leop. Carol. Nat. Cur. Bonn **22**: 605-764.   (English abst. by Busk, Ray Soc. 1853.)

CORTI, B.   1772.   Observationi misc. sulla Tremella e sulla circolazione del fluid in una pianta acquaiola.   Lucca, 1774.

DUJARDIN, F.   1835.   Sur les pretendus estomacs des animalcules infusories et sur une substance appelée Sarcode.   Ann. Sci. Nat. Zool. II **4**: 364-377.

DUTROCHET, H. J. 1837. Mémoires pour servir á l'histoire anatomique et physio-logique des végétaux et des animaux.

FACCHINI. 1845. Ueber die Amici'sche Ansicht von der Befruchtung der Pflanzen, ein Beitrag. Flora **28**: 193–198. See also **27**: 359–360.

FLEMMING, W. 1879*a*, 1880, 1881. Beiträge zur Kenntniss der Zelle und ihre Lebenserscheinungen. I, II, III. Arch. Mikr. Anat. **16, 19, 20**.

1879*b*. Ueber das Verhalten des Kernes bei der Zelltheilung, usw. Virchow's Archiv. **77**.

1882. Zellsubstanz, Kern und Zelltheilung. Leipzig.

1887. Neue Beiträge zur Kenntniss der Zelle. Arch. Mikr. Anat. **29**.

FOL, H. 1873. Die erste Entwicklung des Geryoniden-Eies. Jen. Zeitsch. **7**.

1875. Sur le développement des Pteropodes. Arch. de Zool. **4**.

1876. Sur les phénomènes de la division cellulaire. Comptes Rend. Acad. Sci. Paris **83**: 667–669.

1877. Sur le commencement de l'hénogenie chez divers animaux. Arch. Sci. Nat. et Phys. Geneve **58**.

1879. Recherches sur la fécondation et la commencement de l'hénogenie. Mem. Soc. Phys. et Nat. Geneve **26**: 89–397.

FONTANA. 1781. Sur la structure primitive du corps animal. In Traité sur le vénin de la vipere. Florence.

GAERTNER, J. 1788. De fructibus et seminibus plantarum.

GAERTNER, K. F. 1849. Versuche und Beobachtungen über die Bastardzeugung. Stuttgart.

GEGENBAUR, K. 1861. Ueber den Bau und die Entwicklung der Wirbelthiere. Müller's Archiv f. Anat. u. Physiol. p. 451–529.

GIARD, A. 1877. Sur la signification morphologique des globules polaires.

GOODSIR, J. 1842. On secreting structures. Trans. Roy. Soc. Edinb.

1842. On Peyer's glands. Lond. and Edinb. Mo. Jour.

1845. Anatomical and Pathological Observations. Edinburgh.

1868. Anatomical Memoirs. (Ed. by Turner.) Edinburgh.

GREW, N. 1682. The Anatomy of Plants. London.

GUIGNARD, L. 1899. Sur les antherozoides et la double copulation sexuelle chez les végétaux angiospermes. Comptes Rend. Acad. Sci. Paris **128**: 864–871. figs. 19.

HAECKEL, E. 1866. Generelle Morphologie. Jena.

HANSTEIN, J. 1880. Das protoplasma als Träger der pflanzlichen und thierischen Lebensverrichtungen. Heidelberg.

HARVEY, WM. 1651. Exercitationes de Generatione Animalium. (English transl., Sydenham Society, 1847.)

HENLE, J. 1837. Symbolæ ad anatomiam villorum intestinalium.

1841. Allgemeine Anatomie. Leipzig.

HERTWIG, O. 1875. Beiträge zur Kenntniss der Bildung, Befruchtung, und Theil-ung des tierischen Eies, I. Morph. Jahrb. **1**. See also **2–4**.

1884. Das Problem der Befruchtung und der Isotropie des Eies, eine Theorie der Vererbung. Jenaische Zeitschr. **18**: 276–318.

HOFMEISTER, W. 1849. Die Entstehung des Embryos der Phanerogamen.

1858. Neuere Beobachtungen über Embryobildung der Phanerogamen. Jahrb. Wiss. Bot. **1**: 82–188. pls. 7–10.

HOOKE, R. 1665. Micrographia, or some physiological descriptions of minute bodies made by magnifying glasses. London.

HUXLEY, T. H. 1868. The Physical Basis of Life. (Collected Essays.)

KIESER. 1812. Mémoire sur l'organisation des plantes.

von Kölliker, A. 1845. Die Lehre von der thierischen Zelle. Zeit. Wiss. Bot. **2**.
1885. Die Bedeutung der Zellkerne für die Vorgänge der Vererbung. Zeit. Wiss. Zool. **42**.

Koelreuter, J. G. 1761–1766. Vorläufige Nachricht von einigen Geschlecht der Pflanzen betreffenden Versuchen und Beobachtungen.

Kowalevsky, A. 1871. Embryologische Studien an Würmern und Arthropoden. Mem. Acad. Imp. Sci. de St. Petersburg VII **16**: 13. pl. 4. figs. 24.

Kühne, W. 1864. Untersuchungen über das Protoplasma. Leipzig.

La Valette St. George. 1865. Ueber die Genese der Samenkörper. Arch. Mikr. Anat. **1**. See also Vols. **2** and **3**.

van Leeuwenhoek, A. 1673–1723. Brieven. Leiden, Delft.

Link, H. F. 1807. Grundlehren der Anatomie und Physiologie der Pflanzen.

Malpighi, M. 1675. Anatome Plantarum.

Mark, E. L. 1881. Maturation, fecundation, and segmentation of *Limax campestris*. Bull. Mus. Comp. Zool. Harvard **6**: 173–625. pls. 5.

Meyen, F. J. F. 1826. De Primis Vitæ Phenomenis in Fluidis.
1830. Lehrbuch der Phytotomie. Berlin.
1837–1839. Neues System der Pflanzenphysiologie.

Meyer, A. 1881. Ueber die Struktur der Stärkekörner. Bot. Zeit. **39**: 841–846, 857–864. pl. 9.
1883a. Ueber Krystalloide der Trophoplasten und über die Chromoplasten der Angiospermen. Ibid **41**: 489, 505, 525.
1883b. Das Chlorophyllkorn. pp. 91. pls. 3. Leipzig.

Mirbel, C. F. 1801. Traité d'anatomie et de physiologie végétale.
1808. Exposition et defense de ma théorie de l'organisation végétale.
1833. Recherches sur la *Marchantia*. Mem. French Inst. 1835.

von Mohl, H. 1835, 1837. Ueber die Vermehrung der Pflanzenzelle durch Theilung. Dissert. Tübingen 1835. Flora **45**. 1837.
1844. Einige Betrachtungen über den Bau der vegetabilische Zelle. Bot. Zeit. **2**: 273–277, 289–294, 305–310, 321–326, 337–342. pl. 2.
1845. Vermischte Schriften.
1846. Ueber die Saftbewegung im Inneren der Zellen. Bot. Zeit. **4**: 73–78, 89–94.
1851. Grundzüge der Anatomie und Physiologie der vegetabilische Zelle. (Engl. transl. by Henfrey, London, 1852.)

Moldenhawer, J. J. P. 1812. Beiträge zur Anatomie der Pflanzen.

von Nägeli, C. 1844, 1846. Zellkerne, Zellbildung, und Zellwachsthum. Zeitschr. Wiss. Bot. **1, 3**. (Engl. transl. by Henfrey, Ray Society; London, 1846, 1849.)
1846. On the utricular structures in the contents of cells. Ray Society, 1849. (Engl. Transl. by Henfrey.)
1858. Die Stärkekörner. Zurich.

Nawaschin, S. 1899. Neue Beobachtungen über Befruchtung bei *Fritillaria* und *Lilium*. Bot. Centr. **77**: 62. (Account in Russian, 1898.)

Newport, G. 1851, 1853, 1854. On the impregnation of the ovum in the amphibia. Phil. Trans. Roy. Soc. London.

Payen, A. 1839. Mémoire sur l'amidon, etc. Paris.
1846. Mémoire sur les développements des végétaux, etc. Mem. Acad. Paris **9**.

Prevost and Dumas. 1824. Nouvelle Théorie de la generation. Ann. Sci. Nat. **1**: 1, 167; **2**: 100, 129.

Pringsheim, N. 1855. Ueber die Befruchtung und Keimung der Algen und das Wesen des Zeugungsaktes. Monatsber. K. Akad. Wiss. Berlin **1**.
1856. Ueber die Befruchtung der Algen. Ibid.
1858. Morphologie der Oedogonieen. Jahrb. Wiss. Bot. **1**: 11–81.

REMAK, R. 1841. Ueber Theilung rother Blutzellen beim Embryo. Med. Ver. Zeit. Müller's Archiv f. Anat. u. Physiol. 1858; 177–188. pl. 8.

1852. Ueber extracellulare Entstehung thierische Zellen und über Vermehrung derselben durch Theilung. Müller's Archiv f. Anat. u. Physiol. 1852; 47–57.

RUDOLPHI, K. A. 1807. Anatomie der Pflanzen.

SCHACHT, H. 1850 Entwicklungsgeschichte des Pflanzenembryon. Amsterdam 1850. In Ann. Sci. Nat. Bot. III 15: 80–109. 1851.

1858. Ueber Pflanzenbefruchtung. Jahrb. Wiss. Bot. 1: 193–232. pls. 11–15.

SCHIMPER, A. F. W. 1880–1881. Untersuchungen über die Entstehung der Stärkekörner. Bot. Zeit. 38: 881; 39: 185, 201, 217. pls. 13, 2.

1883. Ueber die Entwicklung der Chlorophyllkörner und Färbkörper. Bot. Zeit. 41: 105, 121, 137, 153. pl. 1.

1885. Untersuchungen über die Chlorophyllkörper und die ihnen homologen Gebilde.

SCHLEIDEN, M. J. 1837. Einige Blick auf die Entwicklungsgeschichte des vegetabilische Organismus bei den Phanerogamen. Wiegmann's Archiv. 1: 289.

1838. Beiträge zur Phytogenesis. Müller's Archiv f. Anat. u. Phys. (English transl., Sydenham Society, 1847.)

1842. Grundzüge zur Wissenschaftlichen Botanik. Zweite Aufl. (English transl. by Lankester, 1849.)

1844. Bemerkung zur Bildungsgeschichte des Veg. Embryo. Flora 27: 787–789.

SCHNEIDER, A. 1873. Untersuchungen über Platelminthen Jahrb. d. Oberhess. Gesell. Natur-Heilkunde 14. Giessen.

SCHMITZ, FR. 1879. Untersuchungen über die Zellkerne der Thallophyten. Verhandl. Naturhist. Ver. Preuss. Rheinl. u. Westf. p. 346.

SCHNEIDER, A. 1883. Das Ei und seine Befruchtung. Breslau.

SCHULTZE, M. 1861. Über Muskelkörperchen und das was man eine Zelle zu nennen hat. Arch. Anat. u. Physiol. 1–27.

SCHWANN, TH. 1839. Mikroskopische Untersuchungen über die Uebereinstimmung in der Struktur und dem Wachsthum der Thiere und Pflanzen. (English transl., Sydenham Soc., 1847). Preliminary statement in Froriep's Notizen. No. 91: 103, 112. 1838.

SCHWEIGGER-SEIDEL, O. 1865. Ueber die Samenkörperchen und ihre Entwicklung. Arch. Mikr. Anat. 1: 309–335. pl. 19.

SPALLANZANI, L. 1786. Expériences pour servir à l'histoire de la génération des animaux et des plantes. Geneva.

SPRENGEL, C. K. 1793. Das neu entdekte Geheimniss der Natur in Bau und Befruchtung der Blumen. Berlin.

SPRENGEL, K. 1802. Anleitung zur Kenntniss der Gewächse.

STRAẞBURGER, E. 1875. Ueber Zellbildung und Zelltheilung. Jena.

1877. Ueber Befruchtung und Zelltheilung. Jenaische Zeitschr. 11.

1879. Die Angiospermen und die Gymnospermen. Jena.

1884. Neue Untersuchungen über die Befruchtungsvorgang bei den Phanerogamen, als Grundlage für eine Theorie der Zeugung. Jena.

1888. Ueber Kern- und Zelltheilung im Pflanzenreich, nebst ein Anhang über Befruchtung. Hist. Beitr. 1.

THURET, G. 1854–1855. Recherches sur la fécondation des Fucées, suivies des observations sur les antheridies des algues. Ann. Sci. Nat. Bot. IV, 2 and 3.

TREVIRANUS, L. C. 1806. Vom inwendigen Bau der Gewächse.

1811. Beiträge zur Pflanzenphysiologie.

TULASNE, L. R. 1849. Études d'embryogénie végétale. Ann. Sci. Nat. Bot. III, 12: 21–137. pls. 3–7.

UNGER, F.   1844.   Ueber das Wachsthum der Internodien, von anatomischer Seite
    betrachtet.   Bot. Zeit.   Meristematische Zellbildung.   Ibid.
VIRCHOW, R.   1855.   Cellular Pathologie.   Arch. Path. Anat. Physiol. **8.**
    1858.   Die Cellularpathologie, usw.   (Transl. by Chance, 1860.)
WEISMANN, A.   1885.   Die Kontinuität des Keimplasmas als Grunglage einer
    Theorie der Vererbung.   Jena.
WHITMAN, C. O.   1878.   The embryology of *Clepsine*.   Quar. Jour. Micr. Sci.
    18:   215–315.   pls. 12–15.
WOLFF, C. F   1759.   Theoria Generationis.

CHAPTER II

PRELIMINARY DESCRIPTION OF THE CELL

In a survey of the evolution of biological science it is noticeable that, while diverging lines of inquiry have broadened the field of view, the attention of investigators, speaking generally, has been directed in turn to successively smaller constituent parts of the organism. For many years plants and animals were studied chiefly as wholes. But very early there were made many scattered observations on the various organs and tissues composing the body, and from these relatively crude beginnings morphology and histology later arose. Again, when the protoplasmic mass which we know as the cell came to be recognized as the unit of structure and of function, it was evident that the problems it presents should be investigated to a certain extent by themselves, and such investigation is the task of modern cytology.

Within the field of cytology itself the focus of attention has gradually shortened. While many workers occupied themselves with a study of the general behavior of the cell nucleus, others devoted their efforts entirely to an investigation of its important constituent elements, the chromosomes. Furthermore, cytologists at present are much interested in knowing whether or not any smaller units, corresponding to the "genes" of the geneticist, can be directly demonstrated, and whether or not the chromatic granules or "chromomeres" are of significance in this respect.

In the course of all such studies there are encountered questions which must be referred ultimately to the chemical molecules and atoms and their interactions within the cell, so that biochemistry may in a measure be looked upon as a department of cytology, just as it is to be regarded in other respects as a subdivision of chemistry. The subject of cytology thus occupies an important position in the system of natural sciences. It stands with chemistry and physics on the one hand and the complex phenomena peculiar to living organisms on the other; and the steady mutual approach of the physico-chemical and biological fields is due in large measure to the results of morphological and physiological studies on the cell.

For the term *cell* we are indebted to Robert Hooke and the other microscopists of the seventeenth century, who applied it to the small cavities in the honeycomb-like structure which they discovered in plant tissues. Today the term denotes primarily the protoplasmic "cell contents," which, strangely enough, the early workers regarded as an unimportant fluid product. The term *protoplast,* proposed by Hanstein (1880), is more appropriate and is coming into more general use, but long usage and brevity have probably insured the permanence of the older term.

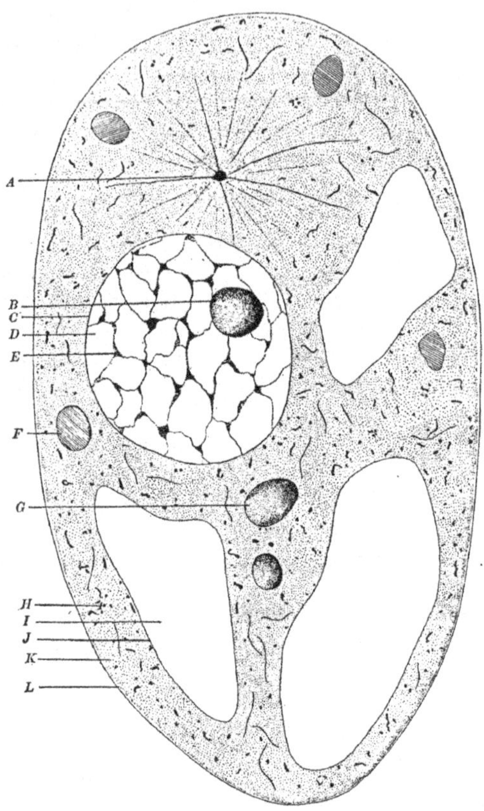

Fig. 1.—Diagram of the cell, showing its principal constituent parts.

A, centrosphere, with centrosome and aster. *B,* nucleolus or plasmosome. *C,* nuclear membrane. *D,* nucleus, filled with karyolymph. *E,* nuclear reticulum, composed of linin and chromatin. *F,* plastid. *G,* metaplasmic inclusion. *H,* chondriosomes. *I,* vacuole. *J,* tonoplast or vacuolar membrane. *K,* cytoplasm. *L,* ectoplast.

Description of the Cell.—The morphology of the cell will here be sketched only in its barest outlines, by way of introduction to the detailed descriptions presented in the subsequent chapters.

The two most constant components of the cell (Fig. 1) are the *cytoplasm,* in which the other cell organs are imbedded, and the *nucleus,* which at least in many respects is the most important of these organs.[1]

The *cytoplasm,* a more or less transparent, viscous, granular fluid, may, with its inclusions, occupy the whole volume of the cell. This is generally true of animal cells and the younger cells of plants. If the cell is vacuolate, as is usually the case in the mature plant cell, the cytoplasm may constitute only a thin layer lining the wall, the central *vacuole* with its cell sap often far exceeding it in volume (Fig. 2, *C*).

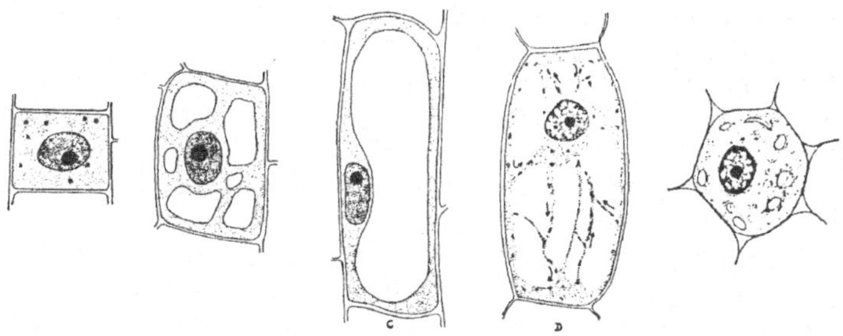

Fig. 2.

A–C, diagram of a plant cell in three stages of development: the vacuoles increase in volume and the protoplasm becomes limited to the parietal region. *D,* cell of stamen hair of *Tradescantia,* showing streaming movements in the cytoplasmic strands. *E,* parenchyma cell from cortex of *Polygonella,* showing nucleus, plastids, and scanty cytoplasm.

In many cases the cytoplasm forms a system of anastomosing strands that often show active streaming (Fig. 2, *D*). Externally the cytoplasm is limited by a layer of different consistency, the *plasma membrane,* or *ectoplast.* Where it comes in contact with the enclosed vacuole it is also limited by a membrane, the *vacuole membrane,* or *tonoplast.*

The *nucleus* is bounded by a *nuclear membrane* and contains an extremely clear fluid, the *nuclear sap,* or *karyolymph.* In the karyolymph is imbedded the *nuclear reticulum,* composed usually of *linin,* an achromatic supporting material. and *chromatin,* the "nuclear substance par excellence." One or more *true nucleoli,* or *plasmosomes,* are commonly present in the nucleus, and may or may not be closely associated with the reticulum. There are often present also *chromatin nucleoli,* or *karyosomes,* which represent accumulations of chromatin at certain points on the reticulum, and should not be confused with the true nucleoli.

[1] According to the older usage the extra-nuclear portion of the protoplast was called "protoplasm," which was unfortunate because we now know that the nucleus also is composed of protoplasm, or living substance in its broader sense. It is now the general custom to avoid this ambiguity by employing Strasburger's terms *cytoplasm* and *nucleoplasm* (*karyoplasm,* Flemming). The older usage, however, has not been entirely superseded.

There are usually *plastids* of one or more kinds in the cytoplasm, the most conspicuous in plant cells being the green chloroplasts.

A *centrosome* is present in the majority of animal cells and in those of certain lower plants. It may occupy the center of a visibly differentiated region, the *centrosphere* or *attraction sphere*, and at the time of cell-division is the focus of a conspicuous system of radiating *astral rays*, collectively known as the *aster*.

Chondriosomes have now been demonstrated in the cells of nearly all plant and animal groups. These are minute bodies having the form of granules, rods, or threads, and apparently constitute a group of materials having various functions.

Metaplasmic inclusions are accumulations of food materials and differentiation products that are relatively passive. These non-proto-plasmic bodies may exist in the form of droplets or crystals, and those which are not transitory or reserve food materials apparently play a very minor rôle in the life of the cell.

Strictly speaking, the *cell wall* as at present understood is not a part of the cell proper, or protoplast, but is rather regarded by many as a secretion of the latter. In many cells, particularly those of animals and the motile cells of algæ and flagellates, it may be absent.

The foregoing is a bare sketch of the general structure of a "typical" cell. It is scarcely necessary to point out that the cell should not be thought of as a static thing with a permanent physical structure: it is rather a dynamic system in a constantly changing state of molecular flux, its constitution at any given moment being dependent upon ante-cedent states and upon environmental conditions. As stated by Moore (1912), "the living cell may be regarded, from the physico-chemical point of view, as a peculiar energy transformer, through which a continu-ally varying flux of energy ceaselessly goes on, and the whole life of the cell is an expression of variations and alterations in rates of flow of energy, and of swings in the balance between various forms of energy." In the words of Harper (1919), the cell is a colloidal system in which the various processes have become progressively localized in certain regions, with the resulting formation of organs, which, with the increasing con-stancy of the processes involved, have come to possess a permanence and individuality of their own. In view of the spatial relationship and definite physiological integration of the various components of the cell, we are to look upon the cell not as a mere mixture of complex substances, but as a definitely organized system.

The Differentiation of Cells.—It is a striking fact that in spite of many minor variations the fundamental structure of the cell is essentially similar in nearly all living organisms, and in all the kinds of tissues which go to make up the body of any one of them. As Harper (1919) remarks, "evolution as we know it has not consisted in the production of new types of protoplasmic structure or cellular organization, but in the development of constantly greater specialization and division of labor between larger and larger groups of cells." One obvious reason for the fundamental similarity of the cells of widely different tissues is found in the fact that all of them are derived by progressive modification from relatively undifferentiated "embryonic" or "meristematic" cells during the course of the ontogeny. In a young vascular plant, for example, definite regions (*meristems*) consisting of such cells are present in the root tip and stem tip, and, at a later stage of development in many cases, in the cambium also. As a general rule these meristematic cells are without large vacuoles or other conspicuous products of differentiation, and are separated by no intercellular spaces. They undergo successive divisions very rapidly (hence the use of root tips for the study of mitosis); and while some of the products of division become greatly modified in structure in connection with their specialization in function, others retain their embryonic or meristematic character and continue to produce new cells from which new tissues are built up throughout the life of the plant.

In the bryophytes and pteridophytes the meristematic activity of the apex (root tip or stem tip, or apical region of thallus) usually centers in a single "apical cell" of definite shape, which cuts off segments (daughter cells) from its various faces with great geometrical regularity. In the Marchantiales and *Anthoceros* the apical cell is cuneate (wedge-shaped) and forms segments from four of its faces; in the anacrogynous Jungermanniales it is sometimes cuneate but more often dolabrate (ax-shaped) and produces segments from its two lateral faces; and in the acrogynous Jungermanniales and mosses it has the form of a triangular pyramid, cutting off segments from its three lateral faces. This last type is found also in the pteridophytes: in the stem tip it produces segments from its three lateral faces, whereas in the root tip, in addition to these three series of segments, it cuts off from its distally directed face a fourth series, which becomes the tissue of the root cap (Fig. 3). In the higher vascular plants there is no single cell characteristically different from the others of the apical meristematic group.

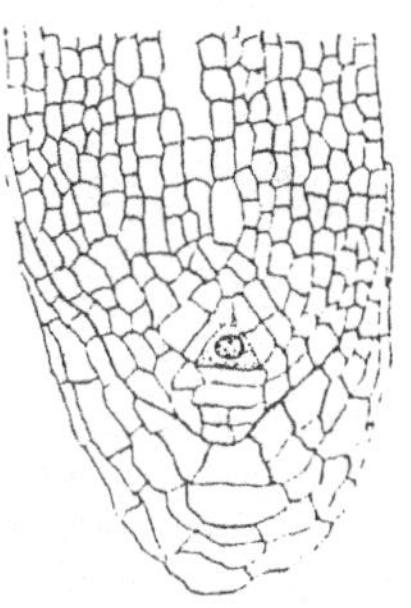

Fig. 3.—Longitudinal section of the root tip of *Osmunda*, showing the triangular pyramidal apical cell. × 144.

Most of the visible characters which ordinarily serve to distinguish the various kinds of differentiated cells of the vascular plant are found in the cell wall rather than in the protoplast itself; strictly speaking, such characters are histological rather than cytological. Thus we have, besides meristematic and little modified parenchymatous cells (Fig. 2,

E), a number of other types, such as tracheids, vessels, wood fibers, sclerenchyma fibers, and sieve tubes (Fig. 4), all of which are characterized by the peculiar ways in which their walls become modified through secondary and tertiary thickenings, and by the form and arrangement assumed by the pits.   (See p. 191.)   The protoplasts may finally disappear completely from wood cells, leaving a tissue or framework composed of lifeless cell walls.

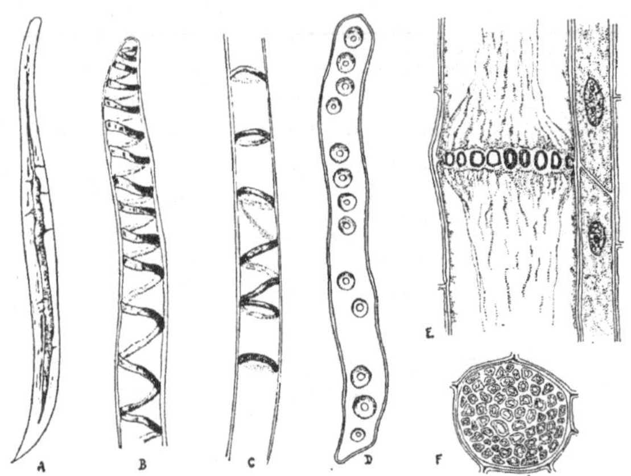

Fig. 4.—Differentiated cells from vascular plants.
A, wood fiber with thickened wall.   *B*, *C*, portions of tracheids with spiral and annular thickenings.   *D*, pitted tracheid.   *E*, portion of sieve tube with adjacent companion cells. *F*, face view of sieve plate shown in section in *E*.

All of this variety of form and structure is conditioned by varied functional activity on the part of different protoplasts: in the process of cell differentiation morphological and physiological changes stand in the closest relationship.   All functional differences are accompanied by chemical or physical differences of some sort in the protoplasm, but it is mainly in the non-protoplasmic inclusions and secretions (including the wall) rather than in any conspicuous structural changes in the protoplast itself that cell differentiation is rendered visible in the case of plants.   Apart from differences in shape, amount of vacuolar material, accumulated food, and other products of differentiation (see p. 133), protoplasts performing widely different functions may appear much alike.

Structural differentiation in connection with division of labor is very striking in animal cells, which are destitute of such walls as plant cells possess.   The muscle cell shows many fine longitudinal fibrillæ which have to do with the cell's power of contractility.   In certain muscles these fibrillæ are so segmented that the whole cell, or muscle fiber, has a transversely striped appearance (Fig. 5, *F*).   The nerve cell (Fig. 5,

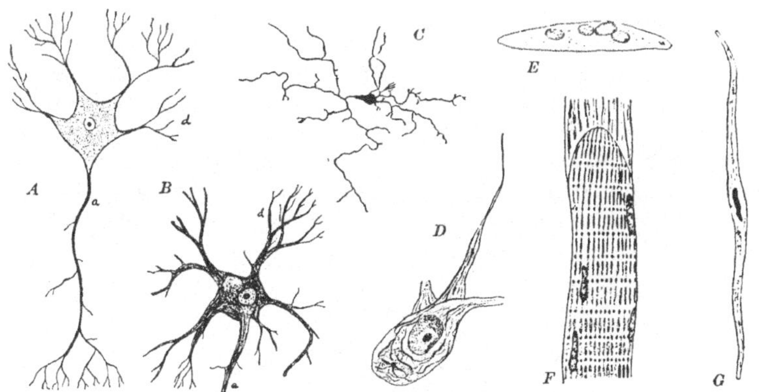

Fig. 5.—Nerve and muscle cells of animals.

A, diagram of a typical neuron: *a*, axis cylinder process or axon, ending in arborescent system; *d*, dendrites. (*After Obersteiner and Hill.*) *B*, cell from human spinal cord, × 75. (*After Obersteiner and Hill.*) *C*, nerve cell from the eye. (*After Lenhossék.*) *D*. Nerve cell from the earthworm. (*After Kowalski.*) *E*, young voluntary muscle cell. *F*, portion of mature voluntary muscle cell, showing striations. *G*, Involuntary muscle cell from intestine. (*E–G after Piersol.*)

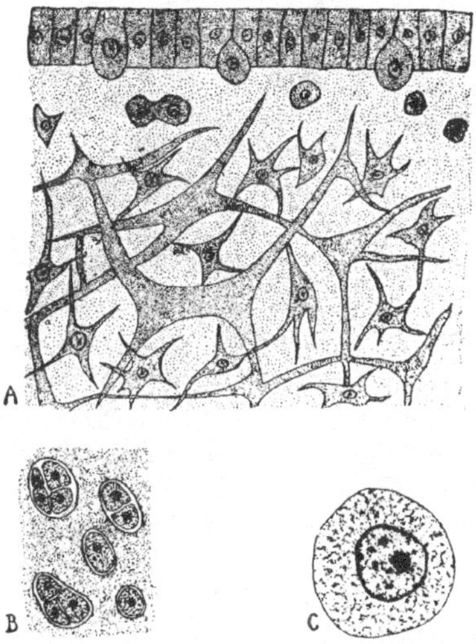

Fig. 6.

A, connective tissue from the jelly fish, showing branching cells and elastic fibers imbedded in gelatinous matrix. (*After Lang.*) *B*, cells of hyaline cartilage imbedded in their secretion. *C*, blood cell from chick embryo.

A–D) typically possesses a single unbranched prolongation (*axon*) and one or more others (*dendrites*) which often become very elaborately branched, especially in the ganglion cells of the spinal cord and brain. The cytoplasm of the nerve cell contains fine fibrils, and also granules of chromatic "Nissl substance."   Cells specialized in connection with motility such as spermatozoa (Fig. 103) and the cells of certain epithelial tissues (Fig. 36), show complex structural modifications not only in the flagellæ, cilia, and cirri which they bear (p. 45), but also in the other cell organs with which the activities of these motile structures are closely

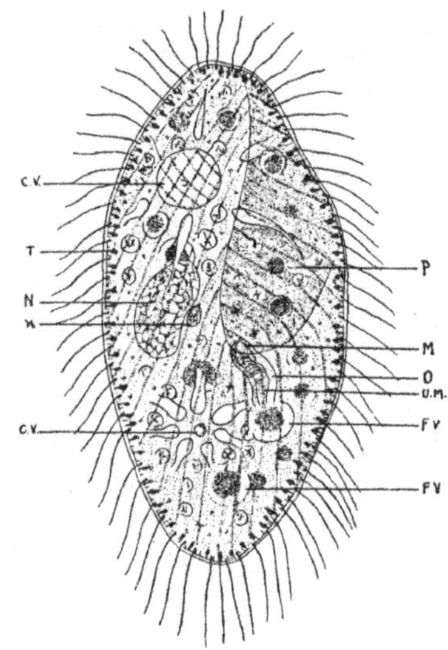

Fig. 7.—*Paramœcium caudatum.* Semidiagrammatic figure showing principal parts. *C. V.*, contractile vacuoles. *T*, trichocyst. *N*, *n*, mega- and micronuclei. *P*, peristomial groove. *M*, mouth. *O*, œsophagus, with undulating membrane, *U. M. F. V.*, food vacuoles. (*After Lang.*)

connected.   (See Chapter IV.)   Secretory cells are often distinguishable not only by the accumulations of secretion products in their cytoplasm, but also by the peculiar forms assumed by their nuclei (Fig. 17, *A, C*). The cells of connective tissue (Fig. 6, *A*) form many long interlacing processes and lie in a supporting matrix which represents their secretions. Cartilage and bone cells (Fig. 6, *B*) are likewise imbedded in their secretions, which are here produced in relatively enormous amounts and, where present, constitute the main supporting framework of the body.

We thus see that the life of the complex multicellular organism is dependent upon the correlated activities of a multitude of cells performing many diverse special functions.   It is a remarkable fact that all of the functions delegated, as it were, to different cells (contractility, motility, mechanical support, the reception and conduction of stimuli, secretion, and excretion), as well as those general functions common to all cells (nutrition and reproduction by division), may in the protozoa and protophyta be carried on within the limits of a single cell. Such a cell as, for example, the body of a *Paramœcium* (Fig. 7), exhibits a corresponding regional differentiation in structure, certain functions being localized in definitely developed organs. Differentiation is therefore something which, fundamentally, does not require multicellular structure for its expression; in fact the most important single step ever taken in differentiation was that which set apart nucleus and cytoplasm, giving the type of organic unit common to all subsequently evolved organisms. It is further evident, however, that the evolution of the higher organisms has unquestionably been very largely conditioned by the multicellular state, and has involved a progressive division of labor in a very real sense. The many functions of a single cell have become distributed among a number of cells in such a way that there has been produced a harmonious whole which is efficient, adaptable, and progressive to a degree not otherwise attainable.

Bibliography 2

See Bibliography 1*A* for general works on the cell and for reviews of early cell literature. For the latter see especially the works of Boveri, Flemming, Koernicke, Mark, Meves, Rückert, Waldeyer, Whitman, and Zimmermann. Other works referred to in Chapter II:

HANSTEIN, J. 1880. Das Protoplasma als Träger der pflanzlichen und thierischen Lebensverrichtungen. Heidelberg.

HARPER, R. A. 1919. The structure of protoplasm. Am. Jour. Bot. 6: 273–300.

MOORE, B. 1912. The Origin and Nature of Life. N. Y. and London.

CHAPTER III

PROTOPLASM

In his famous essay on protoplasm in 1868 Huxley very fittingly referred to it as "the physical basis of life." With a realization of the full significance of this phrase there comes the conviction that protoplasm is the most interesting and important substance to which we can turn our attention, for with it the phenomena of life, in so far as we know them, are invariably associated.

In spite of the enormous amount of work which has been done upon protoplasm during many years, our knowledge of it must still be regarded as very superficial and fragmentary. We can scarcely yet say definitely that a given kind of protoplasm is not a single complex chemical compound, as is held by one prominent school of biochemists: all ordinary analysis seems to indicate that it represents a somewhat looser combination of substances, many of which are in turn very elaborate in composition; and further that these substances probably differ from those found elsewhere not in any fundamental manner, but only in the degree of their complexity. Proteins, fats, crystalloids, water, and other compounds make up protoplasm, but protoplasm is not a mere mixture of these materials; it is organized—it is a *system* of complex substances, the activities of which are fully coördinated. Only if we recognize in protoplasm an organization can we conceive of it as a physico-chemical substratum for those peculiar orderly activities characterizing living substance, namely, synthetic metabolism, reproduction, irritability, and adaptive response.

Physical Properties.—Certain early ideas regarding the physical nature of protoplasm may be briefly reviewed at this point.

Protoplasm appeared to its earliest observers merely as a colorless, viscid substance containing minute granules. Two general opinions soon developed: some held that protoplasm consists of but a single fluid, whereas others regarded it as a combination of two fluids. Brücke (1861), who was one of the first to lay emphasis on the fact that protoplasm is an organized substance, looked upon the cell body as a contractile, semi-solid material through which there streams a fluid carrying granules. Similar to this was the idea of Cienkowski (1863), who believed he saw in the protoplasm of myxomycetes two fluids, one of them hyaline and only semi-fluid (the "ground substance"), and the other a more limpid fluid with granules suspended in it. De Bary (1859,

1864), on the other hand, regarded protoplasm as a single semi-fluid substance, contractile throughout, but showing many local differences due to varying water content. To this general view the work of Hanstein (1870, 1880, 1882) lent support.

Much more prominent have been the structural theories associated with the names of Klein, Flemming, Altman, and Bütschli, and known respectively as the "reticular," "fibrillar," "granular," and "alveolar" theories.

The reticular theory, which was formulated by Fromman (1865, 1876, 1884), was developed especially by Klein (1878–9) and supported by van Beneden, Carnoy, Leydig, and others. These workers saw in protoplasm a reticulum or fine network of a rather solid substance (*spongioplasm*), which holds a fluid and granules in its meshes. This view was adopted for a time by Strasburger.

The fibrillar, or filar, theory, announced by Velten (1873–6) as a result of his observations on *Tradescantia* and other forms, stated that protoplasm is composed of fine fibrils, which, though often branched, do not form a continuous network. This idea was developed mainly by Flemming (1882), who called the substance of the fibrils *mitome* and the fluid bathing them *paramitome*. Some observers asserted that the fibrils are in reality minute canals filled with a liquid, the granules seen by others being merely sections of these canals. An extreme view was that of Schneider (1891), who thought the entire cell might consist of but a single greatly convoluted filament.

To the followers of the reticular and fibrillar theories the fluid held between the fibers was known variously as *ground substance*, *enchylema* (Hanstein 1880), *hyaloplasma* (Hanstein), *paramitome*, and *inter-filar substance*. The granules were known generally as *microsomes* (Hanstein).

According to the granular theory protoplasm is a compound of innumerable minute granules which alone form the essential active basis for the phenomena exhibited; the observed fibrillar and alveolar structures are of secondary importance. Martin (1881) held that fibrils and networks are due entirely to certain arrangements of these granules, or microsomes. Pfitzner (1883) pointed out that the granules are semi-solid and float in a more fluid ground substance. For Altman (1886, etc.), who was the most prominent exponent of the theory, the granules were actual elementary living units, or *bioplasts*, the liquid containing them being a non-living hyaloplasm. The cell was therefore looked upon not as a unit, but as an assemblage of bioplasts, "like bacteria in a zoöglœa," and the bioplasts were believed to arise only by division of others of their kind (*omne granulum e granulo!*).

The alveolar theory, also known as the emulsion, or foam, theory, was elaborated principally by Bütschli (1882, etc.), and is of special interest in view of our present-day notions of protoplasmic structure. According to Bütschli protoplasm consists of minute droplets (averaging 1μ in diameter) of a liquid "alveolar substance" (enchylema) suspended in another continuous liquid "interalveolar substance." The structure is therefore that of an extremely fine emulsion, and the appearances described by other workers are due to optical effects encountered in examining the minute alveolar structure. Bütschli supported his theory by making artificial emulsions with soaps and oils which showed amœboid movement and other striking resemblances to living protoplasm.

The above four theories have been termed "monomorphic theories," for the reason that each of them stated that protoplasm has a single characteristic physical structure. Strasburger in 1892 and thereafter maintained that the protoplast is regularly composed of two portions; an active fibrillar *kinoplasm*, concerned primarily with the motor work of the cell, and a less active alveolar *trophoplasm*, chiefly nutritive in function. It was shown by von Kölliker, Unna (1895), and others, moreover, that one type of structure may be transformed into another. Flemming later adopted the view that no single type characterizes protoplasm, but that the latter may be homogeneous, alveolar, fibrillar, or granular— *i.e.*, it is "polymorphic." Wilson (1899) found that all four states are successively passed through in the echinoderm egg. This observation, which was made upon both living and fixed material, showed in a striking manner the colloidal nature of protoplasm (see below), since it is now known that colloids may assume very diverse structures under the influence of changing environmental conditions. The work of A. Fischer (1899), who treated non-living proteins with cytological fixing reagents and so produced artifacts similar to alveolæ, reticula, and granules, should make one cautious in drawing conclusions regarding protoplasmic structure from fixed material. It should be understood that the only trustworthy observations are those which are made at least in part on living material, for it is not difficult to discover all four types of structure in prepared slides: the protoplasm has there been coagulated by fixing reagents, and we know that in the coagulation of such substances as compose protoplasm an entirely new structure may be assumed.

Protoplasm as a Colloidal System.—For adequate reasons it is now customary to speak of protoplasm in terms of the physics and chemistry of colloids. *Colloids* are those glue-like substances which are uncrystalline, semi-solid, and very slightly or not at all osmotic. They have a high surface tension, coagulate readily, and conduct the electric current very poorly. They are "disperse heterogeneous sytems, *i.e.*, they consist essentially of particles larger than molecules of a substance or substances in a medium of dispersion which may be water or some other fluid" (Child 1915[1]). The particles range in size from those visible to the naked

---

[1] These paragraphs on colloids are based largely upon the convenient summary given by Child (1915, pp. 20 ff.). See also Czapek (1911b), Bayliss (1915), Hatschek (1916), Bechhold (1919), and Robertson (1920).

eye down to single molecules and ions. In the latter case we have a true solution: between colloid and crystalloid the line of demarcation is thus a purely arbitrary one. In many cases the suspended particles are too small to be seen with the ordinary microscope, which will not render visible a body with a diameter less than about 0.20 to 0.25 μ; but with the ultramicroscope, which will reveal particles about one-fortieth of this size, they may be clearly seen. Again, the ultramicroscope is insufficient in the case of certain colloids, in which the presence of suspended particles can still be shown, however, by the Tyndall effect (a milky appearance when a beam of light is passed through them). Most protoplasmic colloids are of this last type.

In a colloidal solution the particles are separate from one another, (*sol*), whereas in the denser "set" condition (*gel*) they are more closely aggregated and hence not free to move upon one another. A colloid may be made to pass from the sol to the gel state or *vice versa;* in some cases this change is reversible, but in others it is not.

Colloids are usually classified as suspensoids and emulsoids. *Suspensoids*, in which the particles are solid, are comparatively unstable; are readily precipitated or coagulated by salts; carry a constant electric charge of definite sign; are not viscous; do not show a lower surface tension than that of the medium of dispersion alone; and are mostly only slightly reversible. *Emulsoids*, in which the suspended particles are fluid, are comparatively stable; are less readily coagulated by salts; are either positively or negatively charged; are usually viscous; have a lower surface tension than the medium of dispersion; form surface membranes; and are highly reversible. Most organic colloids are emulsoids, and there can be no doubt that many of the characteristics of living organisms are due to their presence.

In an emulsion each physically homogenous constituent is known as a *phase*. In mayonnaise dressing, to cite a familiar example, there are three phases: a water phase, consisting of water and substances dissolved in it; an oil phase; and a protein phase (egg). These three physically diverse substances are brought into the emulsified state by beating; one of them is the medium of dispersion (external phase) and the others (internal phases) are suspended in it as liquid particles or droplets. In such an emulsion a given phase usually consists of more than one chemical substance: the water phase, for example, is not pure water, but an aqueous solution of salts and other water-soluble substances. These different chemical substances, including the solvent, which make up a single phase, are known as *components*.

It is shown by certain investigators (Bancroft, Clowes) that the droplets of a suspended phase in a stable emulsion are bounded by films of different constitution: between the phases of an alkaline water-oil emulsion, for example, there appear to be delicate films of a soapy nature.

These films not only prevent the coalescence of the droplets, but also, through alterations in surface tension, influence the transposition or inversion of phases which occurs under certain conditions, whereby the suspended phase becomes the medium of dispersion and *vice versa* (Fig. 8). Such inversion probably plays an important rôle in many cases of transformation of sol into gel and of gel into sol.

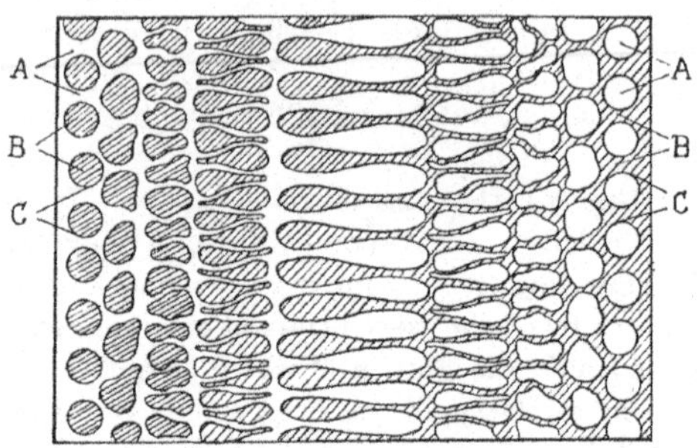

FIG. 8.—Diagram of a colloidal emulsion, illustrating transformation of emulsion of oil in water to emulsion of water in oil.

A, aqueous phase. *B*, oil or other non-aqueous phase. *C*, surface film of soap or other dispersing agent. (*After Clowes*, 1916.)

The properties and behavior of colloidal substances in general appear to be due primarily to the enormous extent of the reacting surface between the constituent phases which results from the finely divided state of one or more of them. In the accompanying table is shown the amount of surface which a given mass of matter may expose when subdivided into successively smaller particles.

TABLE SHOWING THE INCREASE OF SURFACE WITH THE SUBDIVISION OF 1 C.C. OF MATTER IN THE FORM OF A CUBE (DATA PARTLY FROM HATSCHEK, 1919.)

Length of edge of cube	Number of cubes	Total surface exposed
1 cm.	1	6 sq. cm.
1 mm.	1,000	60 sq. cm.
100μ	10^6	600 sq. cm.
$10u$	10^9	6,000 sq. cm.
1μ	10^{12}	6 sq. m.
$100\mu\mu$	10^{15}	60 sq. m.
$10\mu\mu$	10^{18}	600 sq. m.
$1\mu\mu$	10^{21}	6,000 sq. m.

The evidence at hand supports the view that protoplasm is essentially a colloidal solution of the emulsion type. It consists of at least three principal phases: a water phase, containing a number of dissolved components; a fat phase, consisting of fats and fat-soluble components; and a complex protein phase. Since the water phase is here the medium of dispersion protoplasm is classed as a "hydrosol" in its ordinary state, or as a "hydrogel" in the set condition. There are doubtless additional minor phases present, protoplasm being in reality a "complex polyphase colloidal system."

The presence of water in protoplasm is a matter of fundamental importance. As emphatically stated by Henderson (1913), ". . . the physiologist has found that water is invariably the principal constituent of active living organisms. Water is ingested in greater amounts than all other substances combined, and it is no less the chief excretion. It is the vehicle of the principal foods and excretion products, for most of these are dissolved as they enter or leave the body [across the wall of the intestine and across the epithelia of kidneys, lungs, and sweat glands]. Indeed, as clearer ideas of the physico-chemical organization of protoplasm have developed it has become evident that the organism itself is essentially an aqueous solution in which are spread out colloidal substances of great complexity [Bechhold 1912]. As a result of these conditions there is hardly a physiological process in which water is not of fundamental importance" (pp. 75–77).

The amount of water in protoplasm varies greatly under different conditions, but normally it is present in large proportions. It makes up 85 to 95 per cent of the weight of actively streaming protoplasm such as is seen in *Elodea* and *Tradescantia*, and in actively functioning cells it rarely drops below 70 per cent. In dry spores, however, it may be reduced to 10 or 15 per cent, in which case the protoplasm becomes very viscous. The percentage differs constantly in different parts of the cell; nucleus, cytoplasm, and plastids, though all are composed primarily of protoplasm, contain very different amounts of water. Since active protoplasm is a liquid, the phenomena of surface tension and other properties of liquids must enter largely into explanations of its behavior.

The colloidal nature of protoplasm is manifested in many of its properties. Its power of adsorption, which lies at the basis of many cell reactions and certain staining processes, is similar to that of other colloids. Protoplasm, like other colloids, is semi-permeable: a semi-permeable region is probably present wherever protoplasm comes in contact with other substances, such as water; and the permeability of a vacuolate cell is in general the resultant of the permeabilities of the ectoplast, cytoplasm, and tonoplast.

Protoplasm shows most strikingly its colloidal character in the changes of physical state which it undergoes as the effects of variations in the external conditions. The alterations due to changes in temperature will serve for illustration. Above a certain temperature the colloid gelatin exists in the sol state—it is a hydrosol. If the temperature is sufficiently lowered the gelatin "sets"—it becomes a hydrogel. This setting is a reversible process: if the temperature is again raised the sol state is resumed. On the contrary, egg albumen is a hydrosol at ordinary temperatures and becomes a hydrogel when heated; in this case the change is an actual coagulation and is not reversible. Many of the colloids of the cell are of this non-reversible type. "The evidence that in this colloidal condition the transition from liquid to solid, from sol to gel, tends especially to pass into an indefinite series of gradations gave a basis for the explanation of that mixture of the properties of solids and liquids which has puzzled students of protoplasm" (Harper 1919). Protoplasm is thus easily coagulated, not only by a high temperature, but by a variety of chemical substances. The "fixation" of the cell structures by the reagents employed in cytological technique is primarily a coagulation phenomenon, and in the act of coagulation a substance, especially one as complex as protoplasm, undergoes an alteration in physical structure. Although such fixing fluids preserve very well the general structure of the cell, the effects of coagulation should always be borne in mind in interpreting finer details in preparations of fixed cells, and in evaluating the results of those who have made special studies on the ultimate structure of protoplasm.

Microdissection.—Much has been added to our knowledge of the physical nature of protoplasm in recent years through microdissection. Certain workers, notably Barber (1911, 1914), Kite (1913), Chambers (1914, 1915, 1917, 1918), and Seifriz (1918, 1920) have developed a technique (fully described by Barber 1914, and Chambers 1918) whereby they have been able to dissect living cells under the high powers of the microscope, thus opening a most promising field for investigation. Kite, working on the cells of several plants and animals, found that protoplasm exists in the form of sols and gels of varying consistency, that of plant cells being as a rule more dilute and less rigid than that of animals. The cytoplasm is commonly somewhat more viscous than is usually thought, having the consistence of a "soft gel," while the nucleus may often be surprisingly firm. (See Chapter IV.)

Chambers (1917), who gives a convenient bibliography of the subject, states that in the early germ cells and eggs of certain animals the protoplasm is in the sol state with a surface layer in the gel state, whereas adult cells are usually gels. He further asserts that the surface gel is readily regenerated after injury, a new gel film being formed over the injured area. As regards the structure of protoplasm, he finds it to consist of a hyaline fluid carrying microsomes and macrosomes, which measure less than 1μ and from 2–4μ in diameter respectively. Upon disorganization the macrosomes, which are more sensitive to injury than are the microsomes, swell and go into solution, while the hyaline fluid flows out and mixes with water or coagulates, forming a reticular or granular structure.

Seifriz (1920) has investigated with care the viscosity of the protoplasms of a number of myxomycetes, algæ, pollen tubes, protozoans, and echinoderm eggs. He finds the degree of viscosity to vary widely, from a very watery to a fairly rigid gel condition, not only in the different organs of the cell but also in the protoplast as a whole at different stages of its development. He warns against accepting viscosity alone as an index of the gel or sol state of the protoplasm, since it is physical structure and not viscosity which determines these states in an emulsion.

It is to be hoped that the methods employed by the above investigators will be further developed and applied more widely, for through them many misconceptions will undoubtedly be corrected.

It should be evident from all these considerations that there is probably no sinlge visible structure characteristic of protoplasm at all times. Any fundamental structure which it may have remains to be discovered in the ultramicroscopic constitution of the colloids and other materials of which it is composed, and in the physical relations which these bear to one another. It should be pointed out, however, that in the idea of protoplasm as a complex colloidal emulsion we have the best hypothesis yet offered as a basis for the interpretation of the behavior of living substance.

Chemical Nature of Protoplasm.—Chemically, as well as physically, protoplasm is exceedingly complex, and the study of its constitution has opened a field of research which is continually broadening. Only a brief summary of some of the more important chemical facts can be presented here; for more detailed accounts special works on the subject must be consulted.[1]

As already pointed out, the substances of which protoplasm is composed are probably not fundamentally different from those found elsewhere, but show rather a greater complexity and a high degree of organization. Protoplasm is an intricately organized system of water, proteins, enzymes, fatty substances, carbohydrates, salts, and other minor constituents. The often cited analysis by Reinke and Rodewald (1881) of the myxomycete *Æthalium septicum* (*Fuligo*) showed the protoplasm of this form to have the following composition:

PER CENT DRY WEIGHT		PER CENT DRY WEIGHT	
Proteins	40	Cholesterin (lipoid)	2.0
Albumins and enzymes	15	Ca salts (except $CaCO_3$)	0.5
Other N compounds	2	Other salts	6.5
Carbohydrates	12	Resins	1.2
Fats	12	Undetermined	6.5

The protein matter of protoplasm exists in relative complex forms. "The chief mass of the protein substances of the cells does not consist of

<hr>

[1] See especially the books of Hammarsten (1909), Wells (1914), Czapek (1915), Bayliss (1915), Mathews (1916), Palladin (1918), Robertson (1920), and the review by Zacharias (1909).

proteids in the ordinary sense, but consists of more complex phosphorized bodies . . ." (Hammarsten). Such "phosphorized bodies" are the nucleo-proteins, which are "probably the most important constituents of the cell, both in quantity and in relation to cell activity" (Wells). A long series of chemical investigations beginning with the pioneer work of Miescher, Hoppe-Seyler, and Reinke, have shown that these nucleo-proteins are essentially combinations of nucleic acid with proteins, or sometimes with the simpler histones or the still simpler protamines.

The nucleus as a rule is free from or very poor in uncombined carbohydrates, fats, and salts, but is characterized rather by the abundance of a nucleo-protein called *nuclein*, isolated in 1871 by Miescher, who gave it the formula $C_{29}H_{49}N_9P_3O_{22}$. It was shown by Altman (1889) that nuclein, like the other nucleo-proteins, could be split into two substances: nucleic acid and a form of albumin (protein), the two existing in chemical combination like an ordinary salt. Nucleic acid from yeast was given the general formula $C_{40}H_{59}N_{14}O_{22} - 2P_2O_5$, and that from fish sperm $C_{40}H_{56}N_{11}O_{16} - 2P_2O_5$. Nucleic acid was further analysed into phosphoric acid and certain bases. The relation of these simple substances to nuclein, and also the relation of nuclein to more complex nucleo-proteins, are shown in the following scheme (mainly from Wells):

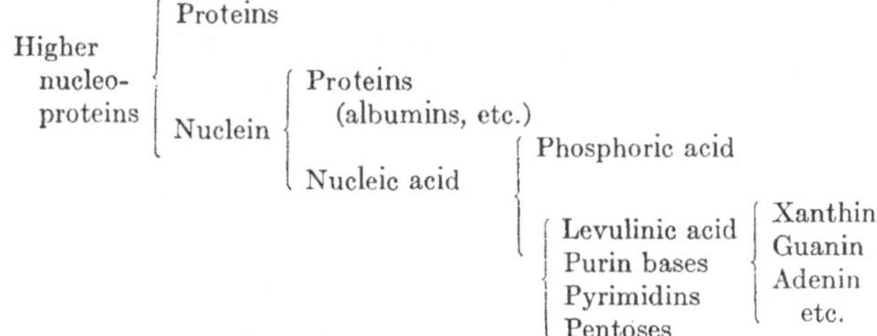

The nucleo-proteins of the nucleus (chromatin) contain very little of the protein constituent and are thus relatively rich in phosphorus. Glaser (1916) accordingly speaks of chromatin as "a conjugated phospho-protein group with a nucleic acid group, the latter group being a complex of phosphoric acid and a nuclein base." Kossel (1889, 1891, 1893) even concluded that in certain instances (during mitosis) chromatin might be simply nucleic acid.

In the cytoplasm, on the contrary, the proportion of the protein constituents is relatively high. The cytoplasm probably has no true nuclein, but is rich in nucleo-albumins, albumins, globulins, and peptones, which, unlike nuclein, have little or no phosphorus. As a result its reaction is alkaline, in contrast to the acidity of the nucleus. According to Hammarsten (1909), "the globulins and albumins are to be considered as nutritive materials for the cell or as destructive products in the chemical transformation of the protoplasm." Granules of "volutin" formed in the cytoplasm are also looked upon as a food substance used by the nucleus in the elaboration of chromatin.

The fatty components of the cell comprise both ordinary fats and lipoids (fat-like bodies not decomposed by alkalis); among the latter lecithin and cholesterin are of great importance, particularly in the cells of animals.

The carbohydrates found in protoplasm are chiefly pentoses and hexoses, which are as a rule combined with proteins and with lipoids. Glycogen exists free in the cells of many tissues and serves as a source of heat and energy. The important rôle played by pentosans in the activity of the plant cell is strongly emphasized by Spoehr (1919) and Macdougal (1920); in fact these authors speak of protoplasm as "an intermeshed pentosan-protein colloid."

Inorganic salts are present in considerable variety, as shown by the presence of the following elements in the ash of *Fuligo* protoplasm: Cl, S, P, K, Mg, Na, Ca, Fe.

Because of their failure to find any new types of chemical compounds in their analysis of protoplasm Reinke and Rodewald (1881) thought it probable that the peculiarities of protoplasm are due to its structure rather than to its chemical composition. It has since been found, however, that certain of the life processes continue for a time after the protoplasm has been ground up mechanically. Moreover, more refined analytical methods have enabled chemists to isolate from protoplasm certain extremely complex and unstable proteins (the "protoplasmids" of Etard), which differ greatly in degree of complexity, if not otherwise, from proteins encountered elsewhere.

Varieties of Protoplasm.—From the foregoing résumé it is plain that in protoplasm, because of the many combinations possible among constituents present in such great variety, we have a substance which may exist in a vast number of different forms. When it is further recalled that many of the constituents exhibit singly the phenomenon of stereoisomerism this number is seen to be incalculable. For example, it was shown by Miescher that an albumin molecule with 40 carbon atoms could have about one billion stereoisomers, and some albumins probably have more than 700 carbon atoms. Albumin, moreover, is only one of many complex substances present in protoplasm. Hence, the statement that all living cells are composed of the same substance, protoplasm, is true only in a general sense. Although they are made up of the same categories of substances existing in the same general type of organization—the hydrocolloidal state—the protoplasms of different organisms vary widely in the relative amounts of these leading constituents. For example, the lipoids are much more abundant in the protoplasm of animals than in that of plants, and the carbohydrate-protein ratio also shows notable differences in the two kingdoms. Analogous differences also exist between the smaller plant and animal groups, and with these differences in chemical constitution are associated many characteristic diversities in metabolic activity. Thus it is not simply with protoplasm but with protoplasms that the working biologist has to deal.

Special emphasis has been placed upon the relation of this great diversity in the constitution of protoplasms to the amazing variety observed among living organisms by Kossel, Reichert, and a number of other writers. As Reichert states, the evidence seems to indicate that "in different organisms corresponding complex organic substances that constitute the supreme structural components of protoplasm and the major synthetic products of protoplasmic activity are not in any case absolutely identical in chemical constitution, and that each substance may exist in countless modifications, each modification being characteristic of the form of protoplasm, the organ, the individual, the sex, the species, and the genus." With regard to the integration of the various protoplasmic constituents, Mathews (1916) says: "Protoplasm, that is the real living protoplast, consists of a gel, or sol, which is composed of the colloids of an unknown nature which include protein, lipin and carbohydrate. Whether these colloidal particles consist of one large colloidal compound in which enzymes, protein, phospholipin and carbohydrate are united to make a molecule which may be called a biogen [Verworn 1895, 1903], cannot be definitely stated, but it seems probable that something of the sort is the case."

The Plasma Membrane.—It was recognized very early that there is at the surface of the protoplast a thin layer of relatively resistant, hyaline protoplasm which Hanstein called *ectoplasm*, distinguishing it thus from the granular *endoplasm* within. Pfeffer (1890) employed the corresponding terms *hyaloplasm* and *polioplasm*. The ectoplasmic envelope, which is best seen on "naked" masses of protoplasm, such as amœbæ, myxomycetes, and the zoöspores and gametes of algæ, has been variously referred to by different writers as the *ectoplast, plasma membrane, Hautschicht,* and *Plasmahaut.*[1]

The proponents of the reticular and fibrillar theories of the structure of protoplasm looked upon this external layer as a region in which the fibrils are more closely compacted or interwoven, whereas Bütschli regarded its relative firmness as due to a compact radial arrangement of alveolæ. Pfeffer (1890) held that such a limiting membrane, which living protoplasm always produces on an exposed surface and which consists mostly or entirely of protein substances, is not itself truly protoplasmic, whereas the majority of cytologists have thought it to be a

---

[1] A discussion of ectoplasm and endoplasm based upon a large number of observations on *Amœba* is given in a new work by Schaeffer (1920).

special protoplasmic layer: Strasburger, for instance, believed it to be composed of kinoplasm.

The microdissection studies of Kite (1913) and Chambers (1917) mentioned above have extended our knowledge of the physical nature of the plasma membrane. Both of these observers describe the ectoplast of an *Amœba* as a concentrated gel. Seifriz (1918), as a result of such studies on the *Fucus* egg and myxomycetes, states that the membrane is a definite morphological structure, very elastic and glutinous, and capable of constant repair. He further asserts that membrane formation is a physical process dependent upon the physical state of the protoplasm and not upon that of the medium, and that it does not occur after death.

That the formation of such a limiting membrane at the surface of protoplasm is the result of the tendency of colloidal particles to accumulate on any interface has been pointed out by physical chemists. Citing, by way of illustration, the film which forms on the surface of cooling milk, Moore (1912) says: "The chief colloid of the milk, on account of its affinities, accumulates on the surface, the accumulation gives increased concentration, the presence of the increased concentration causes the multi-molecules to build together, the larger molecules fall out of solution as particles, and these join to form a close network or film." In a similar manner the unicellular organism or other mass of naked protoplasm develops its resistant envelope, and the enclosed protoplast of the higher plant its ectoplasmic layer and tonoplast.

Permeability.—The physico-chemical nature of the plasma membrane has been a subject of much discussion among physiologists. On the assumption that the permeability of the cell is a case of solubility in the ectoplasm, E. Overton (1895, 1899, 1900) developed a theory of the constitution of the ectoplast. It was pointed out first, that the ectoplast is not miscible with water; second, that in plant and animal cells the only bodies which are not miscible with water in the ordinary state are fats and oils; third, that the ectoplast is more or less permeable to substances according as the latter are more or less soluble in fats and oils; and fourth, that any substance insoluble in another substance will not pass through a membrane composed of the latter. It was therefore concluded that the ectoplast is made up of some lipoid compound, such as lecithin, which acts as a semi-permeable membrane. This theory, though very suggestive, was effectively opposed by Ruhland (1909, 1915) and a number of other investigators, who called attention to many substances which do not behave according to the requirements of the theory stated in so simple a form. A more nearly adequate conception of the constitution of the ectoplast has thus been souhgt.

Of the more recent theories which have been offered in connection with the problem of permeability the most promising are those which interpret the ectoplast as an emulsion. According to Czapek (1910, 1911, 1915) the ectoplast is an emulsion of lipoids, proteins, and other substances, the lipoids forming a suspended phase. "Protoplasm is a colloidal emulsion of lipoids in hydrocolloidal media, the latter containing proteins and mineral salts." Lepeschkin (1910, 1911) advanced the contrary view that the lipoids form the medium of dispersion. In attempting to account for changes in permeability Clowes (1916) points out that inversion of phases probably plays an important rôle, while Spaeth (1916) ascribes changes in permeability to alterations in the degree of dispersion of the colloids, with resulting changes in the viscosity of the membrane. A more definitely stated hypothesis of the latter type is that tentatively suggested by Lloyd (1915) and Free (1918). Colloids are known that "have two liquid phases which differ in composition only in the relative proportion of water and of the substance of the colloid" (Free). It is accordingly possible that alterations in permeability may be due to changes in the distribution of water between two such phases present in the plasma membrane. When water passes from the internal (suspended) to the external (continuous) phase, the droplets of the former would become very small; when the movement is in the opposite direction they would become very large and closely packed. As a result there would be such changes in the physical nature of the membrane as would aid in interpreting the behavior of the latter toward substances entering or leaving the cell. It is held that such a hypothesis accounts more readily for the gradual changes in permeability observed than does the inversion theory of Clowes, according to which the change might be expected to occur suddenly. It is pointed out, however, that both processes are probably involved.

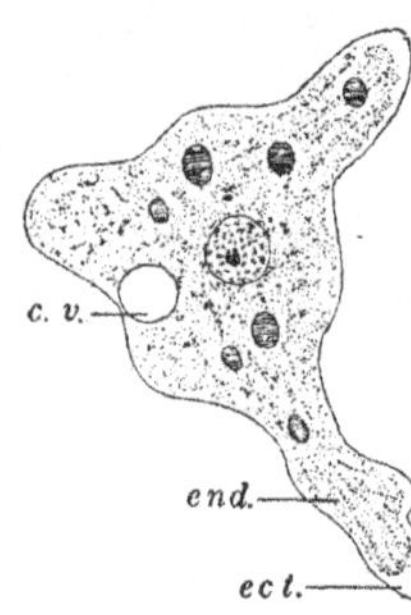

Fig. 9.—*Amœba*, showing ectoplasm, endoplasm, and contractile vacuole.

Whatever the degree of correspondence between the above interpretations and reality may be, it is scarcely open to doubt, especially since the work of Bancroft (1913) and Clowes (1916) on colloids, that in such theories we have our best prospect of reaching an adequate knowledge of the plasma membrane, which, because of its great importance in the life of the cell, is to be regarded as a definite "osmotic organ."

Protozoa.—It is in the Protozoa that the ectoplast shows its most elaborate structural differentiations. (See Minchin, 1912, Chapter V.) Here the ectoplast clearly has several functions: protective, motor, excretory, and sensory. In most forms other than the Sarcodina there is a resistant envelope of some sort. This may represent (*a*) the entire ectoplast modified (the "periplast" of Flagellata); (*b*) a superficial modified layer of the ectoplast (the "pellicle" of Infusoria and some Amœbæ); (c) a secreted layer ("cell membrane") rather than a modification of the ectoplast. In certain cases definite actively protective organs, the *trichocysts*, are differentiated in the ectoplasm.

Among the ectoplasmic structures with a motor function the simplest are the *pseudopodia;* in the larger ones there is a core of endoplasm (Fig. 9), but the more delicate "filose" ones consist entirely of ectoplasm (Fig. 10). The flagellum of *Euglena* was reported by Bütschli to have an elas-

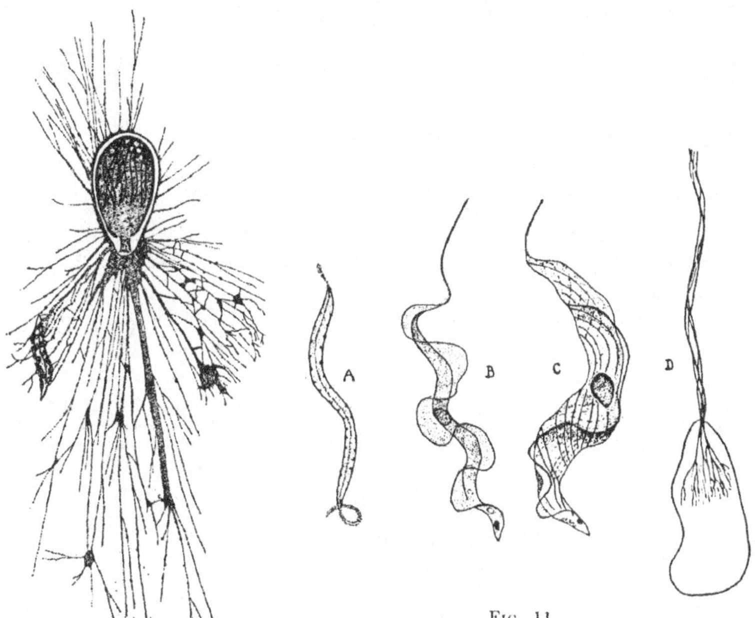

Fig. 10.—*Gromia oviformis,* showing filose-reticulate pseudopodia composed of ectoplasm. (*From Minchin, after Schultze.*)

Fig. 11.

A, flagellum of *Euglena,* showing endoplasmic core and ectoplasmic sheath. (*After Bütschli.*) B, *Trypanosoma tincæ,* with undulating membrane. (*After Minchin.*) C, *Trypanosoma percæ,* showing myonemes. (*After Minchin.*) D, flagellum of *Euglena.* (*After Dellinger.*)

tic endoplasmic core with a contractile ectoplasmic sheath (Fig. 11, *A*), but the later figure of Dellinger (1909) represents it as composed of four twisted filaments ending within the animal as a system of branching rootlets (Fig. 11, *D*). *Cilia,* which are short and numerous and show rythmic pulsation; *cirri,* which are formed of tufts of cilia; *membranellæ,* representing fused rows of cilia; and *undulating membranes,* which are sheet-like extensions of the ectoplasm (Fig. 11, *B*), are all essentially ectoplasmic organs. A further motor differentiation is seen in the minute contractile fibrils known as *myonemes,* which are analogous to a system of muscle fibers (Fig. 11, C). In ciliated forms they run beneath the rows of cilia.

Contractile vacuoles, which exercise an excretory function, originate in the ectoplasm, although later they may lie much deeper.

A sensory function is performed by the "eyespot," which is sensitive to light, and also by the flagellæ and cilia, which are often receptors of tactile stimuli. The eyespot seems in some instances to be plastid-like in character, and will be discussed in Chapter VI.

Protoplasmic Connections.—The fine protoplasmic strands (*Plasmodesmen*) connecting many plant cells through pores in the intervening walls are extensions of the ectoplasm (Fig. 12). Several early workers

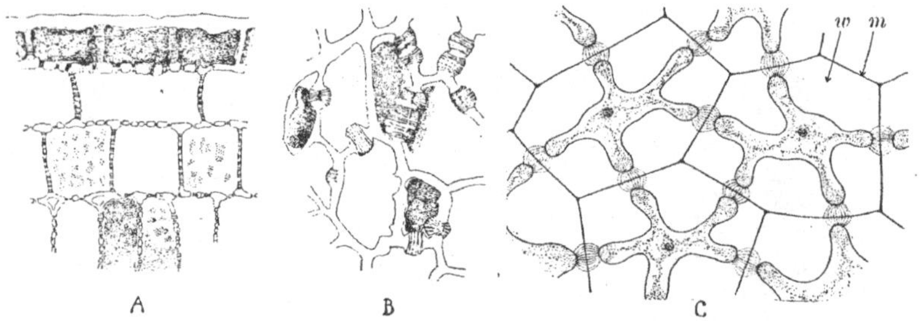

FIG. 12.—Protoplasmic connections in vascular plants.

A, B, Pinus pinea: cells of cotyledon. × 375. (*After Gardiner and Hill,* 1901.) *C, Phytelephas* ("vegetable ivory"): endosperm cells with greatly thickened walls (*w*), showing spindle-shaped bundles of connecting strands. *m*, middle lamella. Semidiagrammatic.

suspected the presence of such connections before they were able to see them, and even the coarse strands passing through the sieve plates of sieve tubes, though often observed, were not well known until the time of Hanstein's work in 1864. The finer strands of other plant tissues where described in a large number of researches between 1880 and 1900. Among these may be mentioned those of Wille (1883) and Borzi (1886) on the Cyanophyceæ; Kohl (1891), Overton (1889), and Meyer (1896) on the Chlorophyceæ; Hick (1885) on the Fucaceæ; Hick (1883), Massee (1884), and Rosenvinge (1888) on Florideæ; and, on vascular plants, those of Tangl (1879), Russow (1882), Strasburger (1882, 1901), Goroschankin (1883), Terletzki (1884), Wortman (1887, 1889), Haberlandt (1890), Kienitz-Gerloff (1891), Jönsson (1892), Kuhla (1900), Gardiner (1884, 1897, 1900), Hill (1900, 1901), and Gardiner and Hill (1901).

With respect to the origin and development of these connecting strands very little is accurately known. Some observers have claimed that the pores through which they pass are present from the time the primary wall is first formed, no wall substance being laid down at these points. Gardiner (1900) believed them to arise directly from the median portion of the fibers of the achromatic figure at the close of mitosis. His observations were made on the endosperm of *Lilium* and *Tamus*. Others, on the contrary, have regarded them as secondarily developed structures. Their absence from the walls between *Cuscuta* and *Viscum* and their hosts (Kienitz-Gerloff, Kuhla, Strasburger 1901), and also from many cells which glide over one another during growth, is a fact opposed to the latter interpretation. Although they have been demonstrated in a number of kinds of tissue they probably do not occur so widely as some have supposed; but it may nevertheless be true that in many cases their apparent absence is due to the fact that the special methods often necessary to their demonstration have not been widely employed.

As to their function, it can scarcely be doubted that they may serve to transmit stimuli of one kind or another from cell to cell (Pfeffer 1896). Noteworthy in this connection is their presence in tissues of plant parts known to be particularly responsive to external stimuli, such as the leaves of *Mimosa* (Gardiner 1884) and *Dionæa* (Gardiner 1884; Macfarlane 1892), the stamens of *Berberis* (Gardiner 1884), and the sensitive labellum of the orchid, *Masdevallia muscosa* (Oliver 1888). Their extensive development in storage tissues, such as the endosperm of seeds (Tangl 1879; Gardiner 1897), would also indicate that they are in part responsible for the readiness with which nutritive materials are translocated in such specialized tissues.

Vacuoles.—Vacuoles in the cytoplasm are more characteristic of plant than of animal cells. They are usually absent in the very young cell, but appear as growth and differentiation progress. In case they are very small and numerous the cytoplasm takes on an alveolar appearance, but more commonly they coalesce to form one large vacuole which may occupy a volume greater than that of the protoplast itself (Fig. 2). This condition is characteristic of many mature cells of plants, but is comparatively rare in animals.

The ordinary vacuole is essentially a droplet of fluid, consisting of water with differentiation products in solution, surrounded by a delicate limiting membrane. DeVries (1885) developed the theory that vacuoles are derived from "tonoplasts." The tonoplasts were believed to be small bodies imbedded in the cytoplasm and multiplying by fission. Through the absorption of water they swell and become vacuoles, the vacuole wall thus being made up of the material of the tonoplast body. We still refer to the vacuole wall as the *tonoplast*. De Vries looked upon the vacuole as a body with an individuality somewhat similar to that of a nucleus, since the tonoplast from which it develops was supposed to arise from a preëxisting tonoplast by division. The theory was supported by certain other workers, but it does not enjoy wide acceptance today

It has been found by Bensley (1910) and others that there is in the cytoplasm of certain comparatively young cells a system of fine canals which later open up to form vacuoles (Fig. 13). The fixing reagents commonly employed in cytological technique destroy these *canaliculæ*; and since Bensley, by using special reagents, demonstrated such canals in the familiar cells of the onion root tip, it is highly probable that they occur very widely. It seems more reasonable oo suppose that the fluid differentiation producos, when they are first forming, gradually come to move along certain paths, forming canals, and later accumulate in the form of vacuoles, than to suppose that the vacuoles originate in such individualized units as the tonoplasts of deVries.

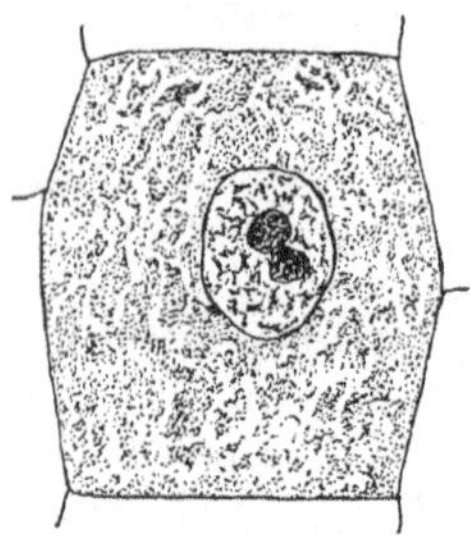

Fig. 13.—Cell from root tip of *Allium cepa*, showing canaliculæ. (*After Chamberlain.*)

Fluids other than water may also occur in the form of vacuoles; oil vacuoles, for example, are not uncommon in oertain cells. If fats, oils, and other products of metabolism take their origin in chondriosomes, as some suppose (see Chapter VI), it is not improbable that something at least analogous to the above mentioned tonoplast behavior may occur in the case of certain substances appearing in the cell. The cell sap and other differentiation products in the cytoplasm will be discussed further in Chapter VII.

PROTOPLASM AS THE SUBSTRATUM OF LIFE

Since the true significance of protoplasm was first recognized in the middle of the last century many suggestions have been ventured regarding the nature of the relation existing between life and its physical basis. A full discussion of this subject obviously cannot be entered upon here, but theories of two types, the micromeric and the chemical, may be cited by way of illustration.

Micromeric Theories.—Many years ago there were developed certain speculative "micromeric theories" of the constitution of protoplasm; these became particularly prominent during the latter half of the nineteenth century. According to these "atomic theories of biology" the principle of life was held to reside in ultimate fundamental particles. The particles were supposed to be for the most part of ultramicroscopic size, capable of independent growth and reproduction, and associated like members of a vast colony in protoplasm. Such vital units were compared by some to chemical molecules, but they were generally regarded as something much more complex. Examples of such units were the "organic molecules" of Buffon, the "microzymes" of Béchamp, the "physiological units" of Spencer, the "plastidules" of Maggi and Haeckel. the "bioplasts" of Altman, the "vital particles" of Wiesner, the "gemmules" of Darwin, the "biophores" of Weismann, the "pangens" of de Vries, and the "ergatules" and "generatules" of Hatschek.

In a somewhat similar manner a number of the later investigators occupied with the study of the ultimate structure of protoplasm have often been led to inquire which of the constituents of protoplasm are the actually living elements. Among those who viewed protoplasm as a reticular structure some held the material of the reticulum to be the true living substance, the liquid ground substance being lifeless, whereas others held the reverse to be true. Many of those who saw in protoplasm a granular structure regarded the granules as the ultimate living units, and more recently there has even been a tendency on the part of some investigators (Beijerinck, Lepeschkin), who have emphasized the emulsion nature of protoplasm, to view the droplets of the suspended phase in a similar light. To Bütschli the continuous phase was the essential substance.

By most modern biologists such attempts to assign the principle of life to any particular constituent unit of protoplasm or of the cell, whether this unit be an observed structural component or a purely imaginary one, are regarded as not in harmony with an adequate modern conception of the term "living." It has been repeatedly emphasized that life should be thought of not as a property of any particular cell constituent, but as an attribute of the cell system as a whole (Wilson 1899); or, as Brooks (1899) put it, not merely as a property but as a relation or adjustment between the properties of the organism and those of its environment. This recalls Herbert Spencer's characterization of life as a "continuous adjustment of internal relations to external relations." As Sachs (1892, 1895) and others urged, the various elements in the cell should be referred to as active and passive rather than living and lifeless. These elements play various rôles in the cell's activity: each contributes to the orderly operation of the whole. When any part fails to function properly, or when the proper adjustment is not maintained, the whole system of correlated reactions, the resultant of which we call life, must become disorganized. As Child (1915) remarks, the theories postulating vital units only transfer the problems of life from the organism to something smaller; the fundamental problem of coördination is no nearer solution than before, and the whole question is placed outside the field of experimentation. Harper (1919) also points out that modern cytology no longer looks upon protoplasm as a substance with a single specific structure, or as one made up of ultimate fundamental units of some kind, but rather as a colloidal system or group of systems of varying structure and composition. "The fundamental organization of living material is expressed in the structure of the cell." The cell itself, and not some hypothetical corpuscle, is the unit of organic structure. Protoplasm is accordingly not made up of structural units arranged in various ways to form the cell organs, but is rather a colloidal system in which special processes and functions have become localized and fixed in certain regions; and this in turn has resulted in the evolution of organs possessing more or less permanence.

Chemical Theories.—Much more suggestive, if not conclusive, have been certain attempts to place the phenomena of the organism upon a purely chemical basis. With the development of organic chemistry from the time of Wöhler's (1828) synthesis of urea onward there has grown up the idea that life processes and chemical reaction not only resemble each other but are actually the same fundamentally. When protoplasm was subjected to chemical analysis and found to consist chiefly of water and proteins, and when these substances became more intimately known, the task of explaining the activity of protoplasm in terms of the chemistry of proteins was undertaken. One group of workers developed the hypothesis that peculiarly labile protein molecules are responsible for the organism's reactions, "death" being primarily a change from the labile to the stable condition on the part of these molecules. Such molecules were called "biogens" by Verworn (1903). The molecule itself was not thought of as alive, but its constitution was held to be the basis of life, which "results from the chemical transformations which its lability makes possible." Accordingly, "life itself consists in chemical change, not in chemical constitution" (Child 1915).

Adami (1908, 1918) contends that life is thus "the function, or sum of functions, of a special order of molecules." These ultimate molecules of living matter he calls *biophores* (not to be confused with the biophores of Weismann, which were molecular complexes), and he locates them in the nucleus, the cytoplasm having merely "subvital" functions. They are proteidogenous in nature, *i.e.*, they compose an active substance which takes the form of relatively inert proteins when subjected to chemical analysis. The biophore is conceived by Adami to have the form of a ring or a ring of rings of the benzene type—a ring of amino acid radicles with many unsatisfied affinities or bonds. The biophore grows in a manner analogous to that of the inorganic crystal: ions and radicles from the surrounding medium become attached as side chains to the free bonds of the central ring and take on a grouping similar to that of the latter; in this way the biophoric molecules are multiplied. Since side chains can be detached and new ones of other kinds added, the biophore is changeable and may exist in many different forms. Although the central ring is thought to be relatively stable and fixed, the variety of side chains and their many possible arrangements probably give to each species a distinct kind of biophore. On this hypothesis the molecule of living matter (biophore) is one "of extraordinary complexity, and in a state of constant unsatisfaction, built up by linking on other simple molecules, and as constantly, in the performance of function, giving up or discharging into the surrounding medium these and other molecular complexes which it has elaborated" (Adami 1918, pp. 251–2). "All vital manifestations are manifestations of chemical change in proteidogenous matter, are, in short, the outcome of arrangement of that matter with the necessary liberation or storing up of energy" (p. 225). Accordingly, life is "a state of persistent and incomplete recurrent satisfaction and dissatisfaction of . . . certain proteidogenous molecules" (1908, Vol. I, p. 55).

Pictet (1918) also associates the phenomena of life with a special structure of the organic molecule. Only the arrangement of the atoms in open chains, he asserts, permits the manifestation of life and its maintenance; the cyclic structure is that of substances which have lost this faculty; and death results, from the chemical point of view, from a cyclization of the elements of the protoplasm.

To the theory that the vital processes are bound up with a special form of protein or protein-like molecule many have objected. For example, Höber (1911) has contended that there are present in the organism only those kinds of proteins which may be formed in the laboratory. He urges that life should not be thought of as a single process, or as dependent upon any particular kind of molecule, but rather that it should be looked upon as the result of many correlated processes occurring between many substances under certain conditions. "If we accept this idea," says Child (1915, p. 19), "we must abandon the assumption of a living substance in the sense of a definite chemical compound. Life is a complex of dynamic processes occurring in a certain field or substratum. Protoplasm, instead of being a peculiar living substance with a peculiar complex morphological structure necessary for life, is on the one hand a colloidal product of the chemical reactions, and on the other hand a substratum in which the reactions occur and which influences their course and character both physically and chemically. In short, the organism is a physico-chemical system of a certain kind."

Harper (1919) is also opposed to theories based upon the conception of protoplasm as a single complex chemical substance, as well as to those which hold protoplasm to be a relatively simple two-phase colloidal system—the alveolar and granular theories, for example. "The crude simplicity and general inadequacy of these . . . conceptions . . . have done much to bring the whole subject of protoplasmic organization into disrepute. On the other hand the conception of protoplasm as an aggregate of complex compounds, a polyphase colloidal system or system of systems, seems to do much more adequate justice to the observed facts."

Conclusion.—As stated at the opening of the present chapter, it is with protoplasm that the phenomena of life, in so far as we know them, are invariably associated. The complex behavior of the living organism can receive scientific explanation (*i.e.*, be fitted into an orderly scheme of antecedents and consequents), if at all, only on the basis of the constitution and properties of the materials composing protoplasm; the structural organization of protoplasm; the relation of the reactions and responses of protoplasm in the form of organized units or cells to the environmental conditions; the chain of energy changes occurring in connection with all of the organism's activities; and the correlation of all these conditions and events. It is largely the effort to account for organization and regulatory correlation, and the consequent behavior of the complex organism as a versatile and consistent unit or individual—as something more than a cell aggregate—that has led to certain present day vitalistic theories, as opposed to those which would hold life to be dependent upon "nothing but" the correlated physico-chemical reactions and interactions occurring in protoplasm.

Whatever our ultimate judgment in this matter shall be—for any decision at present is premature—it is scarcely to be denied that the hypotheses that have thus far been most stimulating to research in biological science and most valuable in analysing the data afforded by this research are those which seek to formulate vital activity in terms of what the physicist for convenience calls matter and energy; and which hold life to be not the manifestation of a super-organic, non-perceptual entity, or even of a distinct perceptual but hypothetical vital energy, but rather the resultant of the many correlated interactions involving only energies of known kinds. The way must not be closed, however, against possible new categories of energy. The description (reduction to order) of our perceptual experience of organic nature, which is the primary task of biological science and which has been scarcely more than begun, must for the present be made as far as possible in terms applicable also to inorganic nature. It is here that achieved results would seem to justify the judicious use of a "mechanistic" working hypothesis, whereby the attempt is made to "describe the changes in organic phenomena by the same conceptual shorthand of notation as suffices to describe inorganic phenomena" (Pearson). To what extent our ultimate biological theory is to show the need of non-mechanical energies or principles will depend very largely upon what this scientific description (orderly formulation) turns out to be like as investigation proceeds, and also upon the degree of success with which the physicist will resume the phenomena of inorganic nature in mechanical formulæ. Thus, as Professor D'Arcy W. Thompson forcefully says:

"While we keep an open mind on this question of vitalism, or while we lean, as so many of us now do, or even cling with a great yearning, to the belief that something other than the physical forces animates the dust of which we are made, it is rather the business of the philosopher than of the biologist, or of the biologist only when he has served his humble and severe apprenticeship to philosophy, to deal with the ultimate problem.   It is the plain bounden duty of the bioloigst to pursue his course unprejudiced by vitalistic hypotheses, along the road of observation and experiment, according to the accepted discipline of the natural and physical sciences. . .   It is an elementary scientific duty, it is a rule that Kant himself laid down, that we should explain, just as far as we possibly can, all that is capable of such explanation, in the light of the properties of matter and of the forms of energy with which we are already acquainted." (Presidential address before the Zoological Section of the British Association for the Advancement of Science, 1911.)

Bibliography 3

Protoplasm

ADAMI, J. G.   1908.   Principles of Pathology.   1st ed.
  1918.   Medical Contributions to the Study of Evolution.   London.
ALTMANN, R.   1886.   Studien über die Zelle.   I.   Leipzig.
  1887.   Die Genese der Zellen.   Leipzig.
  1889.   Die Struktur des Zellkerns.   Arch. Anat. u. Physiol.   p. 409–411·
  1890, 1894.   Die Elementarorganismen und ihre Beziehung zu den Zellen.   Leipzig.
  1892.   Ueber Kernstruktur und Netzstrukturen.   Arch. Anat. u. Physiol.
    223–230·
  1893.   Die Granulartheorie und ihre Kritik.   Ibid.   p. 55–66·
BANCROFT, W. D.   1913.   The theory of emulsification.   Jour. Phys. Chem. **17**:
    501–520·
  1914.   The theory of colloid chemistry.   Ibid. **18**: 549–558·
BARBER, M. A.   1911*a*.   A technic for the inoculation of bacetria and other sub_
    stances into living cells.   Jour. Infect. Diseases **8**: 348–360.
  1911*b*.   The effect on the protoplasm of *Nitella* of various chemical substances and
    of microörganisms introduced into the cavity of the living cell.   Ibid. **9**: 117–129·
  1914.   The pipette method in the isolation of single microörganisms and in the
    inoculation of substances into living cells.   Philip. Jour. Sci. **9**.
DE BARY, H. A.   1859.   Myxomyceten. Zeitschr. Wiss. Zool. **10**: 88–175·   pls. 5.
  1864.   Die Mycetozoen.   2 Aufl.   Leipzig.
BAYLISS, W. M.   1915.   Principles of General Physiology.   London.
BECHHOLD, H.   1912.   Die Kolloide in Biologie und Medizin.   Dresden.   (Engl.
    transl. by J. G. M. Bullowa, N. Y., 1919.)
BENSLEY, R. R.   1910.   On the nature of the canalicular apparatus of animal
    cells.   Biol. Bull. **19**: 174–194.   figs. 3.
BLACKMAN, F. F.   1912.   The plasmatic membrane and its organization.   New
    Phytol. **11**: 180–195.
BORZÌ, A.   1886.   Le communicazioni intracellulari delle Nostochineæ.   Malpighia
    1: 74, 97, 145, 197.   pl. 3.
BROOKS, W. K.   1899.   The Foundations of Zoology.   New York.
BRÜCKE, C.   1861.   Die Elementarorganismen.   Sitzber. Akad. Wiss. Wien II **44**:
    381–406·
BÜTSCHLI, O.   1892.   Untersuchungen über mikroskopische Schaume und das
    Protoplasma.   Leipzig.   (Engl. transl. by E. A. Minchin, 1894.)
  1898.   Untersuchungen über Strukturen.   Leipzig.
  1901.   Meine Ansicht über die Struktur des Protoplasmas und einige ihrer Kritiker.
    Arch. Entwmech. Org. **11**: 499–584·   pl. 20.
CHAMBERS, R.   1914.   Some physical properties of the cell nucleus.   Science
    40: 824–827·
  1915.   Microdissection studies on the germ cell.   Ibid. **41**: 290–293·

1917a.   Microdissection Studies. I.   The visible structure of the cell protoplasm and death changes. Am. Jour. Physiol. **43**: 1–12. figs. 2.

1917b.   Microdissection Studies. II.   The cell aster. A reversible gelation phenomenon. Jour. Exp. Zool. **23**: 483–504.

1918.   The Microvivisection Method. Biol. Bull. **34**: 121–136.

1919.   Changes in protoplasmic consistency and their relation to cell division. Jour. Gen. Physiol. **2**: 49–68.

CHILD, C. M. 1915. Senescence and Rejuvenescence. Chicago.

CIENKOWSKI, L. 1863. Zur Entwicklungsgeschichte der Myxomyceten. Jahrb. Wiss. Bot. **3**: 325–337.

CLOWES, G. H. A. 1916a. Protoplasmic equilibrium. I. Action of antagonistic electrolytes on emulsions and living cells. Jour. Phys. Chem. **20**: 407–451.

1916b. Antagonistic electrolyte effects in physical and biological systems. Science **43**: 750–757.

CZAPEK, F. 1910a. Ueber Fällungsreaktionen in lebenden Pflanzenzellen. Ber. Deu. Bot. Gesell. **28**: 147–169.

1910b. Ueber Oberfläschenspannung und den Lipoidgehalt der Plasmahaut in lebenden Zellen. Ibid. 480–487.

1911a. Ueber eine Methode zur directen Bestimmung der Oberfläschenspannung der Plasmahaut von Pflanzenzellen. Jena.

1911b. Chemical Phenomena of Life. N. Y.

DELAGE, Y. 1895. La structure du protoplasme et les theories sur l'hérédité. Paris.

DELLINGER, O. P. 1909. The cilium as a key to the structure of contractile protoplasm. Jour. Morph. **20**: 171–209. pls. 4. figs. 12.

EWART, A. J. 1903. On the physics and physiology of protoplasmic streaming in plants. Oxford.

FISCHER, A. 1899. Fixierung, Färbung und Bau des Protoplasmas. Jena.

FLEMMING, W. 1882. Zellsubstanz, Kern und Zelltheilung. Leipzig.

1892. Referate über Zelle. Ergeb. d. Anat. u. Entw. II **1**: 43-82.

FREE, E. E. 1918. A colloidal hypothesis of protoplasmic permeability. Plant World **21**: 141–150.

FROMMAN, C. 1865. Ueber die Struktur der Bindesubstanzzellen des Rückenmarks Centr. f. Med. Sci. **3**.

1875. Zur Lehre von der Struktur der Zellen. Jenaische Zeitschr. **9**. (Other papers listed.)

1884. Untersuchungen über Struktur, Lebenserscheinungen und Reaktionen thierischen und pflanzlicher Zellen. Ibid. **17**: 1–349. pls. 3.

GARDINER, W. 1884. On the continuity of the protoplasm through the walls of vegetable cells. Arb. Bot. Inst. Würzburg **3**: 52–87.

1897. The histology of the cell wall, with special reference to the mode of connexion of cells. Proc. Roy. Soc. London **62**: 100–112. figs. 8. (Prelim. note.)

1900. The genesis and development of the wall and connecting threads of the plant cell (Prelim. Comm.) Ibid **66**. 186–188.

GARDINER, W. and HILL, A. W. 1901. The histology of the cell wall with special reference to the mode of connection of cells. Phil. Trans. Roy. Soc. London B **194**: 83–125. pls. 31–35.

GATENBY, J. B. 1919. Identification of intra-cellular elements. Jour. Roy. Micr. Soc. London **2**: 93–119. figs. 14.

1920. The cytoplasmic inclusions of the germ-cells. VII. The modern technique of cytology. Quar. Jour. Micr. Sci. N. S. **64**: 267–302.

GLASER, O. 1916. The basis of individuality in organisms. Science **44**: 219–224.

GOROSCHANKIN, J. 1883. Zur Kenntniss der Corpuscula bei den Gymnospermen. Bot. Zeit. **41**: 825–831. pl. 7.

HABERLANDT, G. 1890. Das Reizleitende Gewebesystem der Sinnpflanzen. Leipzig.

HAMMARSTEN, O. 1909. A Textbook of Physical Chemistry. 5th ed. N. Y.

HANSTEIN, J. 1864. Die Milchsaftgefasse, usw. Berlin.
  1880a. Das Protoplasma als Trager der pflanzlichen und thierischen Lebensverrichtungen. Heidelberg.
  1880b. Einige Züge aus der Biologie des Protoplasmas. Hanstein's Bot. Abhandl. **4**: 1–56. figs. 10. Also Sitzber. Nieder. Ges. Bonn 1870.

HARPER, R. A. 1919. The structure of protoplasm. Am. Jour. Bot. **6**: 273–300.

HATSCHEK, E. 1916. An Introduction to the Physics and Chemistry of Colloids. 2d ed. Philadelphia.

HENDERSON, L. J. 1913. The Fitness of the Environment. New York.

HICK, T. 1883. Protoplasmic continuity in the Florideæ. Nature **28**: 581.
  1885. Protoplasmic continuity in the Fucaceæ. Jour. Bot. **23**: 97–102, 354–357. pl. 255.

HILL, A. W. 1900. Distribution and character of connecting threads in the tissues of *Pinus sylvestris* and other allied species. Proc. Roy. Soc. London **67**: 437–439.
  1901. The histology of the sieve tubes of *Pinus*. Ann. Bot. **15**: 575–611. pls. 31–33.

HÖBER. 1911. Physikalische Chemie der Zelle und der Gewebe. 3d ed. Leipzig.

HOPPE-SEYLER. 1871. Ueber die chemische Zusammensetzung des Eiters. Med.-Chem. Unters. Lab. ausgew. Chem. zu Tübingen **4**. Berlin.
  1881. Physiologische Chemie. Berlin.

JÖNSSON, B. 1892. Siebähnliche Poren in den trachealen Xylemelementen der Phanerogamen, hauptsächtlich der Leguminosen. Ber. Deu. Bot. Ges. **10**: 494–513. pl. 27.

KASSOWITZ, M. 1899. Allgemeine Biologie. Vienna.

KIENITZ-GERLOFF, F. 1891. Die Protoplasmaverbindungen zwischen benachbarten Gewebselementen in der Pflanzen. Bot. Zeit. **49**: 1, 17, 33, 49, 65. pls. 2. (Bibliography.)

KITE, G. L. 1913. Studies on the physical properties of protoplasm. I. Am. Jour. Physiol. **32**: 146–164.

KLEIN, E. 1878–1879. Observations on the structure of cells and nuclei. Quar. Jour. Micr. Sci. **18**: 315–339, pl. 16; **19**: 125–175, pl. 7.
  1879. Ein Beitrag zur Kenntniss der Struktur des Zellkerns. Centr. f. Med. Wiss. Berlin.

KOHL, F. G. 1891. Protoplasmaverbindungen bei Algen. Ber. Deu. Bot. Ges. **9**: 9–17. pl. 1.

KOSSEL, A. 1880. Ueber das Nuclein der Hefe. Zeit. Physiol. Chem. **3, 4**.
  1889. Ueber die Nucleine. Centr. Med. Wiss. pp. 417, 593.
  1893. Ueber die Nucleinsäure. Ibid.

KUHLA, F. 1900. Die Plasmaverbindungen bei *Viscum album*. Bot. Zeit. **58**: 29–58. pl. 3.

KÜHNE, W. 1864. Untersuchungen über das Protoplasma und die Contractilitat. Leipzig.

LECOMTE, H. 1889. Contribution a l'étude du liber des angiospermes. Ann. Sci. Nat. Bot. VII **10**: 193–324. pls. 21–24.

LEPESCHKIN. 1910. Zur Kenntniss der Plasmamembran. Ber. Deu. Bot. Ges. **28**: 91–103, 383–393.
  1911. Zur Kenntniss der chemischen Zusammensetzung der Plasmamembran. Ibid. **29**: 247–260. See also pp. 349–351.

LIVINGSTON, B. E. 1903. The rôle of diffusion and osmotic pressure in plants. Chicago.

LLOYD, F. E. 1915. The behavior of protoplasm as a colloidal complex. Yearbook Carnegie Inst. Washington **14**: 66–69.

MACDOUGAL, D. T. 1920. Hydration and Growth. Carnegie Inst. of Wash., Publ. 297. pp. 176.

MACDOUGAL, D. T. and SPŒHR, H. A. 1920. The components and colloidal behavior of plant protoplasm. Proc. Am. Phil. Soc. **59**: 150–170.

MACFARLANE, J. M. 1892. Contributions to the history of *Dionæa muscipula* Ellis. Contrib. Bot. Lab. Univ. Pa. **1**: 7–44. pl. 4.

MALFATTI, H. 1891–1892. Zur Chemie der Zellkerns. Ber. Naturwiss. Ver. zu Innsbruck **20**.

MANN, G. 1906. Protoplasm, its definition, chemistry, and structure. Oxford.

MARTIN. 1881. Zur Kenntniss der indirekten Kerntheilung. Arch. f. Path. Anat. u. Physiol. **86**: 57–67. pl. 4.

MASSEE, G. 1884. On the formation and growth of cells in the genus *Polysiphonia*. Jour. Roy. Micr. Soc. II **4**: 198–200. pl. 6.

MATHEWS, A. P. 1916. Physiological Chemistry.

MEYER, A. 1896. Die Plasmaverbindungen und die Membranen von *Volvox globator*, usw. Bot. Zeit. **54**: 187–215. pl. 8.

MIESCHER, F. 1871. Ueber die chemische Zusammensetzung der Eiterzellen. Med.-Chem. Unters. Lab. ausgew. Chem. zu Tübingen **4**. Berlin.

——— 1874. Die Spermatozoen einiger Wirbelthiere. Verhandl. d. Naturforsch. Gesell. in Basel **6**.

MINCHIN, E. A. 1912. An Introduction to the Study of the Protozoa. London.

MOORE, B. 1912. The Origin and Nature of Life. New York.

OLIVER, F. W. 1888. On the sensitive labellum of *Masdevallia muscosa* Rchb. f. Ann. Bot. **1**: 237–253. pl. 12.

OVERTON, E. 1889. Ein Beitrag zur Kenntniss der Gattung *Volvox*. Bot. Centr. **39**: 65, 113, 145, 177, 209, 241, 273. pls. 1–4.

——— 1895. Ueber die osmotischen Eigenschaften der lebenden Pflanzen- und Thierzelle. Vierteljahrsschr. Nat. Ges. Zurich **40**: 159–184.

——— 1899. Ueber die allgemeinen osmotischen Eigenschaften der Zelle, ihre vermuthlichen Ursachen und ihre Bedeutung für die Physiologie. Ibid. **44**: 85–135.

——— 1900. Studien über die Aufnahme der Anilinfärben durch die lebende Zelle. Jahrb. Wiss. Bot. **34**: 669–701. See Pflüger's Archiv **92**.

PALLADIN, V. I. 1918. Plant Physiology. (Engl. transl., ed. by Livingston.)

PEARSON, K. 1892. The Grammar of Science.

PICTET, A. 1919. Life and the structure of molecules. (Geneva Arch. Phys. et Nat. Sci.) Engl. Trans. in Sci. American Suppl. **87**: 50–51, 59.

PFEFFER, W. 1885. Zur Kenntniss der Kontaktreize. Unters. Bot. Inst. Tübingen **1**: 483–535. fig. 1.

——— 1890. Zur Kenntniss der Plasmahaut und der Vacuolen, nebst Bemerkungen über den Aggregatzustand der Protoplasmas und über osmotische Vorgänge. Abh. Math.-Phys. kl., Sächs. Ges. Wiss. **16**.

——— 1896. Ueber den Einfluss des Zellkerns auf die Bildung der Zellhaut.

PFITZNER, W. 1883. Beiträge zur Lehre vom Baue des Zellkerns und seinen Theilungserscheinungen. Arch. Mikr. Anat. **22**: 616–688. pl. 25.

PRINGSHEIM, N. 1854. Untersuchungen über Bau und Bildung der Pflanzenzelle. Berlin.

REICHERT, E. T. 1914. The germ plasm as a stereochemic system. Science **40**: 649–661.

Reinke und Rodewald. 1881. Die chemische Zusammensetzung des Protoplasmas von *Æthalium septicum*. Unters. Bot. Lab. Gottingen.

Reinke, F. 1882. Protoplasma-Probleme. Ibid.

1883. Ein Beitrag zur physiologischen Chemie von *Æthalium septicum*. Ibid.

Robertson, T. B. 1920. Principles of Biochemistry. Philadelphia and New York.

Rosenvinge, L. K. 1888. Sur la formation des pores secondaires chez les *Polysiphonia*. Bot. Tidskr. **17**: 10.

Roskine, G. 1917. La structure des myonèmes. Compt. Rend. Soc. Biol. Paris **69**: 363–364.

Ruhland, W. 1909. Beiträge zur Kenntniss der Permeabilität der Plasmahaut. Jahrb. Wiss. Bot. **46**: 1–54. figs. 2.

1915. Zur Kritik der Lipoid- und Ultrafiltertheorie der Plasmahaut, usw. Biochem. Zeitschr. **54**: 59–77.

Russow. 1883. Ueber die Perforation der Zellwand und den Zusammenhang der Protoplasmakörper benachbarten Zellen. Sitzber. Naturfor. Ges. Univ. Dorpat **6**: 562.

von Sachs, J. 1892. Physiologische Notizen. II. Flora **75**: 57–67.

1895. Physiologische Notizen. IX. Weitere Betrachtungen über Energiden und Zellen. Ibid. **81**: 405–434. (See Wilson 1900, p. 30.)

Schæffer, A. A. 1920. Ameboid Movement. Princeton Univ. Press.

Schneider, C. 1891. Untersuchungen über die Zelle. Arb. Zool. Inst. Wien **9**: 179.

Schwarz, F. 1887. Die morphologische und physiologische Zusammensetzung des Protoplasmas. (Rev. in Bot. Zeit. **45**: 576–583.)

Seifriz, W. 1918. Observations on the structure of protoplasm by aid of microdissection. Biol. Bull. **34**: 307–424. figs. 3. (Bibliography.)

1920. Viscosity values of protoplasm as determined by microdissection. Bot. Gaz. **70**: 360–386.

Spaeth, R. A. 1916. The vital equilibrium. Science **43**: 502–509.

Spœhr, H. A. 1919. The carbohydrate economy of cacti. Carnegie Inst. of Wash., Publ. 287. pp. 79.

Strasburger, E. 1882. Ueber den Bau und das Wachsthum der Zellhäute. Jena.

1892. Schwärmsporen, Gameten, pflanzliche Spermatozoiden und das Wesen der Befruchtung. Histol. Beitr. **4**.

1901. Ueber Plasmaverbindungen pflanzlicher Zellen. Jahrb. Wiss. Bot. **36**: 493–610. pls. 14, 15.

Tangl, E. 1879. Ueber offene Communicationen zwischen den Zellen des Endosperms einiger Samen. Ibid. **12**: 170–190. pls. 4–6.

Terletzki, P. 1884. Anatomie der Vegetationsorgane von *Struthiopteris germanica* und *Pteris aquilina*. Jahrb. Wiss. Bot. **15**: 452–501. pls. 24–26.

Unna, P. 1895. Ueber die neueren Protoplasmatheorien, und das Spongioplasma. Deu. Med. Zeit. 1895. p. 98.

Velten, W. 1876. Die physikalische Beschaffenheit des pflanzlichen Protoplasmas. Sitzber. Akad. Wiss. Wien, Math.-Nat. Kl., **73**: I 131–151.

Verworn, M. 1895. Allgemeine Physiologie. Jena.

1903. Die Biogenhypothese. Jena.

de Vries, H. 1885. Plasmolytische Studien über die Wand der Vacuolen. Jahrb. Wiss. Bot. **16**: 465–598.

Wells, H. G. 1914. Chemical Pathology. 2d edition. Philadelphia. Chapter I.

Wille, N. 1883. Ueber die Zellkerne und die Poren der Wände bei den Phycochromaceen. Ber. Deu. Bot. Ges. **1**: 243–246.

WILSON, E. B. 1899. On protoplasmic structure in the eggs of echinoderms and some other animals. Jour. Morph. **15**: Suppl. 1–23.

WORTMANN, J. 1887. Zur Kenntniss der Reizbewegungen. Bot. Zeit. **45**: 785, 801, 817, 833.

—— 1889. Ueber die Beziehungen der Reizbewegungen wachsender Organe zu den normalen Wachsthumerscheinungen. Ibid. **47**: 453, 469, 485.

ZACHARIAS, E. 1881–1893. Ueber die chemische Beschaffenheit des Zellkerns. Bot. Zeit. **39**: 169. Ueber den Zellkern. Ibid. **40**: 611. Ueber Eiweiss, Nuclein, und Flastin. Ibid. **41**: 209. Ueber den Nukleolus. Ibid. **43**: 257. Beiträge zur Kenntniss des Zellkerns und der Sexualzellen. Ibid. **45**: 281. Ueber Chromatophilie. Ber. Deu. Bot. Ges. **11**: 188–195.

ZIMMERMANN, A. 1893. Sammel-Referate 2, 3. Beih. Bot. Centr. **3**: 209–217 321–328.

CHAPTER IV

THE NUCLEUS

It is now half a century since the modern period of cytology was ushered in by a series of researches revealing the remarkable behavior of the nucleus during the critical stages of the life cycle. Because of the peculiarly intimate relation which this behavior has been shown to have to many outstanding biological problems, including that of heredity, it is largely in nuclear phenomena that cytological interest has continued to center throughout the period. The most striking of these phenomena form the subjects of several subsequent chapters: at this point we shall consider the nucleus only as it appears in the "resting" cell, *i.e.*, in the cell not undergoing division.

Occurrence.—The most conspicuous, and in some respects the most important of the cell organs is the nucleus. Whether or not we shall say that every living cell contains a nucleus will depend upon what we are to include under the term. If the chromatin or chromatin-like substances, no matter whether distributed throughout the cell in the form of granules or aggregated to form a well defined organ, be regarded as constituting a nucleus, then it follows that all plant and animal cells normally have nuclei. If, however, as certain protozoölogists prefer, the term nucleus be employed only with reference to a distinctly delimited organ, we must regard those lowly organized cells with scattered chromatic material as devoid of nuclei, although they possess, as all cells apparently do, material which performs at least the nutritive functions of a nucleus. This latter type of organization, which is found in certain members of the Protozoa and Bacteria, and also some Cyanophyceæ, will be discussed later on in connection with nuclear structure (p. 66) and cell-division (Chapter X). In myxomycetes, where simple and primitive conditions might be expected, Jahn (1908' 1911) and Olive (1907) have demonstrated the presence of definite nuclei showing mitosis and the phenomenon of chromosome reduction.

General Characters.—The vast majority of cells have one nucleus each. A few exceptions may be noted. In tapetal cells, laticiferous vessels, the internodal cells of Characeæ, and certain other cells there are often several nuclei arising by the division of one. In the Siphoneæ among green algæ (Fig. 14, *V*) and the Phycomycetes among fungi there are no cross walls in the filamentous and much branched vegetative body, so that large numbers of nuclei are associated in one extensive mass of cytoplasm. Such a body is called a cœnocyte, and the cœnocytic condition is found in a number of the lower organisms. In the Uredineæ (rusts) the typical life history is made up of two phases, with uninucleate and binucleate cells respectively. In certain Infusoria two kinds of nuclei are regularly present. Thus in *Paramœcium caudatum* (Fig. 15) there is one small micronucleus which divides by a peculiar form of mitosis, and one large meganucleus (macronucleus) which divides amitotically. In *P. aurelia* there are two micronuclei and one meganucleus, whereas in

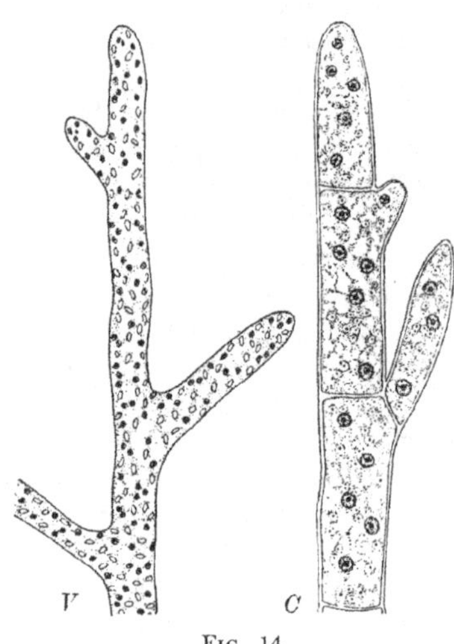

Fig. 14.

V, portion of body of *Vaucheria*, showing cœnocytic condition; nuclei dark and plastids in outline. *C*, portion of body of *C l a d o p h o r a*, showing semi-cœnocytic condition.

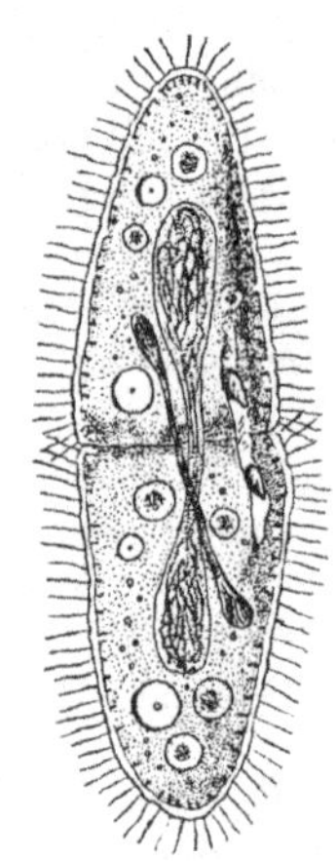

Fig. 15.—*Paramœcium caudatum* undergoing fission; mega- and micronuclei dividing. (*From Minchin, after Bütschli and Schewiakoff.*)

Stentor there may be one meganucleus and several micronuclei. In general the meganucleus seems to be a storage organ of the cell; it may disappear and be replaced by a new one. The micronucleus performs the usual nuclear functions. The mammalian red blood corpuscle begins its life as a nucleated cell, but later on the nucleus is lost.

The position of the nucleus in the cell is determined largely by physical causes, such as surface tension, the position of the vacuoles, and the relative density of the cytoplasm in different portions of the cell. In a non-vacuolated cell it ordinarily occupies the center of the cytoplasmic mass, whereas in a cell with vacuoles it is imbedded in the cytoplasm even when the latter is reduced to a thin parietal layer; it never lies free in the vacuole. In the Cladophoraceæ (Carter 1919) it is regularly imbedded, at least partially, in the chloroplast, and this is true even in cells possessing a considerable amount of colorless cytoplasm. Its position is also related to the functions of the cell: in general it lies in the region characterized by the most active metabolism. For example, in young growing root hairs (Fig. 16, *B*) and pollen tubes it is commonly found where elongation is taking place, and in thickening epidermal cells (Fig. 16, *A*) it frequently, though not always, lies near the wall upon which the thickening material is being deposited. This relation of position to function was emphasized in the works of Haberlandt (1887) and Gerassimow (1890, 1899, 1901).

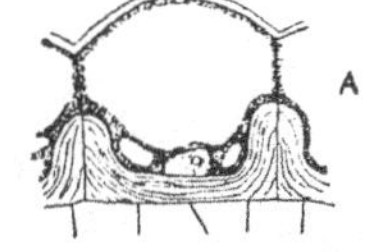
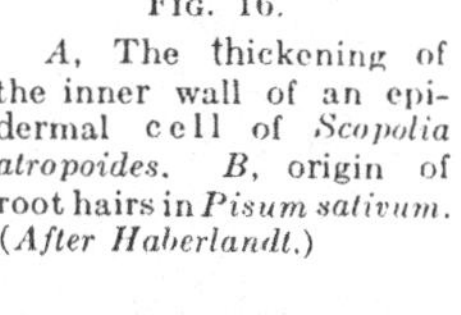

FIG. 16.

A, The thickening of the inner wall of an epidermal cell of *Scopolia atropoides*. *B*, origin of root hairs in *Pisum sativum*. (*After Haberlandt.*)

In form the nucleus is typically spherical or ellipsoidal, its shape being determined by a number of physical factors. Under comparatively uniform conditions, as obtain where a small nucleus lies in a relatively large amount of non-vacuolated cytoplasm, a spherical shape is assumed because of the phenomena of surface tension. Exceptions are often seen in cells with specialized functions.

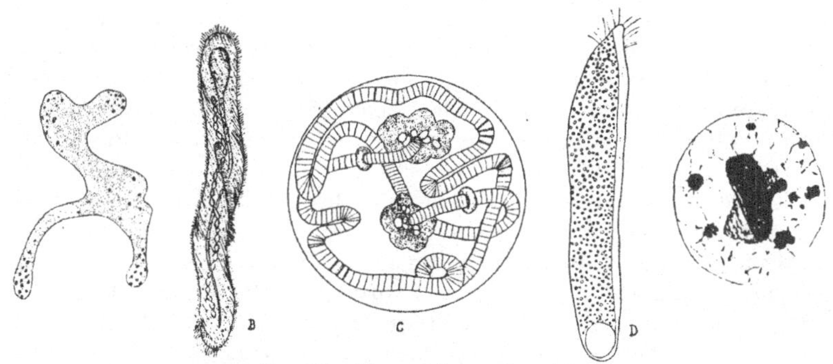

FIG. 17.—Unusual forms of nuclei.

A, portion of nucleus from spinning gland of *Vanessa urticæ*, showing irregular form and finely divided state of the chromatin. (*After Korschelt*, 1896.) *B*, *Spirostomum ambiguum*, with moniliform nucleus. (*After Stein*.) *C*, Nucleus from salivary gland of *Chironomus*: the chromatic material exists as a series of discs in a convoluted thread, which ends in two nucleoli. (*After Balbiani*, 1881.) *D*, *Chœnia teres*, with chromatic granules scattered throughout the body. (*After Gruber*, 1884.) *E*, Nucleus from root tip of *Marsilia*, showing concentration of chromatic material in the nucleolus. (*After Berghs*, 1909.)

In the cells of the spinning glands of *Pieris* and *Vanessa* (butterflies) the physiological conditions result in the assumption of very irregular forms whereby the nuclear surface is considerably increased (Fig. 17, *A*). Nuclei seem rather commonly to undergo amœboid changes in shape;

such active movement can be directly observed in the nucleus of the living cycad spermatozoid. In the long, narrow cells of vascular bundles the nuclei, which are not free to grow in all dimensions, come to be correspondingly elongated. The nucleus may also be passively forced into very irregular shapes by the dense accumulation of starch grains and the diminution in the amount of cytoplasm, as in the endosperm cells of maize. In *Stentor* and *Spirostomum* the nucleus has the form of a string of beads (Fig. 17, *B*).

In size the nucleus shows a wide variation, ranging in plants from the extremely minute nucleus of *Mucor*, 1μ or less in diameter, to the relatively gigantic nucleus of the *Dioon* egg, with a diameter of 600μ. A similar range is seen in animal nuclei. Although the nuclei of the fungi are characterized by small size, most of them being less than 5μ in diameter, they may grow to a large size at certain stages. The primary nucleus of *Synchytrium*, for instance, reaches a diameter of over 60μ. The majority of nuclei, however, fall between 5μ and 25μ. In spite of the wide range in the size of nuclei of different organisms, in a given tissue it is comparatively uniform.

With respect to the physical nature of the nucleus as a whole, the researches of Kite (1913) and Chambers (1914, 1917) have shown that it ordinarily consists at least in part of a gel of higher viscosity than the cytoplasm, often being so firm that it can easily be handled without injury by means of the dissecting instrument. This obviously would be impossible were the nucleus merely a watery droplet or vesicle in the cytoplasm. The germinal vesicle (nucleus) of the animal egg Chambers (1917) finds to be a sol droplet with a gel membrane; if it is pinched in two by the dissecting instrument the two halves will reunite if they come in contact.

The chemical nature of the nucleus has been dealt with in the preceding chapter. With regard to its electrical properties, the nucleus is apparently negative to the cytoplasm. R. S. Lillie (1903) found that free nuclei and the heads of spermatozoa, which are almost entirely nuclear material, pass to the anode in an isotonic cane sugar solution, whereas cells rich in cytoplasm, such as large leucocytes, pass to the cathode. These results have been confirmed by Hardy (1913).

Nucleoplasmic Ratio.—Of more importance than the absolute size of the nucleus is the relation of its volume to that of the cytoplasm—the so-called *nucleoplasmic* or *Kernplasma relation*. Many years ago it was held by Sachs (1892, 1893, 1895) and by Strasburger (1893) that the size of a meristematic cell in a plant, owing to a supposed limitation of the sphere of influence of the nucleus, maintains a very definite relation to the size of its nucleus. This conception has recently been emphasized anew by Winkler (1916), and parallel views have been expressed by several zoölogists (*e.g.*, Hegner on *Arcella*, 1919). In the case of certain terminal meristems of plants such a rule may well hold true within limits, but the condition reported by Bailey (1920) in the lateral meristem (cambium) shows clearly that it cannot have universal application. The cambial initials may vary enormously in size with no corresponding variation in the size of their nuclei: two such initials, one of them having many hundreds of times the volume of the other, may possess nuclei of approximately equal size.

The nucleoplasmic ratio has figured prominently in discussions of the problem of senescence. R. Hertwig in 1889 advanced the theory that senescence and natural death are associated with an increase in the relative size of the nucleus. He later asserted (1903, 1908) that the nucleoplasmic relation is self-regulatory within certain limits for each kind of cell, exercising thereby a control over many cell activities, including cell-division. Minot (1891, 1908, 1913), on the contrary, believed that the increase in the relative volume of the cytoplasm, in addition to its differentiation, is a fundamental factor in senescence and death. Conklin (1912), as a result of his work on *Crepidula*, denied the existence of a constant and self-regulatory nucleoplasmic relation, holding rather that changes in this relation are not causes of such cell activities as cell-division, but are results of the metabolic processes by which such cell activities are brought about. Child (1915) points out that in most animal tissues there is an increase in the relative amount of cytoplasm during senescence, whereas in plants, although the cell enlarges through vacuolation, the relative volume of cytoplasm often does not increase. He therefore concludes that the nucleoplasmic relation cannot be regarded as a universal factor in senescence; it is rather an indication of the kind and rate of metabolism. The differentiation of the cytoplasm, apart from its mere change in volume, Child, with many other workers (Minot, Delage, Jennings, etc.), regards as a matter of the greatest importance in senescence. Further discussion of this subject is deferred to Chapter VII.

Not only has it been held that there is a certain relation between the mass of the nucleus and that of the cytoplasm, whatever the significance of this relation may be, but there also seems to be a size relationship between the nucleus and its contained chromosomes. In 1896 Boveri showed that the size of the nuclei in merogonic echinoderm larvæ (see p. 325) is dependent upon the number of chromosomes each contains. In a more extended study (1905) he demonstrated that it is the surface of the nucleus that is proportional to the chromosome number, and also that the size of the cell is proportional to both. Gates (1909), however, adduced evidence to show that this rule is by no means universal.

Structure.—Having reviewed the general features of the nucleus as a whole, we may next give attention to its structure, as seen in typical cases.

The nucleus is bounded by a distinct *nuclear membrane*. The nature of this membrane has been a subject of much controversy. Some have regarded it as a precipitation membrane laid down when the newly formed karyolymph comes in contact with the cytoplasm at the time the daughter nuclei are reconstructed during the closing phases of mitosis, while others (Lawson 1903) have interpreted it as merely a denser limiting layer of the cytoplasm. The above cited work of Kite and Chambers, however, leaves no doubt that the membrane is a definite morphological structure belonging to the nucleus: although it is at times very delicate, it remains intact when the nucleus is pushed and pulled about by the dissecting instrument, and is thrown into folds when the karyolymph is withdrawn with a pipette.

Within the nuclear membrane is a series of gels of varying consistency. The *nuclear sap*, or *karyolymph*, is a highly transparent substance which is generally looked upon as homogeneous, although it has been thought by some workers (Reinke 1894) to be made up of large, pale "œdamatin granules." It may be in the sol or gel state. Imbedded in the karyolymph is a network or *reticulum*, which may be relatively uniform throughout the nucleus or only fragmentary and incomplete. It is usually said to be composed of a gel substance known as *achromatin* (Flemming 1879) or *linin* (Schwarz 1887). Supported on the linin reticulum is the *chromatin* (Flemming 1879). This highly stainable substance may exist in the form of small granules or droplets at the nodes of the reticulum, or apparently in many nuclei as a fluid thin enough to distribute itself more or less uniformly throughout the achromatic substance. In the latter case the whole reticulum appears to be composed of a single unevenly stained material, careful examination showing the "chromatic granules" and "achromatic support" to be its thicker and finer portions respectively (Fig. 51) (Grégoire and Wyagerts 1903; Grégoire 1906; Sharp 1913, 1920). According to Kite (1913) the granules in the living nucleus consist of a very concentrated gel, the supporting reticulum of a somewhat more dilute but not at all fibrous gel, and the karyolymph of a gel which is the most dilute of all.

Heidenhain (1894) found imbedded in the colorless linin net two sorts of chromatin in the form of granules: *oxychromatin*, consisting largely of plastin, poor in phosphorus, and staining with the acid dyes; and *basichromatin*, composed mainly of nuclein, rich in phosphorus, and staining with the basic dyes. These two forms of chromatin apparently may change into each other by the addition or loss of phosphorus. The periodic changes in the staining reactions of many nuclei therefore indicate changes in the chemical composition of the chromatin, and these in turn point to the intimate association of the nucleus with the periodic physiological processes of the cell. As used by many writers the term oxychromatin includes also the linin, so that in much cytological literature linin and oxychromatin are more or less interchangeable terms, while "chromatin" refers to the basichromatin. Oxychromatin appears to be closely similar in composition to the achromatic structures in the cytoplasm, such as spindle fibers and centrosomes. The prominent place occupied by the nucleus in cytology is due in large measure to the conspicuous behavior of its chromatic substance at the time of cell-division and fertilization, topics which are to receive detailed consideration in subsequent chapters.

In many nuclei basichromatin accumulates at certain points in the reticulum, forming *karyosomes*, also called "net knots" and *chromatin nucleoli*. These seem to be masses of surplus chromatin elaborated by the nucleus during the resting phase or in some cases chromatin which has flowed to these points from the other parts of the reticulum. During the next mitosis they are distributed with the rest of the chromatin. As Rosen (1892) long ago showed by his studies of their staining reactions, they differ decidedly in composition from true nucleoli, although they may closely resemble the latter after treatment with certain stains (iron-alum-hæmatoxylin).

One or more *true nucleoli*, or *plasmosomes* (Ogata 1883), are usually present in the nucleus. A single nucleolus is probably characteristic of most nuclei; there are rarely many, and in some cases there is none. The nucleolus may be in close organic connection with the nuclear reticulum or it may lie entirely apart from it. In composition it consists largely of such oxychromatic substances as plastin and pyrenin, or of nuclein well saturated with protein (Zacharias). It usually stains with the acid dyes: by a proper selection of stains it may, therefore, be distinguished from the karyosomes, which, being composed of basichromatin, take the basic dyes as a general rule. In structure the nucleolus may appear to be homogeneous throughout, like an oil globule; in other cases it has an outer envelope of different consistency and staining reaction. Very often vacuoles, occasionally containing granules, are present in the interior. Crystalloid bodies are also frequently observed in the nucleolus (Digby on *Galtonia*, 1910; Reed on *Allium*, 1914; Kuwada on *Zea*, 1919). In the epithelial cells of the frog intestine Carleton (1920) finds one or more intranucleolar bodies which he calls "nucleolini." These appear to divide and pass to the daughter cells at the time of mitosis, and may possibly initiate the formation of new nucleoli in the daughter nuclei. Montgomery (1899)[1] concluded that the nucleolus grows in size by the apposition of smaller particles of nucleolar material on its surface, and by the intussusception of vacuolar substance arising outside the nucleolus.

Function of Nucleolus.—Various opinions have been entertained regarding the function of the nucleolus. By many workers it has been looked upon as chiefly a passive by-product of no further use in the life

---

[1] An exhaustive review of the literature dealing with the nucleolus up to 1899 is given in this paper.

of the cell (Haecker). Strasburger (1895, 1897), who observed the disappearance of the nucleolus at about the time the spindle fibers appear during the prophases of mitosis, concluded that it is a mass of reserve kinoplasm which gives rise indirectly to the achromatic figure. While some have agreed in the main with this conclusion, many have denied the relationship of nucleolus and spindle, contending that the former is rather a reserve constituent for the linin reticulum (Eisen 1900) or the chromatin (Schürhoff 1918). Frequently the bulk of the basichromatic material of the nucleus is lodged in the nucleolus at certain stages. In the somatic nuclei of *Marsilia* (Fig. 17, *E*), for example, Berghs (1909) shows that it is transferred to the nucleolus during the telophases of mitosis, and returned to the reticulum in the following prophases. This phenomenon, which has an important bearing on the rôle of the chromatin and the individuality of the chromosomes, will be referred to again in Chapter VIII.

In many cells, as shown by the work of the zoölogists the nucleolus appears to be concerned in the elaboration of secretion and storage products. In the eggs of certain animals Macallum (1890) showed that the nucleolar material, which appears to differentiate from the chromatin, passes into the cytoplasm and there combines with another substance to form the yolk globules. In the cells of the pancreas he further found that material often present in the form of nucleoli functions in a similar manner in the production of zymogen. Many other observations of this general nature have been reported. In the silk-gland cells of certain insects it has recently been shown by Nakahara (1917) that some of the nucleoli, which may originally be passive by-products, later pass into the cytoplasm and contribute to the formation of the secretion products. An extreme view of the importance of the nucleolus is that of Derschau (1914), who regards the nucleolus as the real center of the life of the cell. Granules of oxychromatin, he asserts, pass out from the nucleolus through the cytoplasm in the form of chondriosomes, carrying basichromatin as a building material to the places where it is required.

It is highly probable that the nucleolus has various functions in different cells, but in general we may conclude that it is a mass of accumulated material which is usually, though not always, utilized in the metabolic processes of the nucleus

The Nuclei of Bacteria and Other Protista.—The question of the nucleus in bacteria is one that it appears to be particularly difficult to settle satisfactorily. This is due not only to the minute size of these organisms, which makes special methods necessary and observation very difficult, but also to the fact that a variety of conditions seems to be present in the group. That the bacterial cell is devoid of a nucleus has been held by several investigators including Fischer (1894, 1897, 1899,

1903), who looked upon the observed granules as reserve materials rather than nuclear substance. Migula (1894, 1897, 1904) regarded the existence of nuclei in bacteria as very doubtful. The majority of workers, on the contrary, have held that a nucleus or at least nuclear material is present in some form. The most striking view is that which regards the whole cell in some cases as a naked nucleus (Hüppe 1886; Zettnow 1891, 1897, 1899; Růžička 1908, 1909; and, in the case of small bacteria, Bütschli 1890, 1892, 1896, 1902). The evidence advanced in support of this hypothesis, however, is of very doubtful value

In many bacteria, particularly the larger forms, there is present a granular substance which has certain characteristics of chromatin, and which in some species exists as a single well defined mass. The "central body" of the sulphur bacterium Bütschli regarded as the homologue of a nucleus, the peripheral portion of the cell being cytoplasm. In a careful study of the entire life cycle of *Bacillus Bütschlii* Schaudin (1902) found that the chromatic material present during most of the cycle as chromidia unites at certain stages to form peculiar spiral figures; in the spores it takes the form of dense masses. Such scattered chromidia and "spiral filament nuclei" were also observed by Guilliermond (1908, 1909), who has given a review of the subject (1907). Nakanishi (1901), who employed both intra-vitam methods and fixed material, reported the presence of nuclei in the vegetative cells and spores of a number of species.

The nucleus of the large *Bacterium gammari* was studied by Vejdowský (1900), who in 1904 described its division by mitosis. Mencl (1904, 1905, 1907, 1909) demonstrated by careful methods the nuclei in many species and also reported mitotic division in *Bacterium gammari*. Doubt concerning the systematic position of this form, however, has been raised by some investigators, who think it not improbable that it is a yeast-like fungus rather than a bacterium.

Dobell (1908, 1909, 1911), whose review of the subject has been of service in the preparation of this summary, has studied with much care many species of bacteria in their natural culture media. His conclusions are summarized in the following quotation (1911):

"All bacteria which have been adequately investigated are —like all other Protista—nucleate cells.

"The form of the nucleus is variable, not only in different bacteria, but also at different periods in the life cycle of the same species.

"The nucleus may be in the form of a discrete system of granules (chromidia); in the form of a filament of various configuration; in the form of one or more relatively large aggregated masses of nuclear substance; in the form of a system of irregularly branched or bent short strands, rods, or networks; and probably also in the vesicular form characteristic of the nuclei of many animals, plants, and protists.

"There is no evidence that enucleate bacteria exist."

The apparent discrepancy between this view of bacterial organization and that of Minchin, stated below, will be seen to be largely a matter of terminology.

It is therefore among the Proitsta that the widest departures from the usual type of nuclear structure are found, certain of them in all probability representing relatively primitive stages in the evolution of the true nucleus. Such an interpretation is evidently to be placed upon the "distributed nuclei" seen in certain bacteria, protozoans, flagellates, and Cyanophyceæ (p. 202), which consist of granules of a material akin to chromatin scattered throughout the cell, sometimes with a limiting membrane of some sort but often with none. It is doubtful if granules scattered with no definite limitations throughout the cell, as in *Chænia teres* (Fig. 17, *D*) or *Chroococcus turgidus* (Fig. 72, *A*), should be spoken of collectively as a nucleus. As pointed out at the beginning of this chapter, it seems preferable to certain workers to limit the term to those chromatic aggregations which actually have the characters of a definitely localized organ. In discussing the advisability of so restricting the application of the term Minchin (1912, Chapter VI) points out that "the word 'chromatin' connotes an essentially physiological and biological conception . . . of a substance, far from uniform in its chemical nature, which has certain definite relations to the life history and vital activities of the cell. The word 'nucleus,' on the other hand . . . is essentially a morphological conception, as of a body, contained in the cell, which exhibits a structure and organization of a certain complexity, and in which the essential constituents, the chromatin particles, are distributed, lodged, and maintained, in the midst of achromatinic elements which exhibit an organized arranegment, variable in different species, but more or less constant in the corresponding phases of the same species." According to this interpretation the term "nucleus" would not be applicable to a mass of granules (chromidia) scattered throughout the cell. Minchin states further that ". . . as soon as a mass or a number of particles of chromatin begin to concentrate and separate themselves from the surrounding protoplasm, with formation of distinct nuclear sap and appearance of achromatinic supporting elements, we have the beginning at least of that definite organization and structural complexity which is the criterion of a nucleus as distinguished from a chromidial mass."

Those Protista of the lower (bacterial) grade, in which there are only scattered grains of chromatic material, are looked upon as "non-cellular" in organization by Minchin, who believes that from such a primitive state the "strictly cellular grade of organization has been evolved by concentration of some or all of the chromatin to form a nucleus." In its simplest condition such a nucleus consists of one or more chromatin granules in a sort of vacuole, and is known as a "protokaryon." In other cases the chromatin forms a single large mass at the center of the nucleus ("vesicular nucleus"). Since "the chromatic particles are the only constituents of the cell which maintain persistently and uninterruptedly their existence throughout the whole life cycle of living organisms universally," Minchin (1916) believes that the earliest living things, which he calls " Biococci," were minute particles of a chromatin-like substance. These were the ancestors of the present chromatin grains and find their nearest modern representatives in certain pathogenic Chlamydozoa. According to this view the cytoplasm was differentiated later in the evolution of the cell, whereas the more general view probably is that chromatin and cytoplasm were coexistent as two substances in cells from the earliest known stages (Wilson).[1]

The Function of the Nucleus.—It may be said without reservation that the nucleus dominates the morphological and physiological changes in the cell. Although the type of organization formed by a nucleus in combination with cytoplasm is required for the carrying on of cell activity, it is nevertheless evident from a huge mass of accumulated observations that in the nucleus is to be found the center of control for both the functional activities and for cell reproduction (cell-division). Many years ago Claude Bernard (1878) pointed out that while the cytoplasm is the seat of destructive metabolism, the nucleus is the seat of constructive metabolism, this physiological rôle offering "the key to its significance as the organ of development, regeneration, and inheritance" (Wilson). The inability of a cell deprived of its nucleus to carry on synthetic metabolism in any complete manner has often been noted, though such a cell may not perish for some time. The mammalian erythrocyte, for example, loses its nucleus at an early stage and may continue to exist in the enucleate state for from 15 to 30 days (Hunter, Quincke). Klebs found that enucleate cells of *Spirogyra* may continue for some time to form starch. But such cells are apparently unable to divide or to increase their bulk by the elaboration of new cell substance. Many ordinary activities, such as cell wall formation (Townsend 1897; Gerassimow 1899, 1901), fail to occur. From such observations it is concluded that the nucleus is necessary for the synthetic processes associated with growth and reproduction. This conclusion is supported by the facts of regeneration.

The rôle of the nucleus in regeneration was strikingly shown by the well known experiments of Gruber (1885) and F. R. Lillie (1896) on *Stentor*. This unicellular organism, which has a nucleus like a string of beads. may be cut into fragments: any fragment containing a portion of the nucleus has the power of regenerating a complete new animal, whereas enucleate fragments, although they may live for a little time, undergo no regeneration and eventually perish.

[1] For more complete descriptions of the nuclei of Protista the works of Wilson (1900) and Minchin (1912) should be consulted. The behavior of such nuclei at the time of cell-division is briefly described in Chapter X of this book.

A number of biologists (Gruber 1886, Hertwig 1898, Heidenhain 1894, Henneguy 1896, Conklin 1902) concluded that in general the chromosomes (basichromatin) are concerned chiefly with differentiation and regulation, while the achromatin (oxychromatin) has to do with metabolism (Conklin 1917). Metabolism is in reality a great complex of reactions: the reactions are not independent of one another but are closely correlated, and thus constitute an intricately adjusted reaction system. Among these many reactions, according to modern physiology, the most important is *oxidation*, for the energy utilized by the organism is derived immediately from the union of protoplasm or of its constituent elements with oxygen. Oxidation has been called the "independent variable" (Loeb and Wasteneys 1911) upon which the other reactions largely depend: oxidation is the dominant factor in cell activity, and it is therefore of the greatest importance to understand as well as possible the relation of the parts of the cell to this process.

Following the experiments of Spitzer (1897), who observed that nucleoproteins extracted from certain animal tissues have the same oxidizing power as the tissues themselves, it was advocated by Loeb (1899) that the nucleus is the center of oxidation in the cell. Loeb pointed out that this would explain the inability of enucleated cell-fragments to undergo regeneration. This conclusion was supported by R. S. Lillie (1903), who later (1913) showed that rapid oxidation occurs both at the surface of the cell and at the surface of the nucleus, and also by Mathews (1915). Other workers (Wherry 1913, Schultze 1913, Reed 1915), however, have failed to agree. Osterhout (1917), who briefly summarizes the subject, found that "injury produces in the leaf-cells of the Indian Pipe (*Monotropa uniflora*) a darkening which is due to oxidation. The oxidation is much more rapid in the nucleus than in the cytoplasm and the facts indicate that this is also the case with the oxidation of the uninjured cell."

The rôle of the nucleus in development and inheritance, which has been a subject of so much discussion in recent years, will be dealt with in later special chapters (XIV–XVIII), after the behavior of the nucleus in somatic cell-division, maturation, and fertilization has been described.

Bibliography 4

Nucleus

BAILEY, I. W. 1920a. The cambium and its derivative tissues. II. Size variations of cambial initials in gymnosperms and angiosperms. Am. Jour. Bot. **7**: 355–367. figs. 3.

1920b. The cambium and its derivative tissues. III. A reconnaissance of cytological phenomena in the cambium. Ibid. **7**: 417–434. pls. 26–29.

BALBIANI, E. G. 1881. Sur la structure du noyau des cellules salivaire chez les larves de *Chironomus*. Zool. Anz. **4**: 637–641. figs. 7.

BERGHS, J. 1909. Les cinèses somatiques dans le *Marsilia*. La Cellule **25**: 73–84. 1 pl.

BERNARD, C. 1878. Leçons sur les phénomènes de la vie. Paris.

BOVERI, TH. 1895. Ueber die Befruchtungs- und Entwicklungstahigkeit kernloser Seeigeleier und über die Moghchkeit ihrer Bastardcicrung. Arch. f. Entw. 2: 394–443. pls. 24, 25.

1905. Zellen-Studien. V. Ueber die Abhängigkeit der Kerngrösse und Zellenzahl der Seeigel-Larven von der Chromosomenzahl der Ausgangszellen. Jena.

BÜTSCHLI, O. 1890. Ueber den Bau der Bakterien und verwandter Organismen. Leipzig.

1892. Untersuchungen über mikroskopische Schaume und das Protoplasma. Leipzig.

1896. Weitere Ausführungen über den Bau der Cyanophyceen und Bakterien. Leipzig.

1902. Bemerkungen über Cyanophyceen und Bakterien. Arch. f. Protistenk. 1.

CARLETON, H. M. 1920. Observations on the intranucleolar body in columnar epithelium cells of the intestine. Quar. Jour. Micr. Sci. 64: 329–344. pl. 17.

CARTER, N. 1919. The cytology of the Cladophoraceæ. Ann. Bot. 33: 467–478. pl. 27. figs. 2.

CHAMBERS, R. 1914, 1917. See Bibliography 3.

CHILD, C. M. 1915. Senescence and Rejuvenescence. Chicago.

CONKLIN, E. G. 1902. Karyokinesis and Cytokinesis in the maturation, fertilization and cleavage of Crepidula and other Gasteropoda. Jour. Acad. Nat. Sci. Phila. 12: 5–121. pls. 1–6. figs. 33.

1912. Cell size and nuclear size. Jour. Exp. Zool. 12: 1–98. figs. 37.

1913. The size of organisms and of their constituent parts in relation to longevity, senescence, and rejuvenescence. Pop. Sci. Mo. 83: 178–198.

1917. Mitosis and Amitosis. Biol. Bull. 33: 396–436. pls. 10. (Literature.)

DELAGE, Y. 1899a. La fécondation mérogonique et ses resultats. Comptes rendus Acad. Sci. Paris.

1899b. Études sur la mérogonie. Arch. Zool. Exp. III 7: 383–417.

1899c. Sur l'interpretation de la fécondation mérogonique et sur une theorie nouvelle de la fécondation normale. Ibid. 512–527.

1903. L'hérédité et les grandes problemes de la biologie. Paris.

VON DERSCHAU, M. 1908. Beiträge zur pflanzlichen mitose: Centern, Blepharoplasten. Jahrb. Wiss. Bot. 46: 103–118. pl. 6.

1914. Zum Chromatindualismus der Fflanzenzelle. Arch. Zellf. 12: 220–240. pl. 17.

DIGBY, L. 1910. The somatic, premeiotic, and meiotic nuclear divisions of Galtonia candicans. Ann. Bot. 24: 727–757. pls. 59–63.

DOBELL, C. C. 1909. Chromidia and the binuclearity hypothesis. Quar. Jour. Micr. Sci. 53: 279. See other papers at pp. 183 and 579; also vol. 52. p. 121.

1911. Contributions to the cytology of the bacteria. Ibid. 56: 395–506. pls. 16–19. fig. 1.

FISCHER, A. 1894. Untersuchungen über Bakterien. Jahrb. Wiss. Bot. 27: 1.

1897. Untersuchungen über den Bau der Cyanophyceen und Bakterien. Jena.

1899. Fixierung, Färbung und Bau des Protoplasmas. Jena.

1903. Vorlesungen über Bakterien. 2 Aufl. Jena.

FLEMMING, W. 1879. Beiträge zur Kenntniss der Zelle und ihre Lebenserscheinungen. Arch. Mikr. Anat. 16: 302–436. pls. 15–18.

GATES, R. R. 1909. The stature and chromosomes of Œnothera gigas, De Vries. Arch. Zellf. 3: 525–552.

GERASSIMOW, J. J. 1890. Einige Bemerkungen über die Funktion des Zellkerns. Bull. Soc. Sci. Nat. Moscow. 548-554.

1892. Ueber die kernlosen Zellen bei einigen Conjugaten. Ibid. 109-131.

1896. Ueber ein Verfahren kernlose Zellen zu erhalten. Ibid. 477-480.

1899. Ueber die Lage und die Funktion des Zellkerns. Ibid. 220-267.

1901. Ueber den Einfluss des Kerns auf das Wachsthum der Zellen. Ibid. 185-220.

1904. Ueber die Grösse des Zellkerns. Beih. Bot. Centr. **18**: 45-118. pls. 2.

GRUBER, A. 1885. Ueber künstliche Teilung bei Infusorien. Biol. Centralbl. **4**: 717-722; **5**: 137-141.

1886. Beiträge zur Kenntniss der Physiologie und Biologie der Protozoen. Ber. Naturf. Gesell. Freiburg **1**.

GUILLIERMOND, A. 1907. La cytologie des Bacteries. Bull. Inst. Pasteur **5**: 273.

1908. Contribution a l'étude cytologique des Bacilles endosporés. Arch. f. Protistenk. **12**: 9.

1909. Observations sur la cytologie d'un Bacille. Comptes Rendus Soc. Biol. Paris **67**: 102.

1910. A propos de la structure des Bacilles endosporés. Arch. f. Protistenk. **19**: 6.

HABERLANDT, G. 1887. Ueber die Beziehungen zwischen Funktion und Lage des Zellkerns bei den Pflanzen. Jena.

1914. Physiological Plant Anatomy. 4th ed. Transl. by Drummond.

HARDY, W. B. 1913. Note on differences in electrical potential in the living cell. Jour. Physiol. **47**: 108-111.

HEGNER, R. W. 1919. Quantitative relations between chromatin and cytoplasm in the genus *Arcella*, with their relations to external characters. Proc. Nat. Acad. Sci. **5**: 19-22.

HEIDENHAIN, M. 1894. Neue Untersuchungen über die Centralkörper und ihre Beziehungen zum Kern und Zellprotoplasma. Arch. Mikr. Anat. **43**: 423-758. pls. 25-31.

HENNEGUY. 1896. Leçons sur la cellule. Paris.

HERTWIG, R. 1889. Ueber die Kernkonjugation der Infusorien. Abhandl. Bayer. Akad. Wiss. II **17**.

1898. Ueber Kernteilung, Richtungskörperbildung und Befruchtung von *Actinospherium eichornii*. Ibid. **19**.

1903. Ueber Korrelation von Zell- und Kerngrösse und ihre Bedeutung für die Differenzierung und die Theilung der Zelle. Biol. Centralbl. **23**: 49-62. 108-119.

1908. Ueber neue Probleme der Zellenlehre. Arch. Zellf. **1**: 1-32. figs. 9.

HÜPPE, F. 1886. Die Formen der Bakterien und ihre Beziehungen zu Gattungen und Arten. Wiesbaden.

JAHN, E. 1904. Myxomycetenstudien. 3. Kernteilung und Geisselbildung bei den Schwärmern von *Stemonitis flaccida* Lister. Ber. Deu. Bot. Gesell. **22**: 84-92. pl. 6.

1908. Myxomycetenstudien. 7. *Ceratiomyxa* Ibid. **26a**: 342-352.

1911. Myxomycetenstudien. 8. Der Sexualakt. Ibid. **29**: 231-247. pl. 11.

JENNINGS, H. S. 1912. Age, death and conjugation in the light of work on lower organisms. Harvey Lectures 1911-1912. Pop. Sci. Mo. **80**: 563-577.

1913. The effect of conjugation in *Paramœcium*. Jour. Exp. Zool. **14**: 279-392. figs. 2.

KITE, G. L. 1913. See Bibliography 3.

KORSCHELT, E. 1896. Ueber die Structur der Kerne in den Spinndrüsen der Raupen. Arch. Mikr. Anat. **47**: 500-549. pls. 26-28.

Kuwada, Y. 1919. Die Chromosomenzahl von *Zea Mays* L. Jour. Coll. Sci. Tokyo **39**: 1–148. pls. 2.

Lawson, A. A. 1903. On the relationship of the nuclear membrane to the protoplast. Bot. Gaz. **35**: 305–319. pl. 15.

Lillie, F. R. 1896. On the smallest parts of the *Stentor* capable of regeneration. Jour. Morph. **12**: 239–249.

Lillie, R. S. 1902. On the oxidative properties of the cell nucleus. Am. Jour. Physiol. **7**: 412–421. fig. 1.

1903. On differences in the direction of the electrical connection of certain free cells and nuclei. Ibid. **8**: 273–283.

1913. The formation of indophenol at the nuclear and plasma membranes of frogs' blood corpuscles and its acceleration by induction shocks. Jour. Biol. Chem. **15**: 237–248. pl. 1.

Loeb, J. 1899. Warum ist die Regeneration kernloser Protoplasmastücken unmöglich, usw? Arch. Entw. **8**: 689–693.

Loeb, J. and Wasteneys. 1911. Sind die Oxidationsvorgänge die unabhängige Variable in den Lebenserscheinungen? Biochem. Zeitschr. **36**: 345–356.

Mathews, A. P. 1915. Physiological Chemistry. p. 180.

Mencl, E. 1904. Einige Beobachtungen über die Struktur und Sporenbildung bei symbiotischen Bakterien. Centralbl. Bakt. II **12**: 559.

1905. Cytologisches über die Bakterien der Prager Wasserleitung. Ibid. **15**: 544.

1907*a*. Eine Bemerkung zur Organisation der Periplaneta-Symbionten. Arch. f. Protistenk. **10**: 188.

1907*b*. Nachträge zu den Strukturverhältnissen von *Bakterium gammari* Vejd. Ibid. **8**: 259.

1909. Die Bakterienkerne und die "cloisons transversales" Guilliermonds. Ibid. **16**: 62.

Migula, W. 1894. Ueber den Zellinhalt von *Bacillus oxalaticus* Zopf. Arb Bakt. Inst. Karlsruhe **1**.

1897, 1900. System der Bakterien. Jena.

1904. Der Bau der Bakterienzelle. Lafar's Handb. Techn. Mykol. **1**: 48.

Minchin, E. A. 1912. An Introduction to the Study of the Protozoa.

1916 The evolution of the cell. Am. Nat. **50**: 106–118.

Minot, C. S. Senescence and Rejuvenation. Jour. Physiol. **12**: 97–153.

1908. The problem of age, growth, and death. New York.

1913. Moderne Probleme der Biologie. Jena.

Montgomery, T. H. 1899. Comparative cytological studies, with special regard to the morphology of the nucleolus. Jour. Morph. **15**: 265–582. (Bibliography of about 450 titles.)

Nakahara, W. 1917. On the physiology of the nucleoli as seen in the silk-gland of certain insects. Jour. Morph. **29**: 55–74. pls. 2.

1918. Some observations on the growing oocytes of the stonefly, *Perla immarginata*, Say, with special regard to the origin and function of the nucleolar structures. Anat. Rec. **15**: 203–216. figs. 9.

Nakanishi, K. 1901. Ueber den Bau der Bakterien. Centr. Bakt. 1 **30**: 97.

Ogata, M. 1883. Die Veranderung der Pancreaszellen bei der secretion. Arch. Anat. Physiol. (Physiol. Abt.) 405–437. pl. 6.

Olive, E. W. 1907. Cytological Studies on *Ceratiomyxa*. Trans. Wis. Acad. Sci. **15**: 754–773. pl. 47.

Osterhout, W. J. V. 1917. The rôle of the nucleus in oxidation. Science **46**: 367–369.

REED, G. B. 1915. The rôle of oxidases in respiration. Jour. Biol. Chem. **22**: 99–111. pl. 1.

REED, T. 1914. The nature of the double spirem in *Allium cepa*. Ann. Bot. **28**: 271–281. pls. 18, 19.

REINKE, FR. 1894. Zellenstudien. I. Arch. Mikr. Anat. **43**: 377–422. pls. 22–24.

ROSEN, F. 1892. Beiträge zur Kenntniss der Pflanzenzelle. I. Ueber tinctionelle Unterscheidung verschiedener Kernbestandteile und der Sexualkerne. Cohn's Beitr. z. Biol. d. Pfl. **5**: 443–458 pl. 16.

RůŽIČKA, V. 1908. Sporenbildung und andere biologische Vorgänge bei dem *Bact. anthracis*. Arch. f. Hyg. **64**: 219–294. pls. 1–3.

— 1909. Die Cytologie der Sporenbildenden Bakterien und ihr Verhältnis zur Chromidienlehre. Centr. Bakt. II **23**: 289–300. figs. 8.

SACHS, J. 1892. Physiologische Notizen II. Beiträge zur Zellentheorie. Flora **75**: 57–67.

— 1893. Physiologische Notizen VI. Ueber einige Beziehungen der specifischen Grösse der Pflanzen zu ihrer Organisation. Ibid. **77**: 49–81.

— 1895. Physiologische Notizen IX. Weitere Betrachtungen über Energiden und Zellen. Flora (Ergänzungsband) **81**: 405–434.

SCHAUDINN, F. 1902. Beiträge zur Kenntniss der Bakterien und verwandten Organismen. 1. *Bacillus Bütschlii*, n. sp. Arch. Protistenk. **1**: 306.

— 1903. II. *Bacillus sporonema*, n. sp. Ibid. **2**: 421.

SCHULTZE, W. H. 1913. Die Sauerstofforte der Zelle. Verh. Deu. Path. Ges. **16**: 161–168.

SCHÜRHOFF, P. N. 1918. Die Beziehungen des Kernkörperchens zu den Chromosomen und Spindelfasern. Flora **110**: 52–66. figs. 3.

SCHWARZ, FR. 1887. Die morphologische und chemische Zusammensetzung des Protoplasmas. Breslau.

SPITZER. 1897. Die Bedeutung gewisser Nukleoproteide für die oxidative Leistung der Zelle. Arch. f. Ges. Physiol. **67**: 615–656.

STRASBURGER, E. 1893. Ueber die Wirkungssphäre den Kerne und die Zellgrösse. Histol. Beitr. **5**: 97–124. Jena.

— 1895. Karyokinetische Probleme. Jahrb. Wiss. Bot. **28**: 151–204. pls. 2, 3.

— 1897. Ueber Cytoplasmastrukturen, Kern- und Zelltheilung. Ibid. **30**: 375–405. figs. 2.

TOWNSEND, C. O. 1897. Der Einfluss des Zellkerns auf die Bildung der Zellhaut. Jahrb. Wiss. Bot. **30**: 484–510. pls. 20, 21.

VEJDOWSKÝ, F. 1900. Bemerkungen über den Bau und Entwicklung der Bakterien. Centr. Bakt. II **6**: 577–589. 1 pl.

— 1904. Ueber den Kern der Bakterien und seine Teilung. Ibid. **11**: 481–496. 1 pl.

WHERRY, E. T. 1913. On the metamorphosis of an amœba, *Vahlkampfia* sp., into flagellates and vice versa. Science **37**: 494–496.

WILSON, E. B. 1900. The Cell in Development and Inheritance. 2d ed.

WILSON, E. B. and MATHEWS, A. P. 1895. Maturation, fertilization, and polarity in the echinoderm egg. New light on the "Quadrille of the Centers." Jour. Morph. **10**: 319–342. figs. 8.

WINKLER, H. 1916. Ueber die experimentelle Erzeugung von Pflanzen mit abweichenden Chromosomenzahlen. Zeitschr. f. Bot. **8**: 417–531.

YATSU, N. 1905. The formation of centrosomes in enucleated egg-fragments. Jour. Exp. Zool. **2**: 287–312. figs. 8.

ZACHARIAS, E. 1881–1893. See Bibliography 3.

ZETTNOW, E. 1891. Ueber den Bau den Bakterien. Centr. Bakt. **10**: 690.

1897. Ueber den Bau der Grossen *Spirillum*. Zeit. Hyg. **24**: 72.

1899, 1900. Romanowski's Farbung bei Bakterien. Ibid. **30**: 1, and Centr. Bakt. 1 **27**: 803.

1908. Ueber Schwellengrebel's Chromatinbänder in *Spirillum volutans*. Centr. Bakt. 1 **46**: 193.

ZIMMERMANN, A. 1893–1894. Sammel-Referate. 5, 7, 8. Beih. Bot. Centr. **3**: 333–342, 401–436; **4**: 81–89. (No. 7 reviews 153 papers on nuclei of various plant groups.)

CHAPTER V

THE CENTROSOME AND THE BLEPHAROPLAST

THE CENTROSOME

For a full description of the morphology and behavior of the centrosome, based upon the large amount of work done on animal cells prior to 1900, reference should be made to Wilson's book on the cell. In the present account attention will be devoted mainly to centrosome structures in plants. The centrosomes of animal cells will be described only in general terms, their rôle in cell-division being dealt with in later chapters.

Occurrence and General Characters.—The centrosome is an organ which is characteristic chiefly of the cells of animals: in the great majority of these cells it has been found, at least during certain stages. . In plants centrosomes are limited to the cells of algæ and fungi and the spermatogenous cells of certain bryophytes and pteridophytes. If the blepharoplast be regarded as a modified centrosome, a question which will be discussed further on, all motile cells (spermatozoids) of bryophytes, pteridophytes, and gymnosperms must be looked upon as possessing centrosomes. During the last decade of the nineteenth century several botanists reported the presence of centrosomes in the cells of a number of angiosperms, but these cases have all failed to stand the test of subsequent more critical research.[1]

It is scarcely possible to give a description which will apply to all centrosomes, since to any rule there are apparently exceptions. The "typical" centrosome, as seen in animal cells, is a very minute granule, which stains intensely with certain dyes. It is usually situated in the cytoplasm, but in some cases it is found within the nucleus (Fig. 18). It commonly lies in a more or less hyaline mass of material, called the *centrosphere* (Strasburger 1892), *attraction sphere* (van Beneden 1883), *astrosphere* (Fol 1891), or *hyaloplasm sphere* (Wilson 1901).[2] This centrosphere may often show two or more concentric zones differing somewhat in structure and appearance (Fig. 60). At certain stages, especially during nuclear division, the centrosome becomes the focus of a system of delicate rays known collectively as the *aster* (Fol 1877). The aster will

[1] For a review of these cases see Koernicke (1903, 1906).

[2] There has been much confusion in the application these terms. (See Wilson 1900, p. 324.)

receive consideration in the chapter on the achromatic figure.[1] Very often there are two centrosomes lying side by side in the centrosphere,

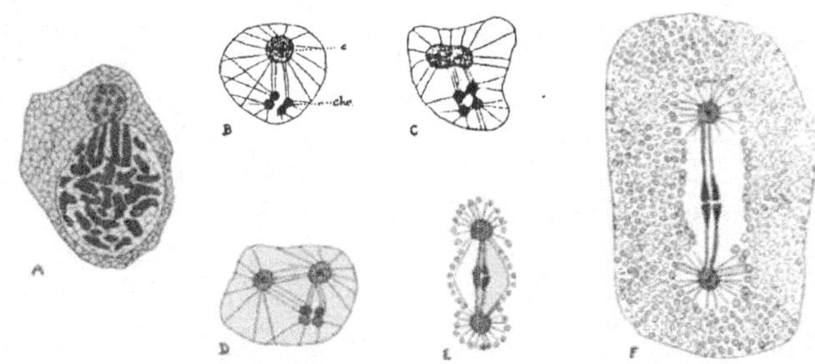

FIG. 18.—Centrosomes in animal cells.

A, attraction sphere above nucleus in spermatocyte of *Salamandra*. (*After Rawitz; see also Fig.* 59.)   *B-F*, intranuclear centrosome in spermatocyte of *Ascaris megalocephala* and its behavior during the prophases of mitosis; *c*, centrosphere; *chr*, chromosome tetrad. (*After Brauer,* 1893.)

the two having arisen by the division of one, apparently in preparation for the next cell-division (Fig. 19). Von Winiwarter (1912) noticed that in interstitial testicular cells, which may have one, two, or four nuclei, there are respectively two, four, and eight rod-like centrosomes lying in midst of a granular mass ("idiosome"). In some cells there may be a larger number of smaller "centrioles" rather then one centrosome, and occasionally there are one or more concentric series of granules about the central centrosome. Several such types described by various writers are shown in Wilson's Fig. 152. It is questionable how far these are normal appearances, for Chambers (1917) asserts that several of them may be produced in the animal egg by subjecting the latter to abnormal environmental conditions.

Individuality.—The centrosome was discovered and described by Flemming (1875) and independently by van Beneden (1876). In 1887 van Beneden and Boveri, as a result of their researches on the thread-worm, *Ascaris megalocephala*, independently concluded that the centrosome, like the nucleus, is a permanent cell organ maintaining its individuality throughout

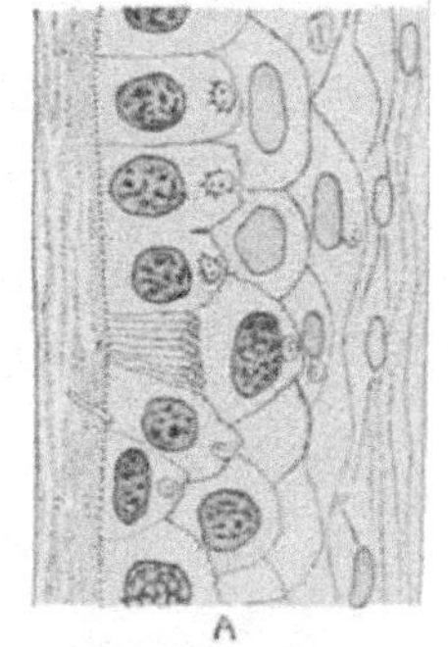

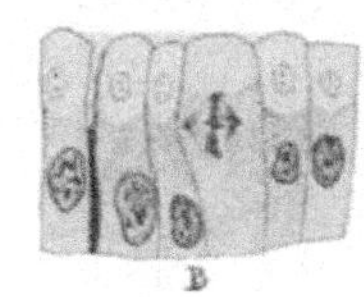

Fig. 19.—Centrosomes in epithelial cells.

A, from cornea of monkey. *B*, from gastric gland of man. (*After Zimmermann.*)

[1] Because of the relation of the centrosome to the achromatic figure it will be necessary to make constant reference to the latter. Chapter IX should be consulted in this connection.

successive cell generations. They observed that, prior to cell-division, the centrosome divides to form two daughter centrosomes, which move apart to opposite sides of the cell and form the poles between which the mitotic figure is established; and further, that after cell-division is completed the centrosome included in each daughter cell does not disappear, but remains visible in the cytoplasm through the ensuing resting stage. Because of this striking behavior at the time of cell-division (see further p. 177) the centrosome soon came to be known as "the dynamic center of the cell."

The above facts seemed to constitute ample ground for the conception of the centrosome as a permanent cell organ, but many obstacles have been found in the way of its acceptance as a theory of universal application. At certain stages in the history of many animal cells its presence

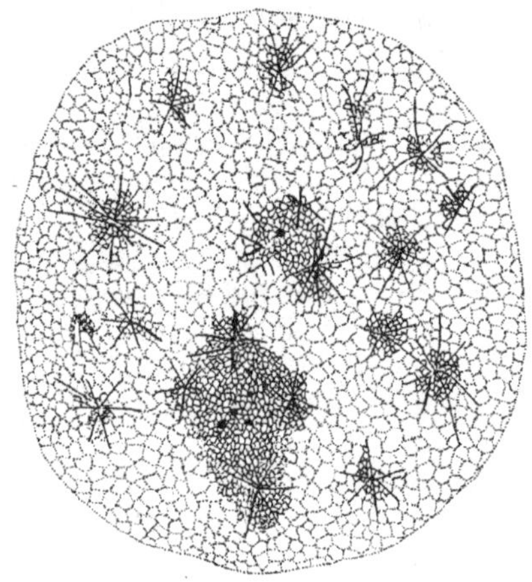

Fig. 20.—Artificial cytasters in the egg of *Arbacia*. (*After Morgan*, 1899.)

cannot be demonstrated, and it is entirely absent from the cells of higher plants. Furthermore, Mead (1898) and Morgan (1896. 1898) found that the formation of centrosomes with asters may be induced in the eggs of certain animals by artifical means (treatment with NaCl and $MgCl_2$ solutions) (Fig. 20), and it has been claimed that centrosomes so formed may function normally in the ensuing division (cleavage) of the egg. Contrary to the opinion of Boveri (1901), Wilson (1901) regarded such "artificial cytasters" as true asters with true centrosomes. Conklin (1912), however, contends that they do not function in mitosis. It is probable that no single conclusion can be drawn concerning this matter which will apply to all cases. There seems to be good evidence for the view that the centrosome in some tissues exists as a permanent cell organ, dividing at each mitosis and remaining visible through the resting stages, at least for a number of cell generations. In other cases it disappears at the close of mitosis, a new one being apparently formed just before the next mitosis. The fact that the formation of centrosomes may be brought about by artificial means suggests that the regular appearance of the centrosome in successive mitoses is closely associated with regularly recurring physiological conditions in the cell; and that its presence in successive cell-divisions does not require an uninterrupted morphological continuity through the intervening stages. Its constant presence in some tissues probably indicates the continuity of some physiological function.

Centrosomes in Algæ.[1]—One of the earliest known centrosomes in plants was that of the diatom *Surirella*, discovered by Smith (1886–7) and Bütschli, and fully described by Lauterborn (1896) and Karsten (1900). It lies near the nucleus, becomes surrounded by radiations, and divides to form the central spindle of the mitotic figure in a very peculiar manner.

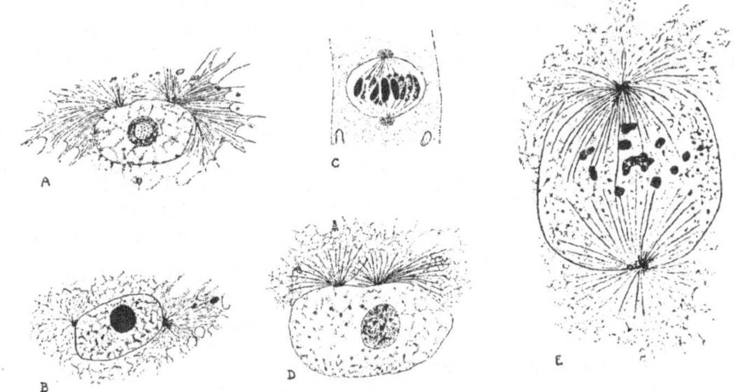

FIG. 21.—Centrosomes in algæ.

A, Stypocaulon. (*After Swingle,* 1897.) *B, Stypocaulon.* (*After Escoyez,* 1909.) *C,* centrosphere-like bodies in *Polysiphonia.* (*After Yamanouchi,* 1906.) *D, E, Dictyota dichotoma.* (*After Mottier,* 1900.)

Centrosomes in the Sphacelariaceæ have been described by Humphrey (1894), Swingle (1897), Strasburger (1900), and Escoyez (1909). In the vegetative cells of *Sphacelaria*, according to Strasburger, the centrosome is situated in a centrosphere at the focus of an aster. Previous to mitosis it divides into two which take up positions at opposite poles of the spindle. In *Stypocaulon* (Swingle) essentially the same condition exists (Fig. 21). Escoyez later concluded, however, that the asters of *Stypocaulon* are formed independently rather than by division, and that the central corpuscles are probably not true centrosomes, but cytoplasmic microsomes.

[1] This review of plant centrosomes and also that of the blepharoplast in subsequent pages are based upon similar reviews given by the author in his paper on Spermatogenesis in *Equisetum* (1912).

In the oögonium and segmenting oöspore of *Fucus* Farmer and Williams (1896, 1898) described two centrospheres containing granules and arising independently at opposite sides of the nucleus. Strasburger (1897) reported definite centrosomes with asters all through mitosis. In the sporeling he observed appearances indicating the division of the centrosome, and concluded that the latter represents a permanent cell organ. In a very detailed investigation Yamanouchi (1909) demonstrated in the antheridium and oögonium two very definite centrosomes, which appear independently of each other, become surrounded by conspicuous asters, and occupy the spindle poles during mitosis (Fig. 61, *C*). He further showed that when the sperm reaches the egg nucleus a new centrosome appears on the nuclear membrane at the point where the sperm enters.

In *Dictyota dichotoma* Mottier (1898, 1900) states that in the two divisions in the tetrasporocyte, in at least the first three or four cell generations of the sporeling, and in all the vegetative cells of the tetrasporic plant curved rod-shaped centrosomes with asters occur at the spindle poles, the two having arisen by the division of one during the early phases of mitosis (Fig. 21, *D, E*). Williams (1904) further reports that the entrance of the sperm causes a centrosome to appear in the egg cytoplasm. Centrosomes in *Nemalion* were described by Wolfe (1904).

In *Polysiphonia violacea* (Yamanouchi 1906) there are present during the prophases of every mitosis two centrosome-like bodies in the kinoplasm at opposite sides of the nucleus. A little later the small bodies disappear, while the kinoplasm takes the form of two large centrosphere-like structures (Fig. 21, *C*); during the later stages of mitosis these fade from view. Yamanouchi believes that these structures do not represent permanent cell organs, but are formed *de novo* at the beginning of each mitosis. Somewhat similar temporary centrospheres, with radiations but no centrosomes, are present in the tetrasporocyte of *Corallina* (Davis 1898; Yamanouchi).

Fungi.—Among the fungi the best known centrosomes are those of the Ascomycetes (Fig. 22). Harper (1895, 1897, 1899, 1905) described granular disc-shaped centrospheres surrounded by asters at the poles of the spindle in the asci of *Peziza, Ascobolus, Erysiphe, Lachnea, Phyllactinia,* and other genera. He regarded them as permanent organs of the cell. In a recent paper (1919) he speaks of the ascomycete centrosome as a structure differentiated "as a region of connection between nucleus and cytoplasm and for the formation of fibrillar kinoplasm." Harper believed the ascospore walls to be formed by the lateral fusion of the curved astral rays focussing upon the centrosome, a point disputed by Faull (1905) and others. Centrosomes in additional genera were figured by Guilliermond (1904, 1905). In *Gallactinia succosa* (Marie 1905; Guilliermond 1911) a single centrosome, which arises within the nucleus with a cone of fibrils extending toward the chromatin, divides into two which take up positions opposite each other at the nuclear membrane, at which time asters develop in the cytoplasm. Faull (1905) found centrosomes in *Hydnobolites, Neotiella,* and *Sordaria;* in the last named genus they appear to be discoid while the cell is in the resting condition but round and smaller during mitosis. In *Humaria rutilans* Miss Fraser (1908) observed two centrosomes lying near each other, each at the apex of a cone of fibers and surrounded by a faint aster. These move apart and

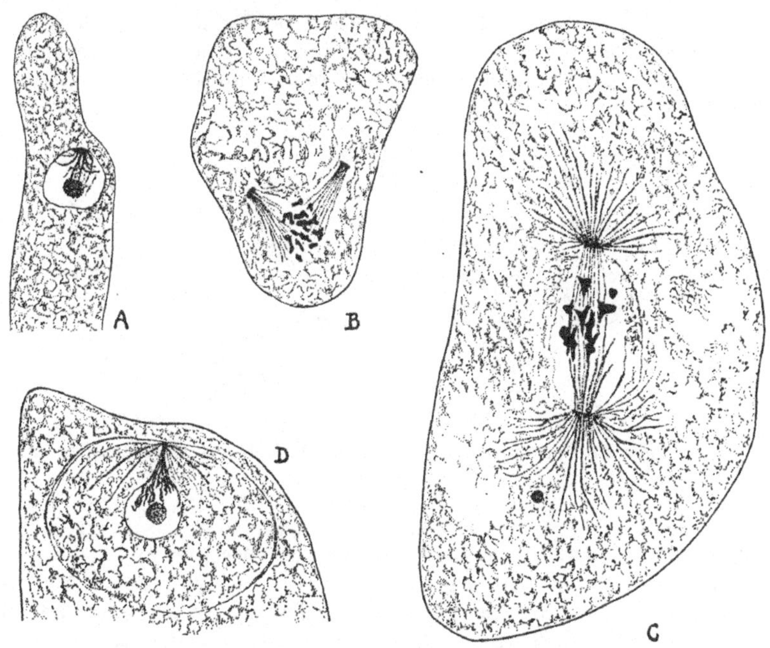

Fig. 22.—Centrosomes in ascomycetes.

A–C, Phyllactinia corylea: division of nucleus in ascus, showing behavior of centrosomes. *D, Erisiphe cichoracearum:* formation of ascopore wall. (*After Harper,* 1905.)

establish the spindle in the usual manner. Centrosomes are also figured in *Ascobolus* and *Lachnea* by Fraser and Brooks (1909); in *Otidea* and *Peziza* by Fraser and Welsford (1908); in *Microsphæra* by Sands (1907); and in *Pyronema* by Claussen (1912).

In the Basidiomycete *Boletus* (Levine 1913) the centrosomes present during the last mitosis in the basidium attach themselves to the basidium wall, and in close connection with them the daughter nuclei are reconstructed. They mark the points of origin of the sterigmata and eventually pass into the spores.

Bryophytes.—The first centrosome known in the liverworts was that of *Marchantia* described by Schottländer (1893), according to whom the centrosome in the spermatogenous cells divides during the anaphases of mitosis, so that each daughter nucleus is accompanied by two (Fig. 27). In the gametophytic cells certain minute bodies with radiations at the poles of the elongated nucleus and of the spindle were believed by Van Hook (1900) to represent centrosomes. Centrospheres with conspicuous radiations but without true centrosomes were described in the mitoses of the germinating spore of *Pellia* by Farmer and Reeves (1894), Davis (1901), and Chamberlain (1903). Grégoire and Berghs (1904), however, pointed out that the centrospheres observed by the foregoing writers in *Pellia* are in reality only appearances due to the intersection of numerous astral rays, and are not distinct bodies.

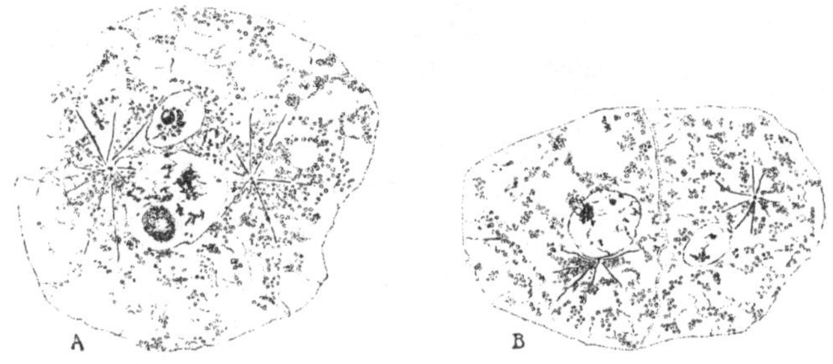

FIG. 23.—Centrosomes in *Preissia quadrata*.

A, in fertilized egg just prior to nuclear fusion, *B*, in cells of young embryo. (*After Graham*, 1918.)

In the cells of *Preissia quadrata* Miss Graham (1918) has more recently made some observations of much interest. She describes and figures two distinct centrosomes with a few astral rays in the cytoplasm of the fertilized egg, at the time when the sexual nuclei are approaching each other and in contact (Fig. 23, *A*). This, together with Yamanouchi's observation on *Fucus* and that of Williams on *Dictyota*, cited above, suggests that in certain plants, as in animals, the formation of centrosomes and asters in the egg cytoplasm is in some way induced by the entrance of the sperm. Similar appearances have been noted by Meyer (1911) in *Corsinia* and by Florin (1918) in *Riccardia* (*Aneura*). Centrosomes were also observed by Miss Graham in the four-celled embryo of *Preissia* (Fig. 23, *B*), this being one of the only cases in which centrosomes have been seen in non-spermatogenous cells in plants above the algæ.

Conclusion.—With regard to centrosomes in plants, it may be concluded from the above review that although there is no adequate evidence for their existence in the cells of angiosperms, they are clearly present in many algæ, fungi, and probably certain bryophytes, where they perform definite functions in the life of the cell. The question of centrosomes in the spermatogenous cells of bryophytes, pteridophytes, and gymnosperms is dealt with in the following discussion of the blepharoplast.

THE BLEPHAROPLAST

Occurrence.—The blepharoplast, as indicated by the name given to it by Webber (1897), is the cilia-bearing organ of the cell. Blepharoplasts of one kind or another are found generally in the motile cells of plants and animals, such as motile unicellular organisms (Flagellata, Ciliata, etc.), swarm spores, spermatozoa, and spermatozoids; and also in cells which, though not freely motile themselves, have motile organs performing other functions, as in the case of ciliated epithelium. In plants blepharoplasts are most conspicuously displayed in the spermatogenous cells of bryophytes, pteridophytes, and gymnosperms (cycads and *Ginkgo*), where their striking resemblance to ordinary centrosomes has led to much controversy over their nature. Some cytologists have regarded the blepharoplast as a more or less modified centrosome, whereas others have contended that it is a special kinoplasmic or cytoplasmic organ distinct from the centrosome. In recent years the evidence has tended strongly to support the former view.

In the following pages the blepharoplasts of various organisms and the manner in which they function in the development of the motor apparatus will be described in some detail. Attention will be given chiefly to the situations found in plants; the corresponding phenomena in animals will be more briefly considered.

Flagellates.—In the flagellates several types of flagellar apparatus are found (see Minchin 1912, pp. 82 ff., 262–3): in one series of forms the cell contains a single nucleus and "centriole," the latter functioning both as a centrosome and as a blepharoplast. The centriole may lie either against or within the nucleus, so that the flagellum which grows from it appears to arise directly from the nucleus (*Mastigina*); in other forms (*Mastigella*) the centriole is quite independent of the nucleus.

In a second series of forms a single nucleus and centrosome are present, and in addition one or more blepharoplasts. Three conditions have been distinguished here: (*a*) at the time of cell-division the blepharoplasts and flagella are lost, new blepharoplasts arising from the centrosomes during or after mitosis; (*b*) the blepharoplasts may persist, dividing to form daughter blepharoplasts from which new flagella arise (*Polytomella*); (*c*) the centrosome divides to furnish a blepharoplast which subdivides to two: a distal blepharoplast or basal granule of the flagellum, and a proximal blepharoplast or "anchoring granule" at the surface of the nucleus, the two being connected by a delicate strand known as the *rhizoplast, rhizonema,* or *centrodesmose* (*Peranema trichophorum*). Entz (1918) has recently reinvestigated the structure of *Polytoma uvella,* first described by Dangeard (1901), and finds the elaborate organization shown in Fig. 24.

In a third series of forms two nuclei are present: a principal or *trophic nucleus* and an accessory or *kinetic nucleus*. Here there are apparently three conditions: (*a*) a single centrosome, associated with the kinetonucleus, acts both as a blepharoplast and as a division center; (*b*) usually both nuclei have centrosomes associated with them, the

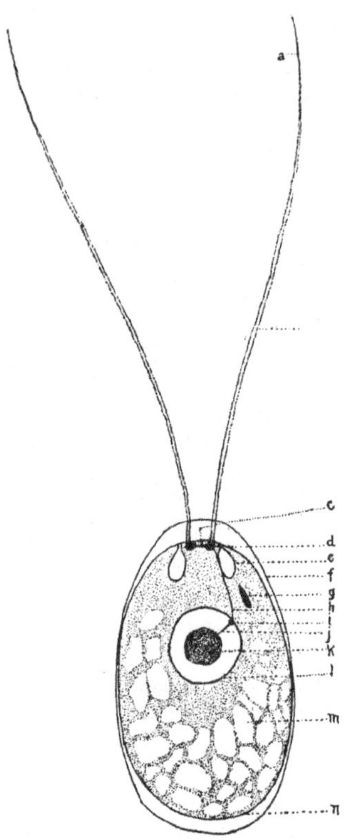

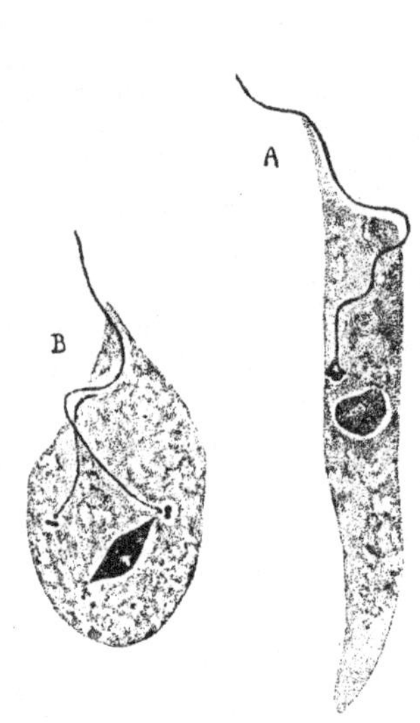

Fig. 24.   Fig. 25.

Fig. 24.—Diagram of structure of *Polytoma uvella*. (*After Entz*, 1918.)

a, end-piece of flagellum. *b*, uniform portion of flagellum. *c*, lateronema. *d*, basoplast or basal granule. *e*, contractile vacuole. *f*, cell envelope. *g*, eyespot. *h*, rhizonema. *i*, karyoplast or anchoring granule. *j*, centronema. *k*, nucleolus. *l*, nuclear membrane. *m*, starch. *n*, surface of protoplast.

Fig. 25.—*Trypanosoma theileri*.

A, flagellum inserted on basal granule. *B*, formation of new flagellum from daughter basal granule after division; nucleus dividing. (*After Hartmann and Nöller*, 1918.)

one lying near or within the kinetonucleus acting as the blepharoplast; (*c*) it is possible that in some cases there is a blepharoplast distinct from the centrosomes accompanying the two nuclei.

In the trypanosomes (Fig. 25) the recent researches of Kuczynski (1917) and Hartmann and Nöller (1918) have shown that the flagellum is inserted on a "basal granule" (centrosome?) very near the "blepharoplast" (kinetonucleus?).   At the time of cell-division the trophic nucleus, blepharoplast, and basal granule all divide, the division of the blepharoplast showing certain features suggesting mitosis.   Although earlier investigators thought the flagellum also split, the above named workers find that the old flagellum remains attached to one of the daughter basal granules while a new flagellum grows out from the other daughter granule.

In flagellate organisms, therefore, the centrosome and the blepharoplast clearly stand in very intimate relationship with one another: in some of the forms they are one and the same organ.

Thallophytes.—Among the earliest investigations of the blepharoplast in algæ were those of Strasburger (1892, 1900).   During the development of the zoöspores of *Œdogonium, Cladophora,* and *Vaucheria* Strasburger

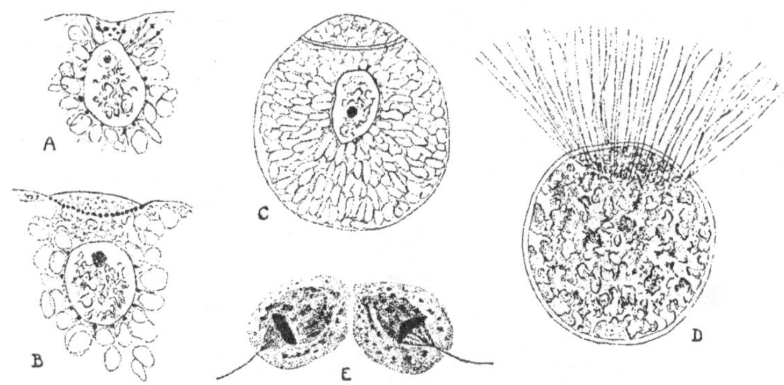

Fig. 26.—Blepharoplasts in Thallophytes.

A–D, formation of the cilia-bearing ring in the zoöspore of *Derbesia*.   (*After Davis,* 1908.)   *E, Stemonitis flaccida:* cilia growing from centrosomes during late stage of division in the formation of swarmers.   (*After Jahn,* 1904.)

found that the nucleus approaches the plasma membrane, which at that point forms a lens-shaped thickening.   From this structure grow out the cilia, and at the base of each a small refractive granule is present.   The blepharoplasts of the higher groups were believed by Strasburger to have been derived from such swollen ectoplasmic organs of the algæ, and that all of them are morphologically distinct from centrosomes.   Dangeard (1898) likewise found a deeply staining granule at the base of the cilia in *Chlorogonium.*

In *Hydrodictyon* (Timberlake 1902) the cilia are inserted on a small body lying in contact with the plasma membrane and joined with the nucleus by a delicate protoplasmic strand.   The possible relationship of this body with the granules seen occupying the spindle poles during the formation of the spore cells was not determined.   In the young zoöspore cell of *Derbesia* (Davis 1908) the nucleus migrates toward the plasma membrane, and from it many granules, which are not centrosomes, move out along radiating strands of cytoplasm to the surface of the cell, where by fusion they form a ring-shaped structure from which the cilia develop (Fig. 26, *A–D*).   In the developing spermatozoid of *Chara* (Belajeff 1894; Mottier 1904) the blepharoplast arises as a differentiation of the plasma membrane and bears two cilia.   No centrosomes or other granules were seen at the base of the cilia, although Schottländer (1893) had previously reported centrosomes in the cells of the spermatogenous filament.

In the zoöspore of the fungus *Rhodochytrium* (Griggs 1904) there is a deeply staining body at the insertion point of the cilia; this is connected by fine cytoplasmic fibers with the nucleus.   In the myxomycete *Stemonitis* Jahn (1904) made an observation that is highly suggestive in connection with the question of the relationship of the centrosome and the blepharoplast.   At the last mitosis in the formation of the swarmers the spindle poles are occupied by centrosomes, and during the anaphases the flagella of the resulting swarmers grow out directly from these centrosomes (Fig. 26, *E*), just as in the spermatocytes of certain insects (p. 95).

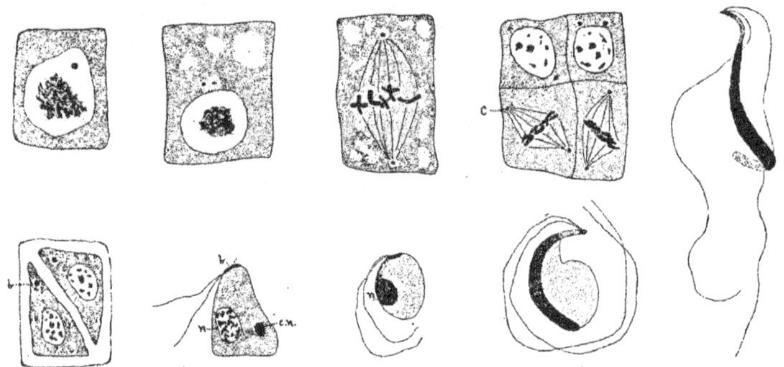

FIG. 27.—Spermatogenesis in *Marchantia*.
b, blepharoplast; *c*, centrosome; *c. n.*, "chromatoider Nebenkörper;" *n*, nucleus. (*After Ikeno,* 1903.)

Bryophytes.—Among the bryophytes the blepharoplasts of *Marchantia* and *Fegatella* (*Conocephalus*) have received much attention. According to Ikeno (1903) a centrosome comes out of the nucleus at each spermatogenous division in *Marchantia* and divides to form two which separate to opposite sides of the cell, occupy the spindle poles, and disappear at the close of mitosis: it is possible that they are included in the daughter nuclei.   After the last (diagonal) division, however, they remain in the cytoplasm as the blepharoplasts, elongating and bearing two cilia (Fig. 27).   Another body, the *chromatoider Nebenkörper*, is also present in the cytoplasm. Similar in most points is the account of Schaffner (1908). Miyake (1905), as the result of his studies on *Marchantia, Fegatella, Pellia, Aneura,* and *Makinoa,* believes that such liverwort centrosomes are merely centers of cytoplasmic radiation, and inclines toward the view of Strasburger that the blepharoplast and the centrosome are not homologous structures. Escoyez (1907) finds two "corpuscles" appearing in contact with the plasma membrane in each cell of the penultimate generation in the antheridia of *Marchantia* and *Fegatella;* they occupy the spindle poles and function as blepharoplasts in the spermatids (the cells which transform directly into spermatozoids). Bolleter (1905) believes the centrosome-like body in *Fegatella* to arise within the nucleus.

In the antheridium of *Riccia* Lewis (1906) reported centrosome-like bodies in both the early and diagonal divisions. They apparently arise *de novo* in the cytoplasm prior to each mitosis, showing no continuity through succeeding cell generations except after the last mitosis, when they persist and become the blepharoplasts.

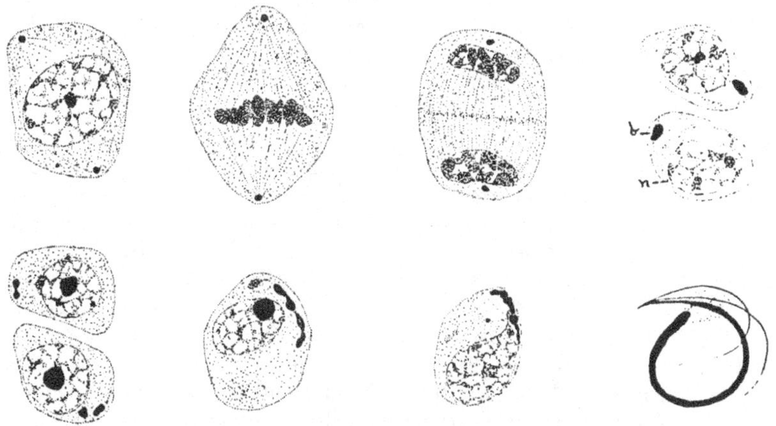

Fig. 28.—Spermatogenesis in *Blasia.*
b, blepharoplast; *n*, nucleus.   × 4200.   (*After Sharp,* 1920.)

Woodburn (1911, 1913, 1915) has given accounts of spermatogenesis in *Porella, Asterella, Marchantia, Fegatella, Blasia,* and *Mnium.* He finds that the blepharoplast is first distinguishable as a special granule in the cytoplasm of the spermatid, and holds that it represents, as Mottier (1904) had formerly suggested, an individualized portion of the kinoplasm arising *de novo* in certain spermatogenous cells. In a more recent contribution (Sharp 1920) it has been shown that in *Blasia* (Fig. 28) a blepharoplast is present at each spindle pole during all stages of the last spermatogenous mitosis, and that in the spermatid it fragments as it becomes transformed into the cilia-bearing thread, after the manner of the blepharoplast of *Equisetum*, described below.

Spermatogenesis in *Pellia, Atrichum*, and *Mnium* has been described by M. Wilson (1911). In *Mnium* and *Atrichum* the spermatogenous divisions show no centrosomes, whereas in *Pellia* centrospheres, and probably centrosomes, are present during the later mitoses. The origin of the blepharoplast as here described is very peculiar. In the spermatid of *Mnium* a number of bodies are said to separate from the nucleolus and pass out into the cytoplasm where they coalesce to form a *limosphere*. The nucleolus then divides into two masses, both of which pass into the cytoplasm; one functions as the blepharoplast and the other gives rise to an *accessory body*. In *Atrichum* the first body separated from the nucleolus becomes the blepharoplast, a second forms the limosphere, and a third the accessory body. In all three plants the blepharoplast goes to the periphery of the cell and grows out into a thread-like structure along the plasma membrane. The nucleus then moves against this thread and the two grow together to form the spirally coiled spermatozoid. Two cilia grow out from the anterior end of the blepharoplast.

The most detailed and critical of all researches on the motile cells of bryophytes are those of C. E. Allen (1912, 1917) on *Polytrichum* (Fig. 29). The first of these papers contains a description of the cytological phenomena accompanying the multiplication of the spermatogenous cells (*androgones*) up through the last mitosis, which differentiates the spermatids (*androcytes*). In the cytoplasm of all the androgones there is a deeply staining kinoplasmic mass; in the early androgones this has the form of a flat plate, while in the later ones it consists of a group of granules (kinetosomes). Prior to each mitosis the plate or group divides to daughter plates or groups which pass to the daugher cells. In the cells of the penultimate generation (androcyte mother-cells) there are no kinetosomes, but instead a spherical "central body" with radiations. This body divides into two which move apart and occupy the spindle poles during the last mitosis. Each resulting androcyte therefore has one such body, which functions as the blepharoplast. Allen does not regard the kinetosomes as definite morphological entities, but rather as masses of reserve kinoplasm. The blepharoplast, however, is a definite cell organ, and although Allen inclines toward the view that it is the homologue of a centrosome he regards the question as an open one. Sapehin (1913) looks upon these bodies as plastids.

Allen's second paper deals with the transformation of the androcyte (spermatid) into the spermatozoid. The blepharoplast elongates to form a uniform rod and develops two cilia from near its anterior end. The nucleus moves against the middle portion of the blepharoplast and the two elongate together in close union to form the body of the spermatozoid, the blepharoplast projecting beyond the anterior end of the nucleus. At about the time the blepharoplast begins to elongate a *limosphere* appears in the cytoplasm and takes up a position near the anterior end of the blepharoplast. Here it divides, giving rise to a small *apical body* that remains visible at the end of the blepharoplast until a comparatively late stage. The remaining portion of the limosphere may be seen lying against the nucleus until the maturity of the spermatozoid

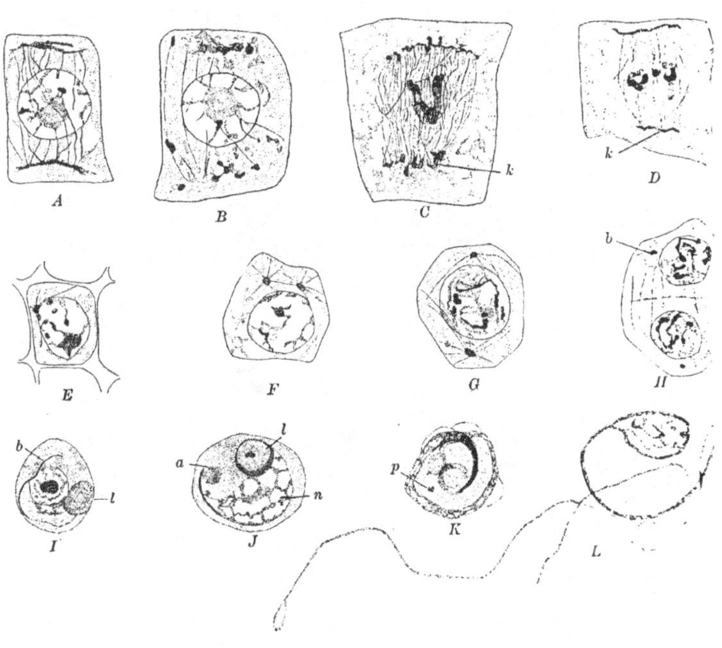

FIG. 29.—Spermatogenesis in *Polytrichum*.

A–D, androgones, showing behavior of kinoplasmic plates and kinetosomes (*k*) during imitoss. *E–G*, androcyte mother-cells, showing division of central body. *H*, telophase of last mitosis; each androcyte has a blepharoplast (*b*). *I–K*, stages in the transformation of the androcyte into the spermatozoid· *a*, apical body; *l*, limosphere; *n*, nucleus; *p*, percnosome. *L*, mature spermatozoid. × 2535. (*After Allen, 1912, 1917.*)

Another body, the *percnosome*, is also seen in the cytoplasm at certain stages. In the opinion of Allen the limosphere is probably identical with the *chromatoider Nebenkörper* described by Ikeno in *Marchantia*, and the percnosome with what M. Wilson (1911) terms the accessory body. The apical body is here described for the first time by Allen.

Pteridophytes.—The early papers dealing with the spermatozoid of pteridophytes, such as those of Buchtien (1887), Campbell (1887), Belajeff (1888), Guignard (1889), and Schottländer (1893), give but little information concerning the development of the blepharoplast. Our more definite knowledge of this subject dates from 1897, when Belajeff published three short papers. In the first of these (1897*a*) it was stated that the fern spermatozoid consists of a thread-shaped nucleus and a plasma band. with a great many cilia growing out from the latter. In the plasma band is enclosed a thin thread which arises by the elongation of a small body seen in the spermatid. In the second paper (1897*b*) the blepharoplast of *Equisetum* was first described as a crescent-shaped body lying against the nucleus of the spermatid; this body stretches out to form the cilia-bearing thread. The third contribution (1897*c*) is a short account of the transformation of the spermatid into the spermatozoid in *Chara, Equisetum,*

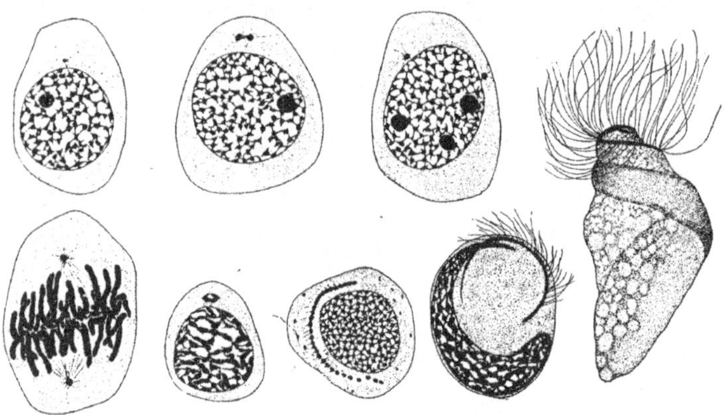

Fig. 30.—Spermatogenesis in *Equisetum arvense*, showing behavior of blepharoplast (centrosome) in last spermatogenous mitosis and in transformation of spermatid into spermatozoid. × 1900. (*After Sharp*, 1912.)

and ferns. In all these forms a small body elongates to form a thread upon which small swellings arise and grow out into cilia. In a comparison with animal spermatogenesis Belajeff here homologized the blepharoplast, the thread to which it elongates, and the cilia of the plant, with the centrosome, middle piece, and tail (perhaps only the axial filament), respectively, of the animal. In the following year (1898) he figured the details made out in *Gymnogramme* and *Equisetum*. In *Gymnogramme* the two blepharoplasts appear at opposite sides of the nucleus in the spermatid mother-cell, whereas in *Equisetum* a single blepharoplast is first figured lying close to the nucleus of the spermatid. More recently it has been shown (Sharp 1912) that the blepharoplast of *Equisetum* (Fig. 30) appears first in the cells of the penultimate generation; there it divides to two which separate and establish between them the achromatic figure after the manner of animal centrosomes. At the close of mitosis the blepharoplast in each spermatid fragments into a number of pieces; these latter join to form a continuous beaded thread from which the cilia grow out. In *Equisetum* the elongating nucleus and blepharoplast do not become closely joined, but are held together only by the rather abundant cytoplasm. The spermatozoid is multiciliate like that of all other pteridophytes with the exception of *Lycopodium, Phylloglossum,* and *Selaginella:* in these three genera the spermatozoids are biciliate like those of the bryophytes.

The most careful work on the blepharoplast of homosporous ferns is that of Yamanouchi (1908) on *Nephrodium* (Fig. 31). In this form no centrosomes are found. The two blepharoplasts, which arise *de novo* in the cytoplasm of the spermatid mother-cell, take no active part in nuclear division, merely lying near the poles of the spindle. In the spermatid the blepharoplast elongates spirally in close union with the nucleus to form the body of the spermatozoid. In *Adiantum* and *Aspidium* Miss R. F. Allen (1911) and Thom (1899) see the blepharoplast first in the spermatid.

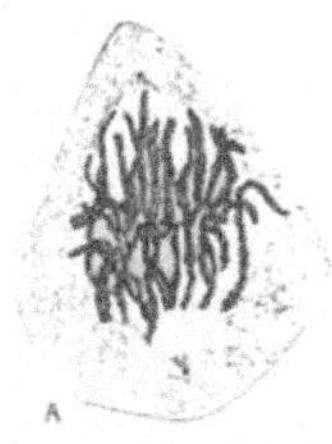

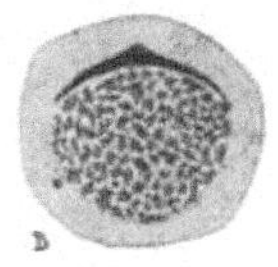

One of the most interesting blepharoplasts is that of *Marsilia* (Fig. 32), first described by Shaw (1898). According to Shaw a small granule, or "blepharoplastoid," appears near each daughter nucleus of the mitosis which differentiates the grandmother-cell of the spermatid (the second of the four spermatogenous mitoses). During the next (third) division the blepharoplastoid divides but soon disappears, and a blepharoplast appears near each spindle pole. In the next cell generation (spermatid mother-cell) the blepharoplast divides into two which are situated at the spindle poles during the final mitosis. In the spermatid the blepharoplast gives rise to several granules by a sort of fragmentation; these together form a thread which elongates spirally in close union

Fig. 31.—Two stages in the spermatogenesis of *Nephrodium*.

A, blepharoplasts near poles of spindle in last spermatogenous mitosis. *B,* elongation of blepharoplast near nucleus. Nebenkern at left. (*After* Yamanouchi, 1908.)

with the nucleus and bears many cilia. The spermatozoid is of the usual fern type, with several coils and a cytoplasmic vesicle. Shaw saw in the foregoing facts no ground for the homology of the blepharoplast and the centrosome. Belajeff (1899) found that centrosomes occur at the spindle poles during all, excepting possibly the first, of the four divisions which result in the 16 spermatids. He reported that after each mitosis the centrosome divides into two which occupy the spindle poles during the succeeding mitosis, and in the spermatids perform the usual functions of blepharoplasts. Belajeff regarded this as a strong confirmation of his theory that the blepharoplast and centrosome are homologous organs.

The results of Shaw were in the main confirmed by the later work of
Sharp (1914), who, however, saw in the achromatic structures accompanying the blepharoplast striking evidence in favor of Belajeff's view
of its homology.   In line with this conclusion the suggestion has recently
been made (Sharp 1920) that the fragmentation of the blepharoplast
in *Blasia, Equisetum, Marsilia,* and the cycads may be homologized with
the normal division exhibited by ordinary centrosomes.

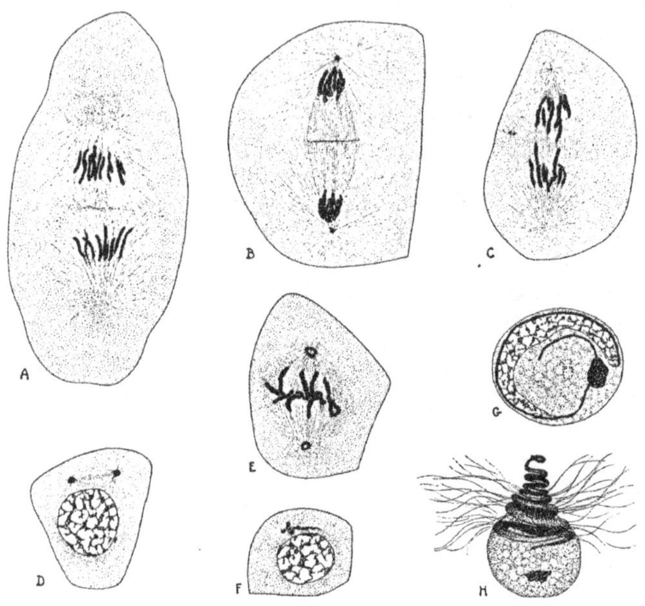

Fig. 32.—Spermatogenesis in *Marsilia quadrifolia.*
   A, first spermatogenous mitosis; no centrosomes.   *B*, second mitosis, centrosomes
present.   *C*, third mitosis; centrosomes present; old centrosome divided and degenerating
in cytoplasm.   *D*, penultimate spermatogenous cell; daughter centrosomes separating.
E, last spermatogenous mitosis; blepharoplasts (centrosomes) becoming vacuolate.   *F*, fragmentation of blepharoplast in spermatid.   *G*, transformation of spermatid into spermatozoid.   *H*, free swimming spermatozoid.   × 1400.   (*After Sharp,* 1914.)

Gymnosperms.—The first known blepharoplast in plants above the
algæ was discovered in *Ginkgo* by Hirasé in 1894.   He observed two, one
on either side of the body cell nucleus, and because of their great similarity to certain structures in animal cells he believed them to be attraction spheres.   In 1897 Webber observed the same bodies, and noted their
cytoplasmic origin.   On account of certain differences between these
organs and ordinary centrosomes he expressed the opinion that they are
not true centrosomes, but distinct organs of spermatogenous cells.
The blepharoplast of *Ginkgo* was later investigated by Fujii (1898, 1899,
1900) and Miyake (1906).

In 1897 and 1901 Webber described the blepharoplast of *Zamia* (Fig. 33). Up to the time of the division of the body cell the two blepharo_ plasts, which arise *de novo* in the cytoplasm, are surrounded by radiations, but they have no part in the formation of the spindle, which is entirely intranuclear. During mitosis they lie opposite the poles, increase greatly in size, become vacuolate, and break up to many granules: these in the spermatid coalesce to form a spirally coiled cilia-bearign band lying just inside the cell membrane. In his full account (1901) Webber gives an extensive discussion of the homology of the blepharoplast.

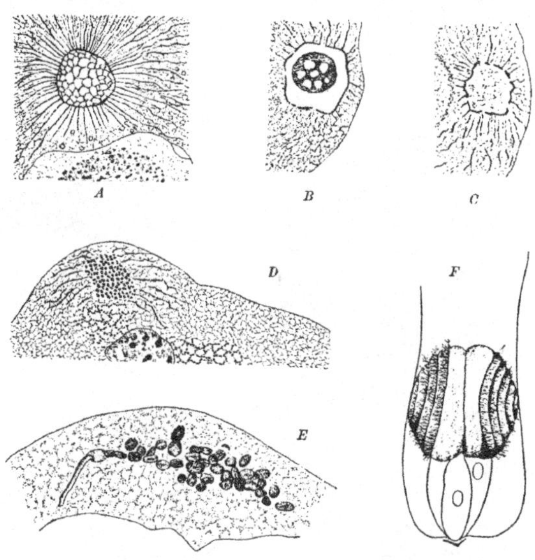

Fig. 33.—Spermatogenesis in *Zamia*.

A–E, five stages in the vacuolation and fragmentation of the blepharoplast during the mitosis differentiating the spermatids. *F*, the two spermatozoids in the end of the pollen tube; prothallial and stalk cells below. Compare Fig. 34. *A–D*, × 350; *E*, × 1200. (*After Webber*, 1901.)

In Ikeno's (1898) account of gametogenesis and fertilization in *Cycas* it was shown that the blepharoplasts appear in the body cell, lie opposite the spindle poles during mitosis, and break up to granules which fuse to form the spiral band in a manner similar to that described by Webber for *Zamia*. The behavior of the blepharoplast in *Microcycas* (Caldwell 1907) is essentially the same.

Chamberlain (1909) observed in the cytoplasm of the body cell of *Dioon* (Fig. 34) a number of very minute "black granules" which he was inclined to believe originate within the nucleus. Very soon two undoubted blepharoplasts are present, and are apparently formed by the enlargement of two of the black granules. Very conspicuous radiations develop about them, and after mitosis they form ribbon-like cilia-bearing bands in the spermatids as in the other cycads.

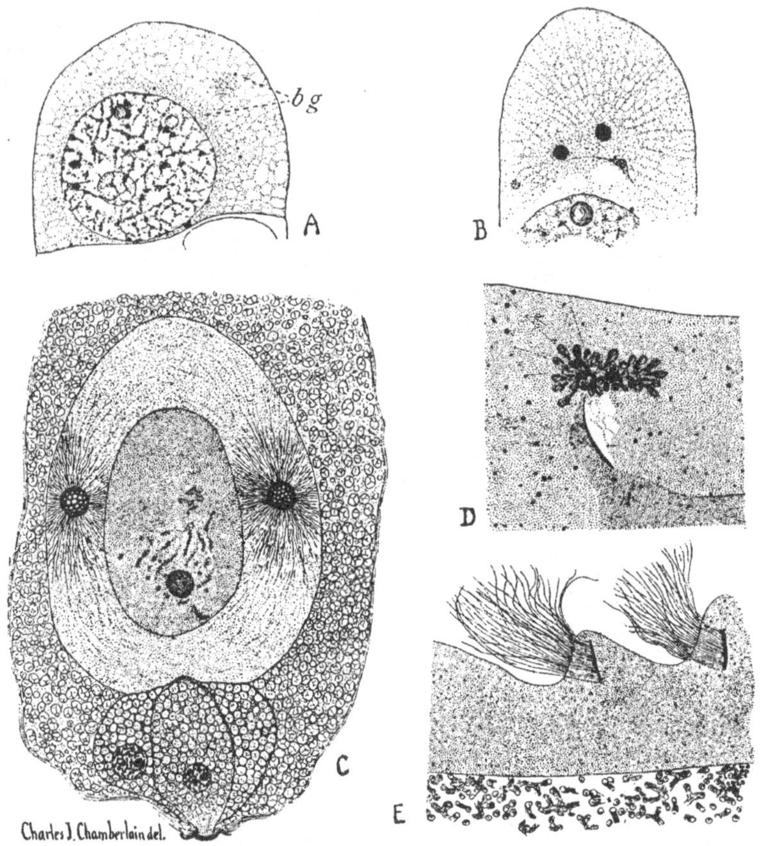

FIG. 34.—Spermatogenesis in *Dioon edule*.

A, "body cell," with black granules in cytoplasm. × 1890. *B*, two blepharoplasts differentiated. × 1890. *C*, body cell with two blepharoplasts; prothallial and stalk cells below. × 237. *D*, fragmentation of blepharoplast in spermatid as spiral band begins to form. × 1890. *E*, portion of edge of spermatozoid, showing spiral band cut at two points and cilia growing from it. × 945. (*After Chamberlain*, 1909.)

Ikeno in 1898 expressed the opinion that the blepharoplast of *Ginkgo* and the cycads is a true centrosome, a view shared by Chamberlain (1898) and Guignard (1898). Two additional papers dealing with this subject were published by Ikeno (1904, 1906). In the first of these he made comparisons with analogous phenomena in animals which he believed to sustain the homologies suggested by Belajeff. He pointed out that in *Marchantia* centrosomes are present in all the spermatogenous divisions, whereas in other liverworts they appear much later, and from this he argued that the bryophytes show various stages in the elimination of the centrosome. He strongly reasserted his belief that blepharoplasts are centrosomes, and spoke of the "transformation of a centrosome into a blepharoplast" in the development of a spermatid into a spermatozoid. The ectoplasmic blepharoplasts of the algæ were also held to be derived from centrosomes. In the second paper he insisted less strongly upon the morphological identity of all blepharoplasts, separating them into three categories: (1) centrosomatic blepharoplasts, including those of the myxomycetes, bryophytes, pteridophytes, and gymnosperms; (2) plasmodermal blepharoplasts, including those of *Chara* and some Chlorophyceæ; (3) nuclear blepharoplasts, found only in a few flagellates.

For a further discussion of this question the student is referred to the present author's papers on *Equisetum* and *Marsilia*. The main conclusions reached may be stated in two extracts from the former paper:

Although limited to a single mitosis in the antheridium, the blepharoplast [of *Equisetum*] retains in its activities the most unmistakable evidences of a centrosome nature, and at the same time shows a metamorphosis strikingly like that in the cycads. In thus combining the main characteristics of true centrosomes with the peculiar features of the most advanced blepharoplasts, it reveals in its ntogeny an outline of the phylogeny of the blepharoplast as it is seen developing through bryophytes, pteridophytes, and gymnosperms, from a functional centrosome to a highly differentiated cilia-bearing organ with very few centrosome resemblances.

The activities of the blepharoplast in *Equisetum* [*Marsilia*, and *Blasia*], taken together with the behavior of recognized true centrosomes in plants and analogous phenomena in animals, are believed to constitute conclusive evidence in favor of the theory that the blepharoplasts of bryophytes, pteridophytes, and gymnosperms are derived ontogenetically or phylogenetically from centrosomes.

Animals.—The early researches of Moore (1895), Meves (1897, 1899), Korff (1899), Paulmier (1899), and many other more recent investigators have established the fact that the centrosome (or centrosomes) of the animal spermatid plays an important rôle in the formation of the motor apparatus of the spermatozoön, the axial filament of the tail growing out directly from it (Fig. 35). Henneguy (1898) even saw flagella attached to the centrosomes of the mitotic figure in the spermatocyte of an insect, an observation which has been often repeated. Wilson (1900, p. 175) concludes that "the facts give the strongest ground for the conclusion that the formation of the spermatozoids agrees in its essential features with that of the spermatozoa . . . ." and that the blepharoplast is without doubt to be identified with the centrosome.

Although there is comparatively little question that the granule at the base of the flagellum in the flagellates, like the body from which the axial filament of the spermatozoön grows, is of centrosomic nature, the nature of the basal granules in Ciliata and in the ciliated epithelial cells of higher animals is much more difficult to determine.   It was held by Henneguy (1897), Lenhossék (1898), Hertwig (1902), and others that these granules, like the basal granules of flagella, are modified centrosomes; whereas certain other investigators (Maier 1903, Studnicka 1899, Schuberg 1905) have found evidence in favor of a contrary interpretation.   An extensive

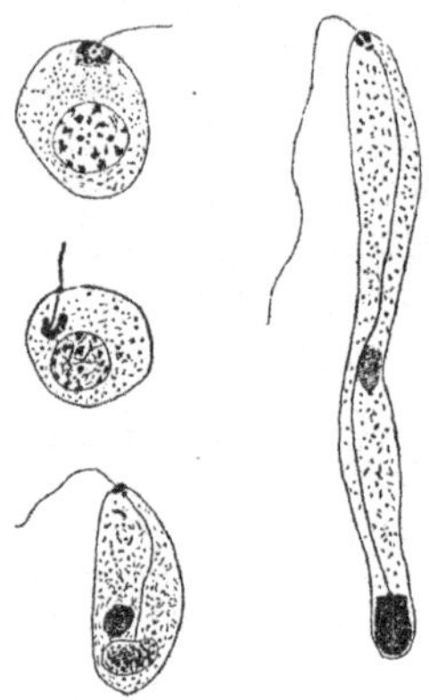

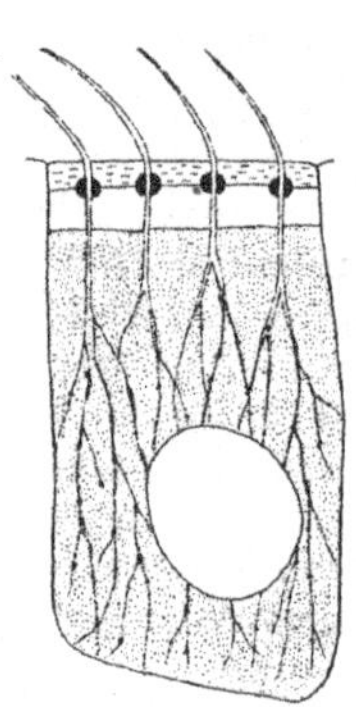

Fig. 35.—Spermatogenesis in *Helix pomatia*, showing growth of flagellum from outer centrosome, and elongation of inner centrosome to form axial filament of middle piece. (*After Korff*, 1899.)

Fig. 36.—Diagram of a ciliated epithelial cell. (*Constructed from figures of Saguchi*, 1917.)

discussion of this question is given by Erhard (1911), who concludes that although the basal corpuscles arise from the nucleus in a manner similar to that of the centrosomes of such cells, the evidence is on the whole unfavorable to the theory of Henneguy and Lenhossék.

Still more recent are the researches of Saguchi (1917), who describes in great detail the insertion of the cilia in epithelial cells.   At the base of each cilium, which itself shows no internal structural differentiation, there is always a basal corpuscle (Fig. 36).   These corpuscles, and hence the cilia, are in parallel rows; and beneath each row there is a transparent zone in which the rootlets of the cilia are anchored, and through which they pass and become continuous with strands of the cytoplasmic recticulum.   Cilium, corpuscle, rootlet, and cytoplasmic strand form one continuous structure.   Saguchi believes that neither the cilium nor the rootlet causes the ciliary movement, but that the kinetic center of this movement is in the basal corpuscles, as Henneguy and Lenhossék thought.   Contrary to the opinion of those authors, however, he regards the ciliary apparatus as entirely independent of centrosomes, holding rather that it is produced by the differentiation of chondriosomes, and that the resemblance of ciliated cells to spermatids, in which centrosomes do produce the motor apparatus, is an accidental one.

Conclusion.—In conclusion it may be said that it is highly probable that cilia-bearing structures are not homologous in all plant and animal groups. It is beyond question that in animals the centrosomes of the spermatid produce the motor apparatus of the spermatozöon. That a similar interpretation is to be placed upon the blepharoplasts in the spermatids (androcytes) of bryophytes, pteridophytes, and gymnosperms appears to be equally well demonstrated. The blepharoplasts of the flagellates are also probably centrosomic in nature, at least in certain cases. In the "plasmodermal blepharoplasts" of motile alga cells we have organs which, in the light of our present knowledge, do not appear to belong to the centrosomic category, but final disposition of them must await further information concerning thòse algæ which possess both centrosomes and blepharoplasts. It can scarcely be doubted that the basal corpuscles of ciliated cells represent organs belonging to various categories. It must be left for further research to determine just how far these structures, which are functionally analogous, are homologous with each other and with other cell organs.

Bibliography 5

Centrosome and Blepharoplast

ALLEN, C. E. 1912. Cell structure, growth and division in the antheridia of *Polytrichum juniperinum* Willd. Arch. Zellforsch. **8**: 121–188. pls. 6–9.

1917. The spermatogenesis of *Polytrichum juniperinum*. Ann. Bot. **31**: 269–292. pls. 15, 16.

ALLEN, R. F. 1911. Studies in spermatogenesis and apogamy in ferns. Trans. Wis. Acad. Sci. **17**: 1–56. pls. 1–6.

BELAJEFF, W. 1888. Ueber Bau and Entwicklung der Spermatozoiden der Gefässkryptogamen. Ber. Deu. Bot. Gesell. **7**: 122–125.

1894. Ueber Bau und Entwicklung der Spermatozoiden der Pflanzen. Flora **79**: Ergänzb.: 1–48. pl. 1.

1897a. Ueber den Nebenkern in spermatogenen Zellen und die Spermatogenese bei den Farnkräutern. Ber. Deu. Bot. Ges. **15**: 337–339.

1897b. Ueber die Spermatogenese bei den Schachtelhalmen. Ibid. **15**: 339–342.

1897c. Ueber die Aehnlichkeit einiger Erscheinungen in der Spermatogenese bei Thieren und Pflanzen. Ibid. **15**: 342–345.

1898. Ueber die Cilienbildner in spermatogenen Zellen. Ibid. **16**: 140–144. pl. 7.

1899. Ueber die Centrosome in den spermatogenen Zellen. Ibid. **17**: 199–205. pl. 15.

VAN BENEDEN, E. 1876. Recherches sur les Dicyemides, survivants actuelles d'un embranchement des Mesozoaires. Bull. Acad. Roy. Belg. **41**: 1160–1205; **42**: 35–97. pls. 3.

1883. Recherches sur la maturation de l'oeuf, la fécondation et la division cellulaire. Arch. d. Biol. **4**.

1887. (van Beneden et Neyt.) Nouvelles recherches sur la fécondation et la division mitosique chez *l'Ascaride megalocephale*. Bull. Acad. Roy. Belg. III **14**: 215-295. pl. 1.

BOLLETER, E. 1905. *Fegatella* (L.) *corda*. Eine morphologisch-physiologische Monographie. Beih. Bot. Centr. **18**: 327-408. pls. 12, 13.

BOVERI, TH. 1887*a*. Ueber die Befruchtung der Eier von *Ascaris megalocephala*. Sitz-ber. Ges. Morph. Phys. München **3**.

1887*b*. Ueber den Anteil des Spermatozoön an der Teilung des Eies. Ibid.

1888. Zellenstudien. II. Jenaische Zeitschrift **22**: 685-882. pls. 19-23

1901. Zellenstudien. IV. Ueber die Natur der Centrosomen. Jen. Zeitschr. **35**: 1-220. pls. 1-8.

BRAUER, A. 1893. Zur Kenntniss der Spermatogenese von *Ascaris megalocephala*. Arch. Mikr. Anat. **42**: 153-212. pls. 11-13.

BUCHTIEN, O. 1887. Entwicklungsgeschichte des Prothalliums von *Equisetum*. Bibliotheca Botanica **8**: 1-49. pls. 1-6.

BUTSCHLI, O. 1891. Ueber die sogenannten Centralkörper der Zelle und ihre Bedeutung. Verh. Naturhist.-Med. Ver. Heidelb. **4**: 535-538.

CALDWELL, O. W. 1907. *Microcycas calocoma*. Bot. Gaz. **44**: 118-141. pls. 10-13.

CAMPBELL, D. H. 1887. Zur Entwicklungsgeschichte der Spermatozoiden. Ber. Deu. Bot. Gesell. **5**: 120-127. pl. 6.

CHAMBERLAIN, C. J. 1898. The homology of the blepharoplast. Bot. Gaz. **26**: 431-435.

1903. Mitosis in *Pellia*. Decenn. Pub. Univ. Chicago **10**: 329-345. pls. 25-27.

1909. Spermatogenesis in *Dioon edule*. Bot. Gaz. **47**: 215-236. pls. 15-18.

CHAMBERS, R. 1917. Microdissection studies. II. The cell aster. A reversible gelation phenomenon. Jour. Exp. Zool. **23**: 483-504.

CLAUSSEN, P. 1912. Zur Entwicklungsgeschichte der Ascomyceten. Zeitschr. Bot. **4**: 1-64. pls. 1-6. figs. 13.

CONKLIN, E. G. 1912. Experimental studies on nuclear and cell division in the eggs of *Crepidula*. Jour. Acad. Nat. Sci. Phila. **15**: 503-591. pls. 43-49.

DANGEARD, P. A. 1896. Mémoire sur les Chlamydomonadinees ou l'histoire d'une cellule. Le Botaniste **6**: 65-290. figs. 19.

1901. Étude sur la structure de la cellule et ses fonctions. Le *Polytoma uvella*. Ibid. **8**: 5-58. figs. 4.

DAVIS, B. M. 1898. Kerntheilung in der Tetrasporenmutterzelle bei *Corallina officinalis* L. var. *mediterranea*. Ber. Deu. Bot. Gesell. **16**: 266-272. pls. 16, 17.

1908. Spore formation in *Derbesia*. Ibid. **22**: 1-20. pls. 1, 2.

ENTZ, G. 1918. Ueber die mitotische Teilung von *Polytoma uvella*. Arch. f. Protistenk. **38**: 324-354. pls. 12, 13. figs. 5.

ERHARD, H. 1911. Die Henneguy-Lenhosseksche Theorie. Ergeb. Anat. u. Entw. **19**: 893-929. (List of 146 papers.)

ESCOYEZ, E. 1907. Blepharoplaste et centrosome dans le *Marchantia polymorpha*. La Cellule **24**: 247-256. pl. 1.

1909. Caryocinèse, centrosome et kinoplasme dans le *Stypocaulon scoparium*. La Cellule **25**: 181-201. 1 pl.

FARMER, J. B. and REEVES, J. 1894. On the occurrence of centrospheres in *Pellia epiphylla*, Nees. Ann. Bot. **8**: 219-224. pl. 14.

FARMER, J. B. and WILLIAMS, J. L. 1896. On fertilization, and the segmentation of the spore in *Fucus*. Ibid. **10**: 479-487.

1898. Contributions to our knowledge of the Fucaceæ; their life history and cytology. Phil. Trans. Roy. Soc. London B **190**: 623-645. pls. 19-24.

FAULL, J. H. 1905. Development of ascus and spore formation in ascomycetes. Proc. Boston Soc. Nat. Hist. **32**: 77-113. pls. 7-11.

FLEMMING, W. 1875. Studien in der Entwicklungsgeschichte der Najaden. Sitzber. Akad. Wiss. Wien **71.**

FLORIN, R. 1918. Das Archegonium der *Riccardia pinguis* (L) B. Gr. Svensk. Bot
Tidskr. **12**: 464–470. figs. 4.

FOL, H. 1877. Sur le commencement de l'henogenie chez divers animaux. Arch.
Sci. Nat. u. Phys. Geneve **58.**

1891. Die "Centrenquadrille," ein neue Episode aus der Befruchtungsgeschichte.
Anat. Anz. **6**: 266–274. figs. 10.

FRASER, H. C. I. 1908. Contributions to the cytology of *Humaria rutilans*. Ann.
Bot. **22**: 35–55. pls. 4, 5.

FRASER, H. C. I. and WELSFORD, E. J. 1908. Further contributions to the cytology
of the ascomycetes. Ibid. **22**: 465–477. pls. 26, 27.

FRASER, H. C. I. and BROOKS, W. E. ST. J. 1909. Further studies on the cytology
of the ascus. Ibid. **23**: 538–549.

FUJII, K. 1898. (Has the spermatozoid of *Ginkgo* a tail or not?) Bot. Mag.
Tokyo **12**: 287–290. (Japanese.)

1899. (On the morphology of the spermatozoid of *Ginkgo biloba*.) Ibid. **13**: 260–
266. pl. 7. (Japanese.)

1900. (Account of a sperm with two spiral bands.) Ibid. **14**: 16–17. (Japanese.)

GRAHAM, M. 1918. Centrosomes in fertilization stages of *Preissia quadrata* (Scop.)
Nees. Ann. Bot. **32**: 415–420. pl. 10.

GRÉGOIRE, V. et BERGHS, J. 1904. La figure achromatique dans le *Pellia epiphylla*.
La Cellule **21**: 193–238. pls. 1, 2.

GRIGGS, R. F. 1912. The development and cytology of *Rhodochytrium*. Bot. Gaz.
53: 127–173. pls. 11–16.

GUIGNARD, L. 1889. Developpement et constitution des anthérozoides. Rev.
Gén. Bot. **1**: 11–27, 63–78, 136–145, 175–194. pls. 2–6.

1898. Centrosomes in Plants. Bot. Gaz. **25**: 158–164.

GUILLIERMOND, A. 1904. Recherches sur la karyokinèse chez les ascomycètes.
Rev. Gén. Bot. **16**: 129–143. pls. 14, 15.

1905. Remarques sur la karyokinèse des ascomycètes. Ann. Mycol. **3**: 344–361.
pls. 10–12.

1911. Aperçu sur l'évolution nucléaire des ascomycètes et nouvelles observations
sur les mitoses des asques. Rev. Gén. Bot. **23**: 89–121. figs. 8. pls. 4, 5

HARPER, R. A. 1895. Beitrag zur Kenntniss der Kernteilung und Sporenbildung
im Ascus. Ber. Deu. Bot. Ges. **13**: (67)–(68). pl 27.

1897. Kerntheilung und freie Zellbildung im Ascus. Jahrb. Wiss. Bot. **30**: 249–
284. pls. 11, 12.

1899. Cell division in sporangia and asci. Ann. Bot. **13**: 467–525. pls. 24–26.

1905. Sexual reproduction and the organization of the nucleus in certain mildews.
Carnegie. Inst. Publ. **37.** Washington.

1919. The structure of protoplasm. Am. Jour. Bot. **6**: 273–300.

HARTMANN, M. und NÖLLER, W. 1918. Untersuchungen uber die Cytologie von
Trypanosoma theileri. Arch. Protistenk. **38**: 355–374. pls. 14, 15. figs. 6.

HENNEGUY, L. F. 1897. Sur les rapports des cils vibratiles avec les centrosomes.
Arch. d'Anat. Micr. **1**: 481–496. figs. 5.

HIRASÉ, S. 1894. Notes on the attraction spheres in the pollen cells of *Ginkgo biloba*.
Bot. Mag. Tokyo **8**: 359.

HUMPHREY, J. E. 1894. Nucleolen und Centrosomen. Ber. Deu. Bot. Ges. **12**:
108–117. pl. 6.

HUMPHREY, H. B. 1906. The development of *Fossombronia longiseta* Austr
Ann. Bot. **20**: 83–108. pls. 5, 6. figs. 8.

IKENO, S.   1898.   Untersuchungen über die Entwicklung der Geschlechtsorgane und den Vorgang der Befruchtung bei *Cycas revoluta*.   Jahrb. Wiss. Bot. **32**: 557–602. pls. 8–10.

    1903.   Die Spermatogenese von *Marchantia polymorpha*.   Beih. Bot. Centralbl. **15**: 65–88.   pl. 3.

    1904.   Blepharoplasten im Pflanzenreich.   Ibid. **24**: 211–221.   figs. 1–3.

    1906.   Zur Frage nach der Homologie der Blepharoplasten.   Flora **96**: 538–542.

JAHN, E.   Myxomycetenstudien.   3.   Kernteilung und Geisselbildung bei den Schwärmern von *Stemonitis flaccida* Lister.   Ber. Deu. Bot. Gesell. **22**: 84–92. pl. 6

KARSTEN, G.   1900.   Die Auxosporenbildung der Gattungen *Cocconeis*, *Surirella*, und *Cymatopleura*.   Flora **87**: 253–283.   pls. 8–10.

KOERNICKE, M.   1903.   Der heutige Stand der pflanzlichen Zellforschung.   Ber. Deu. Bot. Gesell. **21**: (66)–(134).

    1906.   Zentrosomen bei Angiospermen?   Flora **96**: 501–522.   pl. 5.

VON KORFF, K. 1899.   Zur Histogenese der Spermien von *Helix pomatia*.   Arch. Mikr. Anat. **54**: 291–296.   pl. 16.

KUCZYNSKI, M. H.   1917.   Ueber die Teilung der Trypanosomenzelle nebst Bemerkungen zur Organization einiger nahestehender Flagellaten.   Arch. f. Protistenk. **38**: 94–112.   pls. 3, 4.

LAUTERBORN, R.   1896.   Untersuchungen über Bau, Kernteilung, und Bewegung der Diatomen.   Leipzg.

VON LENHOSSÉK, M. 1898.   Ueber Flimmerzellen.   Verb. Anat. Ges. Kiel **12**: 106.

LEVINE, M.   1913.   The cytology of Hymenomycetes, especially the Boleti.   Bull. Torr. Bot. Club **40**: 137–181.   pls. 4–8.

LEWIS, C. E.   1906.   Embryology and development of *Riccia lutescens* and *Riccia crystallina*.   Bot. Gaz. **41**: 109–138.   pls. 5–9.

MAIER, H. N.   1903.   Ueber die feineren Bau der Wimperapparate der Infusorien. Arch. f. Protistenk. **2**.

MAIRE, R.   1905.   Recherches cytologiques sur quelques ascomycètes.   Ann. Mycol. **3**: 123–154.   pls. 3–5.

MEAD, A. D.   1898.   The origin and behavior of the centrosomes in the annelid egg. Jour. Morph. **14**: 181–218.

MEVES, F.   1897.   Ueber Struktur und Histogenese der Samenfäden von *Salamandra maculosa*.   Arch. Mikr. Anat. **50**: 110–141.   pls. 7, 8.

    1899.   Ueber Struktur und Histogenese der Samenfäden des Meerschweinchens. Ibid. **54**: 329–402.   pls. 19–21.   figs. 16.

MEYER, K.   1911.   Untersuchungen über den Sporophyt der Lebermoose.   I.   Entwicklungsgeschichte des Sporogons der *Corsinia marchantioides*.   Bull. Soc. Imp. Moscou 236–286.

MINCHIN, E. A.   1912.   An Introduction to the Study of the Protozoa.   London.

MIYAKE, K.   1905.   On the centrosome of Hepaticæ.   Bot. Mag. Tokyo **19**: 98–101.

    1906.   The spermatozoid of *Ginkgo*.   Jour. Appl. Micr. and Lab. Methods **5**: 1773–1780.   figs. 10.

MOORE, J. E. S.   1895.   Structural changes in the reproductive cells during spermatogenesis of elasmobranchs.   Quar. Jour. Micr. Sci. **38**: 275–313.   pls. 13–16.

MORGAN, T. H.   1896.   The production of artificial astrospheres.   Arch. Entw. **3**: 339–361.   pl. 19.

    1899.   The action of salt solutions on the unfertilized and fertilized eggs of *Arbacia* and other animals.   Ibid. **8**: 448–539.   pls. 7–10.   figs. 21.

MOTTIER, D. M.   1898.   Das Centrosom bei *Dictyota*.   Ber. Deu. Bot. Ges. **16**: 123–128.   figs. 5.

1900. Nuclear and cell division in *Dictyota dichotoma*. Ann. Bot. **14**: 166–192. pl. 11.

1904. The development of the spermatozoid of *Chara*. Ibid. **18**: 245–254. pl. 17.

PAULMIER, F. C. 1899. The spermatogenesis of *Anasa tristis*. Jour. Morph. **15**: Suppl. 223–272. pls. 13, 14.

RAWITZ, B. 1896. Untersuchungen über Zelltheilung. I. Arch. Mikr. Anat. **47**: 159–180. pl. 11.

SAGUCHI, S. 1917. Studies on ciliated cells. Jour. Morph. **29**: 217–279. pls. 1–4.

SANDS, M. C. 1907. Nuclear structure and spore formation in *Microsphæra*. Trans. Wis. Acad. Sci. **15**: 733–752. pl. 46.

SAPĚHIN, A. A. 1913. Untersuchungen über die Individualität der Plastide. Ber. Deu. Bot. Ges. **31**: 14–66. fig. 1.

SCHAFFNER, J. H. 1908. The centrosomes of *Marchantia polymorpha*. Ohio Nat **9**. 363–388.

SCHOTTLÄNDER, P. 1893. Beiträge zur Kenntniss des Zellkerns und der Sexualzellen bei Kryptogamen. Cohn's Beitr. Biol. Pflanzen **6**: 267–304. pls. 4, 5.

SHARP, L. W. 1912. Spermatogenesis in *Equisetum*. Bot. Gaz. **54**: 89–119. pls. 7, 8.

1914. Spermatogenesis in *Marsilia*. Ibid. **58**: 419–431. pls. 33, 34.

1920. Spermatogenesis in *Blasia*. Ibid. **69**: 258–268. pl. 15.

SHAW, W. R. 1898. Ueber die Blepharoplasten bei *Onoclea* und *Marsilia*. Ber. Deu. Bot. Ges. **16**: 177–184. pl. 11.

SMITH, H. L. 1886-1887. A contribution to the life history of the Diatomaceæ. Proc. Am. Soc. Micr. Pts. I and II.

STRASBURGER, E. 1892. Schwärmsporen, Gameten, pflanzliche Spermatozoiden. und das Wesen der Befruchtung. Hist. Beitr. **4**: 49–158. pl. 3.

1897. Kerntheilung und Befruchtung bei *Fucus*. Jahrb. Wiss. Bot. **30**: 351–374. pls. 27, 28.

1900. Ueber Reduktionstheilung, Spindelbildung, Centrosomen, und Cilienbildner im Pflanzenreich. Hist. Beitr. **6**: 1–224. pls. 1–4.

STUDNICKA, F. K. 1899. Ueber Flimmer- und Cuticularzellen mit besonderer Berücksichtigung der Centrosomenfrage. Sitz-Ber. K. Böhmisch. Ges. Wiss. Math.-Naturwiss. Classe, **35**.

SWINGLE, W. T. 1897. Zur Kenntniss der Kern- und Zelltheilung bei den Sphacelariaceen. Jahrb. Wiss. Bot. **30**: 296–350. pls. 15, 16.

THOM, C. 1899. The process of fertilization in *Aspidium* and *Adiantum*. Trans. Acad. Sci. St. Louis **9**: 285–314. pls. 36–38.

TIMBERLAKE, H. G. 1902. Development and structure of the swarm spores of *Hydrodictyon*. Trans. Wis. Acad. Sci. **13**: 486–522. pls. 29, 30.

VAN HOOK, J. M. 1900. Notes on the division of the cell and nucleus in liverworts. Bot. Gaz. **30**: 394–399. pls. 2, 3.

WEBBER, H. J. 1897a. Peculiar structures occurring in the pollen tube of *Zamia*. Bot. Gaz. **23**: 453–459. pl. 40.

1897b. The development of the antherozoid of *Zamia*. Ibid. **24**: 16–22. figs. 5.

1897c. Notes on the fecundation of *Zamia* and the pollen tube apparatus of *Ginkgo*. Ibid. **24**: 225–235. pl. 10.

1901. Spermatogenesis and fecundation in *Zamia*. U. S. Dept. Agr. Plt. Ind. Bull. **2**. pp. 100. pls. 7.

WILLIAMS, J. L. 1904. Studies in the Dictyotaceæ. II. The cytology of the gametophyte generation. Ann. Bot. **18**: 183–204. pls. 12–14.

WILSON, E. B. 1900. The Cell in Development and Inheritance. (p. 175.)

1901. Experimental Studies in Cytology. I. A cytological study of parthenogenesis in sea urchin eggs. Arch. Entw. **12**: 529–596. pls. 11–17.

Wilson, M. 1911. Spermatogenesis in the Bryophyta. Ann. Bot. **25**: 415–457. pls. 37, 38. figs. 3.

von Winiwarter, H. 1912. Observations cytologiques sur les cellules interstitiélles du testicule humaine. Anat. Anz. **41**: 309–320. pls. 1–2.

Wolfe, J. J. 1904. Cytological studies on *Nemalion*. Ann. Bot. **18**: 607–630. pls. 40, 41. fig. 1.

Woodburn, W. L. 1911. Spermatogenesis in certain Hepaticæ. Ann. Bot. **25**: 299–313. pl. 25.

    1913. Spermatogenesis in *Blasia pusilla*. Ibid. **27**: 93–101. pl. 11.

    1915. Spermatogenesis in *Mnium affine*, var. *ciliaris* (Grev.), C.M. Ibid. **29**: 441–456. pl. 21.

Yamanouchi, S. 1906. The life history of *Polysiphonia violacea*. Bot. Gaz. **42**: 401–449. pls. 19–28.

    1908. Spermatogenesis, oögenesis, and fertilization in *Nephrodium*. Ibid. **45**: 145–175. pls. 6–8.

    1909. Mitosis in *Fucus*. Ibid. **47**: 173–197. pls. 8–11.

Zimmermann, A. 1893–1894. Sammel-Referate. 6. Die Centralkörper und die Kerntheilung. 12. Die Cilien und Pseudocilien. Beih. Bot. Centralbl. **3**: 342–354; **4**: 169–171.

CHAPTER VI

PLASTIDS AND CHONDRIOSOMES

PLASTIDS

Next to the nucleus, the most conspicuous organ held within the cytoplasm of the plant cell is the plastid. Cytologists have long been aware of the important physiological rôles played by plastids of various types in the life of the cell, but it is only recently that an added interest has been given these organs by the discovery that certain peculiar characters showing definite modes of inheritance are closely bound up with their behavior. Such problems are complicated by the relation apparently borne by plastids to chondriosomes. In the present chapter will be set forth some of the more important facts regarding these two classes of cell elements.

General Nature and Occurrence.—Plastids are differentiated portions of the protoplasm, as von Mohl long ago pointed out, and represent regions in which certain processes have become localized (Harper 1919). In view of their power of growth and division and their definite relation to certain important physiological functions they are to be regarded as distinct cell organs.

Although plastids can be found in the cells of both animals and plants they are chiefly characteristic of the latter, where they are present in one form or another in all groups with the possible exception of bacteria, myxomycetes, and certain fungi. They are abundant only in those plant parts which have to do with specialized physiological functions. Within a single cell there may be regularly but one plastid, as in many a'gæ, *Anthoceros*, and the meristematic cells of *Selaginella* (Haberlandt 1888, 1905); or two, as in *Zygnema;* or a ligher number, as in the green tissues of most higher plants. They lie imbedded in the cytoplasm and are often closely associated with the nucleus; they are never found normally in the vacuole. The positions which they assume within the cell are frequently related in a definite manner to certain external conditions: in the palisade cells of green leaves, for example, the chloroplasts are found near the upper surface if the incident light is weak, whereas they react to strong illumination by taking up less exposed positions along the lateral walls.

Plastids may be conveniently classified on the basis of their contained coloring matters. This difference in color, however, is secondary in importance; the fundamental distinction is that based upon the kind of physiological work being done, the various pigments being associated in an intimate manner with different reactions occurring within the plastids.

Leucoplasts.—*Leucoplasts* are relatively small and colorless. They are found commonly in the cells of meristematic tissue, and may be retained in some kinds of differentiated cells, such as the glandular hairs of *Pelargonium*. Küster (1911) states that the leucoplasts of *Orchis* are very fluid in consistency, undergoing amœboid changes of shape and multiplying by irregular fission. The smaller leucoplasts appear to represent juvenile stages in the development of plastids of more highly differentiated types, for under certain conditions they develop into the larger and more highly specialized leucoplasts known as *amyloplasts*, and into the various kinds of chromatophores mentioned below.

Chromatophores.—*Chromatophores*, or *chromoplasts*, are plastids bearing one or more pigments, and having thus a more or less decided color. In green plants the most important of these pigments are chlo-

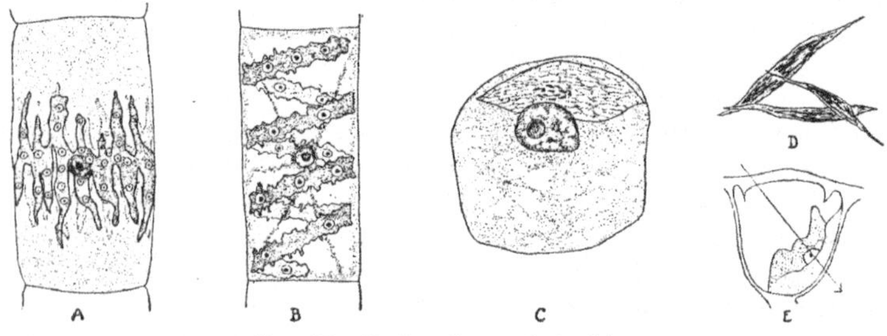

FIG. 37.—Various forms of plastids.

A, Draparnaldia. B, Spirogyra. C, Anthoceros. D, chromoplasts of *Arisæma. E*, cell of *Selaginella*, showing position assumed by plastid in response to light (direction shown by arrow). *A, B*, and *C* show pyrenoids. (*E After Haberlandt*.)

rophyll, carotin, and xanthophyll. Chlorophyll is apparently a combination of two simpler pigments, chlorophyll *a* and chlorophyll *b*. The cells of the Phæophyceæ, Cyanophyceæ, and Rhodophyceæ are characterized respectively by the presence of yellow carotin, blue phycocyanin, and red phycoerythrin, in addition to chlorophyll. The Cyanophyceæ exhibit an especially rich variety of pigments, which in many cases do not appear to be held within definite chromatophores.[1]

Chromatophores are usually spherical, ovoid, or discoid in shape, but many peculiar forms are known, particularly among the green algæ. In *Ulothrix* the chloroplast has the form of a complete or incomplete hollow cylinder; in *Draparnaldia* (Fig. 37, *A*), a hollow cylinder with very

[1] For the literature pertaining to plant pigments see Palladin 1918. See also Haas and Hill (1913), Willstätter and Stoll (1913), Jörgensen and Stiles (1917), Wheldale (1916), and Beauverie (1919). The distribution of carotin is discussed in an earlier paper by Tammes (1900).

irregular ends; in *Ædogonium*, an irregular parietal net; in *Spirogyra* (Fig. 37, *B*), a spirally coiled ribbon; and in the desmids, a series of radiating plates (Carter 1919, 1920). The chromatophore of *Antho- ceros* (Fig. 37, *C*) is spindle-shaped, becoming chain-like in the elongated columella cells (Scherrer 1914). The chromatophore of *Selaginella* may also assume this form (Haberlandt 1888). The chromoplasts of *Arisæma* (Fig. 37, *D*) are frequently sharply angular. In the Clado- phoraceæ (Carter 1919) the cell is completely lined by a thin chromato- phore which may be entire or fenestrated. In many cells irregular strands pass inward through the cell cavity. Indeed it seems not improbable that in some such cases the plastid may be not at all sharply distinct from the rest of the cytoplasm, the two grading one into the other, and the chlorophyll at certain stages permeating all parts of the cytoplasm. The observations of Timberlake and Harper appear to show that such is the condition in the young cells of *Hydrodictyon*. Thus the physiological processes show various degrees of localization in the cell, causing manifold degrees of structural transformation and delimitation of the cytoplasmic regions involved (Harper).

Of all chromatophores the *chloroplasts* stand first in importance, for they bear the green pigment, chlorophyll, which, in the presence of light, enables them to combine water from the soil or other surrounding medium with carbon dioxide from the atmosphere to form carbohydrates, the first visible product being starch. The chloroplasts are therefore the world's ultimate food producers. In addition to chlorophyll other pigments, notably xanthophyll, are usually present. Although the body of the chloroplast can be developed in darkness, the chlorophyll will usually not be elaborated unless light is present. Most young seedlings grown in the absence of light show a pale yellowish color, which is due to a substance known as chlorophyllogen, contained in the plastids. When such "etiolated" plants are placed in the light the plastids become green, apparently through an alteration of the chlorophyllogen to chlorophyll (Monteverde and Lubimenko 1911). Other conditions necessary for the development of chlorophyll are a favorable temperature and the presence of iron, oxygen, and certain carbohydrates.

The structure of the chloroplast is an extremely difficult matter to determine, and has been the subject of some controversy. It is generally thought that the body of the plastid is composed of a finely fibrillar meshwork, the *stroma*, which may be somewhat denser at the periphery, and that the coloring matters are held in the meshes of the stroma in the form of minute droplets. No limiting membrane is definitely known. The included droplets are apparently not composed of the pigments alone: it is probable that they are rather globules of some oily or fatty material containing the pigments in solution. The pigments may easily be dis- solved out with alcohol and other reagents. On the other hand, it has been held by some observers that the stroma is a homogeneous body in which the droplets of chlorophyll solution are imbedded, and that the reticular structure so often reported is an artifact due to the reagent employed in removing the chlorophyll. By others the pigment has been thought to form a layer about the plastid. In any case it seems evident that the chlorophyll is not uniformly distributed throughout the stroma. In chromatophores other than chloroplasts the pigments may at times take the form of solid granules or crystals.

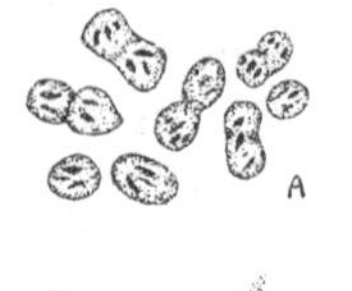

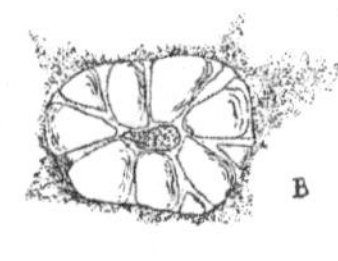

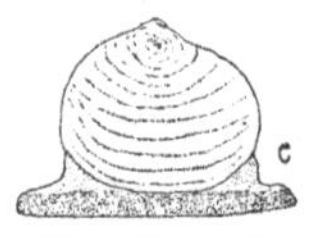

Fig. 38.—Formation of starch by plastids.

A, dividing chloroplasts of *Funaria*, with grains of assimilation starch. × 940. (*After Strasburger.*) *B*, chloroplast of *Zygnema*, with several large s t a r c h grains about a central pyrenoid. (*After Bourquin*, 1917.) *C*, leucoplast (amyloplast) in aerial tuber of *Phajus grandifolius* with grain of reserve starch. (*After Strasburger.*)

Starch.—After a period of photosynthetic activity the chloroplast contains starch, the first visible product of that activity, in the form of minute granules. This "assimilation starch" is formed within the body of the chloroplast, as Meyer originally showed (Fig. 38, *A*, *B*). It is later transformed through the agency of enzymes into some soluble compound, usually a sugar; in this form it may be carried to growing regions, where, after further changes, it is built into the structure of the plant. Or, it may pass to storage organs where it is transformed into the ordinary "reserve starch," or "storage starch." This deposition of reserve starch is brought about through the agency of *amyloplasts*, which are leucoplasts capable of changing already elaborated organic materials, such as glucose, into starch (Fig. 38, *C*).

Reserve starch, upon which we depend so largely for food, is a carbohydrate with a composition expressed by the general formula $(C_6H_{10}O_5)n$, and exists in the form of granules ranging in size approximately from 2μ to 200μ in different plants. Potato-starch grains are usually about 90μ in diameter. The reserve starch grain is formed within the body of the amyloplast, and is made up of a series of concentric layers successively laid down about a center, or "hilum" (Fig. 39, *A*).[1] In case the grain starts to form near the middle of the amyloplast it may develop symmetrically, but commonly the developing grain lies near the periphery of the amyloplast, which becomes greatly distended as the grain grows. Material is thus deposited unevenly upon the grain so that the latter becomes very eccentric; in extreme cases the grain ruptures the amyloplast and remains in contact with it only at one side, where all new material is then deposited. Several grains may start to develop simul-

---

[1] For the structure of the starch grain see the papers of Nägeli, Schimper, Meyer, Binz, Dodel, Salter, and Kramer.

taneously in a single amyloplast, later growing together to form a "compound grain" with more than one hilum. In case the parts making up the compound grain are enveloped in one or more common outer layers the grain is said to be "half-compound." Potato starch is made up of simple, compound, and half-compound grains, whereas in oats and rice all or nearly all of the grains are said to be of the compound type. The successively deposited layers making up the grain differ mainly in water content, the innermost layers being richest and the outermost poorest in water. As a result of this non-uniform composition the grain often splits radially when dried.

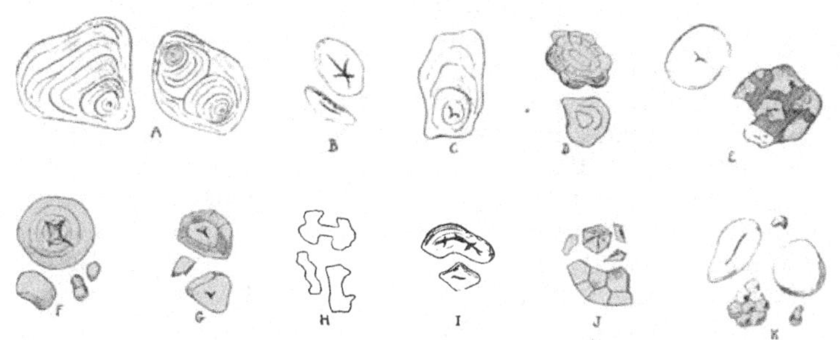

FIG. 39.—Reserve starch grains from various plants.

A, potato; simple and half-compound grains. *B*, colombo starch. *C*, arrowroot. *D*, pea. *E*, maize; intact and partially digested grains. *F*, rye. *G*, maize. *H*, *Euphorbia* *I*, bean. *J*, rice. *K*, wheat. (*After Tschirch.*)

As a result of his classic researches Nägeli (1858) advanced the theory that the starch grain is made up of ultramicroscopic crystalline particles which he called "micellæ," these being surrounded by water films of varying thickness. It was similarly held by A. Meyer (1883, 1895) that the grain is composed of radially arranged needle-shaped crystals known as "trichites;" these are composed of α- and β-amylose which turn blue with iodine. In some starch amylodextrin and dextrin are also present such grains turning red with iodine. Both Nägeli and Meyer held the stratification of the grain to be due to the varying numbers of the crystalline units in the successive layers, and Meyer showed that in certain cases it is correlated with the alternation of day and night, and therefore with a periodic activity on the part of the plastid. This conclusion was confirmed by Salter (1898).

The statement made by Schimper (1880) and Meyer (1883, 1895) that starch is always formed by plastids still holds good in its essential feature: so far as is certainly known no primary product of photosynthesis is formed in the cytoplasm apart from plastids, although in some cases, such as the young cells of *Hydrodictyon*, according to Harper, it is very difficult or even impossible to distinguish the limits of these organs. The granules of paramylum in *Euglena* and those of "Floridean starch" in the red algæ first appear in the cytoplasm; but, although they are the first substances which are visible, it is highly probable that they arise through the transformation of a non-visible product (sugar?) of the photosynthetic activity of the plastids, and are not immediately built up from water and carbon dioxide. A similar interpretation may be placed upon corresponding appearances reported in the case of higher plants. Owing to the great difficulty of determining the true cell structure of the Cyanophyceæ (see p. 202) it is possible to speak of plastid activity in such forms only with great reserve. If, as Olive (1904) and Gardner (1906) hold, these cells are without plastids, the product of photosynthetic activity, commonly glycogen, must be elaborated in the cytoplasm without their aid. If, on the other hand, the peripheral portion of the protoplast represents a large chromatophore (Fischer 1898), or cytoplasm containing a large number of minute chromatophores (Hegler 1901, Kohl 1903, Wager 1903), the photosynthetic process, although it may result in the production of a different substance, is dependent upon the powers of definite protoplasmic organs much the same as in higher plants. Among bacteria and other low forms in which it seems more certain that plastids and the ordinary pigments are absent, widely different types of metabolism are met with. For further discussion of this subject, which lies outside the scope of the present book, more special physiological works should be consulted.

The Pyrenoid.—The term *pyrenoid* was applied by Schmitz (1882) to the refractive kernel-like bodies imbedded in the chromatophores of the algæ. Pyrenoids are characteristic of the Chlorophyceæ especially, being present almost universally in the members of this group. They are known in a few representatives of the Rhodophyceæ (*Nemalion* and the Bangiaceæ), but apparently do not occur in the cells of the Cyanophyceæ, Phæophyceæ, and Characeæ. Very rarely they are present in forms above the algæ: a conspicuous example is the liverwort *Anthoceros*. The chromatophore may contain but one pyrenoid, as in *Zygnema*, or a larger number, as in *Spirogyra, Draparnaldia*, and many other forms (Fig. 37).

As held by de Bary (1858), Schmitz (1884), and Schimper (1885), the pyrenoid appears to be composed of a protein substance with a thick gelatinous consistency. When a single pyrenoid is present in the chromatophore it may multiply by fission along with the latter when the cell divides, while in those forms possessing several pyrenoids this multiplication may be much more extensive. Also, as pointed out by Schmitz and Schimper, and more recently by Smith (1914), the pyrenoid may disappear and arise *de novo* from the cytoplasm or from the plastid protoplasm.

With regard to its function, the early workers referred to above observed that under certain conditions the pyrenoid is closely surrounded by a mass of starch grains, and concluded that it is an organ, or portion of an organ (chromatophore), intimately concerned in the process of starch formation, its action being somewhat similar to that of the amyloplast. The pyrenoid, in fact, has often been likened to a leucoplast imbedded in the chromatophore; Wiesner, for instance, believed the pyrenoid to contain several leucoplast bodies, each of which gave rise to a starch grain. In general, more recent researches have emphasized the close association of the pyrenoid with the starch forming process, although the precise nature of this process remains very much in doubt. According to Timberlake (1901) the pyrenoid in *Hydrodictyon* is differentiated from the cytoplasm and is very active in starch production, segments splitting off from its periphery and forming starch within them. In this way the entire pyrenoid may become a mass of "pyrenoid starch," as distinguished from ordinary, or "stroma starch." McAllister (1913) describes a similar splitting up of the pyrenoid to form several starch grains in *Tetraspora*. Yamanouchi (1913), however, in his description of a new species of *Hydrodictyon*, states that some of the chloroplasts give rise to starch while others give rise to pyrenoids, and that the latter have nothing to do with starch formation.

A similar diversity of opinion exists with respect to the rôle of the pyrenoid in *Zygnema*. Chmielewskij (1896), who looked upon the pyrenoid as a permanent cell organ multiplying only by division, held that starch grains arise wholly from the substance of the pyrenoid, plate-like extensions of the latter being present between and in intimate contact with the developing grains. More recently Miss Bourquin (1917) asserts that the pyrenoid has nothing to do with the appearance of starch, the body of the chromatophore alone being concerned. She observes the starch grains appearing first near the periphery of the chromatophore entirely apart from the pyrenoid, the later formed grains differentiating in positions progressively nearer the pyrenoid (Fig. 38, *B*).

The pyrenoid of *Anthoceros* (Fig. 37, *C*) as described by McAllister (1914) is in reality a group of about 25–300 small "pyrenoid bodies" which are probably composed of a protein substance. The outermost bodies become starch, new ones apparently being formed by the fission of those lying on the interior of the group. McAllister states that no pyrenoid is visible in the young sporogenous tissue, starch being formed without its aid. Somewhat later several small bodies appear and aggregate to form the pyrenoid.

Cleland (1919) recently reports a close association of the pyrenoid of *Nemalion* with the formation of Floridean starch.

Elaioplasts and Oil Bodies.—In 1888 Wakker discovered in the cells of *Vanilla planifolia* and *V. aromatica* certain plastid-like bodies to which he gave the name *elaioplasts*, since they seemed to be concerned in the elaboration of oil (Fig. 40, *A*).   They were soon observed in a number of monocotyledons by Zimmermann (1893), Raciborski (1893), and Küster (1894); and some time later in the flower parts of a dicotyledon, *Gaillardia*, by Beer (1909).   Politis (1914) has found them in monocotyledonous plants belonging to 19 different genera, and in five genera of dicotyledons (Malvaceæ).

There is a considerable lack of agreement in the opinions expressed on the subject of the origin and significance of elaioplasts.   Wakker

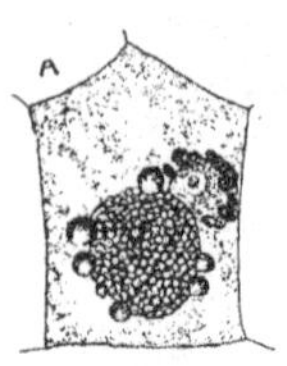

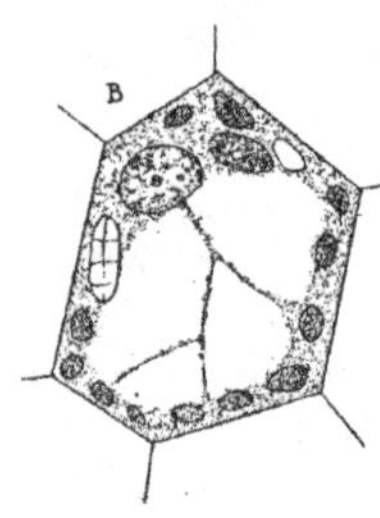

FIG. 40.

A, elaioplast forming oil droplets in epidermal cell of perianth of *Polianthes tuberosa;* nucleus with small plastids at right.   (*After Politis*, 1914.)   *B*, oil bodies in various stages of development in a cell of *Calypogeia*.   (*After Gargeanne*, 1903.)

thought it probable that they represent metamorphosed chloroplasts, which they often closely resemble in structure (Küster), whereas Raciborski asserted that they arise as small refractive granules in the cytoplasm and multiply by budding. In the zygospores of *Sporodinia grandis* and *Phycomyces nitens* Miss Keene (1914, 1919) reports the presence of a number of globular structures with which oil is associated from their earliest stages.   These unite to form one or two large reticulate bodies which Miss Keene believes are related to the elaioplasts of higher plants.   All of these investigators, with Politis, agree that elaioplasts are normal cell organs with a special function, namely, the formation of oily substances having a rôle in nutrition.   Beer, on the contrary, states that in *Gaillardia* they are formed secondarily by the aggregation of many small degenerating plastids and their products at one or more points in the cell, all stages of the process being observed.   Although the bodies so formed may, if green, produce starch, or, if colorless, an oily yellow pigment, Beer thinks it probable that they have no important special function in the life of the plant.

Closely associated with investigations on elaioplasts have been those concerned with the *oil bodies* found in the cells of many liverworts (Fig. 40, *B*).   These bodies, discovered by Gottsche in 1843, were first carefully described by Pfeffer (1874).   Pfeffer stated that they arise by the fusion of many minute droplets of fatty oil appearing in the cytoplasm of very young cells, and later come to lie in the cell sap; he further believed them to possess a special membrane.   Wakker (1888) held them to be analogous to leucoplasts and chloroplasts, multiplying by fission at each cell-division, and pointed out that they lie in the cytoplasm rather than in the cell sap.   He was inclined to view them as products of elaioplasts, which Küster (1894) supposed them to resemble in having a spongy stroma containing oil in the form of minute droplets.

Quite different were the views of Gargeanne (1903). According to him they arise from vacuoles, their limiting membranes thus being the original tonoplasts. While in the juvenile vacuole stage they may multiply by division, but when once fully formed they remain unchanged and divide no further. Gargeanne observed small oil droplets moving about freely within the oil body, and hence concluded that the latter has a fluid consistency rather than a spongy stroma as Küster thought.

The most noteworthy recent observations on oil bodies are those of Rivett (1918), who finds them to be very conspicuous in the cells of *Alicularia scolaris*. Rivett holds that they are in reality only oil vacuoles—that they originate by the coalescence of numerous minute oil droplets secreted by the protoplasm in a manner entirely similar to that in which the ordinary sap vacuole arises (*cf.* Pfeffer). Although they become very large and project well into the sap vacuole, they continue to be surrounded by a thin film of cytoplasm. The oil body, in the opinion of Rivett, is therefore in no sense a plastid, nor is it formed by any special elaioplast: it is simply an accumulation of ethereal and fatty oils together with some protein substance. The "membrane" observed by Pfeffer is the limiting layer of the surrounding cytoplasm, which may be slightly changed by contact with the oil.

Accumulations of oil apparently quite similar to those in liverwort cells have been described in the cells of various angiosperms by a number of writers. To these the term *elaiospheres* was applied by Lidforss (1893).

The published figures of elaioplasts and oil bodies in many cases bear striking resemblance to those of fat- and oil-secreting chondriosomes (see below), and it is not improbable that the problem of their origin and significance will be brought nearer solution by further studies of the latter class of bodies.

The Eyespot.—The so-called *eyespot* present in the flagellate cell and in the zoöspores and gametes of many algæ has certain characteristics in common with plastids, and may therefore receive consideration here. This body, which nearly all workers agree is a light-sensitive organ, is an elongated or circular and flattened structure lying in the anterior region of the cell (flagellates) or near its lateral margin, usually in close association with the chromatophore and the plasma membrane. (Overton 1889; Klebs 1883, 1892; Johnson 1893; Strasburger 1900; Wollenweber 1907, 1908). With respect to its mode of origin, it has been variously reported to arise *de novo* in each newly formed zoöspore in several green algæ (Overton); to develop from a colorless plastid in the young antheridial cell in the case of the spermatozoid of *Fucus* (Guignard 1889); to arise as a differentiated portion of the plastid in the zoöspores and gametes of *Zanardinia* (Yamanouchi); and finally to multiply by fission at the time of cell-division in flagellates (Klebs 1892).

It is generally agreed that the eyespot in many instances consists of a finely reticulate stroma in which an oily red pigment with many of the characteristics of hæmatochrom is held in the form of minute droplets or granules (Schilling 1891; Klebs 1883; Franzé 1893; Wager 1900; Wollenweber 1907, 1908) (Fig. 41, *D*). As shown by the careful researches of Franzé, the stroma may also bear one or more refractive inclusions, which in the Chlamydomonadaceæ and Volvocaceæ consist of starch, and in the Euglenoideæ of paramylum (Fig. 41, *E*). These inclusions were thought by Franzé to increase the sensitivity of the eyespot by concentrating the light at certain points.

The eyespot of the zoöspore of *Cladophora* (Strasburger 1900) appears to arise as a swelling of the plasma membrane, and consists of an external pigmented layer beneath which is a lens-shaped mass of hyaline substance (Fig. 41, *B*). In *Gonium* and *Eudorina* (Mast 1916) the lens-shaped portion lies outside with the cup-shaped opaque portion beneath it (Fig. 41, *A*), an arrangement strongly suggesting the primitive eyes of certain higher organisms. In neither portion could any finer structure be detected. Mast has shown that the orientation of the colony is brought about through changes in the intensity of the light falling upon the light-sensitive substance. As the unoriented swimming colony rotates on its axis, those zoöids turning away from the light have the hyaline portion of their eyespots shaded by the opaque cup; this sudden reduction in the amount of light energy received brings about an increase in the activity of the flagellæ of those zoöids, with the result that the colony as a whole turns more directly toward the source of light.

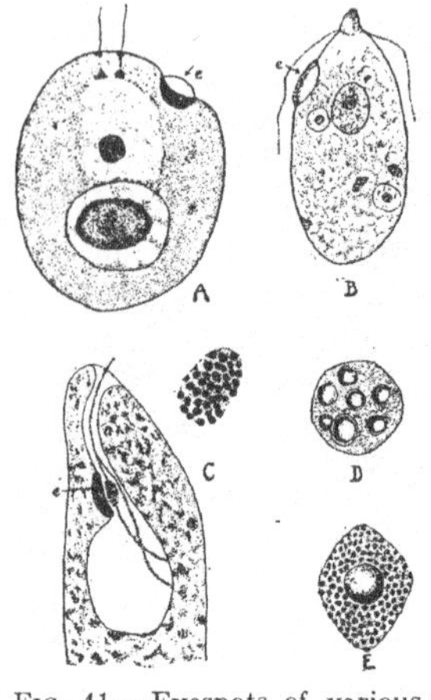

Fig. 41.—Eyespots of various types.

A, zoöid of *Eudorina; e,* eyespot. (*From Mast, After Grave.*) *B*, zoöspore of *Cladophora*, (*After Strasburger*, 1900.) *C*, anterior end of *Euglena viridis*, showing eyespot at surface of œsophagus, and in front of it a swelling on one root of the flagellum; face view of eyespot at right, showing pigment granules. (*After Wager*, 1900.) *D*, eyespot of *Euglena velata*. (*After Franzé*, 1893.) *E*, eyespot of *Trachelomonas volvocina*, with pigment granules and crystalloid body. (*After Franzé*.)

In *Euglena viridis* the morphological connection between the eyespot and the motor apparatus is particularly close. Here Wager (1900) has shown that the eyespot, which is a discoid protoplasmic body containing a layer of large pigment droplets, is situated at the surface bounding the œsophagus in close contact with a swelling on one of the basal branches of the flagellum (Fig. 41, *C*).

In general it may be concluded that the eyespot in some cases bears in its structure, and to a certain extent in its evident function, such a close resemblance to the ordinary plastid that a relationship of some sort between the two seems highly probable; whereas in other cases (*Gonium, Cladophora*) it appears to represent a differentiation of the ectoplast. It is more than likely that light-sensitive organs have arisen more than once in the evolution of the lower organisms, and that they cannot all be placed in the same category.

The Individuality of the Plastid.—It was believed by the early observers, notably Schimper (1883) and Meyer (1883), that plastids never originate *de novo* but always arise from preëxisting plastids by division. Fully differentiated plastids, such as chloroplasts, can readily be seen multiplying in this manner in growing tissues with a frequency sufficient to account for the large number of plastids present in mature plant parts. Since it is known, however, that chloroplasts and other differentiated chromoplasts may arise from leucoplasts through the development of pigments and other characters in the latter, and also that the individual plant arises from sex cells or a spore in which the plastids are usually in a colorless and relatively undifferentiated state, the problem of the individuality of the plastid is mainly one of determining whether these undifferentiated plastids, leucoplasts, or "plastid primordia" later developing into chloroplasts and other types are continuous through the critical stages of the life cycle, multiplying only by division, or arise *de novo* as new differentiations of the cytoplasm. At this point we may review certain cases in which the plastid has been followed through gametogenesis and fertilization.

In *Zygnema* (Kurssanow 1911) each vegetative cell contains one nucleus and two plastids, all of which divide at each vegetative cell-division. In sexual reproduction the entire protoplast, with its nucleus and two plastids, passes through the conjugating tube as a "male" gamete and unites with a similar complete protoplast ("female" gamete) of another filament. The two nuclei fuse, giving the primary nucleus of the new individual (zygospore nucleus), while the two plastids contributed by the "male" gamete degenerate, leaving the two furnished by the "female" gamete as the plastids of the new individual.

In *Coleochæte* (Allen 1905) each vegetative cell and gamete has one nucleus and one plastid: after the sexual union of the gamete nuclei the fertilized egg therefore contains one nucleus and two plastids. These two plastids divide at the first division of the fertilized egg but not at the second, the four resulting cells consequently having one plastid each. In the third cell-division the plastids also divide, so that each cell of the several-celled structure developing from the fertilized egg has its single plastid. Each of the several cells eventually becomes a zoöspore which germinates to produce a new *Coleochæte* body with a single plastid in each cell, the plastid dividing with the nucleus at each cell-division.

A somewhat similar regularity in the behavior of the plastid is shown in *Anthoceros* (Davis 1899; Scherrer 1914). Each gametophytic cell contains a single plastid which divides with the nucleus at each cell-division.   The egg likewise contains a plastid, but the spermatozoid has none: the fertilized egg and sporophyte cells which it later forms are therefore characterized, like the cells of the gametophyte, by the presence of one plastid.   Although it is difficult to demonstrate the plastid in the young sporogenous cells, every sporocyte shows one very clearly.   As shown by Davis (Fig. 42), the sporocyte plastid divides twice during the prophases of the first (heterotypic) division of the sporocyte nucleus, so that each spore of the resulting tetrad receives one.   Upon germination the spore produces a gametophyte with one plastid in each cell, and the cycle is complete.

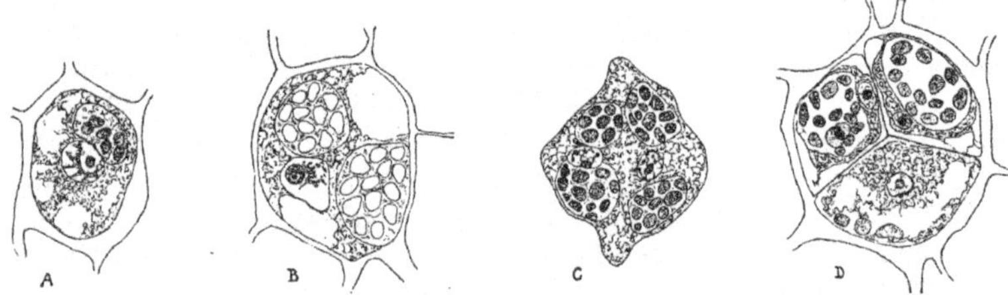

Fig. 42.—The behavior of the plastid in the sporocyte of *Anthoceros*.

A, sporocyte with single nucleus and plastid.   *B*, plastid divided; nucleus in prophase of mitosis.   *C*, plastids divided to four; two nuclei present.   *D*, three of the four spore cells, each of which has a single nucleus and plastid.   (*After Davis*, 1899.)

In all of the foregoing examples it is evident that the plastids, as stated by Scherrer for *Anthoceros*, remain as morphological individuals throughout the whole life cycle, multiplying exclusively by division.   A similar claim is made for the plastids of mosses by Sapĕhin (1915), who has also studied the behavior of the plastids in *Selaginella* and *Isoëtes* (1911, 1913).   In such cases the plastids each possess an individuality comparable to that of nuclei, from which they differ conspicuously, however, in undergoing no fusion at the time of fertilization.   The constancy in number is nevertheless maintained: by the degeneration of the plastids of one gamete in *Zygnema;* by their failure to divide at one cell-division in *Coleochœte;* and because of the fact that the male gamete carries no plastid in *Anthoceros*.   It appears to be generally true that while the eggs in all plant groups contain plastids (usually leucoplasts), the latter are present in male gametes in the algæ only.   Sapĕhin (1913), however, believes that the blepharoplasts of the higher groups represent plastids.

It should be said that only in a comparatively few forms has such a regularity in the behavior of the plastid as that outlined above been demonstrated.   A number of investigators, working on a great variety of cells, have been forced to conclude that plastids are either formed *de novo* as well as by division, or are carried through certain stages of the life cycle in some less conspicuous form. If they represent regional transformations of the cytoplasm resulting from the localization of certain processes, they might well be expected to differentiate anew as these processes begin in the life of the cell, and to preserve varying degrees of permanence depending upon the processes carried on (Harper). Their individual continuity through certain life cycles would accordingly be interpreted to mean that in such forms there is a persistence of certain types of physiological activity through all stages.

In recent years a number of cytologists have described the development of plastids from minute granular primordia in the cytoplasm, and have attempted to show that these primordia are members of the class of cell inclusions known as chondriosomes. A general theory of the individuality of the plastid must therefore involve the question of the relation of plastids to chondriosomes, and the further question of the origin of the chondriosomes themselves. These matters will be taken up in the following pages.

CHONDRIOSOMES

Notwithstanding the large amount of work which has been done upon chondriosomes during recent years, the condition of opinion as to their origin, behavior, and significance is still so unsettled that little more than a review and partial classification of the more prominent views will here be attempted.

Chondriosomes were probably first observed many years ago by Flemming and Altman in the course of their studies on protoplasm. They were first clearly described by La Vallette St. George (1886), who observed them in the male cells of animals and called them "cytomicrosomes." In plants they were first described by Meves (1904) in the tapetal cells of the anthers of *Nymphæa* (Fig. 43, *B*). Benda in 1897 and the following years discovered them in cells of many types, notably in the spermatogenous cells of animals, and applied to them the term "mitochondria." It was not until a decade later, through the researches of Meves, Regaud, Fauré-Fremiet, Lewitski, Guilliermond, and others that they came into prominence. Since that time they have been very intensively studied by both zoölogists and botanists, and a special literature of considerable bulk has developed.[1] It now seems evident that the filaments ("fila") of Flemming, the "bioplasts" of Altman, the "plastidules" of Maggi, the "archoplasmic granules" of Boveri, and the "mitochondria" of Benda are all one and the same thing—chondriosomes (Duesberg 1919).

General Nature and Occurrence.—Chondriosomes occur in the cytoplasm of the cell, commonly in the form of minute granules, rods, and

---

[1] Reviews of the subject are given by Duesberg (1911, 1919), Schmidt (1912), Cavers (1914). and Guilliermond (1919). See also Meves (1918).

threads, but also in a great variety of irregular shapes (Fig. 43). At present it is customary with the majority of workers to refer to all types as *chondriosomes* or *mitochondria*. For those which are definitely rod- and thread-shaped the terms *chondriokonts* and *chondriomites* are also used. It is not to be thought that the various forms constitute distinct classes, for several investigators (N. H. Cowdry; M. and W. Lewis 1915) have observed the chondriosomes undergoing marked changes in shape in living cells, granular ones becoming rod-shaped and filamentous, and *vice versa*. Schaxel (1911) and Kingery (1917) state, moreover, that in fixed material the shape of the chondriosomes is to a certain extent dependent upon the character of the microtechnical methods employed.

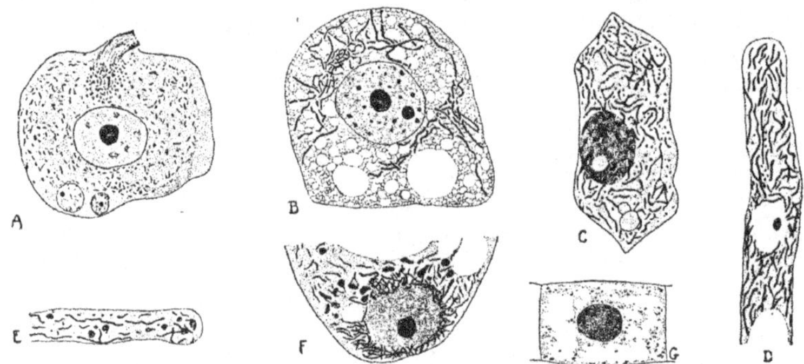

Fig. 43.—Chondriosomes in plant and animal cells.

A, nerve cell from guinea pig. × 480. (*After E. V. Cowdry,* 1914.) *B*, tapetal cell of *Nymphœa alba*. (*After Meves,* 1904.) *C*, living epidermal cell of tulip petal. *D*, ascus of *Pustularia vesiculosa*. *E*, hypha of *Rhizopus nigricans*. *F*, portion of embryo sac of *Lilium;* chondriosomes clustered about nucleus. *G*, cell of root tip of *Allium*—(*C-F. After Guilliermond,* 1918.)

Although when first discovered chondriosomes were believed to be rather limited in distribution, they have now been reported in the cells of plants and animals belonging to nearly all of the larger natural groups. It is asserted by N. H. Cowdry (1917) that "in all forms of animals, from amœba to man, which have been investigated with adequate methods of technique, they occur without exception." They are present, furthermore, in the cells of all tissues. In plants it is probable that they are no less universally present, although it has not yet been possible to demonstrate them with certainty in bacteria, Cyanophyceæ, and certain Chlorophyceæ, such as the Conjugatæ and Confervales (Guilliermond 1915). They are abundant in myxomycetes (N. H. Cowdry 1918), Charales (Mirande 1919), brown and red algæ, fungi, and all the higher groups.

A critical comparison of the chondriosomes of plants with those of animals has been made by N. H. Cowdry (1917), who concludes, contrary to the opinion of Pensa (1914), that there is every reason to regard them as homologous in the two kingdoms. He finds plant and animal chondriosomes to be practically identical in morphology, reaction to fixatives and dyes, and distribution in resting and dividing cells: any conspicuous differences in arrangement seem to be due to the more pronounced polarity of the animal cell. In both cases they are most abundant in the active stages in the life of the cell. As the cell ages and becomes fully differentiated, *i.e.*, as cytomorphosis proceeds, they diminish in number and may completely disappear.

Physico-chemical Nature.—With regard to the chemical and physical nature of chondriosomes, Regaud (1908), Fauré-Fremiet (1910), and Löwschin (1913), working respectively on mammals, protozoa, and plants, agree that they are chemically a combination of phospholipin and albumin. They closely resemble phosphatids, which are combinations of phosphoric and fatty acids, glycerol, and nitrogen bases. Lecithin is such a compound. Since chondriosomes are soluble in alcohol, ether, chloroform, and dilute acetic acid, many of the fixing reagents commonly employed in microtechnique destroy them: this accounts in part for the fact that they were not observed in many familiar tissues until a comparatively recent date. They are well fixed by neutral formalin, potassium bichromate, osmium tetroxid, and chromium trioxid (chromic acid); and these, therefore, are the principal ingredients of the fixing reagents employed in researches upon chondriosomes. Examples of such fluids are those of Altman, Benda, Bensley, Helley, Kopsch, Regaud, and Zenker.[1] Besides staining with hæmatoxylin and several other dyes commonly employed with fixed material, the chondriosomes show a characteristic affinity for certain intra-vitam stains, such as Janus green B, Janus blue, Janus black I, and diethylsafranin, the reaction with the first of these being especially strong. After certain treatments the chondriosomes may closely resemble the "chromidial substance," or granules of nucleoprotein distributed throughout the cytoplasm in some cells. That the two are not to be confused has been emphasized by Duesberg and by E. V. Cowdry. According to the latter author (1916) chondriosomes are "a concrete class of cell granulations," and may be provisionally defined as "substances which occur in the form of granules, rods and filaments in almost all living cells, which react positively to Janus green and which, by their solubilities and staining reactions, resemble phospholipins and to a lesser extent, albumins."

Origin and Multiplication.—The questions of the origin and multiplication of chondriosomes are much debated ones. Certain cytologists

---

[1] For convenient summaries of the effects of various reagents upon chondriosomes the student may refer to Kingsbury's (1912) paper on cytoplasmic fixation, E. V. Cowdry's (1914) on vital staining, and N. H. Cowdry's (1917) on plant and animal chondriosomes.

believe that they have found good evidence for the view that chondriosomes may multiply by division, and some (Guilliermond; Moreau 1914; Terni 1914) have held this to be their sole mode of origin—that they arise only from preëxisting chondriosomes and are therefore permanent cell organs.   Others are convinced that they may arise *de novo* in the cytoplasm, and that the evidence for their division is unsatisfactory (Orman 1913; Löwschin 1913; Scherrer 1914; Miss Beckwith 1914; Chambers 1915; M. and W. Lewis 1915; Twiss 1919; and others).   The investigators of the foregoing group, together with Meves (1900), Lewitski (1910), and Forenbacher (1911), hold that the chondriosomes arise from the cytoplasm, but certain others believe they take their origin from the nucleus. Tischler (1906) and Wassilief (1907), for example, state that they arise from surplus chromatin.   Alexieff (1917) thinks that although cyto-

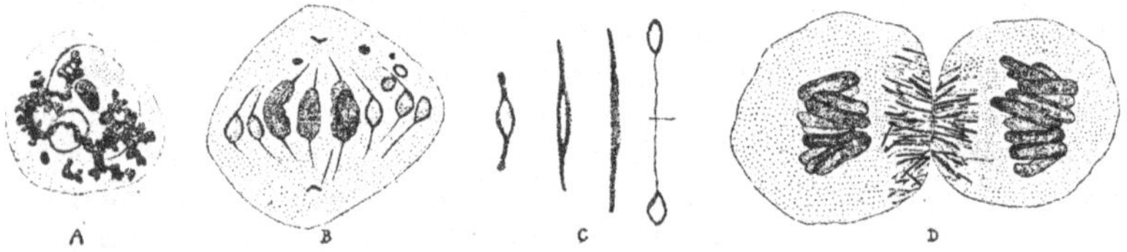

FIG. 44.—Examples of regular behavior of chondriosomes in cell-division.

A–C, spermatocyte of *Gryllotalpa vulgaris*, (*After Voïnov*, 1916): *A*, chondriosomal material in cytoplasm about nucleus; *B*, heterotypic mitosis, showing chondriosomes (at sides) occupying the spindle with the chromosomes (at center); *C*, stages in the division of a chondriosome.   *D*, Dividing cell of *Geotriton fuscus*, showing division of individual chondriosomes as cell constricts at equator.   (*After Terni*, 1914.)

iplasmic dfferentiation is due to them, they are at least in some cases of nuclear origin; and further that they are not fundamentally different from chromosomes and chromidia, a conclusion contradictory to that of Duesberg and E. V. Cowdry, cited above.   Shaffer (1920) believes them to arise as the result of a chemical action of the nucleus upon products of assimilation in the adjacent cytoplasm.   Wildman (1913) classifies the cytoplasmic inclusions present throughout spermatogenesis in *Ascaris* into two main types, both of nuclear origin: "karyochondria," equivalent to the mitochondria of other writers, and "plastochondria," which pass into the cytoplasm, form yolk within them, and fuse to form the food supply ("refractive body") of the spermatozoön.

That the behavior of the chondriosomes at the time of cell-division is a matter of considerable importance has been generally recognized.   In many cases their distribution to the two daughter cells seems to be quite fortuitous, whereas in some tissues more or less definite modes of distribution have been described.   According to Fauré-Fremiet (1910), Terni (1914), Korotneff (1909), and others, the individual chondriosomes divide at the time of mitosis (Fig. 44, *D*), a conclusion with which many others fail to agree (Orman 1913; Miss Beckwith 1914; etc.).   In the cells of the grasshopper, *Dissosteira carolina*, Chambers (1915) finds that the chondriosomal material forms a granular network surrounding the nucleus during the resting stages and the mitotic figure during division. During the later phases of mitosis the strands and granules of this network lengthen into delicate filaments between the two daughter chromosome groups, and finally separate into two granular masses which gradually invest the daughter nuclei.

In the mole cricket, *Gryllotalpa borealis*, the distribution of the chondriosomes to the daughter cells is accomplished with even greater definiteness. According to Payne (1916) they become thread-like and break near the middle, the halves passing to the daughter cells. Voïnov (1916) states that the "mitochondria" in the spermatocyte of *G. vulgaris* fuse to form a thread which then segments into a number (70 or more) "chondriosomes." These are arranged on the spindle along with the chromosomes, which they may closely resemble, and divide to form daughter bodies at both maturation divisions, so that they are equally distributed to the four resulting spermatozoa (Fig. 44, *A–C*).

In certain scorpions also the chondriosomal material is distributed with surprising precision. In a species from Arizona (Wilson 1916) this material in the spermatocyte takes the form of a single ring-shaped body. This ring divides accurately, much like a chromosome, at both maturation divisions, each of the four spermatids, and hence each spermatozoön of the tetrad, receives a quarter of its substance. In a California species (Wilson) there is no ring formed, but instead about 24 hollow spherical bodies. At the two maturation divisions these show no evidence of division, but are passively separated into four approximately equal groups, each spermatid receiving six (occasionally five or seven). A European species described by Sokolow (1913) agrees essentially with this.

Function.—Our knowledge of chondriosomes is yet too incomplete to warrant any categorical statements regarding their functions, but a number of opinions have been expressed, some of them based upon observational evidence and others upon conjecture. Certain of the more prominent opinions may here be reviewed.

It was in 1897 that Benda suggested that chondriosomes might be distinct cell organs with a special function. In a series of papers which began to appear ten years later Meves (1907 etc.) put forth and emphasized the theory that they play an important rôle in heredity—that they carry the hereditary characters of the cytoplasm. Evidence supporting this view was seen by Meves and Benda in certain experiments of Godlewski which seemed to show that the appearance of certain hereditary characters is dependent upon something present in the cytoplasm rather than in the nucleus. (See Chapter XIV.) This theory has had the support of a number of investigators, among whom are the botanists Cavers (1914) and Mottier (1916). Voïnov (1916) also believes that the regular distribution of the chondriosomal substance in *Gryllotalpa* strongly favors the view that this substance is of some significance in heredity. It is probable, however, that the majority of cytologists regard the evidence brought forward in support of the view as very inadequate. Wildman (1913) points out that the chondriosomes may be largely lost during spermatogenesis, and others have recalled cases in which the nucleus is the only portion of the male gamete which can be seen to enter the egg at fertilization. Meves (1911, 1915) and Benda, on the other hand, show that chondriosomes also enter, at least in the forms studied by them (Fig. 45). In the animal spermatid the chondriosomes appear most commonly to contribute to the formation of the Nebenkern of the

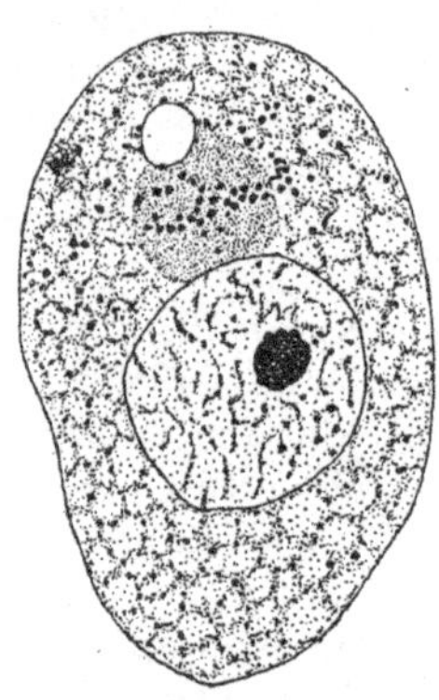

Fig. 45.—Fertilization in *Filaria papillosa*, showing chondriosomes of spermatozoön (at top) distributing themselves in the cytoplasm of the egg. (*After Meves, 1915*.)

spermatozoön (La Vallette St. George 1886; Popoff 1907; Chambers 1915; Shaffer 1920; and others), in some cases later elongating into a sheath around the axial filament of the tail (Shaffer on *Cicada*). Duesberg (1919) states that although the fate of the chondriosomes of the spermatid varies in different animals, they are nevertheless always present in the spermatozoön, and that it has not been clearly shown in any case that they do not enter the egg at fertilization. In many eggs which they do enter, however, they behave with great irregularity during the subsequent cleavage stages (Van der Stricht, etc.). It is not at all improbable that they are in some way concerned in the reactions through which hereditary characters are developed in the individual, but the general opinion is that their apparent variability and indefiniteness in behavior in so many cases are against the view that they take any part in the transmission of factors upon whose presence the development of the characters depends (Gatenby 1918, 1919). The equal distribution of chondriosomes at the time of cell-division is thought to be without any significance in this connection by Harper (1919).

It is obvious that much work remains to be done before the possible relation of chondriosomes to heredity and development can be made clear. For the present it is safest to assume, as will be emphasized in later chapters, that hereditary transmission is the function of the nucleus, chiefly if not entirely, since the chromosomes afford a mechanism of precisely the kind required to account for the observed distribution of hereditary characters.

Meves (1907*ab*, 1909) and Duesberg (1909) have also called attention to the close relation of chondriosomes to muscle fibers in the developing chick embryo. They believe that the chondriosome elongates and directly becomes the young fiber. Gaudissart (1913), on the contrary, shows that the fiber does not arise exclusively from the chondriosome, but that the primary basis is furnished by the plasmatic reticulum with which the chondriosomes coöperate in building up the fiber. Although the chondriosomes thus have a part in the genesis of the muscle fiber, the latter is not a "modified filamentous chondriosome," as Duesberg believed.

Hoven (1910a) and Meves have similarly attempted to show that chondriosomes are concerned in the differentiation of neurofibrils and the collagenous fibers of cartilage. Regaud (1911), Guilliermond (1914),

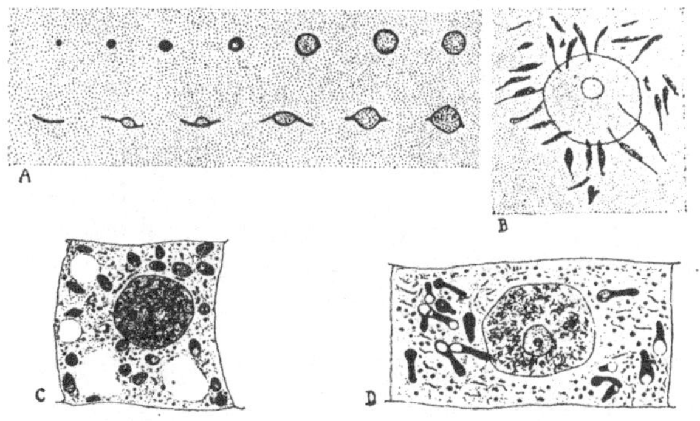

Fig. 46.

A, formation of fat in cell of rabbit by granular and rod-shaped chondriosomes. (*From Guilliermond, after Dubreuil,* 1913.) B, formation of needle-shaped crystals of carotin in chromoplasts derived from chondriosomes in epidermal cell of *Iris* petal. (*After Guilliermond,* 1918.) C, chondriosomes and chloroplasts in young cell of *Pinus banksiana*. ×750. (*After Mottier,* 1918.) D, transformation of plastid primordia into leucoplasts in root cell of *Pisum;* some of the leucoplasts contain starch. (*After Mottier.*)

Hoven (1910b, 1911), and Lewitski (1914) have thought that the chondriosomes may in some cases perform a secretory function, and Dubreuil (1913) has associated them with the production of fat (Fig. 46, A). In the oöcyte of *Cicada* Shaffer (1920) finds them transforming into yolk spherules. The activity of bodies called "plastochondria" by Wildman (1913) in the elaboration of the food supply in the spermatozoön of *Ascaris* has already been mentioned.

Relation of Chondriosomes to Plastids.—One of the most conspicuous views regarding the significance of chondriosomes is that which holds some of them to be the primordia of plastids. After studying the cells of *Pisum* and *Asparagus* Lewitski (1910) concluded that the chondriosomes are essential constituents of the cytoplasm, and that they develop into chloroplasts and leucoplasts in the cells of the stem and root respectively.

Evidence in support of this conception was contributed by Forenbacher (1911), Pensa (1914), Cavers (1914), and others. Guilliermond (1911–1920) in particular was led by the results of his extensive researches on the subject to the view that the chondriosomes, arising only from preëxisting ones by division, persist through the egg and embryonic cells and later become amyloplasts, chloroplasts, and chromoplasts. In this he saw strong evidence for the individuality of the plastid. In 1915 he advanced the opinion that in fungi the chondriosomes function like the amyloplasts of higher plants, forming reserve products as the latter form starch. In this development of chondriosomes into plastids Guilliermond (1913–1915) and Moreau (1914) were able to show that the chondriosomes produce within them certain phenolic compounds which either appear at once as anthocyanin pigments, or as colorless products which may acquire color later through chemical alteration (Fig. 46, *B*).

Among the most recent researches in this field are those of Mottier (1916, 1918) on the cells of *Zea, Pisum, Elodea, Pinus, Adiantum, Anthoceros, Pallavicinia,· Marchantia,* and several algæ. He finds that leucoplasts and chloroplasts are derived from small rod-shaped primordia (Fig. 46, *C, D*) which he regards as permanent cell organs of the same rank as the nucleus. Both primordia and mature chloroplasts multiply by fission. In the cells of *Marchantia, Anthoceros,* and the seed plants he finds also a second series of bodies, which he calls chondriosomes: these like the plastid primordia, are permanent cell organs multiplying by division, but they do not become chloroplasts or leucoplasts. Furthermore, both chondriosomes and primordia are thought by Mottier to be concerned in the transmission of certain hereditary characters.

It is also reported by Emberger (1920*ab*) that in the roots and sporangia of ferns two kinds of granular elements may be recognized at all times, one of them representing the initial stage of plastid development. Contrary to Mottier's opinion, however, he regards both kinds as true mitochondria. Guilliermond (1920) likewise distinguishes two such types in *Iris germanica.*

P. A. and P. Dangeard (1919, 1920), as a result of their researches on the cells of barley, *Selaginella, Larix, Taxus,* and *Ginkgo,* distinguish three classes of cytoplasmic structures differing in reaction to reagents and in function. In their initial stages all have the granular form. The *plastidomes* first appear as minute "mitoplasts," which gradually enlarge and develop into plastids. The *spheromes* are at first recognizable as "microsomes," some of which may be seen to give rise to fat and oil globules while others appear to undergo no change. The *vacuomes* begin their history as "metachromes;" these elongate and form a peculiar network which later develops into a system of vacuoles. Guilliermond (1920) denies the metachromatic nature of this third class of bodies, and holds them to be quite distinct from mitochondria.

The existence of such a genetic relationship between chondriosomes and plastids as that described above has been denied by many writers, among whom may be mentioned Lundegårdh (1910), Meyer (1911), Rudolph (1912), Löwschin (1913, 1914), Scherrer (1914), Miss Beckwith (1914), Derschau (1914), von Winiwarter (1914), Sapĕhin (1915), Chambers (1915), M. and W. Lewis (1915), and Harper (1919). These workers for the most part hold that chondriosomes are not distinct cell organs at all, but regard them rather as more or less transient visible products of protoplasmic activity. Derschau asserts that they arise neither *de novo* nor by fission, but that they are merely small masses of plastin and nuclein concerned in nutrition, arising from basichromatin at the surface of the nucleus. Miss Beckwith speaks of them as differentiation products of the cytoplasm. Löwschin, who made some experiments in the production of artificial chondriosomes, believes them to be due to the emulsified state of the protoplasm and in some instances to the action of fixing agents upon it. To Chambers they appear in living cells not as persistent structures but as temporary physical states of the colloidal substances composing protoplasm. M. and W. Lewis have studied them in tissue cultures and observe that they are continually being formed and used up, and that they show no sharply distinct types. Fauré-Fremiet (1910a) distinguishes "mitochondria," which have an individuality of their own and are permanent cell organs, from "liposomes," which are temporary accumulations of reserve substance.

The almost universal occurrence of chondriosomes in the cells of living organisms, and their frequent alterations in number and appearance, suggest a connection with some fundamental process going on almost constantly and common to all living matter. That this process may be oxidation, the chondriosomes being a "structural expression of the reducing substances concerned in cellular respiration" (Kingsbury), has been regarded as highly probable by Kingsbury (1912), Mayer, Rathery, and Schaeffer (1914), N. H. Cowdry (1917, 1918), and others. Evidence favoring this interpretation is seen in the fact that the chondriosomes occur so widely in the cytoplasm, which acts as a reducing substance; and also in the close similarity between their chemical composition and that of phosphatids, which appear to be capable of auto-oxidation.

Conclusion.—From the foregoing review it should be more than plain that the state of our knowledge of chondriosomes is such that almost no definite final statements can be made regarding their origin and function. The evidence at hand apparently indicates that the class of cell inclusions known as chondriosomes comprises a variety of bodies which play different rôles in the life of the cell. It is scarcely open to doubt that some of them are temporary accumulations of substances involved in metabolism, appearing and disappearing in the cell in a manner somewhat analogous to that of starch. The most plausible hypothesis concerning the specific physiological rôle of such changeable types of chondriosomes is that they have to do with the processes of oxidation and reduction—with cellular respiration. It is also becoming increasingly apparent that other chondriosomes represent the juvenile stages in the development of plastids of various kinds, and that they are in some way concerned in the formation of chlorophyll and other pigments. If this is true they are clearly of the highest importance.

Whether or not any of the chondriosomes are to be considered as permanent cell organs is a question to which, in view of the conflicting testimony of competent observers, no final answer can at present be given. To determine whether these minute bodies arise *de novo* or always multiply by division is a matter of extreme practical difficulty. Until this question is settled it is obviously impossible to come to a decision regarding the individuality of those plastids which appear to take their origin from chondriosomes, or to know what may be the possible relation of chondriosomes to inheritance. With respect to the latter point, the chondriosomes, like all other structures concerned in metabolism, may be indirectly associated with the development of hereditary characters, but the view that they transmit or represent differential factors for such characters is as yet unsupported by adequate evidence.

From the fact that the chondriosomes may not preserve their individuality at all times, however, it does not follow that they must be denied the rank of cell organs. Their great variability, indifferent behavior at the time of cell-division in so many cases, and their unknown mode of origin are, as Kingsbury (1912) states, against the view that they are cell organs; and it is doubtless true that many chondriosomes should for such reasons be denied such rank. On the other hand, those chondriosomes which seem clearly to perform important and specific functions in the life of the cell should, like centrosomes appearing *de novo* at each cell-division, be looked upon as cell organs, though not as permanent ones with an uninterrupted continuity.

In spite of the fact that the study of chondriosomes has so far raised more problems than it has solved, it has already proved of much value, for it has turned to the cytoplasm some of the attention so long directed almost exclusively to the nucleus, and it appears that many problems of much importance to cytology pertain to the cytoplasm. It has also been of great service in bringing about a closer scrutiny of the effects of fixation and a renewed emphasis upon the importance of the study of living protoplasm. Much has already been learned as the result of this study, but the solution of the principal problems involving chondriosomes must await the results of further research.

Bibliography 6

Plastids and Chondriosomes

ALEXIEFF, A. 1917*a*. Mitochondries et corps parabasal chez les Flagelles. Comptes Rend. Soc. Biol. Paris **69**: 358–361.

1917*b*. Mitochondries et rôle morphogene du noyau. Ibid. 361–363.

1917*c*. Nature mitochondiale du corps parabasal des Flagelles. Ibid. 499–502.

1917*d*. Sur les mitochondries a fonction glycoplastique. Ibid. 510–512.

ALLEN, C. E. 1905. Die Keimung der Zygoten bei *Coleochæte*. Ber. Deu. Bot. Ges. **23**: 285–292. pl. 13.

DE BARY, H. A. 1858. Untersuchungen über die Familien der Conjugaten. Leipzig.

BEAUVERIE. 1919. The anthocyanin pigments of plants. Sci. Am. Suppl. **87**: No. 2244, pp. 2–3, 7. (Transl. from Rev. Gen. Sci. Paris.)

BECKWITH, C. J. 1914. The genesis of the plasma structure in the egg of *Hydractinia echinata*. Jour. Morph. **25**: 189–251. pls. 8.

BEER, R. 1909. On elaioplasts. Ann. Bot. **23**: 63–72. pl. 4.

BENDA, C. 1897. Neuere Mitteilungen über die Histogenese des Saugetierspermatozoön. Verh. d. Phys. Gesell. Berlin.

1898. Ueber die Spermatogenese der Vertebraten und höherer Evertebraten. 2. Die Histogenese der Spermien. Ibid, 1898.

1902. Die Mitochondria. Ergeb. d. Anat. u. Entw. **12**: 743–781. (Review.)

BINZ, A. 1892. Beiträge zur Morphologie und Entwicklung der Stärkekörner. Flora **76**: 34–91. pls. 5–7.

BOURQUIN, H. 1917. Starch formation in *Zygnema*. Bot. Gaz. **64**: 426–434. pl. 27.

CARTER, N. 1919*ab*. Studies on the chloroplasts of desmids. I, II. Ann. Bot. **33**: 215–254, pls. 14–18; 295–304, pls. 12, 20.

1919*c*. The cytology of the Cladophoraceæ. Ibid. **33**: 467–478. pl. 27.

1920*a*. Studies on the chloroplasts of desmids. III. The chloroplasts of Cosmarium. Ann. Bot. **34**: 265–285. pls. 10–13.

1920*b*. Studies on the chloroplasts of desmids. IV. Ibid. **34**: 303–320. pls. 14–16.

CAVERS, F. 1914. Chondriosomes (mitochondria) and their significance. New Phytol. **13**: 96–106. (Review of subject.)

CHAMBERS, R. 1915. Microdissection studies on the germ cell. Science **41**: 290–293.

CHMIELEWSKIJ. 1896. Ueber Bau und Vermehrung der Pyrenoide bei einigen. Algen. (Russian, 10 pp.) See Bot. Centr. **69**: 277–278. 1897.

CLELAND, R. E. 1919. The cytology and life history of *Nemalion multifidum* Ag. Ann. Bot. **33**: 323–352. pls. 22–24. figs. 3.

COWDRY, E. V. 1914*a*. The vital staining of mitochondria with Janus green and diethylsafranin in human blood cells. Internat. Monatsschr. f. Anat. u. Physiol. **31**: 267–286.

1914*b*. The comparative distribution of mitochondria in spinal ganglion cells of vertebrates. Am. Jour. Anat. **17**: 1–29. pls. 3.

1916. The general functional significance of mitochondria. Am. Jour. Anat. **19** 423–446. (Review of subject.)

COWDRY, N. H. 1917. A comparison of mitochondria in plant and animal cells. Biol. Bull. **33**: 196–228. figs. 26.

1918. The cytology of the myxomycetes with special reference to mitochondria Biol. Bull. **35**: 71–94. 1 pl.

1920. Experimental studies on mitochondria in plant cells. Ibid. **39**: 188–206. pls. 3.

DANGEARD, P. A. 1919. Sur la distinction du chondriome des auteurs en vacuome, plastidome, et sphérome. Compt. Rend. Acad. Sci. Paris **169**: 1005–1010.

1920*a*. Plastidome, vacuome et sphérome dans *Selaginella Kraussiana*. Ibid. **170**: 301–306. 1 pl.

1920*b*. La structure de la cellule végétale et son metabolisme. Ibid. **170**: 709–714.

DANGEARD, P. 1920. Sur l'évolution du système vacuolaire chez les gymnospermes. Ibid. **170**: 474–477. figs. 8.

DAVIS, B. M. 1899. The spore mother cell of *Anthoceros*. Bot. Gaz. **28**: 89–109. pls. 9, 10.

VON DERSCHAU, M. 1914, Zum Chromatindualismus der Pflanzenzelle. Arch. f. Zellf. **12**: 220–240. pl. 77.

DODEL, A. 1892. Beitrag zur Morphologie und Entwicklung der Stärkekörner von *Pellionia Daveauana*. Flora **75**: 267–280. pl. 5–6.

DUBREUIL, G. 1913. Le chondriome et le dispositif de l'activité secretaire. Arch. d'Anat. Micr. **15**: 53–151.

DUESBERG, J. 1907. Der Mitochondrialapparat in den Zellen der Wirbeltiere und Wirbellosen. Arch. Mikr. Anat. **71**: 284–296. pl. 24.

1909. Les chondriosomes des cellules embryonnaires du poulet et leur rôle dans la genese des myofibrilles, avec quelques observations sur le developpement des fibres musculaires striées. Arch. Zellf. **4**: 602–671. pls. 28–30. figs. 10.

1911. Plastosomes, "apparato reticolaro interno," und Chromidialapparat. Ergeb. d. Anat. u. Entw. **20**: 567–916. (Extensive review.)

1917. Chondriosomes in the cells of fish embryos. Am. Jour. Anat. **21**: 465–493. pls. 1–3. figs. 8.

1918. Chondriosomes in the testicle-cells of *Fundulus*. Am. Jour. Anat. **23**: 133–154. pls. 2. figs. 21.

1919. On the present status of the chondriosome-problem. Biol. Bull. **36**: 71–81.

DUESBERG, J. et HOVEN, H. 1910. Observations sur la structure du protoplasme des cellules végétales. Anat. Anz. B **36**: 96–100. figs. 5.

EMBERGER, L. 1920*a*. Évolution du chondriome chez les cryptogames vasculaires. Compt. Rend. Acad. Sci. Paris **170**: 282–284. figs. 5.

1920*b*. Évolution du chondriome dans la formation du sporange chez les fougéres. Ibid. **170** : 469–471. figs. 7.

FAURÉ-FREMIET, M. E. 1910*a*. Mitochondries et liposomes. Comptes Rend. Soc. Biol. Paris **62**: 537–539.

1910*b*. La continuité des mitochondries a travers des generations cellulaires et le rôle de ces elements. Anat. Anz. **36**: 186–191. figs. 3.

FISCHER, A. 1905. Die Zelle der Cyanophyceen. Bot. Zeit. **63**: 51–129. pls. 4, 5.

FORENBACHER, A. 1911. Die Chondriosomen als Chromatophorenbildner. Ber. Deu. Bot. Ges. **29**: 648–660. pl. 25.

FRANZÉ, R. 1893. Zur Morphologie und Physiologie der Stigmata der Mastigophoren. Zeit. Wiss. Zool. **56**: 138–164. pl. 8.

GARDNER, N. L. 1906. Cytological studies in Cyanophyceæ. Univ. Calif. Publ. Bot. **2**: 237–296. pls. 21–26.

GARGEANNE, A. J. M. 1903. Die Oelkörper der Jungermanniales. Flora **92**: 457–482. figs. 18.

GATENBY, J. B. 1918. Cytoplasmic inclusions of the germ cells. III. The spermatogenesis of some other Pulmonates. Quar. Jour. Micr. Sci. **63** . 197–258. pls. 16–18. figs. 3.

1919. The cytoplasmic inclusions of the germ cells. V. The gametogenesis and early development of *Limnæa stagnalis* (L.) with special reference to the Golgi apparatus and the mitochondria. Quar. Jour. Micr. Sci. **63**: 445–492. pls. 27, 28. figs. 6. (See also Parts I and II in vol. **62**, and Part IV in **63**.)

GAUDISSART, P. 1913. Reseau protoplasmique et chondriosomes dans la genese des myofibrilles. La Cellule **30**: 29–43. pls. 1, 2.

GOTTSCHE. 1843. Anatomische-physiologische Untersuchungen über *Haplomitrium Hookeri.* Verb. Leop. Carol. Akad. **12**: I, 286.

GUIGNARD, L. 1889. Developpement et constitution des antherozoides. Rev. Gén. Bot. **1**: 11, 63, 136, 175. pls. 2–6.

GUILLIERMOND, A. 1911. Sur les mitochondries des cellules végétales. Comptes Rend. Acad. Sci. Paris **153**: 199–201. figs. 4.

1912. Recherches cytologiques sur le mode de formation de l'amidon et sur les plastes végétaux. Arch. d'Anat. Micr. **14**: 309–428. pls. 13–18.

1913. (*a*) Sur la signification du chromatophore des algues. Comp. Rend. Soc. Biol. Paris **75**: 85–87. (*b*) Quelques remarques nouvelles sur la formation des pigments anthocyaniques au sein des mitochondries. Ibid. 478–481. (*c*) Nouvelles observations sur le choudriome de l'asque de *Pustularia vesiculosa.* Ibid. 646–649. (*d*) Nouvelles remarques sur la signification des plastes de W. Schimper par rapport aux mitochondries actuelles. Ibid. 437–440.

1914*a*. Etat actuel de la question de l'evolution et du rôle physiologique des mitochondries. Rev. Gén. Bot. **26**: 129–149, 182–210. figs. 16.

1914*b*. Bemerkungen über die Mitochondrien der vegetativen Zellen und ihre Verwandlung in Plastiden. Ber. Deu. Bot. Ges. **32**: 282–301. figs. 2.

1915*a*. Nouvelles observations vitales sur le chondriome des cellules epidermiques de la fleur d'*Iris germanica.* Comp. Rend. Soc. Biol. Paris **67**: 241–249.

1915*b*. Recherches sur le chondriome chez les champignons et les algues. Rev. Gén. Bot. **27**: 193, 236, 271, 297, 315. pls. 12.

1917*a*. Sur la nature et le rôle des mitochondries des cellules végétales. Comp. Rend. Soc. Biol. Paris **69**: 916–924.

1917*b*. Observations vitales sur le chondriome de la fleur de Tulipe. Comp. Rend. Acad. Sci. Paris **164**: 407–409.

1917*c*. Contributions a l'étude de la fixation du cytoplasme. Ibid. 643–646.

1917*d*. Recherches sur l'origine des chromoplastes et le mode de formation de pigments du groupe des xanthophylles et des carotins. Ibid. 232–234.

1917*e*. Sur les alterations et les caractères du chondriome dans les cellules epidermique de la fleur de Tulipe. Ibid. 609–612.

1917*f*. Sur les phénomènes cytologiques de la degénérescence de cellules epidermiques pendant la fanaison des fleurs. Compt. Rend. Soc. Biol. Paris **69**: 726–729.

1918. Sur l'origine mitochondriale des plastids. Compt. Rend. Acad. Sci. Paris **167**: 430–433.

1919. Observations vitales sur le chondriome des végétaux et recherches sur l'origine des chromoplastides et le mode de formation des pigments xanthophylliens et carotiniens. Rev. Gén. Bot. **31**: 372–413, 446–508, 532–603, 635–770. pls. 60. figs. 35.

1920*a*. Sur l'évolution du chondriome dans la cellule végétale. Compt. Rend. Acad. Sci. Paris **170**: 194–197. figs. 4.

1920*b*. Sur les éléments figurés du cytoplasme. Ibid. **170**: 612–615. figs. 5.

1920*c*. Nouvelles recherches sur l'appareil vacuolaire dans les végétaux. Ibid. **171**: 1071–1074 figs. 25.

HAAS, P. and HILL, T. G. 1913. An introduction to the chemistry of plant products. London.

HABERLANDT, G. 1888. Die Chlorophyllkörper der Selaginellen. Flora **71**: 291–308. pl. 5.

1905. Ueber die Plasmahaut der Chloroplasten in den Assimilationzellen von *Selaginella Martensii* Spring. Ber. Deu. Bot. Ges. **23**: 441–452. pl. 20.

HARPER, R. A.   1919.   The structure of protoplasm.   Am. Jour. Bot. **6** : 273–300.

HEGLER, R.   1901.   Untersuchungen über die Organization der Phycochromzelle. Jahrb. Wiss. Bot. **36** : 229–354·   pls. 5, 6.   figs. 5.

HOVEN, H.   1910*a*.   Sur l'histogenese du systeme nerveux peripherique et sur le rôle des chondriosomes dans la neurofibrillation.   Arch. de Biol. **25** : 427–494. pls. 15, 16.

   1910*b*.   Contribution a l'étude du fonctionnement des cellules glandulaires.   Du rôle du chondriome dans la secretion.   Anat. Anz. **37** : 343–351·   figs. 7.   (Note prelim.)

   1911.   Du rôle du chondriome dans l'elaboration des produits de la glande mammaire.   Anat. Anz. **39** : 321–326·   figs. 4.

JANSSENS, F. A., VAN DE PUTTE, E., et HELSMORTEL, J.   1913.   Le chondriosome dans les champignons.   (Notes prelim.)   La Cellule **28** : 447–452·   pls. 1, 2.

JOHNSON, L. M.   1893.   Observations on the zoöspores of *Draparnaldia*.   Bot. Gaz. **18** : 294–298·   pl. 32.

JÖRGENSEN, I. and STILES, W. 1917.   Carbon assimilation.   New Phytol.   Reprint No. 10.   London.

KEENE, M. L.   1914.   Cytological studies of the zygospores of *Sporodinia grandis*. Ann. Bot. **28** : 455–470·   pls. 35, 36.

   1919.   Studies of zygospore formation in *Phycomyces nitens* Kunze.   Trans. Wis. Acad. Sci. **19** : 1195–1220·   pls. 16–18·

KINGERY, H. M.   1917.   Oögenesis in the white mouse.   Jour. Morph. **30** : 261–316· pls. 2.

KINGSBURY, B. F.   1912.   Cytoplasmic fixation.   Anat. Record **6** : 39–52·

KLEBS, G.   1883.   Ueber die Organization einiger Flagellaten-Gruppen und ihre Beziehungen zu Algen und Infusorien.   Unters. Bot. Inst. Tübingen **1** : 233–262. pls. 2, 3.

   1892.   Flagellaten-Studien.   1, II.   Zeit. Wiss. Zool. **55** : 265–445·   pls. 13–18·

KOHL, F. G.   1903.   Ueber die Organization und Physiologie der Cyanophyceenzelle, und die mitotische Teilung ihres Kerns.   Jena.

KOROTNEFF, A.   1909.   Mitochondrien, Chondriomiten, und Faserepithel der Tricladen.   Arch. Mikr. Anat. **74** : 1000–1016·   pls. 2.

KRAMER, H.   1902.   The structure of the starch grain.   Bot. Gaz. **34** : 341–354. pl. 11.

KURSSANOW, L.   1911.   Ueber Befruchtung, Reifung und Keimung bei *Zygnema*. Flora **104** : 65–84·   pls. 1–4.

KÜSTER, W.   1894.   Die Oelkörper der Lebermoose und ihr Verhältniss zu Elaioplasten.   Inaug. Dissert., Basel.

   1911.   Ueber amöboide Formveränderung der Chromatophoren höher Pflanzen. Ber. Deu. Bot. Ges. **29** : 362–369·   figs. 4.

LA VALETTE ST. GEORGE, A. 1886.   Spermatologische Beiträge.   2.   Arch. Mikr. Anat. **27** : 1–12·   pls. 1, 2.

LEWIS, M. R. and LEWIS, W. H.   1915.   Mitochondria (and other cytoplasmic inclusions) in tissue cultures.   Am. Jour. Anat. **17** : 339–401·   figs. 26.

LEWITSKI, G.   1910.   Ueber die Chondriosomen in pflanzlichen Zellen.   Ber. Deu. Bot. Ges. **28** : 538–546·   pl. 17.

   1911.   Die Chloroplastenanlagen in lebenden und fixierten Zellen von *Elodea canadensis*, Rich.   Ibid. **29** : 697–703·   pl. 28.

   1914.   Die Chondriosomen als Secretbildner bei den Pilzen.   Ibid. **31** : 517–528· 1 pl.

LIDFORSS, B.   1893.   Studien öfver elaiosferer i örtbladens mesophyll och epidermis. Inaug. Dissert., Lund; also in Kgl. Fysiogr. Sällsk. i Lund Handl. **4**.

Löwschin, A. M. 1913 "Myelinformen" und Chondriosomen. Ber. Deu. Bot. Ges. **31**: 203–209.

1914. Vergleichende experimental-cytologische Untersuchungen über Mitochondrien in Blättern der höheren Pflanzen. (Vorl. Mitt.) Ibid. **32**: 266–270. pl. 5.

Lundegårdh, H. 1910. Ein Beitrag zur Kritik zweier Vererbungshypothesen. Jahrb. Wiss. Bot. **48**: 285–378.

Mast, S. O. 1911. Light and the behavior of organisms. pp. 410. N. Y.

1916. The process of orientation in the colonial organism, *Gonium pectorale*, and a study of the structure and function of the eyespot. Jour. Exp. Zool. **20**, 1–17. figs. 6.

Mayer, Rathery and Schaeffer. 1914. Les granulations ou mitochondries de la cellule hepatique. Jour. Physiol. Path. Gén. **16**: 607–622.

McAllister, F. 1913. Nuclear division in *Tetraspora lubrica*. Ann. Bot. **27**: 681–696. pl. 56.

1914. The pyrenoid of *Anthoceros*. Am. Jour. Bot. **1**: 79–95. pl. 8.

Meves, Fr. 1904. Ueber das Vorkommen von Mitochondrien bezw. Chondriomiten in Pflanzenzellen. Ber. Deu. Bot. Ges. **22**: 284–286. pl. 16.

1907*a*. Ueber Mitochondrien bzw. Chondriokonten in den Zellen junger Embryonen. Anat. Anz. **31**: 399–407.

1907*b*. Die Chondriokonten in ihrem Verhältnis zur Filarmasse Flemmings. Ibid. **31**: 561–569.

1908. Die Chondriosomen als Träger erblicher Anlagen. Cytologische Studien am Hühnerembryo. Arch. Mikr. Anat. **72**: 816–867. pls. 39–42.

1909. Ueber Neubildung quergestreifter Muskelfäsern nach Beobachtungen am Hühnerembryo. Anat. Anz. **34**: 161–165. figs. 3.

1914. Was sind die Plastosomen? Ibid. **85**: 279–302. figs. 17.

1915. Ueber Mitwirkung der Plastosomen bei der Befruchtung des Eies von *Filaria papillosa*. Arch. Mikr. Anat. **87**: II 12–46. pls. 1–4. See also p. 286.

1918. Die Plastosomentheorie der Vererbung. Eine Antwort auf verschiedener Einwände. Ibid. **92**: II, 41–136. figs. 18. (Bibliography.)

Meyer, A. 1881. Ueber die Struktur der Stärkekörner. Bot. Zeit. **39**: 841–846, 857–864. pl. 9.

1883*a*. Ueber Krystalloide der Trophoplasten und über die Chromoplasten der Angiospermen. Ibid. **41**: 489–498, 505–514, 525–531.

1883*b*. Das Chlorophyllkorn. pp. 91. pls. 3. Leipzig.

1895. Untersuchungen über die Stärkekörner. Jena.

1911. Bemerkungen zu G. Lewitski: Ueber die Chondriosomen in pflanzlichen Zellen. Ber. Deu. Bot. Ges. **29**: 158–160.

Mirande, M. 1919. Sur le chondriome, les chloroplastes et les corpuscules nucleolaires du protoplasme des *Chara*. Comptes Rend. Acad. Sci. Paris **168**: 283–286. figs. 7.

von Mohl, H. 1835, 1837. Ueber die Vermehrung der Pflanzenzellen durch Theilung. Dissert. Tübingen 1835. Flora **45**, 1837.

1851. Gründzüge der Anatomie und Physiologie der vegetabilische Zelle. Engl. Transl. by Henfrey, London 1852.

Monteverde, N. A. and Lubimenko, V. N. 1911. Recherches sur la formation de la chlorophylle chez les plantes. [Russian.] Bull. Acad. Imp. Sci. St. Petersbourg VI **5**: 73–100.

Moreau, F. 1914. Le chondriosome et la division des mitochondries chez les *Vaucheria*. Bull. Soc. Bot. France **61**: 139–142.

Mottier, D. M. 1918. Chondriosomes and the primordia of chloroplasts and leucoplasts. Ann. Bot. **32**: 91–114. pl. 1.

VON NÄGELI, C. 1858.   Die Stärkekörner.   Zurich.

OLIVE, E. W.   1904.   Mitotic division of the nuclei of the Cyanophyceæ.   Beih. Bot. Centr. **18**: 9–44.   pls. 1, 2.

ORMAN, E.   1913.   Recherches sur les differenciations cytoplasmiques dans les végétaux.   1. Le sac embryonnaire des Liliacées.   La Cellule **28**: 365–441.   pls. 1–4.

OVERTON, E.   1889.   Beitrag zur Kenntniss der Gattung *Volvox*.   Bot. Centr. **39**: 65–72.   113–118, 145–150, 177–182, 209–214, 241–246, 273–279.   pls. 4.

PALLADIN, V. I.   1918.   Plant Physiology.   (Engl. edition by Livingston.)   Philadelphia.

PAYNE, F.   1916.   A study of the germ cells of *Gryliotalpa borealis* and *Gryllotalpa vulgaris*.   Jour. Morph. **28**: 287–327.   pls. 4.   figs. 5.

PENSA, A.   1914.   Ancora a proposito di condriosomi e pigmento antocianico nelle cellule vegetali.   Anat. Anz. **46**: 13–22.

PFEFFER, W.   1874.   Die Oelkörper der Lebermoose.   Flora **57**: 2–6, 17–27, 33–43.   pl. 1.

POLITIS, I.   1914.   Sugli elaioplasti nelle mono- e dicotiledoni.   Atti Inst. Bot. Univ. Pavia II **14**: 335–361.   pls. 13–15.

POPOFF, M.   1907.   Eibildung bei *Paludina vivipara* usw.   Arch. Mikr. Anat. **70**: 43–129.   pls. 4–8.   fig. 1.

RACIBORSKI, M.   1893.   Ueber die Entwicklungsgeschichte der Elaioplasten der Liliaceen.   Anz. Akad. Wiss.   Krakau.   p. 259.

REGAUD, C.   1908.   Sur les mitochondries de l'epithelium seminal.   Comptes Rend. Soc. Biol.   Paris **65**: 660–662.

   1910.   Étude sur la structure des tubes seminiferes, etc.   Arch. d'Anat. Micr. **11**: 291–433.

   1911.   Les mitochondries, organites du protoplasma considerés comme les agents de la fonction eclectique et pharmacopexique de la cellule.   Revue de Medicin.

RIVETT, M. F.   1918.   The structure of the cytoplasm in the cells of *Alicularia scolaris*, Cord.   Ann. Bot. **32**: 207–214.   pl. 6.   figs. 3.

RUDOLPH, K.   1912.   Chondriosomen und Chromatophoren.   Beitrag zur Kritik der Chondriosomentheorien.   Ber. Deu. Bot. Ges. **30**: 605–629.   pl. 18.   1 fig. (Review.)

SALTER, J. H.   1898.   Zur naheren Kenntniss der Stärkekörner.   Jahrb. Wiss. Bot. **32**: 117–166.   pls. 1, 2.

SAPĔHIN, A. A.   1911.   Ueber das Verhalten der Plastiden im sporogonen Gewebe. (Vorl. Mitt.).   Ber. Deu. Bot. Ges. **29**: 491–496.   figs. 5.

   1913*a*.   Untersuchungen über die Individualität der Plastide.   [Russian.]   pp. 133.   pls. 17.   Odessa.

   1913*b*.   Untersuchungen über die Individualität der Plastide.   (Zweite Vorl. Mitt.)   Ber. Deu. Bot. Ges. **31**: 14–16.   1 fig.

   1915.   Untersuchungen über die Individualität der Plastide.   Arch. Zellf. **13**: 319–398.   pls. 10–26.

SCHAXEL, F.   1911.   Plasmastructuren, Chondriosomen und Chromidien.   Anat. Anz. **39**: 337–353.   figs. 16.

SCHERRER, A.   1914.   Untersuchungen über Bau und Vermehrung der Chromatophoren und das Vorkommen von Chondriosomen bei *Anthoceros*.   Flora **107**: 1–56.   pls. 1–3.

SCHILLING, A. J.   1891.   Die Süsswasser-Peridineen.   Flora **74**: 220–299.   pls. 8–10.

SCHIMPER, A. F. W.   1880–1881.   Untersuchungen über die Entstehung der Stärkekörner.   Bot. Zeit. **38**: 881–902, pl. 13; **39**: 185–194, 201–211, 217–227.   pl. 2.

1883. Ueber die Entwicklung der Chlorophyllkörner und Färbkorper. Bot. Zeit. **41**: 105–112, 121–131, 137–146, 153–162. 1 pl.

1885. Untersuchungen über die Chlorophyllkorper und die ihnen homologen Gebilde.

SCHMIDT, E. W. 1912. Pflanzliche Mitochondrien. Prog. Rei Bot. **4**: 163–181. figs. 6.

SCHMITZ, FR. 1882. Die Chromatophoren der Algen. Verh. Naturhist. Ver. Preuss. Rheinl. u. Westf. **40**. (See also Sitz-Ber. Med. Gesell. f. Nat. u. Heilk. Bonn, 1880.)

1884. Beiträge zur Kenntniss der Chromatophoren. Jahrb. Wiss. Bot. **15**: 1–177. pl. 1.

SENN, G. 1908. Die Gestalts-und Lageveränderung der Pflanzen-Chromatophoren. Leipzig.

SHAFFER, E. L. 1920. The germ-cells of *Cicada* (*Tibicen*) *septemdecim* (Homoptera). Biol. Bull. **38**; 404–474. pls. 9.

SMITH, G. M. 1914. The cell structure and colony formation in *Scenedesmus*. Arch. f. Protistenk. **32**: 278–297. pls. 16, 17.

SOKOLOW, I. 1913. Untersuchungen über die Spermatogenese bei den Arachniden. I. Ueber die Spermatogenese der Skorpione. Arch. f. Zellf. **9**: 399–432. pls. 22, 23. 1 fig.

STRASBURGER, E. 1900. Ueber Reduktionstheilung, Spindelbildung, Centrosomen, und Cilienbildner im Pflanzenreich. Hist. Beitr. **6**: 1–227. pls. 4. Jena.

1903. Text-Book of Botany (Strasburger, Noll, Schenck, and Schimper). 2d. English edition from 5th German ed.

TAMMES, T. 1900. Ueber die Verbreitung des Carotins im Pflanzenreich. Flora **87**: 205–247. pl. 7.

TERNI. T. 1914. Condriosomi, idiozoma e formazioni periidiozomiche nella spermatogenesi degli anfibii. (Ricerche sul *Geotriton fuscus*.) Arch. Zellf. **12**: 1–96. pls. 1–7.

TIMBERLAKE, H. G. 1901. Starch formation in *Hydrodictyon utriculatum*. Ann. Bot. **15**: 619–635.

1903. The nature and function of the pyrenoid. Science N. S. **17**: 460.

TISCHLER, G. 1906. Ueber die Entwicklung des Pollens und der Tapetenzellen bei *Ribes*-Hybriden. Jahrb. Wiss. Bot. **42**: 545–578. pl. 15.

TWISS, W. C. 1919. A study of plastids and mitochondria in *Preissia* and corn. Am. Jour. Bot. **6**: 217–234. pls. 23, 24.

VOÏNOV, D. 1916. Sur l'existence d'une chondriodierese. Comptes Rend. Soc. Biol. Paris **68**: 451–454. figs. 4.

WAGER, H. 1900. On the eyespot and flagellum of *Euglena viridis*. Jour. Linn. Soc. **27**: 463–481. pl. 32.

1903. The cell structure of the Cyanophyceæ. (Prelim. note.) Proc. Roy. Soc. London **72**: 401–408. figs. 3.

WAKKER, J. H. 1888. Studien über die Inhaltskörper der Pflanzenzelle. Jahrb. Wiss. Bot. **19**: 423–496. pls. 12–15.

WASSILIEF, A. 1907. Die Spermatogenese von *Blatta germanica*. Arch. Mikr. Anat. **70**: 1–42. pls. 1–3. 1 fig.

WHELDALE, M. 1916. The anthocyanin Pigments of Plants. Cambridge.

WIESNER, J. 1892. Die Elementarstructur und das Wachstum der lebenden Substanz. Wien.

WILDMAN, E. E. 1913. The spermatogenesis of *Ascaris megalocephala* with special reference to the two cytoplasmic inclusions, the refractive body and the "mitochondria;" their origin, nature and rôle in fertilization. Jour. Morph. **24**: 421–457. pls. 3.

WILLSTÄTTER, R. und Stoll, A. 1913. Untersuchungen über Chlorophyll. Berlin.

WILSON, E. B. 1916. The distribution of the chondriosomes to the spermatozoa of scorpions. Science **43**: 539. (Refers also to work of Sokolow.)

VON WINIWARTER, H. 1912. Observations cytologiques sur les cellules interstitielles du testicule humain. Anat. Anz. **41**: 309–320. pls. 1, 2.

WOLLENWEBER, W. 1907. Das Stigma von *Hæmatococcus*. Ber. Deu. Bot. Ges. **25**: 316–320. pl. 11.

    1908. Untersuchungen über die Algengattung *Hæmatococcus*. Ibid. **26**: 238–298. pls. 12–16. figs. 12.

YAMANOUCHI, S. 1913. *Hydrodictyon Africanum*, a new species. Bot. Gaz. **55**: 74–79.

ZIMMERMANN, A. 1893. Ueber die Elaioplasten. Beitr. z. Morph. u. Physiol. der Zelle **1**: 185–197.

    1894. Sammel-Referate. 9. Die Chromatophoren. 10. Die Augenfleck. 11. Elaioplasten, Elaiosphären und verwandte Körper. 15. Die Stärkekörner und verwandten Körper. Beih. Bot. Centr. **4**: 90–101, 161–165, 165–169, 329–335.

CHAPTER VII

METAPLASM; POLARITY

In the foregoing chapters we have described successively the various organs of the cell. Our account of the resting cell will now be completed by passing in brief review some of its more conspicuous non-protoplasmic inclusions. We shall also call attention to another characteristic but imperfectly understood attribute of the protoplast, namely, its polarity.

Metaplasm.—In addition. to their definite cell organs—nucleus, cytoplasm, centrosomes, plastids, and possibly chondriosomes—cells which have undergone any amount of differentiation usually contain a variety of other materials representing products of metabolism. Many of these substances are held in solution in the cell sap, itself a differentiation product, while others are present in insoluble form in the cytoplasm, often in special vacuoles. All such non-protoplasmic inclusions, particularly those existing in some visible form, are referred to as *metaplasm*, a term introduced by Hanstein. Although it has been held by some (Kassowitz 1899) that metaplasm is always inactive and to be sharply set apart from active protoplasm, it is more probable, as Child (1915) contends, that no absolute distinction can be made between the two. Most of the products of differentiation, however, are clearly non-protoplasmic and relatively inactive.

In cells of many types, even in the comparatively undifferentiated cells of the root meristem, there often occur accumulations of chemically complex substances in the form of small globules or irregular masses in the cytoplasm. In many cases these more or less transient bodies, which often stain intensely with the nuclear dyes and are therefore referred to as "chromatic bodies," show reactions indicating a composition closely approaching that of the extra-nuclear granules of nucleo-protein (chromidia) which R. Hertwig and Goldschmidt interpret as granules of escaped chromatin concerned in cell differentiation. Others resemble the fatty chondriosomes in form and composition. It is therefore a matter of some difficulty to distinguish between these various substances, which, as a matter of fact, probably do not represent sharply distinct classes.

The most conspicuous non-protoplasmic inclusions represent food materials in transitory form or in the storage condition; they are consequently abundant in cells carrying a supply of reserve foods, such as spores and eggs, and in storage organs, such as many roots and the endosperm and cotyledons of seeds.   In the animal egg the storage material commonly exists in the form of "yolk globules," or "deutoplasm spheres," which consist for the most part of relatively complex protein compounds. Fat or oil globules are usually present with them.   In plants the most characteristic storage product is starch, the origin and characters of which were described in Chapter VI along with the plastids by which they are formed.   In some organisms, including the fungi, glycogen appears to carry on the function performed by starch and sugar in the higher plants. Fats and oils, usually in the form of droplets but sometimes of soft grains or even crystals (nutmeg), comprise another important class of storage substances: these are especially prevalent in seeds and spores, where light weight is of advantage.   In many cases oil may be produced anywhere in the cell, but in certain forms it has been found that special elaioplasts, and

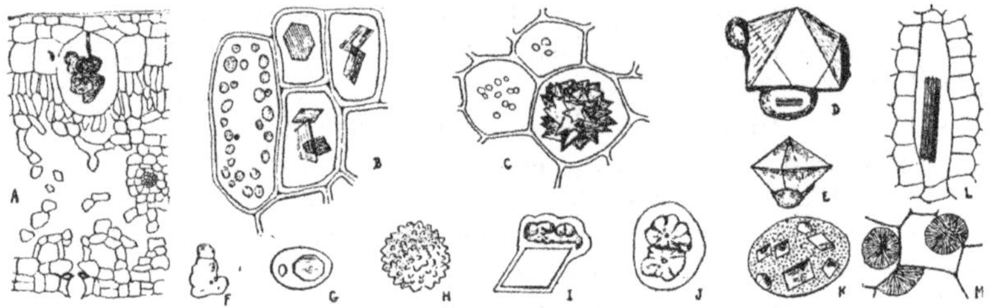

Fig. 47.—Crystalline and other inclusions in the cells of various plants.

A, cystolith in subepidermal cell of *Ficus* leaf.   B, crystal cells in *Arctostaphylos*.   C, druse in cell of *Rheum palmatum*.   D–K, aleurone grains: D, E, from *Myristica*; F, from *Datura stramonium*; G, from *Ricinus communis*; H, from *Amygdalus communis*; I, from *Bertholletia excelsa*; J, from *Fœniculum*; K, from *Elœis guiniensis*.   L, raphides in leaf of *Agave*.   M, inulin crystals in preserved cells of artichoke.   (*B–K after Tschirch*.)

possibly also chondriosomes, are concerned in this process.   The peculiar oil bodies found in the cells of certain liverworts appear to represent oil vacuoles: these also have been discussed, together with elaioplasts, in Chapter VI.   Large masses of intranuclear metaplasm are found characteristically in the eggs of gymnosperms.

Aleurone grains occur in small vacuoles in the cells of many seeds, particularly in such oily ones as those of *Ricinus, Juglans,* and *Bertholletia.*   In maize and wheat grains they are limited to a single layer of cells, the "aleurone layer."   The aleurone grain varies much in structure and form, several types being described by Pfeffer in 1872.   The grain consists primarily of an amorphous protein substance, often with an outer, somewhat more opaque shell.   Some examples show no greater differentiation than this, but many are much more elaborate (Fig. 47, D–K). Those of *Ricinus* contain within them a single angular crystal of protein (albumen), often referred to as the "crystalloid," and a globule of a double phosphate of calcium and magnesium with certain organic substances called the "globoid" (Fig. 47, *G*). The crystalline inclusions sometimes grow to be very large. It has been thought by certain workers that aleurone grains are self-perpetuating bodies with an individuality comparable to that of nuclei and certain plastids. That this view is correct has been rendered very improbable by the researches of East and Hayes (1911, 1915) and Emerson (1914, 1917) on the inheritance of aleurone characters in maize, and also by the work of Thompson (1912), who succeeded in producing artificial aleurone grains in all essential respects similar to those elaborated by the plant.

Crystals occur in great variety in the differentiated cells of plants. They may lie in the cytoplasm, in vacuoles, attached to or imbedded in the cell wall, and even in special cells. They are usually salts of calcium, calcium oxalate being especially prevalent. The bundles of needle-shaped crystals known as "raphides" (Fig. 47, *L*) found in the leaves of a number of plants are composed of the latter salt, as are also the spherical aggregations called "druses," or "sphærraphides" (Fig. 47, *C*). The curious clustered "cystoliths" of the *Ficus* leaf (Fig. 47, *A*) are made up of cellulose and calcium carbonate. Crystals of silica are very abundant in the thickened walls of wood cells and in many other tissues, such as the outer cells of the *Equisetum* stem. Crystals of albumen, aside from those found in aleurone grains, are frequently present in the cytoplasm of cells poor in starch, as in the outer portion of the potato tuber. The leucoplast of *Phajus* often contains a rod-shaped albumen crystal. Protein crystals of various shapes are occasionally observed within the nucleus (Stock 1892; Zimmermann 1893).

Cellulose is a common storage material, existing as a rule in the form of laminæ deposited upon the original cell wall.

As already pointed out, the *sap* of vacuolated cells may contain a number of differentiation products in solution. The cell sap is usually slightly acid in reaction, owing to the presence or organic acids (malic, formic, acetic, oxalic) and their salts. Inogranic salts are probably always present. Amides, such as glutamin and asparagin, glucosides, sugars, proteins, tannin, and many other substances are of frequent occurrence in the cell sap of various plants. The carbohydrate inulin may be precipitated out of the sap by alcohol: this accounts for the presence of nodules of radiating inulin crystals frequently encountered in preserved material (Fig. 47, *M*). Rubber is present in the form of a suspension of minute droplets in the cell sap of *Ficus elastica* and several other plants. Gutta-percha occurs in a similar state in *Isonandra gutta*. The cell sap in such cases has a characteristic milky appearance. The cell sap is often colored by red, blue, and yellow anthocyanin pigments (Wheldale 1916; Palladin 1918; Beauverie 1919), some of which change color when the reaction of the sap is altered from acid to basic and *vice versa*. The striking colors of flowers are due to "(1) the varying color of the sap, (2) the distribution of the cells containing it, and (3) combinations of colored sap with chloro- and chromoplasts." Autumnal coloring is due to the formation of pigments as disorganization products: when cytoplasm and chlorophyll are the main disorganizing substances a yellowish color results, whereas if sugars are present in considerable amounts in the cell sap the brighter pigments are formed.

Extruded Chromatin.[1]—The actual extrusion of chromatin from the nucleus into the cytoplasm has been reported in a number of instances: in the microsporocytes of various angiosperms by Digby (1909, 1911, 1914), Derschau (1908, 1914), West and Lechmere (1915), and others; in ferns by Farmer and Digby (1910); and in the Ascomycete *Helvella crispa* by Carruthers (1911). The extruded chromatin commonly takes the form of deeply staining globules or irregular masses in the cytoplasm; often a clear area suggesting a nuclear vesicle is present about them. In some cases, such as *Galtonia candicans* (Digby 1909) and *Lilium candidum* (West and Lechmere 1915), the chromatin may pass through the wall into an adjacent cell, where it forms a rounded mass connected by a chromatic strand with the nucleus from which it originated.

The significance of this phenomenon is by no means apparent. It is not at all unlikely that nutritive materials passing from nucleus to cytoplasm during the normal metabolism of the cell occur at times as visible globules at the nuclear surface. The extrusion of chromatin into neighboring cells, on the other hand, in many cases has every appearance of a phenomenon associated with degeneration or some other abnormal physiological condition. West and Lechmere, however, view the process as one which occurs normally at certain stages, and which will probably be found to be more general in plants. Sakamura's (1920) extensive researches on chloralized cells have led him to regard the extrusion of large masses of chromatin as an abnormal phenomenon which occurs as a result of a disturbance of the metabolism of the cell. Its more frequent occurrence in sporocytes than in other cells is attributed to the unusual sensitiveness of the former to disturbing influences.

The Senescence of the Cell.—The accumulation of products of metabolism ("differentiation products") has a direct bearing on the problem of protoplasmic senility. As its life progresses the cell gradually "ages," and if nothing occurs to prevent it the process eventually terminates in death. What shall be taken as an index of the degree of senescence has been the subject of much discussion. We have already called attention to the attempts which have been made to correlate senescence with a progressive change in the nucleoplasmic relation, concluding that no constant correlation of the kind has been shown to exist (p. 63).

Child (1915) has brought forward much evidence to show that the

---

[1] Extruded chromatin is not metaplasm, but it has been found convenient to treat it at this point along with other inclusions of the cytoplasm.

relative rate of metabolism is the main criterion of the cell's physiological age, "young" cells having a high rate and "old" cells a relatively low rate, and a gradual decline in this rate occurring throughout the life of the cell. In embryonic (physiologically young) cells the cytoplasm appears to be comparatively homogeneous and undifferentiated. Older cells, on the contrary, are ordinarily marked by the presence of products of differentiation in the cytoplasm. The true measure of age is therefore not time, but physiological differentiation.

In many cells a rejuvenating process may occur, whereby a high metabolic rate is restored and the products of differentiation lost: this is regarded as a "return to the embryonic state"—a real physiological rejuvenescence. "Senescence is primarily a decrease in rate of the dynamic processes conditioned by the accumulation, differentiation, and other associated changes of the material of the colloid substratum. Rejuvenescence is an increase in rate of dynamic processes conditioned by changes in the colloid substratum in reduction and dedifferentiation" (Child, p. 58). Such a rejuvenescence occurs in connection with regeneration, vegetative and other asexual reproduction, and sexual reproduction. In each case the cell which begins the new life cycle—the meristematic regenerating cell, the zoöspore, or the zygote—has a high metabolic rate and is comparatively free from the products of differentiation.

In the lower organisms cell differentiation in this sense is not so great but that almost any cell may retain the power to "dedifferentiate" and begin the development of a new individual vegetatively. In these forms asexual reproduction may occur repeatedly and keep the organism as a whole (in protozoa and protophyta) or the protoplasm of the race (in lower metazoa and metaphyta) physiologically young. Only when the metabolic rate falls very low does sexual reproduction, the most effective of all the rejuvenating agencies, ensue.

In the higher plants the retention of the power of dedifferentiation is strikingly shown in the well known cases of *Begonia* and *Bryophyllum*, which can regenerate complete new individuals from a few leaf cells. In the higher animals cell differentiation is usually so great that the somatic cells can no longer dedifferentiate and reproduce the organism asexually. Here rejuvenation occurs only after the union of two gametes, which are themselves, unlike the zoöspores of algæ, physiologically old. Although local rejuvenescence may occur, as in secretory cells which are "younger" after secretion, and also in wound tissue, the differentiation of the body cells is carried so far that their metabolic rate falls low enough to make a recovery or rejuvenescence no longer possible. Thus it is only the functioning reproductive cells that endure: the ultimate cessation of all life processes in the body cells is the price which is inevitably paid by the complex multicellular organism for the advantages conferred by its high degree of differentiation.

Of the highest importance in this connection are the results of attempts to maintain the cells and tissues of higher animals in the living condition in artificial culture media outside the body. It has been shown by the remarkable experiments of Carrel, Leo Loeb, Burrows, H. V. Wilson and others that cells may be isolated from any of the highly differentiated essential tissues of the body and kept actively growing and multiplying *in vitro* for a length of time frequently far exceeding that to which they would have lived in the body. They do not appear to grow old: indeed it is not improbable that in such a constantly favorable environment somatic cells are as "potentially immortal" as the germ cells (see p. 403). In the words of Pearl (1921), "It is the differentiation and specialization of function of the mutually dependent aggregate of cells and tissues which constitutes the metazoan body which brings about death, and not any inherent or inevitable mortal process in the individual cells themselves."

POLARITY

Polarity is a feature which is exhibited in some form by the cells of all higher organisms, and in at least many of the simpler ones, as shown by Tobler (1902, 1904) for certain algæ; indeed it is probable that it is possessed in some form and degree by all cells. Harper (1919) calls attention to the fact that "in the presence of polarity and the various symmetry relations we have a fundamental distinction between cell organization and that of polyphase colloidal systems as they are commonly produced *in vitro*."

This polarity has two aspects, the morphological and the physiological. In the first place, the various constituents of the cell may be arranged symmetrically about one or more ideal axes, so that the cell has more or less distinctly differentiated anterior and posterior ends. This structural aspect of polarity has been the one chiefly emphasized by certain workers: van Beneden (1883), for instance, looked upon polarity as "a primary morphological attribute of the cell," the axis passing through the nucleus and the centrosome. Later writers, among them Heidenhain (1894, 1895), made this conception of morphological polarity the basis for interpretations of many of the phenomena of cell behavior. (See Wilson 1900, pp. 55–56.) However, as Harper (1919) points out, polarity "is apparently independent of the uni- or multinucleated condition of the cell, which shows that it is in some cases at least a more generalized characteristic of the cell as a whole rather than a mere expression of the space relations of the nucleus and cytoplasm . . ." Other investigators (Hatschek 1888; Rabl 1889, 1892) early laid emphasis upon the physiological expression of polarity. The cell shows a polar differentiation in physiological labor: the processes in one portion of the cell differ from those in another, this difference in the case of tissue cells being due to different environments in the tissue. For these workers this physiological differentiation is the essential element of polarity; any morphological polarity is due secondarily to it.

Metabolic Gradient.—The most suggestive physiological conception recently developed in this connection is that of Child (1911–1916). Child has shown in the case of *Planaria* and other lower animals, as well as in certain algæ, that along each of the axes of symmetry there exists a "metabolic gradient," or "axial gradient:" the rate of the physiological processes is highest at one end of the axis and diminishes progressively toward the other end. The anterior end of a planarian, for example, has a higher metabolic rate than the posterior portions. Furthermore, the portions of higher rate dominate and control the development of those portions having a lower rate, with the result that the young individual soon develops and maintains a definite physiological correlation of anterior and posterior parts. Similarly in individuals with more than one axis of symmetry, there may be a corresponding dorsal-ventral, as well as an axial-marginal, correlation. That polarity is here primarily a physiological matter is indicated by the fact that experimental alterations in the metabolic rate in different parts is followed by abnormalities in structural development.

As to the means by which the dominance of certain regions over others is exercised, correlating the activities of the various parts of the organism, there are two principal theories in the field. According to one theory chemical substances (hormones) are produced at certain places and transmitted through the body. Although the circulation of such hormones clearly has much to do with correlation in higher complex organisms, Child adduces good evidence in support of the second theory, namely, that the fundamental relations of polarity "depend primarily upon impulses or changes of some sort transmitted from the dominant region, rather than upon the transportation of chemical substances" (p. 224).

It cannot at present be said to what extent this conception of polarity is applicable to the single cell. The work of Child shows in a very definite manner the coincidence of the morphological and physiological axes of polarity, which indicates that the two are but different aspects of one and the same polar differentiation. A similar coincidence exists very generally in the case of the single cell. In the cell, as in the organism as a whole, functional and structural differentiation are inseparably connected. In the present state of our knowledge the attempt to determine the real essence of polarity raises questions which cannot yet be answered. Does physiological polarity depend upon a polarized structure which is a fundamental attribute of the cell's ultimate organization? Or does a polarized morphological arrangement follow and depend upon a physiological division of labor arising as a difference in intensity or rate in processes originally common to all parts of the cell?   If so, to what internal or external factors is the establishment of this difference due in cells having no initial polarity?   Analogies with electrical polarity have been resorted to in this connection, concerning which Harper (1919) says: "To provide an adequate basis for understanding the observed facts of polarity, however, it seems to me that the conception of compound aggregate polyphase systems is more suggestive than these attempted analogies . . . In the spatial arrangement and interactions of these systems polar differences of the most diversified types are bound to arise in the mass as a whole and express themselves in the form and relative rigidity and surface tension of different parts, as well as in the interrelations between the cells of a group in contact."

The polarity of the multicellular organism as a whole is closely bound up with the polarities of its constituent cells.   Harper has clearly shown (1918) that in *Pediastrum* the position of the swarm-spores in the colony which they unite to form is directly dependent upon their polarity. This does not mean, however, that the polarity of the multicellular organism is nothing more than the sum ʻof the polarities of its constituent cells, unless we return to Schwann's simple conception of the organism as merely an aggregate of independent cells.   (See p. 12.)   The higher individuality, the colony, has its own polarity, which may be related to, but is not the same as, that of its individual cells.   In the ordinary multicellular organism the polarity is an outgrowth of the polarity of the fertilized egg cell rather than of the polarities of the many adult tissue cells.

In polarity, then, we encounter another problem which must be brought nearer a solution before we can have any adequate understanding of the relation of the cell to the multicellular organism as a whole, and of the perplexing matter of organic individuality.

Bibliography 7

Metaplasm—Senescence—Polarity

van Beneden, E.   1883.   Recherches sur la maturation de l'oeuf, la fécondation et la division cellulaire.   Arch. de Biol. **4.**

Boveri, Th.   1901*a*.   Ueber die Polarität des Seeigeleies.   Verh. Phys.-Med. Ges. Würzburg **34.**

1901*b*.   Die Polarität von Ovocyte, Ei und Larve des *Strongylocentrotus lividus.* Zool. Jahrb. (Anat. Abt.) **14:** 630–653.   pls. 48–50.

Carruthers, D.   1911.   Contributions to the cytology of *Helvella crispa* Fries. Ann. Bot. **25:**   243–252.   pls. 18, 19.

Child, C. M.   1911.   Studies on the dynamics of morphogenesis and inheritance in experimental reproduction: I.   The axial gradient in *Planaria dorotocephala* as a limiting factor in regulation.   Jour. Exp. Zool. **10:** 265–320.   figs. 7.

1912.   Studies, etc.   IV.   Certain dynamic factors in the regulatory morphogenesis of *Planaria dorotocephala* in relation to the axial gradient.   Ibid. **13:** 103–152. figs. 46.

1913. Studies, etc. VI. The nature of the axial gradients in *Planaria* and their relation to antero-posterior dominance, polarity and symmetry. Arch. Entw. **37**: 108–158. figs. 13.

1915. Senescence and Rejuvenescence. Chicago.

1916. Axial susceptibility gradients in algæ. Bot. Gaz. **62**: 89–114.

von Derschau, M. 1908. Beiträge zur pflanzlichen mitose: Centern, Blepharoplasten. Jahrb Wiss. Bot. **46**: 103–118. pl. 6.

1914. Zum Chromatindualismus der Pflanzenzelle. Arch. Zellf. **12**: 220–240. pl. 17.

Digby, L. 1909. Observations on "chromatin bodies" and their relation to the nucleolus in *Galtonia candicans* Decsne. Ann. Bot. **23**: 491–502. pls. 33, 34.

1910. The somatic, premeiotic, and meiotic nuclear divisions in *Galtonia candicans*. Ann. Bot. **24**: 727–757. pls. 59–63.

1914. A critical study of the cytology of *Crepis virens*. Arch. Zellf. **12**: 97–146. pls. 8–10.

Duesberg, J. 1911. Plastosomen, "Apparato reticolaro interno," und Chromidialapparat. Ergebn. Anat. Entw. **20**: 567–916. (Review.)

East, E. M. and Hayes, H. K. 1911. Inheritance in maize. Conn. Exp. Sta. Bull. No. 167.

1915. Further experiments in inheritance in maize. Ibid No. 188.

Emerson, R. A. 1914. The inheritance of a recurring somatic variation in variegated ears of maize. Am. Nat. **48**: 87–115.

1917. Genetical analysis of variegated pericarp in maize. Genetics **2**.

Harper, R. A. 1918*a*. Organization, reproduction and inheritance in *Pediastrum*. Proc. Am. Phil. Soc. **57**: 375–439. pls. 2. figs. 35.

1918*b*. The evolution of cell types and contact and pressure responses in *Pediastrum*. Mem. Torr. Bot. Club **17**: 210–240. figs. 27.

1919. The structure of protoplasm. Am. Jour. Bot. **6**: 273–300.

Hatschek, B. 1888. Lehrbuch der Zoologie.

Heidenhain, M. 1894. Neue Untersuchungen über die Centralkörper und ihre Beziehungen zum Kern und Zellprotoplasma. Arch. Mikr. Anat. **43**: 423–758. pls. 25–31.

1895. Cytomechanische Studien. Arch. Entw. **1**: 473–577. pl. 20. figs. 17.

Kassowitz, M. 1899. Allgemeine Biologie. Vienna.

Pearl, R. 1921. The biology of death. II–The conditions of cellular immortality. Sci. Mo. **12**: 321–335; figs. 6.

Pfeffer, W. 1872. Untersuchungen über die Proteinkörner und die Bedeutung des Asparagins beim Keimen der Samen. Jahrb. Wiss. Bot. **8**: 429–574. pls. 36–38.

Rabl, C. 1885. Ueber Zelltheilung. Morph. Jahrb. **10**: 214–330. pls. 7–13. figs. 5.

1889. Ueber Zelltheilung. Anat. Anz. **4**: 21–30. figs. 2.

Sakamura, T. 1920. Experimentelle Studien über die Zell- und Kernteilung mit besonderer Rücksicht auf Form, Grosse und Zahl der Chromosomen. Jour. Coll. Sci. Imp. Univ. Tokyo **39**: pp. 221. pls. 7.

Stock, G. 1892. Ein Beitrag zur Kenntniss der Proteinkrystalle. Cohn's Beitr. Biol. Pflanzen **6**: 213–235. pl. 1.

Thompson, W. P. 1912. Artificial production of aleurone grains. Bot. Gaz. **54**: 336–338. 1 fig.

Tobler, F. 1902. Zerfall und Reproduktionsvermogen des Thallus einer Rhodomelaceæ. Ber. Deu. Bot. Ges. **20**: 357–365. pl. 18.

1904. Ueber Eigenwachstum der Zelle und Pflanzenform. Jahrb. Wiss. Bot. **39**: 527–580. pl. 10.

Tschirch, A. 1889. Angewandte Pflanzenanatomie. Wien u. Leipzig.

West, C. and Lechmere, A. E. 1915. On chromatin extrusion in pollen mother-cells of *Lilium candidum*, Linn. Ann. Bot. **29**: 285–291. pl. 15.

Wheldale, M. 1916. The anthocyanin pigments of plants. Cambridge.

Wilson, E. B. 1900. The Cell in Development and Inheritance.

Zimmermann, A. 1893*a*. Ueber die Proteinkrystalloide. Beitr. z. Morph. u. Physiol. d. Pflanzenzelle **1**: 54–79. pls. 2.

1893*b*. Ueber Proteinkrystalloide. Ibid. **2**: 112–158.

1894. Sammel-Referate. 11. Elaioplasten, Elaiosphären und verwandte Körper. 13. Die Aleurone- oder Proteinkörner, Myrosin- und Emulsionkörner. 14· Die Proteinkrystalloide, Rhabdoiden und Stachelkugeln. 15. Die Stärkekörnrr und verwandten Körper. Beih. Bot. Centr. **4**: 165–169, 321–335.

CHAPTER VIII

SOMATIC MITOSIS AND CHROMOSOME INDIVIDUALITY

SOMATIC MITOSIS

Since the time when the cell was pointed out as the unit of structure and function it has been recognized that the mode of origin of new cells is a matter of fundamental importance. We have seen in our historical sketch that cells were believed by the founders of the Cell Theory to arise *de novo* from a mother liquor, or "cytoblastema," a misconception removed by later investigations in which it was shown beyond question that cells arise only by the division of preëxisting cells. By several early observers the nucleus was seen to have a more or less prominent part in the process, its division preceding that of the cell, but "it was not until 1873 that the way was opened for a better understanding of the matter. In this year the discoveries of Anton Schneider, quickly followed by others in the same direction by Bütschli, Fol, Strasburger, Van Beneden, Flemming, and Hertwig, showed cell-division to be a far more elaborate process than had been supposed, and to involve a complicated transformation of the nucleus to which Schleicher (1878) afterward gave the name *karyokinesis*. It soon appeared, however, that this mode of division was not of universal occurrence; and that cell-division is of two widely different types, which Van Beneden (1876) distinguished as *fragmentation*, corresponding nearly to the simple process described by Remak, and *division*, involving the more complicated process of karyokinesis. Three years later Flemming (1879) proposed to substitute for these terms *direct* and *indirect* division, which are still used. Still later (1882) the same author suggested the terms *mitosis* (indirect or karyokinetic division) and *amitosis* (direct or akinetic division), which have rapidly made their way into general use, though the earlier terms are often employed. Modern research has demonstrated the fact that amitosis or direct division, regarded by Remak and his followers as of universal occurrence, is in reality a rare and exceptional process;. . . it is certain that in all the higher and in many of the lower forms of life, indirect division or mitosis is the typical mode of cell-division" (Wilson 1900, pp. 64–65).[1]

[1] The following additional historical data are of interest. The chromosomes, though they appeared in the figures of Schneider (1873), were first adequately drawn by Strasburger in 1875. Longitudinal splitting was described by Flemming in 1882. The terms *prophase, metaphase,* and *anaphase* were introduced by Strasburger in

143

In view of the fact that the phenomena of growth, differentiation, reproduction, and inheritance are now known to be intimately bound up with the process of cell-division, it is obvious that a detailed knowledge of this process is an absolute prerequisite to a solution of many of the problems which confront us. In the present chapter the essential features of vegetative or somatic nuclear division will be described. After a preliminary sketch of the process of mitosis we shall take up in some detail the behavior of the chromosomes and the question of their individuality. In the following chapter attention will be devoted to other features of cell-division: the achromatic figure, the mechanism of mitosis, cytokinesis (the division of the extra-nuclear portion of the cell), and the formation of the cell wall.

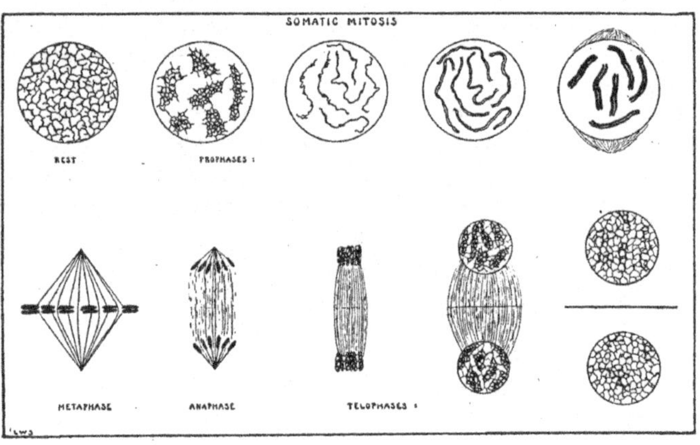

Fig. 48.—Diagram of a typical case of somatic mitosis in plants.

Preliminary Sketch of Mitosis.—The main steps in a typical case of somatic mitosis in plants may be very briefly outlined as follows (Fig. 48):

The chromatic material of the "resting" nucleus, as described in Chapter IV, exists in the form of a more or less irregular *reticulum*. As the process of mitosis begins this reticulum resolves itself into a definite number of slender threads which represent *chromosomes*. These in

1884, and Heidenhain in 1894 first used the term *telokinesis* (*telophase*). Lundegårdh (1912b) added *interphase*. The chromosome was named by Waldeyer in 1888. Hermann in 1891 distinguished connecting fibers (central spindle) and mantle fibers. That the halves of each split chromosome go to opposite poles was shown by van Beneden for animals and by Heuser for plants in 1884. The achromatic spindle was first figured by Kowalevsky (1871) and Fol (1873), and first carefully described by Bütschli (1875ab).

many nuclei are distinct from each other from the first, whereas in other cases they may be arranged end-to-end in a more or less continuous thread, or *spireme*, which later segments transversely into independent chromosomes. The slender threads (chromosomes) now split longltndinally throughout their entire length. A progressive shortening and thickening of the split threads ensues, so that the nucleus is eventually seen to have a certain number of chromosomes which have become double through a longitudinal cleavage.

While the above changes are occurring, fine fibrils are differentiated in the cytoplasm near the nucleus and become arranged in two opposed groups. The nuclear membrane now disappears and the fibers extend into the nuclear region, where some of them (the "mantle fibers") attach themselves to the double chromosomes, while others (the "connecting fibers") pass through from one pole to the other. The double chromosomes quickly become arranged in a single plane at the equator of the cell, the fibers meanwhile forming the *achromatic figure*, or *spindle*. This stage is known as the *metaphase*; all the steps leading up to it, beginning with the initial changes in the resting reticulum, constitute the *prophase*.

The daughter chromosomes (the halves of the longitudinally split chromosomes) now move apart toward the poles of the achromatic figure, where they soon form two closely packed groups with the central spindle of connecting fibers extending between them. The period during which the daughter chromosomes are thus moving apart is known as the *anaphase*. The two groups of daughter chromosomes now reorganize the daughter nuclei, in each of which the chromosomes again form a reticulum like that of the original mother nucleus. This reorganization period is called the *telophase*. During the telophase there is formed upon the connecting fibers (central spindle) a separating wall, which completes the division of the cell. The nucleolus as a rule plays no conspicuous part in mitosis: it usually disappears during the late stages of the prophase, new nucleoli being formed in the daughter nuclei in the telophase. In rapidly dividing cells the period between two successive mitoses is called the *interphase*.

Mitosis in animals (Fig. 49) is closely similar to that in plants as regards the behavior of the chromosomes. It normally differs in two conspicuous features, namely, the presence of centrosomes and the mode of cytokinesis following the division of the nucleus.

During the prophases the centrosome with its aster, if not already double, divides. The two daughter centrosomes, each with its own aster, move apart, and a small bundle of fibers extends between them; all these structures together form the *amphiaster*. The rays on the side toward the nucleus extend into the latter when the membrane dissolves and become attached to the chromosomes, often before the two centro-

10

somes have reached polar positions. The centrosomes, surrounded by asters, remain at the poles of the achromatic figure during metaphase, anaphase, and telophase, and after mitosis has been completed they may disappear or remain through the resting stage to function in the next mitosis. (See p. 78.)

The division of the cell following nuclear division is commonly brought about in animals by the formation of a cleavage furrow, which grows inward from the periphery as described in the following chapter, rather than by the formation of a wall on the spindle fibers as in plants.

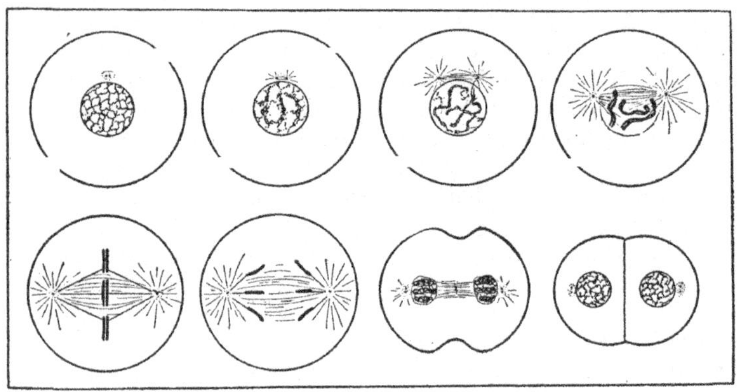

Fig. 49.—Diagram of a typical case of somatic mitosis in animals.

Although the two above points serve in general to distinguish mitosis in animals from that in plants, the distinction is not a sharp one: centrosomes are regularly present in the cells of many lower plants, while cytokinesis by furrowing also occurs in certain cases, as will later be shown. The essential point to be borne in mind is that the significant feature of mitosis—the division of the chromatin and its distribution to the daughter nuclei—is fundamentally the same in both plants and animals.

The relative duration of the various phases of mitosis has been studied in a few cases. As an example may be taken the observations of M. and W. Lewis (1917) on the mesenchyme cells of the chick growing in tissue cultures. These investigators summarize the researches of others upon the subject and give the following figures for the chick cells: prophase, 5 to 50 minutes, usually more than 30; metaphase, 1 to 15, usually 2 to 10; anaphase 1 to 5, usually 2 to 3; telophase up to cytokinesis, 2 to 13, usually 3 to 6; telophasic reconstruction of daughter nuclei, 30 to 120; total, 70 to 180 minutes.

Detailed Description of the Behavior of the Chromosomes in Somatic Mitosis.[1]—In the present account we shall depart from the order usually followed in descriptions of mitosis. Instead of commencing with the resting nucleus and tracing the steps leading to the formation of two daughter resting nuclei, we shall begin the description with the fully formed chromosomes as they appear at the metaphase and follow them through anaphase, telophase, resting stage, and prophase to the next metaphase, when they are again clearly seen. This is done in order that the account of the telophasic transformation of the chromosomes to form a resting reticulum, and the prophasic condensation of the latter to form chromosomes, may be given without interruption, which seems advisable in view of the nature of certain questions which are later to be discussed in the light of chromiosome behavior.

Metaphase (Fig. 50, *A*).—As the chromosomes arrange themselves upon the spindle preparatory to their anaphasic separation their double nature is clearly evident. The two halves may lie very close together and in the case of long chromosomes may be somewhat twisted about each other. When they lie a little apart they may often show small connecting strands or anastomoses; in the immediately preceding stages (late prophase) the halves are usually pressed tightly together, so that these anastomoses appear to be due to mutual coherence at certain points when the halves move slightly apart after the disappearance of the nuclear membrane. As the double chromosomes take their places on the spindle, the spindle fibers become attached to them, not to all parts but to a particular portion of each. In the case of long chromosomes the point of attachment is often at about the middle, whereas in shorter ones it is commonly near one end. At their points of attachment to the spindle the double chromosomes all lie with their halves superposed (one half toward each spindle pole) and in a single plane; those portions to which no fibers are attached may extend in various directions with no regular arrangement.

Anaphase (Fig. 50, *B-D*).—The daughter chromosomes (the halves of the double chromosomes seen at metaphase) now begin to separate, first at the point of insertion, and gradually move away from the equatorial plane. Owing to the different locations of the points of fiber attachment, and also to the fact that the free ends of the chromosomes occupy various positions, the chromosomes, unless they are very short, may now

[1] This description is based on the author's accounts of somatic mitosis in *Vicia faba* (1913) and *Tradescantia virginiana* (1920). In these papers, especially in the first, there is presented a more extensive comparison of the results of other investigators than can be given here. Comparative studies have shown that in general the present description is widely applicable to mitotic phenomena in plants and animals, although many modifications in detail are known, particularly in forms with small chromosomes. A useful list of works on mitosis in angiosperms is given by Picard (1913).

be drawn into a number of peculiar shapes.   In the case of long chromosomes the portions to which the fibers are attached may have reached the poles of the spindle while the other portions are not yet separated at the equatorial plane.   As soon as the daughter chromosomes become entirely free from one another they quickly draw apart and contract into two dense masses, which are often actually farther apart than were

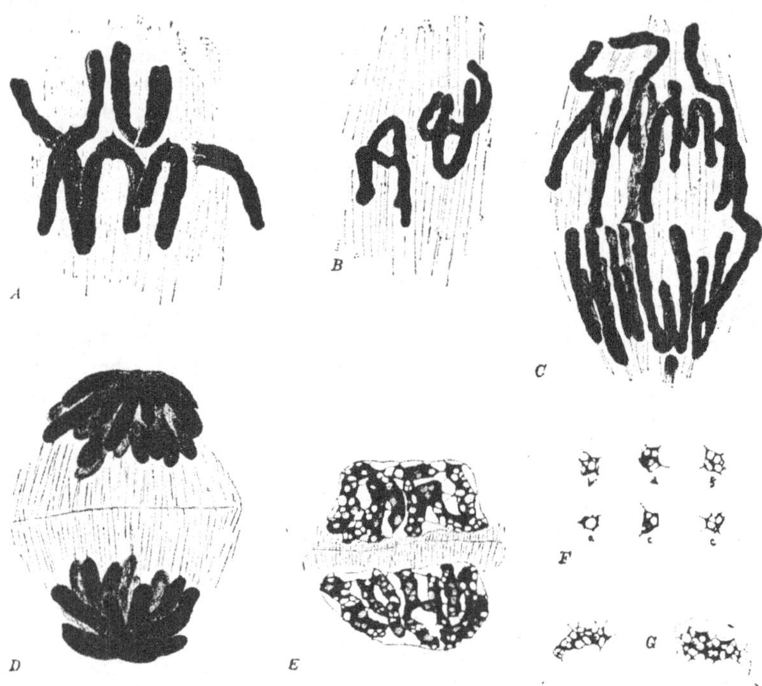

FIG. 50.—Somatic mitosis in *Tradescantia virginiana:* metaphase (*A*), anaphase (*B-D*), and telophase (*E-G*).   At *F* are shown cross sections of chromosomes in the stage shown at *E*.   × 1900.   (*After Sharp*, 1920.)

the poles of the spindle at metaphase.   In these masses the individual chromosomes can be distinguished only with great difficulty or not at all. With this stage, which has been referred to by Grégoire and Wygaerts (1903) as the *tassement polaire*, the anaphase ends and the telophase begins.

   Telophase (Fig. 50, *E*—Fig. 51, *I*).—After remaining tightly pressed together for a short time the chromosomes of each daughter group begin to separate, their individual boundaries again becoming visible.   As they do so they cohere at various points where their substance becomes drawn out to form anastomoses.   It seems clear that the main connections between the chromosomes of the reorganizing telophase nucleus are formed in this way, at least in mitoses showing a *tassement polaire* stage; but it is also probable, as several investigators have pointed out, that other anastomoses may grow out from one chromosome to another in the manner of pseudopodia (Boveri 1904; Gates 1912; Strasburger 1905; Dehorne 1911; Müller 1912; Lundegårdh).

Reactions taking place between the various chromosomes and especially between them and the cytoplasm now result in the production of the nuclear sap, or karyolymph. Between the outermost chromosomes and the cytoplasm and also within the chromosome group droplets of clear karyolymph appear, and where these come in contact with the cytoplasm a nuclear membrane is formed. As the karyolymph increases in amount the nucleus enlarges and the chromosomes become more widely separated.

The telophasic alveolation of the chromosomes, although it may in exceptional cases begin much earlier, usually commences at about the time the chromosomes first separate from one another in the early reorganization stages of the daughter nucleus. Within each chromosome vacuoles appear, first as obscure though rather sharply delimited circular or elongated areas. They lie not only along the axis but also near and against the periphery. This point is of importance in evaluating the claim advanced by certain investigators (Lundegårdh 1910, 1912; Fraser and Snell 1911; Fraser 1914; Digby 1919) that the vacuolation is median and results in a splitting of the chromosomes during the telophase, rather than in the prophase. While the vacuoles develop into open spaces through the breaking down of the thin portions bounding them the nucleus increases rapidly in volume, so that each chromosome appears as an irregular net-like band joined to its neighbors by fine anastomoses. Careful study of the details of these telophasic changes (see cross sections of chromosomes in Fig. 50, *F*) shows that the alveolation proceeds with little regularity, and that each chromosome becomes an alveolar and then reticulate body with nothing which can properly be called a longitudinal split.

In certain cases these internal changes, which result in the transformation of the chromosomes into a reticulum, and which as a rule do not begin until the telophase, may be initiated during the anaphase. In *Allium*, for instance, Miss Merriman (1904), Lundegårdh (1910, 1912*b*), and Němec (1910) all report that the vacuolation of the chromosomes begins at this time. Even more striking is the case of *Trillium* (Grégoire and Wygaerts 1903), in which the unusually large chromosomes may show vacuoles as early as the metaphase (Fig. 54, *A*). Internal changes of other types have also been described in anaphase chromosomes, but not with sufficient clearness to warrant their use in general interpretations.

According to Bonnevie (1908, 1911) the chromosomes of *Allium*, *Ascaris*, and *Amphiuma* each give rise to an endogenous spiral thread during the telophases, this spiral thread persisting through the resting stages until the next prophase, when it again condenses to form a chromosome (Fig. 54, *B*). In his work on *Salamandra* Dehorne (1911) asserted that each chromosome is represented at telophase by two interlaced spirals arising from an anaphasic split, and further that these double structures are associated in pairs and persist in this condition through the resting stages. These two conceptions have been criticized by Grégoire (1912), Sharp (1913), and de Smet (1914), who have interpreted such appearances as occasional aspects of the alveolized chromosomes without the significance attributed to them by Bonnevie and Dehorne.

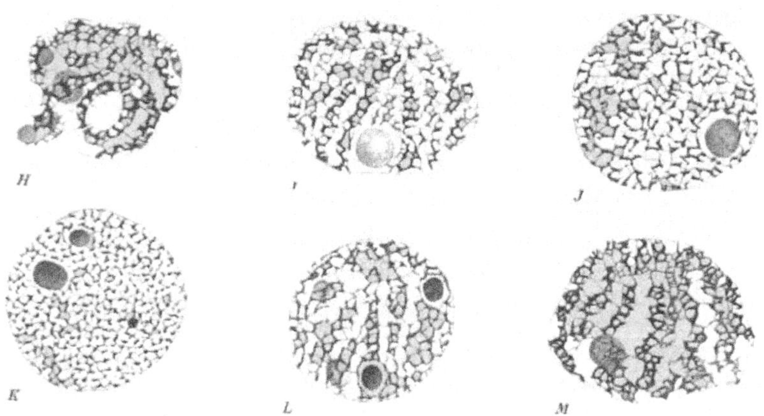

Fig. 51.—Somatic mitosis in *Tradescantia virginiana:* late telophase (*H, I*), interphase and resting stage (*J, K*), and early prophase (*L, M*). × 1900. (*After Sharp,* 1920.)

In the young telophase nucleus the chromosomes may become arranged in the form of a more or less continuous daughter spireme which is then transformed into the resting reticulum. This spireme stage, however, is not a necessary one; its absence is being reported with sufficient frequency to throw much doubt upon the view that it is a phenomenon of even general occurrence.

The nucleolus usually makes its appearance during the early telophase as a small droplet or as several such droplets which may later flow together. It seems to have little direct connection with the chromosomes, but there can be no doubt that its appearance is closely associated with their physiological activities.

As the telophasic changes proceed the chromosomes with their anastomoses gradually form a more and more uniform reticulum, in which, however, the limits of the component chromosomes can be distinguished until a very late stage.

Interphase (Fig. 51, *J*).—It often happens that in rapidly growing tissue, such as the meristem of the root tip, the mitoses succeed one another so rapidly that the telophasic changes may not proceed far enough to obscure the limits of the chromosomes in the reticulum before the changes of the ensuing prophase begin. In such tissue it is not always possible to tell whether a given nucleus will undergo further telophasic change or will at once enter upon the prophases. Such interphasic nuclei develop nucleoli, but karyosomes (in species which have these bodies) are usually not formed until a more advanced stage.

Resting Stage (Fig. 51, *J, K*).—In slowly growing tissue the successive mitoses do not follow one another with very great rapidity, and the telophasic changes are carried on until the condition characteristic of the typical resting nucleus is reached: the interphase here becomes the prolonged resting stage. The structure of the resting nucleus has been fully described in Chapter IV. In the reticulum the limits of the constituent chromosomes usually become indistinguishable, although it is known that in certain cases such nuclei, if properly sectioned and stained, may reveal heavier and lighter areas in the reticulum which represent respectively the chromosomes and the regions of anastomosis between them. The importance of these facts will be apparent in our treatment of the individuality of the chromosomes.

Prophase (Fig. 51, *L*–Fig. 52).—The first indication that the prophasic changes have begun is seen in the breaking down of the reticulum in certain regions. In the case of nuclei which show heavier and lighter areas in their reticula this breaking down occurs along the light portions. In view of what has been said concerning the origin of the reticulum at telophase it is apparent that the breaking up of the reticulum in the prophase represents in such cases the separation of the constituent chromosomes from each other along the lines of their telophasic union, and it has been inferred that a similar interpretation applies to those nuclei in which the reticulum is perfectly uniform or in which the nuclear material assumes more irregular forms. In this way there are developed from the resting reticulum a number of more or less distinct reticulate units, which, in view of their subsequent behavior, we know to be the chromosomes (Fig. 51, *L, M*). That these units are essentially the same as those which went to make up the reticulum at the preceding telophase seems highly probable; there can be little doubt on this point when the interphase is short.

The material of each reticulate unit (chromosome) now gradually condenses in a very irregular fashion about its open spaces and cavities. The thinner regions bounding these spaces and cavities become broken down, and the thicker portions remain as a very irregular zigzag thread of

uneven thickness, which soon begins to straighten out (Fig. 52, *P*).   At
the same time the material composing the thread becomes more evenly
arranged throughout its length, so that the chromosome eventually takes
the form of a single slender thread.   All of these changes—condensation, straightening, and equalization in thickness—may be seen going
on simultaneously in different chromosomes of the same nucleus, or even
in different portions of a single chromosome.

The formation of the slender prophase chromosomes from the reticulum in the above manner was first described in detail by Grégoire and
Wygaerts (1903) and Grégoire (1906), and new cases have since been
added.   The above writers, together with Němec (1910), Digby (1910),

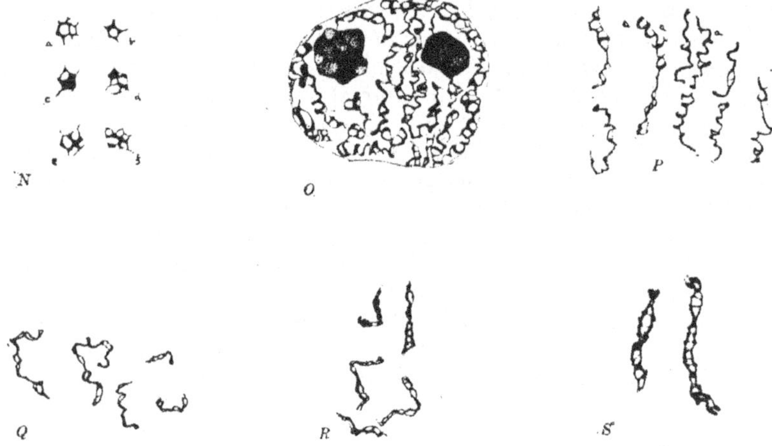

Fig. 52.—Somatic mitosis in *Tradescantia virginiana:* prophases.

At *N* are shown cross sections of chromosomes in the stage shown in Fig. 51, *M.*
× 1900.   (*After Sharp, 1920.*)

and Müller (1912), believe that the separated portions of the reticulum
may also condense directly into the slender threads without passing
through the very irregular zigzag stage above described.   It is probable
that both methods are followed in different cases, direct condensation
possibly being the rule in small nuclei.   The view of Bonnevie (1908,
1911) concerning the origin of the zigzag threads has already been
mentioned in the paragraphs on the telophase.   According to this worker
and to certain others (Wilson 1912*ab*) the chromatic material forms a
spiral thread within the chromosome during the telophase, this thread
uncoiling and emerging from the chromosome in the following prophase.
It is true that the zigzag threads occasionally have a strikingly regular
spiral aspect, but in view of the many other aspects observed and the
process which is known to give rise to them, it is probable that the

formation of spirals in the manner described by Bonnevie is at least very exceptional.

The true *longitudinal splitting of the chromosomes* is initiated in the slender threads of the early prophase (Figs. 52 and 53, Q). As soon as a thread becomes sufficiently equalized in diameter small vacuoles appear along its axis and rapidly develop into a more or less continuous split. Not all the threads, nor even all portions of the same thread, undergo the change at the same time. If we consider the whole nucleus at once, the processes of condensation, straightening, equalization, and

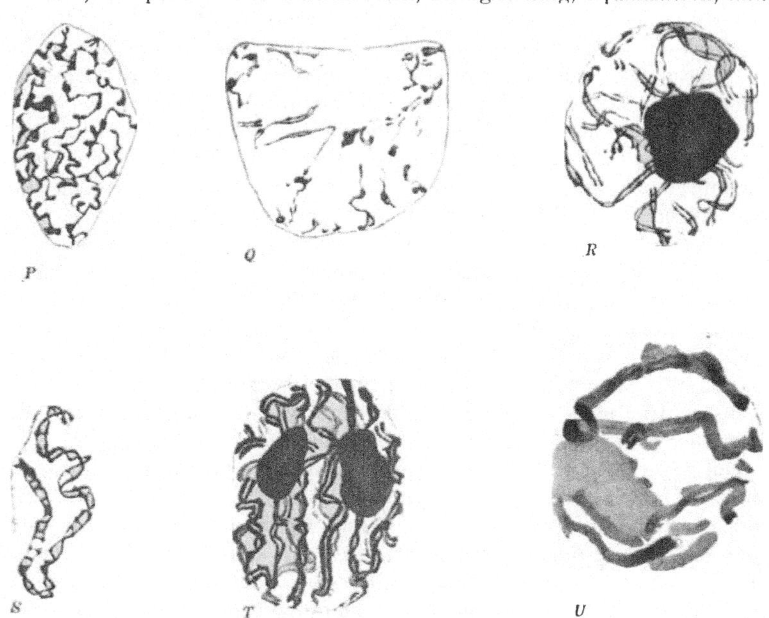

Fig. 53.—Somatic mitosis in *Vicia faba:* prophases.
Stages *P*, *Q*, and *S* correspond with *P*, *Q*, and *S* of Fig. 52. × 1650. (*After Sharp*, 1913.)

splitting are all going on simultaneously; only in a given small portion of a thread do they follow in definite sequence. Furthermore, as soon as the threads become equalized they at once begin to shorten and thicken, so that when vacuolation and splitting are a little delayed they occur in somewhat heavier threads.

The manner in which the vacuoles develop into the complete split should be carefully noted. The vacuoles quickly form openings which extend completely through the chromosome, so that the latter soon takes the form of two parallel strands connected by heavy cross pieces representing the portions between the original vacuoles (Figs. 52 and 53, *S*).

The material constituting the cross pieces gradually moves to the two side strands, the center portion of the cross piece becoming progressively thinner and the material accumulating on the side strands as a pair of chromatic lumps. Although some of the cross pieces may persist until a relatively late stage most of them soon disappear completely, and the material in the two chromatic lumps is gradually distributed more or less evenly along the parallel strands, which represent the daughter chromosomes resulting from the split.

The double chromosomes now shorten and thicken, forming the "thick spireme" so conspicuous in prophase nuclei (Fig. 53, *T*, *U*). As pointed out in the preliminary sketch of mitosis, the chromosomes in the prophase may form a more or less continuous spireme, but it is becoming increasingly apparent that this is not a universal phenomenon. It is certain that in many cases the chromosomes are separate from the first, and it seems therefore that any association in the form of a continuous spireme is a matter of secondary importance. As the shortening and thickening proceed the split may become obscured by the close association of the halves, but suitable methods reveal its presence.

While indications of spindle formation are appearing in the cytoplasm the nucleolus disappears and the nucleus begins to contract, so that the thick double chromosomes become very closely packed together. While the contraction is at its height the nuclear membrane disappears, after which the chromosomes loosen up as an irregularly arranged group. This contraction stage evidently does not occur in many mitoses: the membrane may disappear while the nucleus has its full size. However, when it does occur it is of very short duration, so that it may take place in more cases than has been supposed. After the disappearance of the nuclear membrane the spindle fibers establish connection with the chromosomes, which quickly become arranged with their halves in superposition at the equatorial plane, as described in the paragraph on the metaphase. This brings us to the point with which our description began.

It should be added that in many descriptions of mitosis, notably those presented in general text books, the chromosomes are said to split during the metaphase, after they have become arranged upon the spindle. Such a late development of the split may indeed occur in some cases, but it is not improbable that closer examination would often reveal the inception of the process at a much earlier stage. As has been pointed out in the foregoing description, the early formed split frequently becomes obscured during the later prophases owing to the shortening and thickening of the chromatin threads, and becomes conspicuous again only after the metaphase figure has been established.

Chromomeres.—One matter which should receive special attention is that of the chromomeres. It was held by Roux (1883) that the complicated process of mitosis is meaningless unless the chromatin is qualitatively different in the various regions of the nucleus, and that the arrangement of the material of the chromosome in the form of a long thread prior to its splitting is a means whereby all these qualities, arranged in a linear series in the thread, are equationally divided and distributed to the daughter nuclei. The theory of Balbiani (1876) and Pfitzner (1881), that the chromatin granules visible in the nuclear reticulum arrange themselves in a series in the chromosome and by their division initiate its splitting, had much to do with the formulation of this hypothesis. That the chromatic granules, or *chromomeres* (Fol 1891), represent the qualities of Roux is a theory which has been widely accepted by cytologists. It was the opinion of Brauer (1893) and many later workers that the granules or chromomeres, rather than the chromosomes themselves, are the significant units in the nucleus, and that their division is an act of reproduction. The division and separation of chromosomes was accordingly regarded as a means of distributing the daughter granules to the daughter cells. That the chromomere is made up of still smaller "chromioles" was held by Eisen (1899, 1900). Strasburger, Allen (1905), and Mottier (1907) also found the chromomere to be composed of smaller chromatic granules.

Fig. 54.

A, vacuoles in chromosomes at metaphase in *Trillium*. × 1800. (*After Grégoire and Wygaerts*, 1903.) *B*, spiral arrangement of chromatin material within the chromosomes of *Allium*. (*After Bonnevie*, 1911.) *C, D*, stages of chromosome splitting in *Najas marina*, showing chromomeres. × 2250. (*After Müller*, 1912.)

Although a large number of investigators, particularly those interested in the hereditary rôle of the chromatin, have placed much confidence in the importance of the chromomeres (Strasburger 1884, 1888), others have raised serious objections to the theory that they are significant units or individuals. Grégoire and Wygaerts (1903), Martins Mano (1904), Grégoire (1906, 1907), Maréchal (1907), Bonnevie (1908), Stomps (1910), Lundegårdh (1912), Sharp (1913, 1920), and others have found no such definite behavior on the part of the chromatin granules in the dividing chromosomes studied by them, and have suggested other explanations for the appearances observed. According to a modification of the chromomere theory adopted by Müller (1912) the portions of the thread between the chromomeres split first, the division of the chromomeres then following. It has been pointed out (Sharp 1913) that Müller's figures (Fig. 54, *C, D*), which are very similar to the later ones of Strasburger (1907), may be interpreted as steps in the division of a homogeneous chromatic thread by the formation of vacuoles, and that the chromomeres in this case are merely the cross pieces between the halves of the incompletely split chromosome, as described in the foregoing account of the prophase (Fig. 53, *S*).

It is becoming increasingly apparent that the distinction between chromatin granules and supporting thread is not so sharp as has been supposed, since the chromatic substance is often very fluid in consistency; and many have felt that the granules when present are far too inconstant in number and behavior to serve as the ultimate units which students of heredity hope to find. On the other hand, it should be said that the constancy in size and position of the chromomeres described by Wenrich (1916) for the grasshopper, *Phrynotettix* (Fig. 155), argues strongly for the hereditary significance of these bodies, some of which can be seen to retain their identity through the resting stages. But whatever their importance may be, the arrangement of the chromatic material in the form of a long slender thread and its accurate splitting into exactly similar halves are very suggestive in connection with the theory of Roux that many qualities are arranged in a row and all divided at the time of nuclear and cell division. This subject will receive further attention in the chapters dealing with heredity.

Summary.—The chromosomes, after having arrived at the poles of the achromatic figure, become irregularly alveolized during the telophase and form ragged net-like structures. These are joined to each other by fine anastomoses and so make up the continuous reticulum of the resting stage. In the next prophase this reticulum breaks up into separate small nets or alveolar units, each of which represents a chromosome. The units condense in a peculiar manner and become long slender threads. These threads undergo a longitudinal splitting. The double threads so formed shorten and thicken, and become the double chromosomes which are arranged on the spindle at metaphase. The two halves (daughter chromosomes) making up each double chromosome separate and pass to opposite poles during the anaphase.

The outstanding and significant feature of somatic mitosis is this: *each chromosome is accurately divided into two exactly equal longitudinal halves which are distributed to the two daughter nuclei. The two daughter cells thus receive exactly similar halves of the chromatin of the mother cell.* Furthermore, as will be shown below, there is good evidence for the view that the chromosomes maintain an individuality of some sort, so that, since all the nuclei of the body arise by the repeated equational division of a single nucleus, *all the somatic (body) cells are qualitatively similar in chromatin content: they contain representatives or descendants of each and every chromosome present in the first cell of the series.* The great theoretical importance of these facts will be apparent when we take up the subject of chromosome reduction, and the application of cytological phenomena to the problems of heredity.

THE INDIVIDUALITY OF THE CHROMOSOME

In later chapters the question of the significance of the nuclear structures in heredity is to be considered. In connection with this question it is of the highest importance to determine whether or not the chromosomes to which the reticulum gives rise in the prophase are in any real sense the same as those which went to make up the reticulum at the preceding telophase. That they do preserve their identity as individuals through the resting stage, arise only by division, and maintain therefore a genetic continuity throughout the life cycle, was held by van Beneden (1883), Rabl (1885) and Boveri (1887, 1888, 1891) many years ago, and since that time the idea has received the support of a large number of investigators. We shall now briefly review some of the evidences which have led the majority of cytologists to the view that the chromosomes, "if . . . not actually persistent individuals, as Rabl and Boveri have maintained, . . . must at least be regarded as genetic homologues that are connected by some definite bond of individual continuity from generation to generation of cells" (Wilson 1909).

The Frequent Persistence of Visible Chromosome Limits in the Resting Reticulum.—In the foregoing description of the behavior of the chromosomes in mitosis it was pointed out that in rapidly dividing tissue the telophasic alveolation of the chromosomes and their anastomosis to form the reticulum often do not proceed far enough during the interphase to obliterate the boundaries between the chromosomes, which separate again in the ensuing prophase without having lost their visible identity. In such nuclei there can be little doubt that the autonomy of the chromosome is preserved. In other cases, however, the telophasic transformation of the chromosomes is more complete and the resulting reticulum reacts very weakly to the stains, so that the limits of the constituent chromosomes disappear from view completely. Many workers have therefore objected to the statement that here also the chromosomes are present as individuals, although invisible. Haecker (1902) and Boveri (1904) pointed out that this objection may be met by assuming that it is the achromatic framework of the alveolized chromosome, and not necessarily the basichromatic fluid held within it, that maintains a structural independence. This view had the support of the earlier observation made by Boveri (1887a, 1888a, 1891; also 1909) and confirmed by Herla (1893), that the chromosomes in the segmenting egg of *Ascaris* have a certain arrangement when they build up the nuclear reticulum in the telophase and reappear from the reticulum in the same position at the next prophase.

Special emphasis was laid upon this interpretation by Maréchal (1904, 1907) as a result of his studies on the growth stage of animal oöcytes. At this period in the development of the ovum the chromosomes assume a finely branched form (Fig. 86, *C, D*) and their ordinary staining capacity is lost completely. Although the chromatic fluid may flow from the reticulum to the nucleolus and *vice versa*, and may periodically undergo chemical changes which radically alter its staining reactions, the achromatic chromosomal substratum nevertheless maintains an uninterrupted structural continuity. Such a transfer of the basichromatic material from the persistent reticulum to the nucleolus during the telophase, and to the reticulum again during the succeeding prophase, has also been

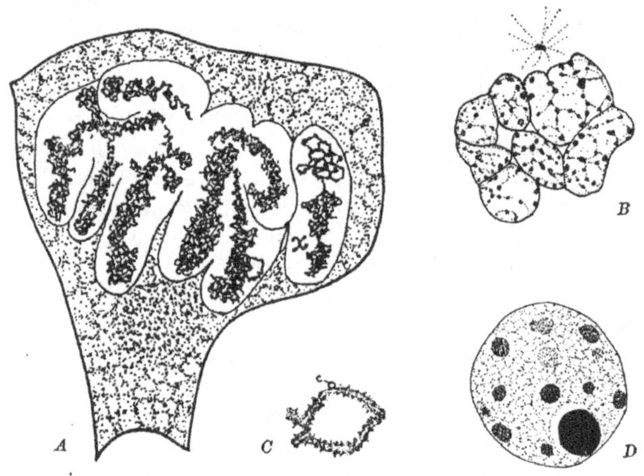

Fig. 55.—Some evidences for chromosome individuality.

A, chromosomal vesicles in *Brachystola magna;* x-chromosome in vesicle at right. (*After Sutton*, 1902.)　*B*, chromosomal vesicles in *Fundulus* embryo.　× c. 1800.　(*After Richards*, 1917.)　*C*, chromomere vesicle (*c*) on chromosome of *Chorthippus*.　× 1500. (*After Wenrich*, 1917.)　*D*, prochromosomes in *Pinguicula*.　× 4200.

observed by Strasburger (1907) and Berghs (1909) in the somatic nuclei of *Marsilia* (Fig. 17, *E*). The chromosome, as Maréchal urges, is not simply a mass of chromatin, but rather "a structure periodically chromatic;" hence the disappearance of stainable substance does not signify the loss of structural continuity on the part of the chromosome.

The "chromosomal vesicles" (Fig. 55, *A, B*) observed by certain investigators constitute valuable evidence in this connection. In the spermatogonia of the grasshopper, *Phrynotettix*, for example, Wenrich 1916) has shown that each of the alveolizing chromosomes forms its own vesicle about it at telophase, the several vesicles joining to form a common nucleus. In some cases the boundaries between the vesicles do not entirely disappear during the resting stages, and at the next prophase the chromatic material of each vesicle organizes in the form of a chromosome. The same condition is found in the nuclei of *Fundulus* (Richards 1917), *Crepidula* (Conklin 1902), and certain fish hybrids (Pinney 1918). From this it is evident that the morphological identity of the chromosomes has not been lost between mitoses, although a very different type of organization has been assumed.

In *Carex aquatilis* Stout (1912) has found a peculiar condition. Here the very small spherical chromosomes, which maintain a serial arrangement, are visible in the resting state, and can be traced continuously through all stages of the somatic and germ cell divisions with the exception of synizesis.

The interpretations of Bonnevie (1908, 1911) and Dehorne (1911), according to whom the chromosomes persist through the resting stage as spirals or double spirals, have been mentioned in the description of mitosis.

Prochromosomes.—Bodies known as prochromosomes have been described in the nuclei of a number of plants: in *Thalictrum, Calycanthus, Campanula, Helleborus, Podophyllum,* and *Richardia* by Overton (1905, 1909); in the Cruciferæ by Laibach (1907); in *Drosera* and other forms by Rosenberg (1909); in *Acer platanoides* by Darling (1914); in *Musa* by Tischler (1910); and in a number of other forms. These prochromosomes appear as small chromatic masses in the reticulum (Fig. 55, *D*), and correspond approximately in number to the chromosomes of the species. They are generally looked upon as portions of chromosomes which have not undergone complete alveolation, and as centers about which the chromosomes again condense at the next prophase. This interpretation is in all probability a valid one in many of the described cases, but in others the significance of such chromatic masses is questionable. In *Crepis virens* de Smet (1914), in harmony with the conclusions of Miss Digby (1914), finds them to be accumulations of material formed during the resting stages. If such is the case they are to be regarded as karyosomes.

Persistence of Parental Chromosome Groups After Fertilization.—In Chapter XII it will be shown that at fertilization there are brought together two sets of chromosomes, one set from each parent; and that in every nucleus of the resulting individual the chromosomes furnished by the two parents are present together, all of them dividing at every mitosis. When the chromosomes of the male parent are similar to those of the female parent it is usually impossible to distinguish them in the nuclei of the offspring. In a number of cases, however, such as *Crepidula* (Conklin 1897, 1901), *Cyclops* (Hæcker 1895; Rückert 1895), and *Cryptobranchus* (Smith 1919) (Fig. 109), the two parental groups are distinguishable on the mitotic spindle, and often at other stages, through several embryonal cell generations. It is in hybrids that this phenomenon is shown most strikingly. In hybrid fishes obtained by crossing *Fundulus* with *Menidia* Mœnkhaus (1904) was able to distinguish easily between the long (2.18 μ) chromosomes of *Fundulus* and the short (1 μ) ones of *Menidia*. Here, as in *Crepidula* and *Cyclops*, the paternal and maternal chromosomes form separate groups in the mitotic figure. A similar condition was seen by Tennent (1912) in hybrid echinoderms obtained by crossing in various ways *Moira, Toxopneustes,* and *Arbacia.* In the later cell-divisions the parental chromosomes mingle more or less, but are nevertheless distinguishable. In *Fundulus* × *Ctenolabrus* hybrids (Morris 1914; Richards 1916), as well as in the normally fertilized *Cryptobranchus* (Smith 1919), the chromatin contributions of the two parents are distinguishable even in the resting nuclei.

Size and Shape of Chromosomes.—One of the most striking evidences favoring the theory of individuality has been found in those plants and animals which show constant differences in size and shape among the various members of each parental chromosome group, so that particular chromosomes are recognizable in the group appearing at each mitosis. Since each parent furnishes a set of chromosomes to the new individual, each kind of chromosome is present in duplicate in the nuclei of this individual: it is therefore customary to speak of them as being present in pairs, although at most stages of the life history there is ordinarily no actual spatial pairing.

Since the description of the chromosomes of *Brachystola* by Sutton in 1902 (Fig. 101) the reported cases in which the different pairs of the chromosome complement possess different characteristic sizes and shapes have become increasingly numerous. This is notably true of insect cytology, as is evident in a review of the extensive researches of McClung (1905, 1914, 1917), Robertson (1916), Harman (1915), Carothers (1917), and many others. In the sea urchin, *Echinus*, Baltzer (1909) found that the 36 chromosomes have constant differences in length and shape, some being hooked and some horseshoe-shaped. In the flatworm, *Gyrodactylus*, (Gille 1914) there are six pairs, all different in length. In *Ambystoma tigrinum* Parmenter (1919) finds 14 pairs of graded sizes. In plants may be cited the cases of *Crepis virens* (Rosenberg 1909; de Smet 1914; M. Nawaschin 1915) (Fig. 56 *bis, A*), which has three pairs of different size; *Vicia faba* (Sharp 1914; Sakamura 1915), with five short pairs and one long pair (Fig. 56); and *Najas* (Tschernoyarow 1914), in which there are seven distinguishable pairs (Fig. 56 *bis, B*). In *Najas* the smallest pair is attached to one of the larger pairs: Sakamura (1920) thinks that these together are really a single pair with pronounced constrictions.

Not only may certain chromosomes be distinguished on the basis of comparative length, but in some cases there may be other characteristics which serve as marks of identification. In the chromosomes of many plants and animals there are pronounced constrictions in some of the

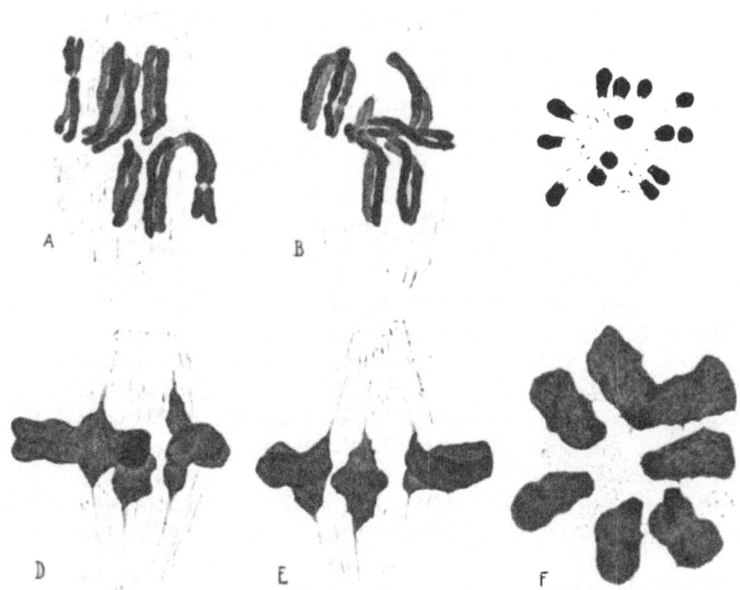

Fig. 56.—The chromosome complement of *Vicia faba.*

A, B, two successive sections of a mitotic figure in the root tip, showing together the 12 split chromosomes, 2 of them about twice as long as the other 10. *C,* cross section of the group of chromosomes at anaphase: each of the long chromosomes, being drawn polewards by the middle, shows both ends, making the number apparently 14. *D, E,* two successive sections of a heterotypic figure in the microsporocyte, showing the 6 bivalents; the large one is at the left. *F,* polar view of heterotypic mitosis at metaphase, showing the 6 bivalents. × 1400.   (*Original.*)

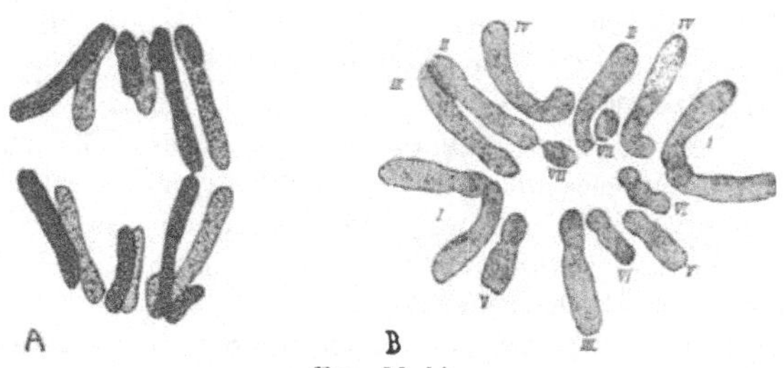

Fig. 56 *bis.*

A, anaphase of somatic mitosis in *Crepis virens,* showing 2 long, 2 medium sized, and 2 short chromosomes passing to each pole. (*After Rosenberg,* 1920.) *B,* the chromosome complement in a somatic cell of *Najas major,* showing the 7 homologous pairs. (*After Tschernoyarow,* 1914.)

members of the group. It has been shown in certain instances that these constrictions have constant positions in the chromosome. A careful study of this phenomenon has been made by Sakamura (1915, 1920). In *Vicia faba*, for example, he finds that each of the two long chromosomes ("M-chromosomes") of the somatic group has two constant constrictions, one at the middle and one near the end ("m-constriction" and "e-constriction") (Fig. 56, *A*). The m-constriction marks the point of attachment of the spindle fibers. There are also end-constrictions in 8 of the 10 short chromosomes. On the basis of the widespread occurrence of constrictions in the chromosomes of both plants and animals Sakamura has interpreted a number of puzzling phenomena, such as the apparent variation in chromosome number within the species (see below) and certain features of the reduction process (Chapter XI)

Such regularly situated constrictions have also been demonstrated in *Fritillaria tenella* by S. Nawaschin (1914). Here they are present at the middle of the largest chromosomes, nearer one end in the medium-sized chromosomes, and close to the end of the smallest ones. In *Crepis virens* (M. Nawaschin 1915) there are constrictions near one end in two of the three chromosomes of the haploid group in the pollen grain, in four of the six chromosomes of the diploid group in the somatic cells, and in six of the nine chromosomes of the triploid group in the endosperm cells. Such a definiteness in the location of constrictions was also seen earlier by Agar (1912) in the chromosomes of the fish, *Lepidosiren.*

Somewhat similar evidence has been brought forward by Wenrich (1916), who finds that the chromatic lumps, or chromomeres, have a striking constancy in position as well as in size in the chromosomes of *Phrynotettix* (Fig. 155). Wenrich (1917) also reports that the small "chromomere vesicles" attached to the chromosomes of certain orthopterans always appear at definite points along the chromosome (Fig. 55, *C*).

It therefore appears that the chromosomes of a given group or complement not only maintain a genetic continuity from cell to cell, but are also in some way qualitatively different from one another. They are consequently said to have a *specificity* as well as an *individuality,* or *continuity.* The relatively constant positions of the constrictions, chromatic lumps, and chromomere vesicles afford further visible evidences that the chromosome may possess some kind of lengthwise differentiation, a fact which, if clearly demonstrated, would be of the highest importance in connection with current views of the rôle of the chromosomes in heredity. (See Chapter XVII.) The significance of chromosome constrictions in this respect has been emphasized by Janssens (1909), S. Nawaschin (1915), and Sakamura (1920).

Chromosome Number.[1]—It was long ago noticed by Boveri, van

[1] For lists of chromosome numbers in plants see Ishikawa (1916) and Tischler (1916). For the numbers in animals see Harvey (1916, 1920).

Beneden, and Strasburger that the number of chromosomes in any given species is relatively constant. It was largely upon this fact that the theory of chromosome individuality was originally based: the fact that the number of chromosomes appearing at every mitosis is almost invariably the same was taken to mean that the structural identity of the chromosomes is never lost. Certain observers (Fick 1905, 1909) have held that the apparent constancy in number is not due to a structural continuity or individuality of any sort, but rather to the fact that the successive nuclei have a relatively uniform amount of nuclear material, the chromosomes "crystallizing out" of this material in each prophase and going into solution at the close of mitosis. This idea was especially developed by Della Valle (1909, 1912ab), who described the formation of chromosomes by the aggregation of fluid crystals during the prophase. These chromosomes he held to be in no sense morphologically continuous individuals, but only temporary chromatic accumulations which are inconstant in number and lose their identity in the telophase. Della Valle's interpretation of chromosome formation has been criticized by a number of writers and his position shown to be untenable by Montgomery (1910), McClung (1917), and Parmenter (1919).

Some of the experiments on echinoderm eggs with which Boveri (1895, 1902, 1903, 1904b, 1905, 1907) and others supported the theory of chromosome individuality may be briefly reviewed.

Boveri found that if the number of chromosomes is increased or decreased by artificial means the altered number appears at every mitosis thereafter. (a) An enucleate egg fragment may be entered by a spermatozoön, and may then develop into a larva with half the normal number of chromosomes in every cell. (b) In another experiment the unfertilized egg of a sea urchin was caused to undergo division by artificial means, after which a spermatozoön was allowed to enter one of the blastomeres (daughter cells). A larva resulted in which one-half of the cells had regularly 18 chromosomes (half the normal number) while the other half had the normal 36. (c) Two spermatozoa occasionally fertilized one egg: the cells of the resulting larvæ had 54 chromosomes, the triploid number. Abnormal mitotic figures were often formed in such dispermic eggs, bringing about an irregular distribution of the chromosomes. For example, a quadripolar spindle was produced, separating the 54 split chromosomes (108 daughter chromosomes) into four groups, with 18, 22, 32, and 36 chromosomes respectively (Fig. 127 bis). The resulting abnormal larva ("pluteus") showed these four chromosome numbers in the cells of four different regions of its body. Boveri (1914) later suggested that malignant tumors might be due to such abnormal chromosome distribution. (d) The number of chromosomes was doubled by shaking the eggs while the chromosomes were split during the early stages of celldivision. In this manner larvæ were produced with 72 chromosomes, the tetraploid number, in all of their cells.   (*e*) In the threadworm, *Ascaris megalocephala*, fertilization of an egg of the variety *bivalens* (two chromosomes) by a spermatozoön of the variety *univalens* (one chromosome) resulted in a larva with three chromosomes in all its cells, the chromosome contributed by the male parent being distinguishable from the other two (Boveri 1888*a*; Herla 1893; Zoja 1895).

Results such as the above led Boveri to the conclusion that the number of chromosomes arising from the reticulum in prophase is directly and exclusively dependent upon the number that went to make it up in the preceding telophase.   If a nucleus is reconstructed in the telophase by an abnormal number of chromosomes as the result of a disturbance of the

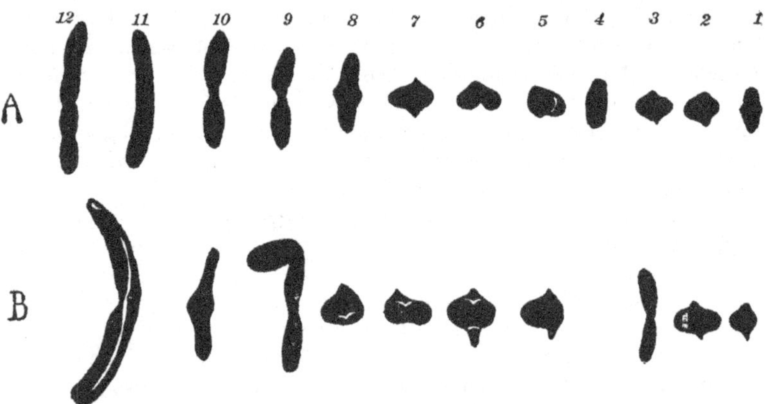

FIG. 57.—The chromosome complement of *Hesperotettix viridis*.
A, the 12 bivalent chromosomes of the spermatocyte, including the accessory chromosome (No. 4.)   *B*, complement from another individual, showing two "multiple chromosomes."   Nos. 11 and 12 have united temporarily, as have also Nos. 4 and 9.   × 1800. (*After McClung*, 1917.)

mitotic process, the altered number invariably appears in the succeeding prophase: if extra chromosomes are present they are not eliminated in any way during the resting stages, and if chromosomes have been lost during abnormal mitosis they are not replaced.   These conclusions have been strikingly confirmed by Sakamura's (1920) work on cells subjected to the influence of chloral hydrate and other agencies causing aberrant chromosome behavior.

Variations in Number.—Although the number of chromosomes in a given species is on the whole remarkably constant, departures from normal numbers are occasionally observed.   Strasburger (1905) believed that the number, though determined by heredity, is not so rigidly fixed that all variation in the vegetative cells is excluded; only in the reproductive cells did he hold constancy in number to be necessary.   Much light has been recently thrown upon such apparent variations in number by McClung (1917) and Miss Holt (1917) in their researches on multiple chromosomes and chromosome complexes.

McClung finds in his analysis of the chromosome groups of the orthopterans *Hesperotettix* and *Mermiria* that temporary associations often occur between various members of a group, with the resulting formation of "multiple chromosomes" and a consequent decrease in the apparent number. In *Hesperotettix*, for instance, the cells normally have 12 pairs of chromosomes, but because of the formation of such multiple chromosomes individuals with apparently 11, 10, or 9 pairs are frequently found (Fig. 57). For a given individual the number so formed is exactly constant, since the members of a multiple remain together in all the cells of the body; but for the species it is variable within certain limits, owing to the varying numbers of chromosomes which may become involved in such multiple combinations. In all cases the full number of chromosome pairs is present, but some of them are so combined that there is an apparent, though not actual, variation in the number. A similar condition is found in other forms by Robertson (1916).

In *Culex* there are three pairs of chromosomes in the somatic cells. During a certain stage in the insect's metamorphosis it has been shown by Miss Holt (1917) that the chromosomes may split repeatedly, giving cells with much larger numbers—up to 72 in some cases. These larger numbers, however, are nearly always multiples of three, indicating that the subdivision of the chromosomes is an orderly process. The daughter chromosomes, moreover, that are formed by the subdivision of each of the original six, remain more or less closely associated as a "multiple complex," which behaves as a single individual in mitosis. It therefore appears that the three pairs of chromosomes "are made up of quite distinct individuals differing from each other to such a degree that chromatin split from one cannot associate itself with that from another pair. . . . Chromosome individuality, alone, can account for these conditions."

Somewhat similar evidence has been brought forward by Hance (1917, 1918ab). Hance finds that the chromosome number in the spermatogonia of the pig is regularly 40, whereas in the somatic cells it varies from 40 to 57. Similarly in *Œnothera scintillans*, which has 15 chromosomes in its microsporocytes, there may be from 15 to 21 chromosomes in the somatic cells. Measurements of the members of the various chromosome groups show that the larger numbers are due to a fragmentation, probably of the larger chromosomes, in the somatic cells. Such fragments divide normally, and it appears probable that the fragments of a single original chromosome are held together by colorless portions and behave as a unit, much as do the multiple complexes of *Culex*.

Sakamura (1920) believes that the chief reason for frequently reported inconstancies in chromosome number is to be found in the chromosome constrictions, which under certain conditions become especially pronounced and temporarily divide one or more of the chromosomes of the group into loosely connected smaller parts. This suggestion, which Sakamura supports with much direct evidence, is probably one of the most fruitful which has been made in this connection.

The theory of chromosome individuality is believed by McClung and Hance to be strengthened, rather than weakened, by such instances of numerical variation as those described above. McClung emphasizes the point that the composition of a given chromosome can be fully understood only if something is known of its genetic history, for what appears as a chromosome may often be either an aggregation of two or three chromosomes, or, on the other hand, only a portion of the true chromosome individual. How widely this interpretation may be applicable to other reported cases of numerical variation and to chromosome structure in general cannot at present be stated, but it promises to lead to significant results.

Discussion and Conclusions.—The author's views on the subject of the individuality of the chromosomes can be most effectively stated in the words of McClung (1917):

" . . . the practical matter before us is to decide whether the metaphase chromosomes of two cells are individually identical organic members of a series because they were produced by the observed reproduction of a similar series of the parent cell, or whether the resemblance is independent of this genetic relation and due to chance association of indifferent materials, or to a reconstituting action of the cell as a whole."

"If it were possible for chromosomes to reproduce themselves and still preserve their physical configuration unchanged, there would probably be little question of their continuity and individuality—the demonstration would be self-evident. But it happens that the necessities of the case require that each newly produced chromosome should take part in the formation of a new nucleus, through whose activities the cell as a whole and each chromosome, individually, is enabled to restore the volume diminished by the act of division. During this process the outlines of the chromosomes become materially changed and in their extreme diffusion can no longer be traced in many cases. Because of our limitations in observational power they appear to be lost as separate individuals and we are thus deprived of the simple test of observed continuity. Later, in the same cell, there reappears a series of chromosomes severally like those which seemed to disappear during the period of metabolic activity. We confront two alternative explanations for this reintegration of the chromosomes; either they actually persist as discrete units of extremely variable form, or they are entirely lost as individual entities and are reconstituted by some extrinsic agency. There is no other possible explanation and we must weigh the facts for one or the other of the alternatives.

All the facts which indicate order and system in chromosome features speak for the former, those which demonstrate variability and indefiniteness, for the latter.   The case for discontinuity is strongest in the absence of any chromosome order, and becomes progressively weaker with the establishment of definiteness and precision in form and behavior."

"So far as I can see there is no half way ground between the assumption that the chromosomes are definite, self-perpetuating organic structures and the other which presents them as mere incidental products of cellular action.   According to one view individual chromosomes are descendents of like elements and possess certain qualities and behavior because of their material descent, the visible mechanism for which is the process of mitosis: according to the other any similarities that may exist in the complexes are the result of chance aggregations of non-specific materials.   It is a choice between organization and non-organization in the last analysis, at least in terms of cellular structures.   To attempt the substitution of a conception of molecular organization, which is beyond the experience of the biologist and which exceeds the present powers of the chemist to analyse, is to cast aside all hope of solving the problem of cellular action, because it is necessary to understand, not only the physical and chemical phenomena involved, but also their different forms in the various parts of the cell."

"That the chromosomes do not maintain a compact and easily recognizable form in the interval between mitoses is accepted by many . . . biologists as proof that they no longer exist as entities.   All the other manifold indications of character and continuity do not weigh against this apparent loss of identity.   Doubtless it would be more satisfying if we could at all times perceive the chromosomes in unchanging form in all stages of cellular activity, but why we should demand this condition as a test of individuality in the chromosomes when we unhesitatingly admit the unity of the organism in all the varied changes of its development from a single cell, through such complexities of change and metamorphosis as to give rise to doubts of even the phyletic position of some stages, it is difficult to see.   Being organic, the chromosomes must change their form, they must suffer division of their substance and they are obliged to restore this loss through metabolic changes.   Since these changes of substance take place at surface contacts there is an obvious advantage in increased superficies and, in common with other, larger structural elements, the chromosomes become extended and their substances are diffused.   In this state their boundaries may not be well defined and this circumstance has been seized upon as a disproof of their continuity."

"Since it is not possible to observe directly the action of the chromosome we are obliged to make use of indirect evidence, seeking parallels between elements of structure and action in the chromosomes, and the mass effect of cellular action as exhibited in the so-called body characters.   Such a method is justified by all other experience in tracing relations between structure and function in organisms, and while it apparently resolves the organism into parts of greater or less independence, has given us our best conceptions of it as a whole."

"What is postulated . . . is that the chromosomes are self-perpetuating entities with individual peculiarities of form and function to identify them. Characteristics of form and behavior we see; certain very definite parallels between these and the manifestations of somatic characters exist beyond question; provision for the perpetuation of the organic unity of the individual chromosomes is found in the process of mitosis; the actual direct result of its operation appears in the uniform conditions of the complex in the individual animal; the extension of this beyond the organism to the group and the means for it in the phenomena of maturation and fertilization are easily established by observation; the age old existence of all these circumstances is revealed by the near approach to uniformity in the chromosome complex of the multitude of species of unnumbered individuals constituting a family. And yet, in the face of this overwhelming mass of evidence indicative of order, system and specific chromosome organization, some conceive only the action of ordinary chemical forces, or the chance association of indifferent substances, while others, over impressed with the thought of a general coördinating force in the organism, deny significance to the orderly play of its cellular parts."

"It is my belief that the observed act of reproduction, by which the organization of the chromosomes is materially transmitted in each mitosis, together with all facts indicating extensive distribution of given conditions, definiteness of organization, uniformity of behavior and consistence of deviation from the normal, are so many clear indications of the individual character of the chromosomes. Transmutation of form, even to an extreme degree, can not be held as a valid argument against a persistent individuality. A consideration of the criteria applied to larger organic aggregates well supports this view. Such objects are said to possess individuality when they exhibit a more or less definite unity which is persistent and characterized by peculiarities of form and function. Most clearly defined is this individuality when it may be perpetuated through some form of reproduction to find expression in new units of similar character. The term does not connote unchangeability, and there may be fusions with more or less loss of physical delimitations, followed by separation, even after exchange of substances. *The test of individuality is material continuity, but it does not necessarily involve complete or entirely persistent contiguity.*[1] An organism may bud off new individuals similar to itself, the substance of its body differs from time to time, movements of parts take place, fragmentation occurs, extreme attenuation or extension of substance is found, even separation and recombination of parts may happen and yet the individual maintains itself. What it may have been in the past, what its possibilities of future development are, what potentialities of multiplied individuality it suppresses do not affect the reality of its individuality. It is, as Huxley says, 'a single thing of a given kind.' If one such thing divides into two, there are two individuals; if two unite into one indistinguishably there is a single individual; if a fusion of two things occurs in part, without loss of physical configuration, there are still two individuals in existence. Only when the substance of one thing disappears or becomes incorporated integrally into the organization of another does its individuality depart.

If all these variations of physical state may occur in the history of an organism without sacrifice of individuality, there can be no reason for urging them against a conception of the individuality of the self-perpetuating chromosomes."

Bibliography 8

Somatic Mitosis. Individuality of Chromosomes

AGAR, W. E. 1912. Transverse segmentation and internal differentiation of chromosomes. Quar. Jour. Micr. Sci. **58**: 285–298. pls. 12, 13.

---

[1] Italics ours.

BALBIANI, E. G.   1876.   Sur les phénomènes de la division du noyau cellulaire. Compt. Rend. Acad. Sci. Paris **83**: 831–834.

BALTZER, F.   1909.   Die Chromosomen von *Strongylocentrotus lividus* und *Echinus microtuberculatus*.   Arch. Zellf. **2**: 549–632.   pls. 37, 38.   fig. 25.

VAN BENEDEN, E.   1876.   Recherches sur les Dicyémides, survivants actuelles d'un embranchement Mésozoaires.   Bull. Acad. Roy. Belg. **41**: 1160–1205; **42**: 35–97.   pls. 3.

  1883.   Recherches sur la maturation de l'oeuf, la fécondation et la division cellulaire.   Arch. de Biol. **4**.

  1884.   (van Beneden et Julin).   La spermatogénèse chez *l'Ascaride mégalocéphale*. Bull. Acad. Roy. Belg.   III **7**: 312–342.

  1887.   (van Beneden et Neyt).   Nouvelles recherches sur la fécondation et la division mitosique chez l'*Ascaride mégalocéphale*.   Ibid. III **14**: 215–295.   pl. 1.

BERGHS, J.   1909.   Les cinèses somatiques dans le *Marsilia*.   La Cellule **25**: 73–84. 1 pl.

BONNEVIE, K.   1908.   Chromosomenstudien.   I. Arch. Zellf. **1**: 450–514.   pls. 11–15.

  1911.   Chromosomenstudien.   II.   Chromatinreifung in *Allium cepa*.   Ibid. **6**: 190–253.   pls. 10–13

BOVERI, T.   1887*a*.   Ueber die Befruchtung der Eier von *Ascaris megalocephala*. Sitzber. Gesell. Morph. Phys. München **3**.

  1887*b*.   Ueber den Anteil des Spermatozoön an der Teilung des Eies.   Ibid.

  1887*c*.   Ueber Differenzierung der Zellkerne während der Furchung des Eies von *Ascaris megalocephala*.   Anat. Anz. **2**: 688–693.

  1887*d*.   Zellen-Studien.   I.   Die Bildung der Richtungskörper bei *Ascaris megalocephala* und *Ascaris lumbricoides*.   Jen. Zeitschr. **21**: 423–515.   pls. 25–28.

  1888*a*.   Zellen-Studien.   II.   Ibid. **22**: 685–882.   pls. 19–23.

  1888*b*.   Ueber partielle Befruchtung.   Sitzber. Ges. Morph. Phys. München **4**.

  1890.   Zellen-Studien.   III.   Ueber das Verhalten der chromatischen Kernsubstanz bei der Bildung der Richtungskörper und bei der Befruchtung.   Jen. Zeitschr. **24**: 314–401.   pls. 11–13.

  1891.   Befruchtung.   Ergebn. d. Anat. u. Entw. **1**: 386–485.   figs. 15.

  1902.   Ueber mehrpolige Mitosen als Mittel zur Analyse des Zellkerns.   Verhandl. Phys.-Med. Ges. Würzburg **35**.

  1904*a*.   Ergebnisse über die Konstitution der chromatischen Kernsubstanz.   Jena.

  1904*b*.   (Boveri und Stevens.)   Ueber die Entwicklung dispermer Ascariseier. Zool. Anz. **27**: 406–417.

  1907.   Zellen-Studien.   VI. Die Entwicklung dispermer Seeigel-Eier.   Jena.

  1909.   Die Blastomerenkerne von *Ascaris megalocephala* und die Theorie der Chromosomenindividualität.   Arch. Zellf. **3**: 181–268.   pls. 7–11.

  1914.   Zur Frage der Entstehung maligner Tumoren.

BRAUER, A.   1893.   Zur Kenntniss der Spermatogenese von *Ascaris megalocephala*. Arch. Mikr. Anat. **42**: 153–212.   pls. 11–13.

BÜTSCHLI, O.   1875*a*.   Vorl. Mitteilung über Untersuchungen betreffend die ersten Entwicklungsvorgänge im befruchteten Ei von Nematoden und Schnecken. Zeit. Wiss. Zool. **25**: 201.

  1875*b*.   Vorl. Mitteilung einiger Resultate von Studien über Conjugation der Infusorien und die Zellteilung.   Ibid. **25**: 426.

CAROTHERS, E. E.   1917.   The segregation and recombination of homologous chromosomes as found in two genera of Acrididæ (Orthoptera).   Jour. Morph. **28**: 445–521.   pls. 14.   figs. 5.

CONKLIN, E. G.   1897.   The embryology of *Crepidula*.   Jour. Morph. **13**: 1–226. pls. 9.

  1919–1920.   The mechanism of evolution in the light of heredity and development. Sci. Mo. **9**: 481–505; **10**: 52–62, 170–182, 269–291, 388–404, 498–515.

DARLING, C. A.   1914.   Prochromosomes in synapsis.   Science **41**: 182–183.

DEHORNE, A.   1911.   Recherches sur la division de la cellule.   I.   La duplicisme constant du chromosome somatique chez *Salamandra maculosa* Lour. et chez *Allium cepa* L.   Arch. Zellf. **6**: 613–639.   pls. 35, 36.   figs. 2.

DELLA VALLE, P.   1909.   L'organizzazione della cromatina studiata mediante il numero dei cromosomi.   Archivio Zool. Ital. **4**: 1–177.   1 pl.

1912*a*.   La morphologia della cromatina dal punto di vista fisico.   Ibid. **6**: 37–321.   pls. 4, 5.   figs. 75.   diagrams 9.

1912*b*.   Die morphologie des Zellkerns und die Physik der Kolloide.   Zeitschr. f. Chemie u. Indus. d. Kolloide **12**: 12–16.

DIGBY, L.   1910.   The somatic, premeiotic, and meiotic nuclear divisions of *Galtonia candicans*.   Ann. Bot. **24**: 727–757.   pls. 59–63.

1914.   A critical study of the cytology of *Crepis virens*.   Arch. Zellf. **12**: 97–146.   pls. 8–10.

1919.   On the archesporial and meiotic mitoses of *Osmunda*.   Ann. Bot. **33**: 135–172.   pls. 8–12.   1 fig.

EISEN, G.   1899.   The chromoplasts and the chromioles.   Bot. Cent. **19**: 130–136.   figs. 5.

1900.   The spermatogenesis of *Batrachoseps*, polymorphous cells.   Jour. Morph. **17**: 1–117.   pls. 1–14.

FARMER, J. B. and SHOVE, D.   1905.   On the structure and development of the somatic and heterotype chromosomes of *Tradescantia virginica*.   Quar. Jour. Micr. Sci. **48**: 559–569.   pls. 42, 43.

FICK, R.   1905.   Betrachtungen über die Chromosomen, ihre Individualität, Reduktion und Vererbung.   Arch. Anat. u. Phys. Suppl. 179–228.

1909.   Bemerkungen zu Boveris Aufsatz über die Blastomerenkerne von *Ascaris* und die Theorie der Chromosomen.   Arch. Zellf. **3**: 521–524.

FLEMMING, W.   1880.   Beiträge zur Kenntniss der Zelle und ihre Lebenserscheinungen.   II.   Arch. Mikr. Anat. **19**.

1882.   Zellsubstanz, Kern und Kernteilung.   Leipzig.

1891–1897.   Zelle.   Ergeb. d. Anat. u. Entw. **1–7**.   (Review.)

FOL, H.   1873.   Die erste Entwicklung des Geryoniden-Eies.   Jen. Zeitschr. **7**.

1891.   Die "Centrenquadrille," ein neue Episode aus der Befruchtungsgeschichte.   Anat. Anz. **6**: 266–274.   figs. 10.

FRASER, H. C. I. and SNELL, J.   1911.   The vegetative divisions in *Vicia faba*.   Ann. Bot. **25**: 845–855.   pls. 62, 63.

FRASER, H. C. I.   1914.   The behavior of the chromatin in the meiotic divisions of *Vicia faba*.   Ann. Bot. **28**: 633–642.   pls. 43, 44.

GATES, R. R.   1912.   Somatic mitoses in *Œnothera*.   Ann. Bot. : 993–1010.   pl. 86.

GILLE, K.   1914.   Untersuchungen über die Eireifung, Befruchtung und Zellteilung von *Gyrodactylus elegans* var. *Nordmann*.   Arch. Zellf. **12**: 415–456.   pls. 32–34.

GRÉGOIRE, V.   1906.   La structure de l'élément chromosomique au repos et en division dans les cellules végétales.   La Cellule **23**: 311–353.   pls. 1, 2.

1912.   Les phénomènes de la métaphase et de l'anaphase dans la caryocinèse somatique.   A propos d'une interprétation nouvelle.   Ann. Soc. Sci. Bruxelles **34**.

GRÉGOIRE, V. et WYGAERTS, A.   1903.   La reconstitution du noyau et la formation des chromosomes dans les cinèses somatiques.   La Cellule **21**: 7–67.   pls. 2.

1907.   La formation des gemini hétérotypiques dans les végétaux.   Ibid. **24**: 369–420.   pls. 2.

HAECKER, V.   1896.   Ueber die Selbständigkeit der väterlichen und mütterlichen Kernbestandteile während der Embryonalentwicklung von *Cyclops*.   Arch. Mikr. Anat. **46**: 579–617.   pls. 28–30.

1902. Ueber das Schicksal der elterlichen und grosselterlichen Kernanteile. Jen. Zeitschr. **37**: 299–400. pls. 17–20. figs. 16.

HANCE, R. T. 1917. The diploid chromosome complexes of the pig (*Sus scrofa*) and their variation. Jour. Morph. **30**.

1918a. Variations in the number of somatic chromosomes of *Œnothera scintillans* de Vries. Genetics **3**: 225–275. pls. 7.

1918b. Variations in somatic chromosomes. Biol. Bull. **35**.

HARMAN, M. T. 1915. Spermatogenesis in *Paratettix*. Science **41**: 440.

HARPER, R. A. 1919. The structure of protoplasm. Am. Jour. Bot. **6**: 273–300.

HARVEY, E. B. 1916, 1920. A review of the chromosome numbers in the Metazoa. Jour. Morph. **28**: 1–63; **34**: 1–68.

HEIDENHAIN, M. 1894. Neue Untersuchungen über die Centralkörper und ihre Beziehungen zum Kern und Zellprotoplasma. Arch. Mikr. Anat. **43**: 423–758. pls. 25–31.

HERLA, V. 1893. Étude des variations de la mitose chez l'*Ascaride megalocephale*. Arch. de Biol. **13**: 423–520. pls. 15–19.

HERMANN, J. 1891. Beiträge zur Lehre von der Entstehung der karyokinetischen Spindel. Arch. Mikr. Anat. **37**: 569–586. pl. 31. 2 figs.

HERTWIG, O. 1875. Beiträge zur Kenntniss der Bildung, Befruchtung, und Theilung des tierischen Eies. I. Morph. Jahrb. **1**; see also **2–4**.

HEUSER, E. 1884. Beobachtungen über Zellkernteilung. Bot. Centr. **17**: 27–32, 57–59, 85–95, 117–128, 154–157. pls. 2.

HOLT, C. M. 1917. Multiple complexes in the alimentary tract of *Culex pipiens*. Jour. Morph. **29**: 607–627. pls. 4.

ISHIKAWA, M. 1916. A list of the number of chromosomes. Bot. Mag. Tokyo **30**: 404–448. (Plants.)

JANSSENS, F. A. et WILLEMS, J. 1909. Spermatogénèse dans les Batrachiens. IV. La Cellule **25**: 151–178. pls. 2.

KOWALEVSKY, A. 1871. Embryologischen Studien an Würmern und Arthropoden. Mem. Acad. Imp. Sci. de St. Petersburg VII **16**: 13.

LAIBACH, F. 1907. Zur Frage nach der Individualität der Chromosomen in Pflanzenreich. Beih. Bot. Centralbl. **22**: 191–210. pl. 8.

LEWIS, W. H. and LEWIS, M. R. 1917. The duration of the various phases of mitosis in the mesenchyme cells of tissue cultures. Anat. Record **13**: 359–368.

LUNDEGÅRDH, H. 1910. Ueber Kernteilung in den Wurzelspitzen von *Allium cepa* und *Vicia faba*. Svensk. Bot. Tidskr. **4**: 174–196. figs. 11.

1912a. Die Kernteilung bei höheren Organismen nach Untersuchungen an lebenden Material. Jahrb. Wiss. Bot. **51**: 236–282. pl. 2. figs. 8.

1912b. Das Karyotin im Ruhekern und sein Verhalten bei der Bildung und Auflösung der Chromosomen. Arch. Zellf. **9**: 205–330. pls. 17–19. figs. 9.

1912c. Chromosomen, Nukleolen, und die Veränderung im Protoplasma bei der Karyokinese. Beitr. Biol. Pflanzen **11**: 373–542. pls. 11–14.

MARÉCHAL, J. 1904. Ueber die morphologische Entwicklung der Chromosomen im Keimbläschen des Selachiereies. Anat. Anz. **25**: 383–398. figs. 25.

1907. Sur l'ovogénèse des Selachiens. La Cellule **24**: 1–239. pls. 10.

MARTINS MANO, TH. 1904. Nucléole et Chromosomes. La Cellule **22**: 57–76. pls. 1–4.

McCLUNG, C. E. 1905. The chromosome complex of orthopteran spermatocytes. Biol. Bull. **9**: 304–340. figs. 21.

1914. A comparative study of the chromosomes in orthopteran spermatogenesis. Jour. Morph. **25**: 651–749. pls. 1–10.

1917. The multiple chromosomes of *Hesperotettix* and *Mermiria* (Orthoptera). Ibid. **29**: 519–604. pls. 8.

MERRIMAN, M. L. 1904. Vegetative cell division in *Allium*. Bot. Gaz. **37**: 178–207. pls. 11–13.

MEVES, FR. 1896, 1898. Zelltheilung. Ergeb. Anat. Entw. **6**: 285–390; **8**: 430–542. (Review.)

——— 1911. Chromosomenlängen bei *Salamandra*, nebst Bemerkungen zur Individualitätstheorie der Chromosomen. Arch. Mikr. Anat. **77**: 273–300. pls. 11, 12.

MOENKHAUS, W. J. 1904. The development of the hybrids between *Fundulus heteroclitus* and *Menidia notatus*, with especial reference to the behavior of maternal and paternal chromosomes. Jour. Anat. **3**: 29–65.

MONTGOMERY, T. H. 1910. On the dimegalous sperm and chromosomal variation in *Euschistus*, with reference to chromosomal continuity. Arch. Zellf. **5**: 121–145. pls. 9, 10.

MORRIS, M. 1914. The behavior of the chromatin in hybrids between *Fundulus* and *Ctenolabrus*. Jour. Exp. Zool. **16**: 501–522. pls. 5.

MÜLLER, H. A. CL. 1911. Kernstudien an Pflanzen I u. II. Arch. Zellf. **8**: 1–51. pls. 1, 2.

NAWASCHIN, M. 1915. "Haploide, diploide, und triploide Kerne von *Crepis virens*, Vill." (Russian; cited by Sakamura, 1915, 1920.)

NAWASCHIN, S. 1914. "Sur quelques indices de l'organisation du chromosome." (Russian.)

NĚMEC, B. 1910. Das Problem der Befruchtungsvorgänge und andere Zytologische Fragen. pp. 532. pls. 5. Berlin.

OVERTON, J. B. 1905. Ueber Reduktionsteilung in den Pollenmutterzellen einiger Dikotylen. Jahrb. Wiss. Bot. **42**: 121–153. pls. 6, 7.

——— 1909. On the organization of the nuclei in the pollen mother cells of certain plants, with especial reference to the permanence of the chromosomes. Ann. Bot. **23**: 19–61. pls. 3.

PICARD, M. 1913. A bibliography of works on meiosis and somatic mitosis in the angiosperms. Bull. Torr. Bot. Club **40**: 575–590.

PINNEY, E. 1918. A study of the relation of the behavior of the chromatin to heredity and development in teleost hybrids. Jour. Morph. **31**: 225–292. figs. 88.

PFITZNER, W. 1881. Ueber den feineren Bau der bei der Zellteilung auftretenden fadenförmigen Differenzierung des Zellkerns. Morph. Jahrb. **7**: 289–311. figs. 2.

RABL, C. 1885. Ueber Zelltheilung. Ibid. **10**: 214–330. pls. 7–13. figs. 5.

REMAK, R. 1841. Ueber Theilung rother Blutzellen beim Embryo. Arch. Anat. Physiol. 1858.

——— 1852. Ueber extrazellulare Entstehung thierischer Zellen und über Vermehrung derselben durch Theilung. Ibid.

RICHARDS, A. 1916. Chromosome individuality in fish eggs. Science **43**: 178.

——— 1917. The history of the chromosomal vesicles in *Fundulus* and the theory of genetic continuity of chromosomes. Biol. Bull. **32**: 249–290. pls. 4.

ROBERTSON, W. R. B. 1916. Chromosome studies. I. Jour. Morph. **27**: 179–332. pls. 1–26.

ROSENBERG, O. 1909a. Zur Kenntniss von den Tetradenteilung der Compositen. Svensk. Bot. Tidskr. **3**: 64–77. pl. 1.

——— 1909b. Ueber den Bau des Ruhekerns. Ibid. **3**: 163–173. pl. 5. fig. 1.

——— 1918. Chromosomenzahlen und Chromosomendimensionen in der Gattung *Crepis*. Ark. f. Bot. **15**: 1–16. figs. 6.

——— 1920. Weitere Untersuchungen über die Chromosomenverhältnisse in *Crepis*. Svensk Bot. Tidskr. **14**: 320–326. figs. 5.

Roux, W. 1883. Ueber die Bedeutung der Kernteilungsfiguren. Leipzig.

Rückert, J. 1895. Ueber das Selbstandigbleiben der vaterlichen und mutterlichen Kernsubstanz während der ersten Entwicklung des befruchteten *Cyclops*-Eies. Arch. Mikr. Anat. **46**: 339–369. pls. 21, 22.

Sakamura, T. 1914. Ueber die Kernteilung bei *Vicia cracca* L. Bot. Mag. Tokyo **28**: 131–147. pl. 2.

    1915. Ueber die Einschnürung der Chromosomen bei *Vicia faba* L. (Vorl. Mitt.). Ibid **29**: 287–300. pl. 13. figs. 12.

    1920. Experimentelle Studien über die Zell- und Kernteilung mit besonderer Rucksicht auf Form, Grosse, und Zahl der Chromosomen. Jour. Coll. Sci Imp. Univ. Tokyo **39**: pp. 221. pls. 7.

Schleicher, W. 1878. Die Knorpelzelltheilung. Arch. Mikr. Anat. **16**: 248–300. pls. 12–14. 1879. (Centr. Med. Wiss., Berlin, 1878.)

Schneider, A. 1873. Untersuchungen über Platelminthen. Jahrb. Oberhess. Gesell. Natur-Heilk. **14**. Giessen.

    1883. Das Ei und seine Befruchtung. Breslau.

Sharp, L. W. 1913. Somatic chromosomes in *Vicia*. La Cellule **29**: 297–331. pls. 2.

    1914. Maturation in *Vicia*. (Prelim. note.) Bot. Gaz. **57**: 531.

    1920. Somatic chromosomes in *Tradescantia*. Am. Jour. Bot. **7**: 341–354. pls. 22, 23.

de Smet, E 1914. Chromosomes, prochromosomes, et nucléole dans quelques Dicotylées. La Cellule **29**: 335–377. pls. 3.

Smith, B. G. 1919. The individuality of the germ nuclei during the cleavage of the egg of *Cryptobranchus alleghaniensis*. Biol Bull. **37**: 246–287. pls. 9.

Stomps, T. J. 1910. Kerndeeling en synapsis bij *Spinacea oleracea*. (German transl. in Biol. Centralbl. **31**: 257–320. pls. 3.)

Stout, A. B. 1912. The individuality of the chromosomes and their serial arrangement in *Carex aquatilis*. Arch. Zellf. **9**: 114–140. pls. 11, 12.

Strasburger, E. 1875. Ueber Zellbildung und Zelltheilung. Jena.

    1884. Die Controversen der indirekten Kernteilung. Arch. Mikr. Anat. **23**: 246–304. pls. 13, 14.

    1888. Ueber Kern- und Zellteilung im Pflanzenreich, nebst einem Anhang über Befruchtung. Hist. Beitr. **1**. pp. 258. pls. 3.

    1900. Ueber Reduktionsteilung, Spindelbildung, Centrosomen und Cilienbildner im Pflanzenreich. Hist. Beitr. **6**. pp. 124. pls. 4.

    1905. Typische und allotypische Kernteilung. Jahrb. Wiss. Bot. **42**: 1–71. pl. 1.

    1907a. Apogamie bei Marsilia. Flora **97**: 123–191. pls. 3–8.

    1907b. Ueber die Individualität der Chromosomen und die Propfhybriden-Frage. Jahrb. wiss. Bot. **44**: 482–555. pls. 5–7.

Sutton, W. S. 1902. On the morphology of the chromosome group in *Brachystola magna*. Biol. Bull. **4**: 24–39. figs. 11.

Tennent, D. H. 1912. Studies in cytology. I, II. Jour. Exp. Zool. **12**: 391–411. figs. 21.

Tischler, G. 1910. Untersuchungen über die Entwicklung des Bananen-Pollens. I. Arch. f. Zellf. **5**.

    1916. Chromosomenzahl, -Form und -Individualitat im Pflanzenreich. Prog. Rei Bot. **5**: 164–284. (Review and list of numbers.)

Tschernoyarow, M. 1914 Ueber die Chromosomenzahl in besonders beschaffene Chromosomen im Zellkerne von *Najas major*. Ber. Deu. Bot. Ges. **32**: 411–416. pl. 10.

WALDEYER, W. 1888. Ueber Karyokinese und ihre Beziehung zu den Befruchtungsvorgängen. Arch. Mikr. Anat. **32**: 1–122. figs. 14. (English transl. in
Quar. Jour. Micr. Sci. **30**: 159–281. 1889.)

WENRICH, D. H. 1916. The spermatogenesis of *Phrynotettix magnus* with special
reference to synapsis and the individuality of the chromosomes. Bull. Mus.
Comp. Zool. Harvard Coll. **60**: 55–136. 10 pls.

    1917. Synapsis and chromosome organization in *Chorthippus* (*Stenobothrus*)
curtipennis and *Trimerotropis suffusa* (Orthoptera). Jour. Morph. **29**: 471–516.
pls. 1–3.

WILSON, E. B. 1900. The Cell in Development and Inheritance. 2d ed.

    1909. The cell in relation to heredity and evolution. In "Fifty Years of Darwin-
ism." Holt & Co.

    1912*a*. Some aspects of cytology in relation to the study of genetics. Am. Nat
46: 57–67.

    1912*b*. Studies on chromosomes. VIII. Jour. Exp. Zool. **13**: 345–449. pls. 9.

ZIMMERMANN, A. 1893. Sammel-Referate 6, 7. Beih. Bot. Cent. **3**: 342–354,
401–436.

ZOJA, R. 1895. Sulla indipendenza della cromatina paterna e materna nel nucleo
delle cellule embrionali. Anat. Anz. **11**: 289–293. figs. 3.

ZUR STRASSEN, O. L. 1898. Ueber Riesenbildung bei *Ascaris*-Eieren. Arch.
Entw. **7**: 642–676. pls. 16, 17. figs. 9.

CHAPTER IX

THE ACHROMATIC FIGURE, CYTOKINESIS, AND THE CELL WALL

THE ACHROMATIC FIGURE

The spindle fibers and asters about the centrosomes (when these are present) are collectively termed the *achromatic figure*, in contradistinction to the *chromatic figure*, or chromosomes. Compared with the chromosomes the achromatic figure is relatively little understood, which makes it a very unsatisfactory subject for discussion. We shall first describe the achromatic figure in its more common forms, and after mentioning certain theories which have been propounded to explain its origin and nature we shall briefly review a few of the suggestions which have been made on the subject of the mechanism of mitosis.

In Higher Plants.—In somatic mitosis in higher plants the achromatic figure is devoid of centrosomes and asters. Ordinarily it arises and behaves as follows: While the prophasic changes are taking place within the nucleus the first indications of spindle formation appear in the cytoplasm in the immediate vicinity of the nucleus. At the two sides of the latter, in the general position of the future spindle poles, there are developed two masses of more or less hyaline material, usually called "kinoplasmic caps." In these two polar caps delicate fibrils soon appear, as if by a process of condensation (Fig. 58, *A*, *B*). The nucleus commonly shrinks at this time, while the fibrous areas increase in size and together form a more definitely spindle-shaped figure. After the nuclear membrane has shrunken more closely about the chromosomes it goes into solution and the ingrowing fibers attach themselves to the longitudinally split chromosomes. In many cases the membrane disappears without shrinking, the fibers growing considerably in length to reach the chromosomes. The latter quickly become regularly arranged in the equatorial plane preparatory to their separation (Chapter VIII). The mitotic figure is now established (Fig. 48). The many fibers composing the spindle may focus at a single sharp point at each pole, or they may end indefinitely without converging to a point, forming in the latter case a broad-poled figure which in extreme cases may be as wide at the poles as at the equator (Fig. 74, *D*). Some of the fibers extend from the poles to the chromosomes, to which they are attached, while others pass through from one pole to the other without being so attached: these two sets of fibers are known respectively as *mantle fibers* and *connecting fibers*.	The latter are also collectively termed the *central spindle*.

It is during the anaphases and telophases that the connecting fibers become most evident; in mitotic figures with many chromosomes it may be impossible to see them at metaphase.	At the beginning of the telophase they may form a bundle no greater in diameter than the daughter chromosome groups, but as the daughter nuclei reorganize the fibers commonly bend outward at the middle, forming a barrel-shaped *phragmoplast* (Fig. 58, *C*) which in plants usually continues to widen by the addition of new fibers until it comes in contact with the lateral walls of the cell.

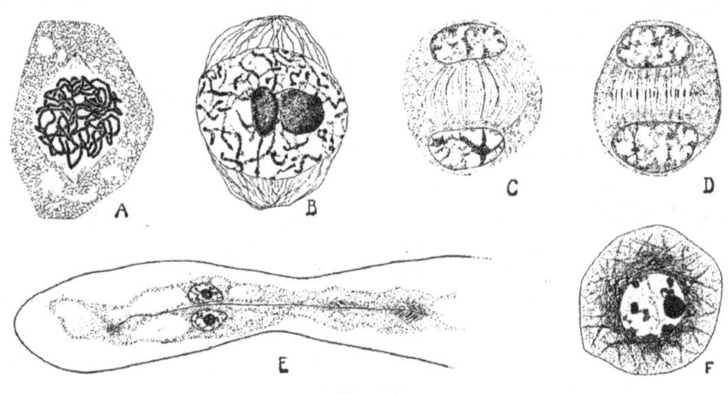

FIG. 58.

A, spindle beginning to differentiate in kinoplasmic caps at poles of nucleus in *Nephrodium*.	(*After Yamanouchi*, 1908.)	*B*, same in *Marsilia*.	(*After Berghs*, 1909.)	*C, D,* the origin of the cell wall in *Pinus:*	*C*, connecting fibers between daughter nuclei at telophase; *D*, thickenings appearing on fibers.	*E*, the continued extension of the cell wall after the completion of mitosis in the endosperm of *Physostegia virginiana*.	× 215.	(*After Sharp*, 1911.)	*F*, multipolar stage of spindle development in microsporocyte of *Acer Negundo*.	× 1125.	(*After Taylor*, 1920.)

While the above changes are occurring the new cell wall which is to be formed between the daughter nuclei begins to differentiate.	As the central spindle widens the fibers become fainter near the nuclei and more prominent at the equatorial region: this appearance seems to be due to the flow of the material composing the fibers toward the latter region. On the thickened fibers there now appear small swellings (Fig. 58, *D*) which increase in size until they fuse to form a continuous plate across the equator of the mother cell, thus dividing the latter into two daughter cells.	As this cell plate undergoes further changes (see p. 190) the fibers disappear completely, first near the two nuclei and ultimately at the equatorial region near the new wall.	If the cell undergoing division is very broad it often happens that wall formation begins near the center of the phragmoplast while the latter is still extending laterally.	In extreme cases wall formation may still be seen in progress at the periphery after the fibers have completely disappeared at the central region (Fig. 58, *E*). Such is notably the case in the tangential divisions of elongated cambium cells (Bailey 1919, 1920).

The spindle in many cases has an origin somewhat different from that described above. The first indication of its differentiation is here the appearance of a weft of fine fibrils in the cytoplasm all around the nucleus (Fig. 84, *E*). As these fibrils increase in size and number they may form several distinct groups extending in various directions, thus giving a multipolar spindle (Fig. 58, *F*). Some of the groups then gradually disappear, while others alter their positions and coalesce, so that a bipolar spindle eventually results. This, in general, is the manner in which the spindle arises in the microsporocytes of angiosperms. For example, Lawson (1898, 1900, 1903) finds that in *Cobœa*, *Gladiolus*, and *Iris* a zone of granular "perikaryoplasm" collects about the nucleus during the prophases of mitosis. When the nuclear membrane dissolves, this substance together with the linin of the nucleus forms a fibrous network which grows out into several cones of fibers, and these later become arranged in two opposed groups.

In Animals.—In the majority of animal cells, and in certain cells of lower plants also, the achromatic figure is a much more elaborate structure than that of the higher plants described above. This is due to the presence of centrosomes, which with their asters are very conspicuous at the time of mitosis. Commonly the aster is not present during the resting stages of the cell, but cases are known in which both centrosome and aster are visible, forming with other materials an "attraction sphere" in the cytoplasm. As the process of mitosis begins (Fig. 59), an aster, if not already present, develops about the centrosome. The centrosome divides, and as the daughter centrosomes move apart each is seen to be surrounded by its own aster, and a small group of fibers ("central spindle") extends between them. The achromatic figure, made up of the asters and the spindle connecting them, is known both at this stage and later as the *amphiaster*. As the daughter centrosomes continue to separate the astral rays increase in prominence. Some of the rays grow into the nucleus when its membrane disappears and become attached as mantle fibers to the chromosomes, while the lengthening central spindle between the asters becomes the central spindle portion (connecting fibers) of the completed mitotic figure (Fig. 49). All the fibers focus upon the centrosomes.

During the anaphase the asters remain very conspicuous, but as the telophases progress they gradually fade from view, except in those forms which have a more or less permanent attraction sphere. Aside from the presence of centrosomes and asters the achromatic figure in animal cells differs most conspicuously from that of higher plant cells in its behavior

12

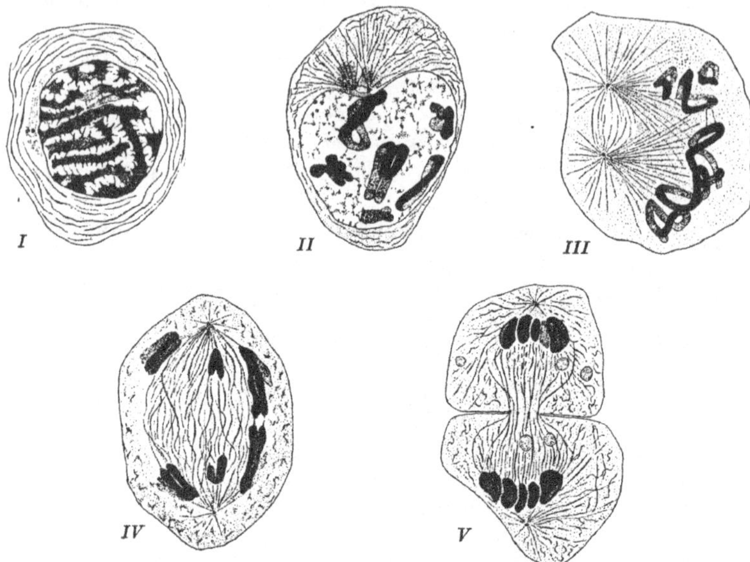

FIG. 59.—Mitosis in the spermatocyte of *Salamandra.*
I, prophase, centrosomes in astrosphere substance; latter spread out on nucleus. *II*, prophase; bivalent chromosomes formed; centrosomes beginning to diverge; central spindle and asters developed. *III*, late prophase: nuclear membrane dissolved; spindle fibers attaching to chromosomes, centrosomes moving apart. *IV*, anaphase: connecting fibers prominent. *V*, telophase: constriction of cell nearly complete; mid-body forming on central spindle or interzonal fibers. (*After Meves*, 1907.)

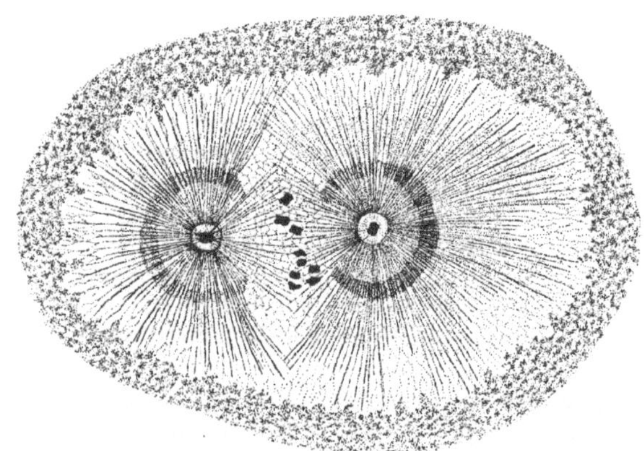

FIG. 60.—Elaborate achromatic figure in oöcyte of *Pisciola.* × 1000.
(*After Jorgensen,* 1913.)

during the telophases. Instead of forming thickenings which become an equatorial cell plate, the connecting fibers play relatively little part in cytokinesis. One or more granules may be differentiated on the fibers at the equatorial region, forming the so-called "mid-body," but the actual division of the cell is brought about by the development of a cleavage furrow, as will be described in the section on cytokinesis.

Intranuclear Figures.—In the above described cases of mitosis in plants and animals the achromatic figure is derived mainly from the cytoplasmic region of the cell, the nuclear materials playing a relatively minor part. In a number of forms, both among animals and plants (fungi, for example), the spindle arises entirely in the nuclear region, forming an intranuclear figure which may be completely established before the

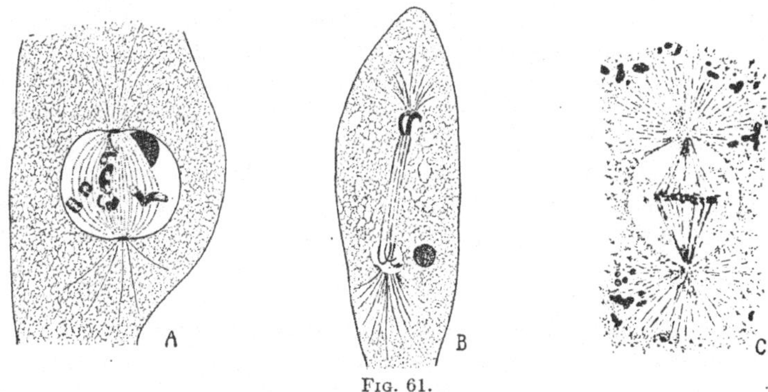

FIG. 61.

A, B, anaphase and telophase of mitosis in ascus of *Laboulbenia chætophora.* × 1350. (*After Faull,* 1912.) (*See also Fig.* 22.) C, intranuclear mitotic figure in oogonium of *Fucus.* (*After Yamanouchi,* 1909.)

nuclear membrane disappears. Cases are known in which the centrosomes themselves are also intranuclear, but usually these bodies lie in the cytoplasm against the nuclear membrane, so that although the spindle portion of the figure is within the nucleus the asters lie in the cytoplasm.

In the division of the nucleus in the ascus of an ascomycete,[1] to take a single example, the process is as follows (Figs. 22; 61 *A, B*): The centrosome, which in ascomycetes is often discoid in shape, lies against the nuclear membrane. As mitosis begins an aster develops in the cytoplasm about the centrosome, and the latter divides to form two daughter centrosomes. The central spindle, if formed at all, does not persist. From each of the daughter centrosomes, which begin to move apart along the nuclear membrane, a group of fibers extends into the nucleus where the chromosomes are being formed from the reticulum. The centrosomes finally reach opposite sides of the nucleus, and their two

[1] For references to the literature of mitosis in ascomycetes see page 290.

groups of fibers become arranged in the form of a sharp poled spindle extending through the nucleus with the chromosomes at the equator. The nuclear membrane commonly remains intact until the chromosomes approach the poles at anaphase; it then disappears, allowing the nucleolus, which has remained unchanged, to escape into the cytoplasm nearby. Between the two densely packed daughter chromosome groups there extends a long strand of chromatic material: this soon disappears and the two daughter chromosome groups reorganize two daughter nuclei not separated by a wall. In those cases in which the division of the fungus nucleus is followed by the development of a separating wall the latter is formed by a cleavage furrow independently of the achromatic figure.

Origin of the Figure.—Having before us the above examples of the achromatic figure, we may now refer very briefly to some of the ideas which have been advanced regarding the details of its origin in the cell.[1]

Early observers looked upon the whole mitotic figure—chromosomes, spindle, and all—as a transformed nucleus, all the structures being formed from the nuclear material at each mitosis. Strasburger, who first held this view, later (1888), with Hermann (1891), believed the spindle to arise wholly from the cytoplasm, whereas O. Hertwig pointed out cases in which the astral rays arise from the cytoplasm and the spindle from the linin reticulum of the nucleus. Flemming (1891) derived the fibers from the linin and the nuclear membrane. It soon became evident that the spindle, although in some cases arising entirely within the nucleus or wholly from the cytoplasm, is commonly made up of materials derived from both regions, as is evident from the examples described in the foregoing paragraphs.

When van Beneden and Boveri announced their view that the centrosome is a permanent cell organ, transmitted by division to daughter cells and directly concerned in the formation of the asters, the theory was adopted that the figure arises from the cytoplasm as a result of the influence of the centrosome. The centrosome therefore came to be known as "the dynamic center of the cell." Although this organ does play a conspicuous rôle when present, its importance in connection with the achromatic figure was somewhat diminished when it became evident that many centrosomes do not persist from one cell generation to the next, and that such bodies are entirely absent from the cells of higher plants.

Rearrangement Theories.—Many attempts have been made to account for the formation of the achromatic fibrils in the cytoplasm. According to some the fibers and astral rays arise as the result of a morphological rearrangement of the preëxistent protoplasmic structure, chiefly under

---

[1] Extensive reviews of the early theories are given by Wilson (1900, pp. 72–86 and 316–329)

the influence of the centrosome. Bütschli (1876), who looked upon protoplasm as alveolar in nature, held that the rays are not really fibers, but only the lamellæ between radially elongated alveolæ about the centrosome. It was the opinion of Wilson (1899) on the other hand, that the rays are actual fibers, though their material is derived from the alveolar walls. Klein (1878) and others who believed protoplasm to be ultimately fibrillar or reticular in structure, regarded the rays as radially arranged fibrillæ. Van Beneden (1883) supposed these fibrillæ to be derived partly from the intranuclear reticulum, and Rabl (1889) pointed out that they are continuous with the unaltered cytoplasmic meshwork and arise by a direct transformation of the latter. In *Passiflora* Williams (1899) found that the nuclear membrane forms a meshwork connecting the linin reticulum with the cytoplasmic reticulum, all three together organizing the spindle.

Special Substance Theories.—According to another group of theories the spindle and asters are not formed merely by the rearrangement of a structure already present, but arise from a special substance in the cell. This substance was held by some to be a constantly present constituent of the cell, forming the achromatic figure at the time of mitosis and remaining in reserve through the resting stages. Boveri's archoplasm hypothesis in its earlier form (1888) was a prominent development of this idea. According to this hypothesis the attraction sphere is composed of a distinct substance called *archoplasm*, which consists in turn of fine granules or microsomes aggregated about the centrosome as a result of the centrosome's attractive force. The entire achromatic figure was held to arise from this mass of archoplasm, the fibers and astral rays growing out from it like roots, to be withdrawn again into the daughter masses of archoplasm at the two poles during the closing phases of mitosis. In this way each daughter cell was thought to receive half of the archoplasm. Although other workers (Watase 1894) also held that the fibers are outgrowths of the centrosome or centrosphere substance, it was made evident later that the material composing the fiber comes from the cytoplasm, being added to the growing fiber at its end. This was the view of Drüner (1894, 1895). Boveri later (1895) modified his archoplasm hypothesis, adopting the view that the fiber is formed from the substance of the cytoplasm and not necessarily from a constantly present archoplasm.

Another theory based on the idea of a special substance in the cell was that of Strasburger (1892, 1897, 1898). Strasburger held that the cell has two kinds of protoplasm: an active fibrillar *kinoplasm* and a less active alveolar *trophoplasm*. The former constitutes the ectoplast, centrosomes, the mitotic fibers, and the contractile substance of cilia and allied structures. The kinoplasm is thus concerned with the motor work of the cell, whereas the trophoplasm has to do chiefly with nutrition.

The nucleolus has been thought by some observers to furnish material for the formation of the spindle, because of the fact that it very commonly disappears from view at about the time the spindle begins to differentiate. It is possible that in some cases there may be a connection of this sort between nucleolus and spindle, but it is clear that this cannot serve as a general interpretation of spindle origin.

That the achromatic figure may arise from a special substance not constantly present in the cell, but formed anew at each mitosis, is a theory which several workers have advanced. The researches of Devisé (1914) and Miss Nothnagel (1916) may be cited for illustration. Devisé, as the result of a careful study of the development of the spindle in the microsporocytes of *Larix*, concluded that the spindle is not formed by the rearrangement of any preëxistent nuclear or cytoplasmic structures, but arises from a substance which develops in the nuclear region during the late prophases (after diakinesis). He was not able to decide whether this substance is of purely nuclear origin or is formed when the karyolymph comes in contact with the cytoplasm. The interaction of karyolymph and cytoplasm is emphasized by Miss Nothnagel in her work on *Allium*. She points out that the contact of newly formed karyolymph with the cytoplasm at telophase brings about the precipitation of the nuclear membrane, and that in an analogous manner an exosmosis of karyolymph through the nuclear membrane into the cytoplasm during prophase causes the precipitation of fine fibrils around the nucleus, these fibrils then developing into the spindle. The achromatic figure therefore arises from a special substance, but this substance, as in the case of *Larix*, is newly formed at each mitosis.

Conclusion.—In general it may be said that although the spindle fibers and the motor and contractile elements of the cell appear to have a substantial relationship with one another, the substance common to them is probably "not to be regarded as being necessarily a permanent constituent of the cell, but only as a phase, more or less persistent, in the general metabolic transformation of the cell substance" (Wilson). Indeed the conspicuous tendency on the part of cytologists at present is to regard the achromatic figure neither as a mere rearrangement of a structure previously present, nor as a form assumed by a special spindle substance, but rather as the result of streaming, gelation, and other temporary alterations in the colloidal substratum. This interpretation is strongly supported by the microdissection studies to be cited in a subsequent paragraph.

The Mechanism of Mitosis.—Since the phenomenon of mitosis was first described there have been put forward a number of theories to account for the operation of the achromatic structures in bringing about the separation of the daughter chromosomes and for the division of the cell. Many of the suggestions undoubtedly contain elements of truth, but it must be admitted that there is no immediate prospect of a satisfactory solution of these problems.

Contractility.—One of the simplest and most widely accepted theories was that of fibrillar contractility suggested by Klein (1878) and van Beneden (1883, 1887), according to which the chromosomes are simply dragged apart by the contraction of two opposed groups of spindle fibers. This theory and its modifications are fully reviewed by Wilson: it will be sufficient here to point out that, whereas many facts were cited in its favor, and elastic models made which simulated the supposed contraction and its results (Heidenhain), the further evidence brought forward by Hermann (1891), Drüner (1894, 1895), Calkins (1898), and others led to the general restriction of the rôle of contractility, until it became apparent that this factor, although it may contribute to the general result, must be of minor importance. The contractility factor appeared again in the more elaborate theory proposed by Rhumbler, which may be briefly stated as follows: The centrosome arises as a local solidification of the walls of the alveolæ; the denser constituents of the protoplasm collect at this point and form an attraction sphere, driving the less dense constituents to the other parts of the cell where the pressure is lower; this migration of fluid affects particularly those strands of the protoplasmic reticulum which radiate more directly from the centrosomes; these strands or rays, in giving up their fluid, shorten, and thus exert a tractive force which draws the daughter chromosomes apart. In this theory, therefore, the main factors are streaming and contractility.

Streaming.—The phenomena of streaming and surface tension have been prominent factors in several attempts to explain both karyokinesis and cytokinesis. The rôle of streaming in karyokinesis has been held to be especially important since Bütschli, Hertwig, and Fol showed many years ago that currents exist in the protoplasm. Rhumbler (1896, 1899), Morgan (1899), Wilson (1901), and Conklin (1902) all held that the astral rays are due at least in part to centripetal currents. This interpretation has recently been confirmed by Chambers (1917) in his microdissection studies on the living cell. With regard to the aster Chambers says: "The formation of the aster consists in the gelation of the hyaloplasm which comes under the influence of the astral center. A hyaline liquid separates out during the gelation and flows in innumerable centripetal paths toward the center where it accumulates to form a sphere. This centripetal flow brings about an arrangement of the gelled hyaloplasm containing the cell-granules into radial strands separated by the hyaline-liquid paths. This produces the astral figure. The strands of gelatinized cytoplasm merge peripherally into the surrounding liquid cytoplasm or reach and anchor themselves in the substance of the gelled surface when the aster is fully formed. The liquid rays merge centrally into the substance of the sphere, the liquid of the rays and of the sphere being thus identical."

Sakamura (1920), although holding the fibers to be important agents in the normal separation of the daughter chromosomes, observes that in abnormal nuclear divisions where no fibers are present the chromosomes still show movements which are probably due to streaming of the cytoplasm and to surface tension phenomena.

The relation of streaming and surface tension to cytokinesis will be discussed in the section dealing more particularly with cytokinesis.

Osmosis.—In a theory of the mechanics of karyokinesis proposed by Lawson (1911) the principal factor involved is osmosis. Lawson's explanation is essentially as follows.   During the late prophase karyolymph passes outward through the nuclear membrane by osmosis, this loss of fluid resulting in a contraction of the nucleus. Owing to the fact that the cytoplasmic reticulum is continuous with the nuclear membrane this contraction sets up radial lines of tension in this reticulum on all sides of the nucleus. As the process continues these lines or "fibers" gradually become arranged in two opposed groups, while the nuclear membrane to which they are attached continues to contract until it actually enwraps each double chromosome. To each double chromosome there are thus attached fibers which represent stretched and distorted regions of the cytoplasmic reticulum extending to the two sides of the cell. When the chromosomes become properly arranged at the equatorial plane the fibers, which are under considerable tension, are able to pull the daughter chromosomes apart and draw them to the poles. As the fibers relax they resume their true reticular state. Although the chromosomes are thus drawn apart by the shortening of "fibers" attached to them, Lawson points out that this is not to be regarded as a case of true active contractility, but only as a release of tension set up in the passive but elastic cytoplasmic reticulum as the result of the exosmosis of karyolymph from the nucleus. This theory has been severely criticized by a number of writers, chiefly on the grounds that such an enwrapping of the chromosomes by the nuclear membrane as Lawson describes cannot be demonstrated in many objects subsequently examined, and that the membrane frequently goes into solution when both it and the growing fibers are still some distance from the chromosomes.

Electrical Theories.—The striking resemblance between the achromatic figure and the lines of force in an electromagnetic field early led to attempts to account for mitosis on the basis of electrical principles. Several investigators, working with various chemical substances, succeeded in modelling fields of force that illustrated graphically the changes supposed to take place in the dividing cell. In later years the electromagnetic interpretation was again brought into prominence by Gallardo, Hartog, and Prenant. At first Gallardo (1896) believed the two spindle poles to be of unlike sign, but later (1906), as the result of the researches of Lillie (1903) (see p. 62) on the behavior of nucleus and cytoplasm in the electromagnetic field, he concluded that the chromosomes and the cytoplasm carry charges of unlike sign: the daughter centrosomes repel each other and move apart because of their like sign, the spindle poles being of like sign also. The movement of the chromosomes to the poles he held to be due to the combined action of two forces: the mutual repulsion of the similarly charged daughter chromosomes, and the attraction between the oppositely charged centrosomes and chromosomes.

The fact that the two centrosomes and hence the two spindle poles are electrically homopolar (Lillie) and alike osmotically at once makes it apparent that the mitotic figure does not represent an ordinary electromagnetic field, for in the latter the poles are of unlike sign—the field is heteropolar. It has consequently been suggested by Prenant (1910) and Hartog (1905, 1914) that the mitotic figure is the seat of a special force, analogous to electrostatic force but not identical with it, which is peculiar to living organisms. This new force they call "mitokinetism."

A large amount of discussion has centered about the possible rôle of electrical forces in mitosis, and many kinds of normal and abnormal mitotic phenomena have been cited as evidence for various views. So far as conclusive statements are concerned, there is disappointingly little of a definite nature that can be said. Meek (1913) asserts that the only generalization which is at present possible is the negative one that "the mitotic spindle is not a figure formed entirely by the action of forces at its poles."

Conclusion.—In conclusion we may emphasize the fact that the achromatic figure depends for its operation upon a variety of interacting factors. Certain investigators have doubtless done good service in emphasizing the importance of one or another of these factors—streaming, surface tension, contractility, gelation, electrical phenomena, and the like—but it has become increasingly evident that in no one of them alone is the key to the problem of mitosis to be found. In spite of the confidence that some progress has been made, at least in the elucidation of certain phenomena which must have a part in any ultimate explanation, it is nevertheless true that the statements made twenty years ago by Wilson (1900, p. 111) may be taken as an essentially accurate expression of the condition of the subject: "When all is said, we must admit that the mechanism of mitosis in every phase still awaits adequate physiological analysis. The suggestive experiments of Bütschli and Heidenhain lead us to hope that a partial solution of the problem may be reached along the lines of physical and chemical experiment. At present we can only admit that none of the conclusions thus far reached, whether by observation or by experiment, are more than the first *naïve* attempts to analyse a group of most complex phenomena of which we have little real understanding."

CYTOKINESIS

In the foregoing pages discussion has been limited largely to karyokinesis. In the present section attention will be directed to cytokinesis, or the division of the extra-nuclear portion of the cell.

In plants the wall separating the two daughter cells is formed by two general methods: cell plate formation and furrowing. The first and more common of these methods, by which a wall is formed in close association with the spindle fibers at the close of mitosis, has been briefly described in the foregoing section on the achromatic figure (p. 176) and will be taken up in greater detail in the following section on the cell wall (p. 190). At this point we shall therefore describe the second method, that of furrowing, which in plants is seen most conspicuously in the thallophytes and in the microsporocytes of the higher plants. The review of the subject given by Farr (1916) will be followed.

Thallophytes.—In *Spirogyra* Strasburger (1875) showed that the wall between the two daughter cells appears as a "girdle" or ring-like ingrowth from the side wall of the parent cell. This wall continues to

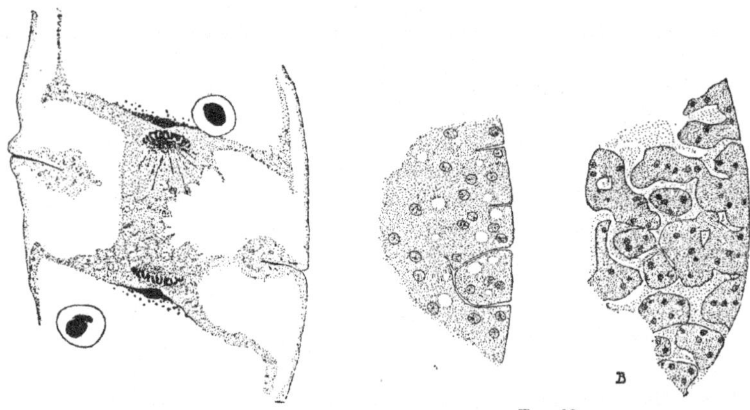

Fig. 62.   Fig. 63.

Fig. 62.—Cytokinesis by furrowing in *Closterium*. Only the central part of the cell is shown. × 700. (*After Lutman, 1911.*)

Fig. 63.

A, Cleavage furrows beginning to form at periphery of sporangium of *Rhizopus nigricans*. × 1500. *B*, Cleavage in the sporangium of *Phycomyces nitens:* intersporal substance in the angular furrows. × 500. (*Both after D. B. Swingle, 1903.*)

grow centripetally by the addition of new material at its inner edge while the protoplast develops a deep cleavage furrow, the process continuing until the separating wall is completed at the center of the cell. A somewhat similar process occurs in *Closterium* (Lutman 1911) (Fig. 62). In the brown algæ *Sphacelaria* (Strasburger 1892; W. T. Swingle 1897) and *Dictyota* (Mottier 1900) the wall develops uniformly across the whole equatorial plane at the same time, and not as a progressive in-growth from the periphery.

In the fungi Harper and others showed that the two daughter cells are separated by the development of a cleavage furrow in which the new wall is laid down. In large multinucleate masses that become broken up into spores this progressive cleavage is a very complicated process. The manner in which the furrows develop is shown in the studies of Timberlake (1902) on *Hydrodictyon,* D. B. Swingle (1903) and Moreau (1913) on *Rhizopus* and *Phycomyces,* Davis (1903) on *Saprolegnia,* Rytz (1907) on *Synchytrium,* and Harper (1899, 1900, 1914) on *Synchytrium, Pilobolus, Sporodinia, Fuligo,* and *Didymium.* In *Rhizopus* (Fig. 63, *A*) the cleavage furrows begin to form both at the peripheral membrane of the sporangium and at the columella and work gradually into the multi-nucleate protoplasm, eventually cutting out multinucleate blocks which become the spores. In *Phycomyces* (Fig. 63, *B*) small vacuoles appear in the midst of the multinucleate protoplasm, enlarge and become stellate, and cut out spore masses with from 1 to 12 nuclei each. In the myxomycete, *Fuligo,* the cleavage is from the surface inward, and the multinucleate blocks are subdivided by further furrowing into uninucleate spores. In *Didymium* the spores are delimited in a similar way by furrows which begin to form along the young capillitium fila-ments in the interior of the multinucleate mass as well as at its periphery.

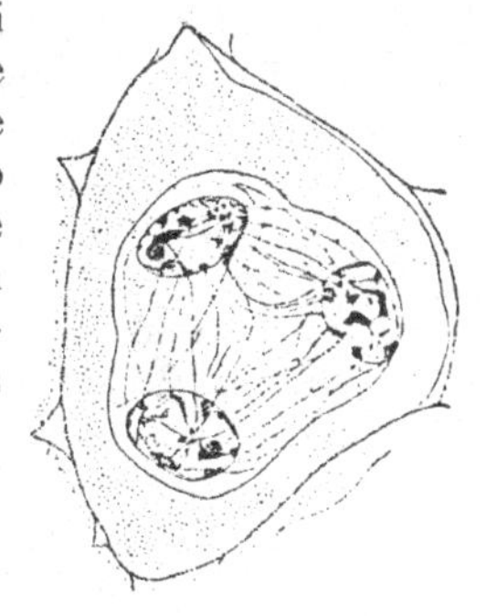

Microsporocytes.—In the microsporocytes of the higher plants it has been shown with great clearness by Farr (1916, 1918) that the quadri-partition to form spore tetrads of the tetrahedral type is brought about by furrowing, previous ac-counts having generally stated that the walls are formed by the cell plate method. Farr finds that after the four microspore nuclei are formed they all become connected by a series of six spindles, or sets of connecting fibers. The two spindles of the second maturation mitosis may persist, four new ones being added, or the two may disappear, six new ones being developed.

Fig. 64.—Cytokinesis by furrowing in the micro-sporocyte of *Nicotiana.* × 1400. (*After Farr,* 1916.)

Although some sporadic thickenings may ap-pear on these fibers they have nothing to do with the formation of the separating walls, there being no centrifugally growing cell plates such as are seen in cells dividing by the cell plate method. Constriction fur-rows appear at the periphery of the cell (Fig. 64) and grow inward until they meet at the center, dividing the protoplast simultaneously into four spores. Any fibers which these furrows encounter as they grow inward are probably incorporated in the new wall, but they play no prominent part in wall formation: the development of the furrows appears to be entirely independent of the fibers present.

In his first paper (1916) Farr states that the microspore tetrads of the bilateral type are usually formed by the cell plate method, a wall being formed across the diameter of the microsporocyte on the connecting fibers after the first maturation mitosis, and the two daughter cells being divided in a similar way after the second mitosis. In his second contribution (1918) he shows that in *Magnolia* such tetrads also are formed by furrowing. After the first mitosis a cleavage furrow starts to form, but its development is arrested until after the second mitosis, when it resumes its growth toward the center and forms a wall across the diameter of the spherical protoplast. At the same time other new furrows subdivide each hemisphere, so that four uninucleate microspores result. Farr states that no case of bipartition by furrowing is known in the higher plants; bipartition begins in *Magnolia*, but the furrow ceases to grow until other furrows are formed after the second mitosis, the eventual division occurring by quadripartition. In the lower plants, however, bilateral tetrads may be formed by the cell plate method. It is the opinion of Farr that furrowing in microsporocytes is due to conditions similar to those which bring it about in animal eggs (see below), since both float freely in a liquid.

Animals.—In animals there is found nothing corresponding to the formation of a cell plate on the spindle fibers and its development into a thick wall such as is seen in plants. As noted in the section on the achromatic figure, there is often a slight differentiation at this region (the "mid-body"), but it has nothing to do with cytokinesis, which is brought about by simple constriction or furrowing. This process is most easily followed in the segmenting egg. In small eggs, such as those of worms, the daughter cells (blastomeres) round up and become more or less spherical, whereas in larger eggs, such as that of the frog, a cleavage furrow appears at one pole and develops through the egg without altering the shape of the latter, so that the first two blastomeres have the form of hemispheres. It is with animal eggs that most of the researches on the mechanism of cytokinesis by furrowing have been carried out.

Mechanism of Furrowing.—Attempts to explain furrowing and the separation of the daughter cells on physico-chemical grounds have been rather numerous. Many years ago Bütschli (1876) advanced the view that as a result of a specific activity on the part of the centrosomes cytoplasmic currents are set up which flow toward the centrosomes and produce a higher surface tension at the equator of the cell, this in turn bringing about furrowing and cell-division. McClendon (1910, 1913) also reported an increase in surface tension at the region of furrowing.

On the contrary, Robertson (1911, 1913) and others attribute furrowing rather to a decrease in the equatorial surface tension, this decrease being due to a diffusion of materials toward that region from the daughter nuclei. Evidence favoring Bütschli's interpretation has been afforded by the studies of Spek (1918). Spek imitated furrowing and division with oil and mercury droplets in water, and showed that by lowering the surface tension at two poles of the droplet the relatively higher surface tension at the equatorial region could be made to bring about the constriction and fission of the droplet. In both droplet and dividing egg he found streamings such as Erlangen (1897) had described in the nematode egg: an axial movement polewards to the regions of low surface tension and a superficial streaming toward the equatorial region of higher surface tension, the streams turning inward at the furrow (Fig. 65).

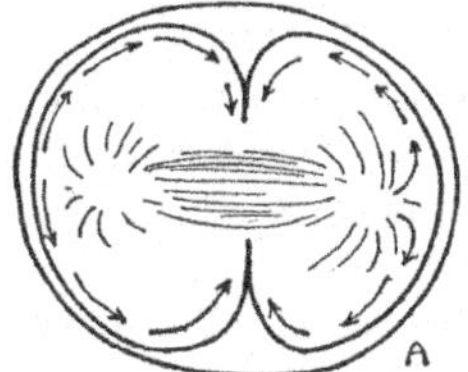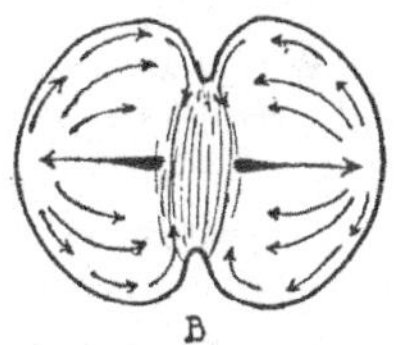

FIG. 65.—Diagram showing streaming and furrowing in the egg of *Rhabditis* (*A*) and an oil droplet (*B*). (*After Spek*, 1918.)

Although the causes of the initial changes in surface tension in the case of the cell are relatively obscure, these experiments of Spek show beyond question that alteration in surface tension and streaming are very important factors in cell-division of this type.

The relation of periodic changes in the viscosity of the egg substance to cytokinesis by furrowing has recently been discussed by Chambers (1919). Immediately after the entrance of the spermatozoön into the echinoderm egg the sperm aster begins to differentiate as a semi-solid region near the sperm head. (See p. 279.) When the aster is most fully developed the egg has its maximum viscosity (Heilbrunn 1915). As the aster disappears the egg again becomes more fluid. Then a second solidification begins at two centers forming the amphiaster, or bipolar figure. The growth of these two semi-solid masses results in the elongation of the egg, and eventually in the development of a cleavage furrow in the more fluid portion of the egg substance separating them. After cleavage is complete the semi-solid masses (asters) revert to a more fluid state. The formation of the cleavage furrow, moreover, may be prevented by mechanical means. At the second mitosis in eggs so treated (binucleate eggs) there are four centers of semi-solidification rather than two, and the egg cleaves simultaneously into four blastomeres. An egg cut into two pieces during the amphiaster stage will, provided it does not return to the fluid state, continue to cleave along the normal plane through the equator of the cell as if nothing unusual had happened. All of these observations indicate a close dependence of cytokinesis upon the temporary differentiation of semi-solid masses in the egg cytoplasm, and throw much light upon the question of the true nature of the achromatic figure.

THE CELL WALL

Probably the most striking difference which meets the eye in a comparison of animal and plant tissues lies in the relative degree of distinctness with which the limits of the individual cells may be made out. Animal cells as a rule are separated only by very thin limiting membranes which in many tissues are so delicate as to be scarcely discernible, whereas the cells of plants usually possess conspicuous firm walls, which in the case of woody plants become greatly thickened and afford mechanical support to large bodies.

The Primary Wall Layer.—Since the time when mitotic cell-division was first carefully studied with the aid of modern methods it has been known that in the cell wall of plants the primary layer, or *middle lamella* (the "intercellular substance" and "cement" of early writers), is formed in most cases in close connection with the spindle fibers at the close of mitosis.[1] The exact manner of its origin, however, has proved to be a very difficult point to determine, and has formed the subject of a long continued controversy. (See papers of Timberlake and Allen, 1900 and 1901.) During the telophases of mitosis the spindle fibers connecting the two daughter nuclei develop thickenings (Fig. 58, *D*), enlarge until they come in contact with one another and fuse to form a *cell plate*, or partition, between the daughter cells. For some time it was thought (Strasburger 1875, 1882, 1884) that the cell plate so formed became at once the middle lamella, upon which secondary and frequently tertiary layers were subsequently deposited by the protoplasts on either side. Strasburger here found support for his theory that the cell wall is essentially a transformed layer of the protoplast, in opposition to Nägeli and von Mohl, who regarded it as primarily a secretion product. As a result of further researches, however, he later (1898) abandoned this view and adopted an interpretation that had been suggested by Treub (1878), namely, that the cell plate formed by the consolidation of the swellings ("microsomes") on the spindle fibers very soon splits to form the plasma membranes of the two daughter cells, and that there is then secreted between these membranes by the protoplasts a substance which becomes the primary layer, or middle lamella. The correctness of this view was confirmed by the careful researches of Timberlake (1900) and Allen (1901). Timberlake pointed out that in the micro-

---

[1] Discussion is here limited to the walls of higher plant tissues. The ectoplast of naked cells has been dealt with in Chapter III.

sporocytes of *Larix* and the root cells of *Allium* the connecting fibers first thicken near the nuclei, then become uniform throughout their length, and finally become swollen at the equatorial region, indicating a transfer toward that region of the material that is to compose the cell plate. Allen was able to show not only that the middle lamella itself may increase in thickness by the addition of new material before the secondary layers begin to be laid down, but also that it consists in reality of two layers representing the secretions contributed by the two daughter protoplasts. Where these two masses of secreted material meet there is developed a median plane of weakness which is ordinarily invisible but along which the lamella invariably splits when inter-cellular spaces are developed by the rounding up of the cells. By the use of proper staining methods it has been found possible to differentiate this "primary cleavage plane." The continuity of the middle lamella is interrupted, if at all, only by the fine pores through which pass the protoplasmic strands connecting adjacent cells. (See p. 46.)

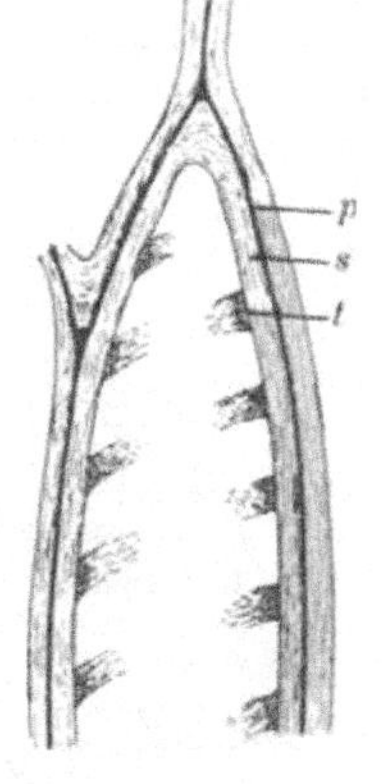

Secondary and Tertiary Wall Layers (Fig. 66).— It is probable that the deposition of the secondary layer begins after the cell has reached nearly or quite its full size, though to this there are apparently certain exceptions. The secondary layer, which seems to be formed with considerable rapidity, differs from the primary layer not only chemically (see below) but also in structure, being interrupted by circular or elongated areas in which no secondary substance is deposited, so that the cells at these places are separated only by the delicate primary membrane. Such a wall is said to be "pitted," the primary lamella extending across the *pit* being termed the *closing membrane.* The central portion of this membrane sometimes

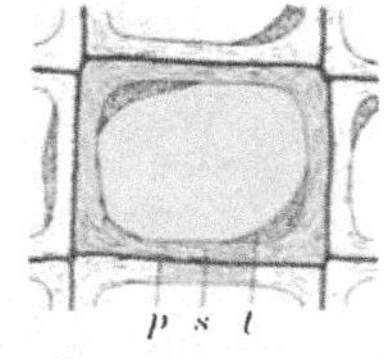

Fig. 66.—Longitudinal and transverse sections of a gymnosperm tracheid; *p,* primary wall or middle lamella; *s,* secondary layer; *t,* spiral tertiary thickening.

(vascular cells of gymnosperms chiefly) has a more or less conspicuous thickening known as the *torus.* The portion of the membrane around the torus is pierced by fine pores: in some cases these may become so large and numerous that the torus appears to be suspended on a meshwork (Fig. 67), while extreme cases are known in which it is held in place only by a few strands. In *bordered pits* (Fig. 68) the secondary wall overarches the margins of the closing membrane. In this type of pit, characteristic chiefly of water-conducting cells of the gymnosperms, the closing membrane is of such a nature that its position in the center of the pit is readily altered. Probably because of changes in pressure it swings to the side of the pit; the torus then lies against the pit opening, or "mouth," and the pit is blocked except for slow diffusion through the rather thick torus. The latter may even be forced tightly into the pit mouth.

The secondary wall layer may be even more limited in extent, only a small portion of the primary wall being covered. Such is the case in protoxylem cells, in which the secondary layer is deposited in the form of rings and spirals (Fig. 4). This form of thickening, together with the

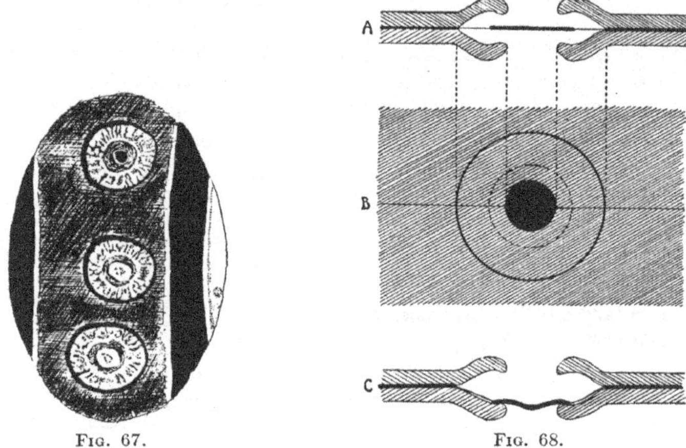

Fig. 67.   Fig. 68.

Fig. 67.—Pits in the wood of *Larix*, showing perforations in pit membrane. × 800. (*After Bailey.*)

Fig. 68.—Diagram of bordered pit of coniferous wood.

A, section of pit showing closing membrane supporting the torus, and secondary layers on each side of middle lamella. *B*, face view of same. *C*, section showing torus forced against mouth of pit. (*After Bailey.*)

peculiarly extensible character of their primary walls, allows for the great increase in length of these cells necessitated by the continued growth of the young organs in which they chiefly function. In some cells, notably the tracheids of certain gymnosperms and the vessels of many angiosperms, a tertiary layer is deposited upon the secondary wall. This tertiary layer takes the form of slender spirals, rings, and other figures resembling the secondary thickenings of protoxylem cells.

The Physical Nature of the Cell Wall.—Hugo von Mohl (1853, 1858) first expressed the idea that the cell wall grows by *apposition*, *i.e.*, by the deposition of material in successive laminæ. Although certain other workers (Wigand 1856) supported this view, it became over-shadowed for a time by the theory of Nägeli. This investigator, as a result of his classic researches on the wall and on starch grains (1858, 1862, 1863), concluded that the wall is made up of ultramicroscopic crystalline *micellæ* surrounded by water films. Growth of the wall in thickness and in area he believed to be due to the intercalation of new micellæ between the old ones, a process termed *intussusception*. Contrasted with this was Strasburger's development of the apposition theory (1882, 1889). Although Strasburger agreed that the wall had both solid and liquid constituents, he held that the latter were not complex micellæ, but only molecules linked together in the form of a reticular framework by their chemical affinities. Growth in area he thought was merely a matter of stretching without the intercalation of additional particles, while increase in thickness was supposed to be accomplished by apposition, or the deposition of layers of new material in the form of small particles, or microsomes. The striations which both he and von Mohl observed in the wall substance were regarded by Strasburger as due to the linear arrangement of these microsomes.

That the cell wall is not merely a lifeless secretion of the protoplast, but contains protoplasm in some form, is a view which has often been upheld, and involves problems which are still far from being solved. Prominence was given to the view by Wiesner (1886), who looked upon the growing cell membrane as a living part of the cell. Following Strasburger's early view, he held the primary layer to be wholly protoplasmic, and supposed the growing wall to be made up of regularly arranged particles, which he called *dermatosomes*, connected by fine fibrils of protoplasm. Growth was accomplished by the intussusception of new dermatosomes. Evidence in support of Wiesner's interpretation was brought forward by Molisch (1888), who showed that when tyloses come into contact pits are formed exactly opposite each other in the two abutting walls, a phenomenon which it would be difficult to explain were the walls without living substance.

The new intussusception theory of Wiesner was accepted by a number of workers including Haberlandt and Zacharias (1891). The apposition, or lamination, theory of Strasburger also had many supporters, among them being Noll (1887), Klebs (1886), Zimmermann (1887), and Askenasy (1890). According to Pfeffer (1892) both processes, the intussusception of new particles or molecules and the apposition of new material in layers, are concerned in the development of the wall. This view was later adopted by Strasburger (1898), and has received general acceptance. But much work must be done before any final conclusion can be drawn regarding many points. Especially obscure is the exact relationship of the protoplasm and the wall. The solution of this difficult problem must await the results of further inquiries by both the cytologist and the biochemist.

The Chemical Nature of the Cell Wall.—Through the researches of Payen (1842), Frémy (1859), Kabsch (1863), Wiesner (1861, 1878), and

13

particularly Mangin (1888–1893) it has been found that the chief constituents of the newly formed cell walls of plants are pectose and cellulose—that the primary wall or middle lamella consists of pectose, the secondary layer of pectose and cellulose, and the tertiary layer of cellulose. These substances however, rarely exist in the wall in pure and unmodified form. The pectose of the primary layer changes later to insoluble pectates, especially the pectate of calcium, while the secondary and tertiary layers very soon become greatly changed in composition, not alone through the addition of a variety of new substances, but also through an actual transformation which in some cases appears to be complete. For example, the secondary and tertiary layers of xylem cells, although at first containing much cellulose, may later become so completely transformed into or replaced by lignin that they show no reaction whatever to cellulose stains. In some cases the primary wall may undergo a certain amount of lignification also. The walls of many cells become heavily impregnated with cutin or suberin, the latter substance being responsible for the peculiar character of corky tissues. Infiltration by cutin, or "cutinization," is to be distinguished from "cuticularization," by which is meant the secretion of a layer of cutin (cuticle) on the outside of the cell. A variety of mineral substances, such as silica, calcium oxalate, and calcium carbonate, as well as more complex organic compounds, such as tannin, oils, and resins, are often deposited in the walls of old cells. The heartwoods of trees owe their qualities largely to the presence of these additional materials.

In spite of these modifications, however, it is still true that cellulose is the substance chiefly characteristic of plant cell walls in general. Although cellulose has been identified in certain animals, the membranes of practically all animal cells are composed of other substances, such as keratin, elastin, gelatin, and chitin. In the fungi also the rôle of cellulose appears to be played in part by chitin.

The Walls of Spores.—Special attention has been given to the development of the elaborate walls, or coats, of the spores of various plants in a number of investigations. Strasburger (1882, 1889, 1898, 1907) concluded that such coats arise by two general methods: (1) by the growth in thickness (by apposition) of the original wall of the spore cell through the activity of the protoplast, as in the pollen grains of *Malva* and other angiosperms, and (2) by a deposition of material upon the original wall by the tapetal fluid in which the young spores lie, as in the case of the megaspore of *Marsilia*.

The highly specialized coats of the megaspore of *Selaginella* have been most intensively studied, particularly by Fitting (1900, 1906) and Miss Lyon (1905), whose accounts disagree in several points. At the close of the tetrad division there is formed about each young spore a thick gelatinous "special wall," at the inner surface of which, according to Fitting, the spore coats begin to differentiate.   The *exospore* first appears, and just outside of it the rough *perispore* soon begins to develop.   Then a second layer, the *mesospore*, is formed within the exospore.   Between the protoplast, which is at this time very small, and the mesospore, and between the exospore and the mesospore, there are developed two cavities filled with a sporangial fluid which furnishes material to the growing coats.   Emphasis is placed on the fact that the coats are able to increase in thickness while they have no immediate contact with the protoplast. The protoplast now expands, after which a third coat, the *endospore*, is formed at its surface.   The mature spore thus has three coats according to Fitting's interpretation, which Denke (1902) and Campbell (1902) confirmed.

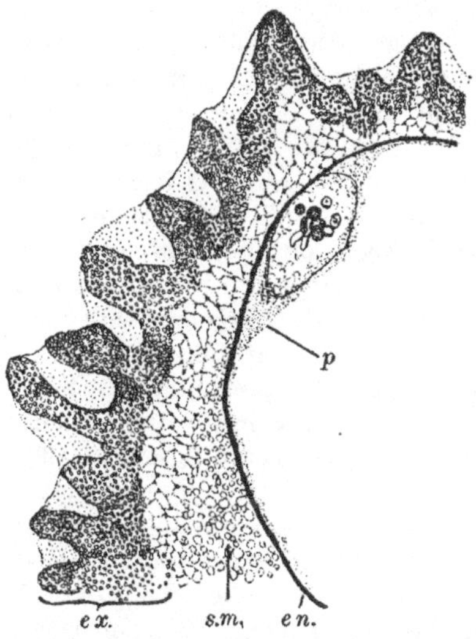

FIG. 69.—The developing megaspore coat of *Selaginella rupestris:*
p, protoplast with nucleus; *en*, endospore; *s.m.*, undifferentiated portion of "spore membrane;" *ex*, exospore: the outer denser portion is the "perinium."   (*After Lyon*, 1905.)

Miss Lyon found that the spore coats (*S. rupestris*) begin to differentiate in the midst of the "spore membrane" ("special wall:" Fitting), rather than at its inner surface as Fitting thought.   The *exospore* first appears as a double zone, the outer part of which becomes the *perinium* (*perispore*: Fitting) (Fig. 69).   The small protoplast gradually expands and pushes back the undifferentiated inner portion of the spore membrane; and while it does so a second coat is formed at its surface and becomes the *endospore* (*mesospore*: Fitting) which increases in thickness by lamination.   In another species (*S. emiliana*) the exospore and endospore form simultaneously.   Miss Lyon thus finds two coats rather than three, but points out that a portion of the spore membrane which may remain in an undifferentiated condition until a late stage may easily be mistaken for a third coat. The two "spaces" in the immature spore wall she holds to be undifferentiated regions in the spore membrane, and not cavities filled with a foreign fluid; and further urges that the protoplast is at all times in contact with the gelatinous spore membrane in which the coats are differentiating, opposing the view that the latter have the power of independent growth in thickness.

Evidence favoring the view that the spore coats can grow while not in contact with the protoplast has been brought forward by Beer (1905, 1911) and Tischler (1908). Beer asserts that although both the primary wall and the secondary thickening layer of the pollen grain (in certain members of the Onagraceæ) originate in intimate connection with the plasma membrane, most of their subsequent growth occurs by intussusception while they are completely separated from the protoplast, which secretes the material used. The development of the pollen wall in *Ipomoea purpurea* has been described in great detail by Beer. Around each young spore immediately after its formation there appears a temporary gelatinous "special wall," upon the inner surface of which the protoplast deposits the *exine*, or outer spore coat. This is at first homogeneous, but soon differentiates into a thin outer lamella and an inner zone made up of a network of thickenings with the rudiments of spines at its nodes. Both the spines and the small rodlets, which develop in a clear space appearing between the outer lamella and the network of thickenings (*mesospore*), undergo most of their development after they are separated from the protoplast. Tischler (1908) reports that the exine of the pollen of sterile *Mirabilis* hybrids may continue to increase in thickness after the protoplast begins to degenerate.

As an example of the formation of spore coats through the activity of a tapetal plasmodium may be taken the case of *Equisetum*, described by Beer (1909) and Hannig (1911). The spores of this form have three coats: an *endospore*, an *exospore*, and a *perispore* consisting of several layers including the one which splits to form the "elaters." The young spore cell at first has a simple membrane, the rudiment of the exospore. The walls of the tapetal cells dissolve, allowing the cell contents to flow freely among the spores as a tapetal plasmodium. Upon the spore membrane the plasmodium deposits successively (1) an inner gelatinous layer, (2) the "middle coat," (3) an outer gelatinous layer, and (4) the elater layer. The exospore develops from the original membrane after the middle coat is formed, and the endospore, or innermost coat, is developed last of all.

From this brief review, to which other examples might be added, it is evident that spore coats may develop in a variety of ways, but too little is known to warrant any statement as to which method may be the most general one. Although cytological interest centers chiefly in other problems, further studies on spore coats would not only contribute to our understanding of cell wall formation, but would also aid in solving the problem of the possible existence of protoplasm in the wall.

Bibliography 9

Achromatic Figure.   Cytokinesis.   Cell Wall.

ALLEN, C. E.   1901.   On the origin and nature of the middle lamella.   Bot. Gaz. **32**: 1-34.

ASKENASY, E.   1890.   Ueber einige Beziehungen zwischen Wachsthum und Temperatur.   Ber. Deu. Bot. Ges. **8**: 61-94.

BAILEY, I. W.   1915.   The effect of the structure of wood upon its permeability. 1. Am. Railway Eng. Assn. Bull. 174.

1919.   Phenomena of cell division in the cambium of arborescent gymnosperms and their cytological significance.   Proc. Nat. Acad. Sci. **5**: 283-285.   1 fig.

1920.   The cambium and its derivative tissues.   III.   A reconnaissance of cytological phenomena in the cambium.   Am. Jour. Bot. **7**: 417-434.   pls. 26-29.

BEER, R.   1904.   The present position of cell-wall research.   New Phytol. **3**: 159-164.

1905.   On the development of the pollen grain and anther of some Onagraceæ. Beih. Bot. Centr. **19**: I 286-313.   pls. 3-5.

1909.   The development of the spores of *Equisetum*.   New Phytol. **8**: 261-266.

1911.   Studies in spore development.   Ann. Bot. **25**: 199-214.   pl. 13.

VAN BENEDEN, E.   1883.   Resherches sur la maturation de l'oeuf, la fécondation et la division cellulaire.   Arch. de Biol. **4**.

1887.   (van Beneden & Neyt).   Nouvelles recherches sur la fécondation et la division mitosique chez *l'Ascaride mégalocéphale*.   Bull. Acad. Roy. Belg. III. **14**: 215-295.   pl. 1.

BOVERI, TH.   1887.   Ueber die Befruchtung der Eier von *Ascaris megalocephala*. Sitz.-Ber. Ges. Morph. Phys. München **3**.

1888.   Zellenstudien.   II. Jen. Zeitschr. **22**: 685-882.   pls. 19-23.

1895.   Ueber das Verhalten der Centrosomen bei der Befruchtung des Seeigeleies, nebst allgemeinen Bemerkungen über Centrosomen und Verwandtes.   Verh. Phys.-Med. Ges. Würzburg, N. F. **29**.

BÜTSCHLI, O.   1876.   Studien über die ersten Entwicklungsvorgänge der Eizelle, die Zelltheilung, und die Kunjugation der Infusorien.   Senckenb. Naturforsch. Ges. **10**.

CALKINS, G. N.   1898.   Mitosis in *Noctiluca miliaris* and its bearing on the nuclear relations of the Protozoa and Metazoa.   Jour. Morph. **15**: 711-770.   pls. 40-42.

CAMPBELL, D. H.   1902.   Studies in the gametophyte of *Selaginella*.   Ann. Bot. **16**: 419-428.   pl. 19.

CHAMBERS, R.   1917.   Microdissection Studies.   II. The Cell aster.   A reversible gelation phenomenon.   Jour. Exp. Zool. **23**: 483-504.

1919.   Changes of protoplasmic permeability and their relation to cell division. Jour. Gen. Physiol. **2**: 49-68.   figs. 14.

CONKLIN, E. G.   1902.   Karyokinesis and cytokinesis in the maturation, fertilization and cleavage of *Crepidula* and other Gasteropoda.   Jour. Acad. Nat. Sci.   Phila. **12**: 5-121.   pls. 1-6.   figs. 33.

DAVIS, B. M.   Oögenesis in *Saprolegnia*.   Bot. Gaz. **35**: 233-249, 320-349.   pls. 9, 10.

DENKE, P.   1902.   Sporenentwicklung bei *Selaginella*.   Beih. Bot. Centr. **12**: 182-199.   pl. 5.

Devisé, R.	1914.	Le fuseau dans les microsporocytes du *Larix*.	(Note prelim.)
Comptes. Rend. Acad. Sci. Paris **158**: 1028–1030.

Drüner, L.	1894.	Zur Morphologie der Centralspindel.	Jen. Zeitschr. **28**: 469–474.

1895.	Studien über den Mechanismus der Zelltheilung.	Ibid. **29**: 271–344.
pls. 4–8.

von Erlangen, R.	1897.	Beobachtungen über die Befruchtung und ersten Teilungen an den lebenden Eiern kleiner Nematoden.	Biol. Centr. **17**: 152–160,
339–346.	figs. 25.

Farr, C. H.	1916.	Cytokinesis of the pollen-mother-cells of certain Dicotyledons
Mem. N. Y. Bot. Gard. **6**: 253–317.	pls. 27–29.

1918.	Cell-division by furrowing in *Magnolia*.	Am. Jour. Bot. **5**: 379–395.
pls. 30–32.

Faull, J. H.	1912.	The cytology of *Laboulbenia chœtophora* and *L. Gyrinidarum*.
Ann. Bot. **26**: 325–355.	pls. 37–40.

Fitting, H.	1900.	Bau und Entwicklungsgeschichte der Makrosporen von *Isoetes*
und *Selaginella* und ihre Bedeutung für die Kenntniss des Wachsthums pflanzlichen Zellenmembranen.	Bot. Zeit. **58**^{1}:107–164.	pls. 5, 6.

1906.	(Review of paper by Lyon, 1905).	Ibid. **64**: 42–43.

Flemming, W.	1891.	Neue Beiträge zur Kenntniss der Zelle.	II. Arch. Mikr.
Anat. **37**: 685–751.	pls. 38–40.

Frémy, E.	1859a.	Recherches chimiques sur la composition des cellules végétales.
Comptes Rend. Acad. Sci.	Paris **48**: 202–212.

1859b.	Recherches chimiqies sur la cuticle.	Ibid. 667–673.

1859c.	Recherches sur la composition chimique du bois.	Ibid. 862–868.

Gallardo, A.	1896a.	La carioquinesis.	Ann. Soc. Cientif.	Argentina **42**.

1896b.	Essai d'interprétation des figures karyokinétiques.	Ann. Mus. Nac. d.
Buenos Aires.

1901.	Les croisments des radiations polaires et l'interprétation dynamique des
figures de karyokinèse.	Soc. de Biol. **53**.

1906.	L'interprétation bipolaire de la division karyokinétique.	Ann. Mus. Nac.
d. Buenos Aires **13**: 259.

1909.	La division de la cellule phénomène bipolaire de caractère électro-colloidal.
Arch. Entw. **28**: 125–154.	Figs. 9.

Hannig, E.	1911.	Ueber die Bedeutung der Periplasmodien.	I. Die Bildung des
Perispors bei *Equisetum*.	II. Die Bildung der Massulæ von *Azolla*.	Flora **102**:
209–278, pls. 13, 14.	figs. 17.

Harper, R. A.	1899.	Cell division in sporangia and asci.	Ann. Bot. **13**: 467–525.
pls. 24–26.

1900.	Cell and nuclear division in *Fuligo varians*.	Bot. Gaz. **30**: 217–251.	pl. 14.

1914.	Cleavage in *Didymium melanospermum* (Pers.) Macbr.	Am. Jour. Bot. **1**:
127–144.	pls. 11, 12.

Hartog, M.	1905.	The dual force of the dividing cell.	I.	The achromatic spindle
figure illustrated by magnetic chains of force.	Proc. Roy. Soc.	London **76**
B: 548–567.	pls. 9–11.

1914.	The true mechanism of mitosis.	Arch. Entw. **40**: 33–64.	figs. 16.

Heilbrunn, L. V.	1915.	Studies in artificial parthenogenesis.	II.	Physical changes
in the egg of *Arbacia*.	Biol. Bull. **29**: 149–203.

Hermann, J.	1891.	Beiträge zur Lehre von der Entstehung der karyokinetischen
Spindel.	Arch. Mikr. Anat. **37**: 569–586.	pl. 31.

Jörgensen, M.	1913.	Zellenstudien.	II. Die Ei- und Nährzellen von *Piscicola*.
Arch. Zellf. **10**: 125–160.	pls. 13–18.

KABSCH, W. 1863. Untersuchungen über die chemische Beschaffenheit der Pflan_
zengewebe; mit Bezug auf die neuesten Arbeiten Frémys über diesen Gegenstand.
Jahrb. Wiss. Bot. **3**: 357–399.

KLEBS, G. 1886. Ueber die Organization der Gallerte bei einigen Algen und Flagel_
laten. Unters. Bot. Inst. Tubingen **2**: 333–418. pls. 3, 4.

KLEIN, E. 1878, 1879. Observations on the structure of cells and nuclei. Quar.
Jour. Micr. Sci. **18**: 315–339, pl. 16; **19**: 125–175. pl. 7.

KÜHN, A. 1920. Untersuchungen zur kausalen Analyse der Zellteilung. I. Arch.
f. Entw. **46**: 259–327. pls. 11, 12. figs. 21.

LAWSON, A. A. 1898. Some observations on the development of the karyokinetic
spindle in the pollen mother cell of *Cobæa scandens*. Proc. Calif. Acad. Sci. Bot.
III **1**: 169–184. pls. 33–36.

1900. Origin of the cones of the multipolar spindle in *Gladiolus*. Bot. Gaz. **30**:
145–153. pl. 12.

1903. Studies in spindle formation. Ibid. **36**: 81–100. pls. 15, 16.

1911. Nuclear osmosis as a factor in mitosis. Trans. Roy. Soc. Edinb. **48**: 1
137–161. pls. 1–4.

LILLIE, R. S. 1903. On differences in the direction of the electrical connection of
certain free cells and nuclei. Am. Jour. Physiol. **8**: 273–283.

LUTMAN, B. F. 1911. Cell and nuclear division in *Closterium*. Bot. Gaz. **51**:
401–430. pls. 22, 23.

LYON, F. 1905. The spore coats of *Selaginella*. Bot. Gaz. **40**: 285–295. pls. 10, 11.

MANGIN, L. 1888. Sur la constitution de la membrane des végétaux. Comptes
Rend. Acad. Sci. Paris. **107**: 144.

1889. Sur la presence des composées pectiques dans les végétaux. Ibid. **109**: 579.

1890*a*. Sur la substance intercellulaire. Ibid. **110**: 295.

1890*b*. Sur les réactifs colorants des substances fondamentales de la membrane.
Ibid. **111**: 120.

1891. Observations sur la membrane cellulosique. Ibid. **113**: 1069.

1893. Sur l'emploi du rouge de ruthénium en anatomie végétale. Ibid. **116**. 653.

McCLENDON, J. F. 1910*a*. On the dynamics of cell-division. I. The electric charge
on colloids in living cells in the root tips of plants. Arch. Entw. **31**: 80-90. pl. 3.
figs. 2.

1910*b*. On the dynamics of cell-division. II. Changes in permeability in developing eggs to electrolytes. Am. Jour. Physiol. **27**: 240–275.

1913. The laws of surface tension and their applicability to living cells and cell
division. Arch. Entw. **37**: 233–247. figs. 10.

MEEK, C. F. U. 1913. The problem of mitosis. Quar. Jour. Micr. Sci. **58**: 567–592.

MEVES, FR. 1896, 1898. Zelltheilung. Ergeb. d. Anat. u. Entw. **6**: 285–390; **8**:
430–542. (Review.)

VON MOHL, H. 1853. Ueber die Zusammensetzung der Zellmembran aus Fäsern.
Bot. Zeit. **11**: 753–762, 769–775.

1858. Die Untersuchungen des Pflanzengewebes mit Hülfe des polarisierten
Lichtes. Ibid. **16**: 373–375. 1 pl.

MOLISCH, H. 1888. Zur Kenntniss der Thyllen, nebst Beobachtungen über Wund-
heilung in der Pflanzen. Sitzber. k. Akad. Wiss. Wien, Math.-Naturw. Cl.
I **97**: 264–299. pls. 1, 2.

MOREAU, F. 1913. Les Karyogamies multiples de la zygospore de *Rhizopus nigricans*. Bull. Soc. Bot. France **60**: 121–123.

MORGAN, T. H. 1899. The action of salt solutions on the unfertilized and fertilized
eggs of *Arbacia*, and of other animals. Arch. Entw. **8**: 448–539. pls. 7–10.
figs. 21.

MOTTIER, D. M.   1900.   Nuclear and cell-division in *Dictyota dichotoma*.   Ann. Bot. **14**: 166–192.   pl. 11.

VON NÄGELI, C.   1858.   Die Stärkekörner.   Zurich.

   1863*a*.   Ueber die chemische Zusammensetzung der Stärkekörner und Zellmembran.   Ibid 1863.   (See also Nägeli's Bot. Mitt. **1, 2**.)

   1863*b*.   Die Anwendung der Polarisationsmikroscops auf die Untersuchungen der organischen Elementartheile.   Beitr. z. Wiss. Bot. **3**: 1–126.   pls. 1–7.

   1864.   Ueber den innern Bau der vegetabilischen Zellmembran.   Sitzber. k. b. Akad.   1864.

NOLL, F.   1887.   Experimentaluntersuchungen über d. Wachsthum d. Zellmembran.

NOTHNAGEL, M.   1916.   Reduction divisions in the pollen mother-cells of *Allium tricoccum*.   Bot. Gaz. **61**: 453–476.   pls. 28–30.   fig. 1.

PAYEN, A.   1842.   Mémoires sur les développements des végétaux.   Paris.

PFEFFER, W.   1892.   Studien der Energetik.

PRENANT, A.   1910.   Théories et interprétations physiques de la mitose.   Jour. de l'Anat. et Phys. **46**.

RABL, C.   1889.   Ueber Zellteilung.   Anat. Anz. **4**: 21–30.   figs. 2.

RHUMBLER, L.   1896.   Versuch einer mechanischen Erklärung der indirekten Zell- und Kerntheilung.   I. Cytokinese.   Arch. Entw. **3**: 527–623.   pl. 26. figs. 39.

   1897.   Stemmen die Strahlen der Astrosphäre oder ziehen sie?   Ibid. **4**: 659–730. pl. 28.   figs. 27.

   1898.   Die Mechanik der Zelldurchschnurung nach Meves' und nach meiner Auffassung.   Ibid. **7**: 535–554.   pl. 12.   figs. 5.

   1899.   Furchung des Ctenophoreneies nach Ziegler und deren Mechanik: usw. Ibid. **8**: 187–238.   figs. 28.

   1903.   Mechanische Erklärung der Aehlichkeit zwischen magnetischen Kraftlinien- system und Zellteilungsfiguren.   Ibid. **16**: 476–535.   figs. 36.

ROBERTSON, T. B.   1909.   Note on the chemical mechanics of cell-division.   Arch. Entw. **27**: 29–34.

   1911.   Further remarks on the chemical mechanics of cell-division.   Ibid. **32**: 308– 313.

   1913.   Further explanatory remarks concerning the chemical mechanics of cell division.   Ibid. **35**: 692–707.   figs. 3.

RYTZ, W.   1907.   Beiträge zur Kenntniss der Gattung *Synchytrium*.   Centr. f. Bakt. II **18**: 635–655, 799–825.   1 pl.   figs. 10.

SAKAMURA, T.   1920.   Experimentelle Studien über die Zell- und Kernteilung mit besonderer Rücksicht auf Form, Grosse und Zahl der Chromosomen.   Jour. Coll. Sci. Imp. Tokyo **39**: pp. 221.   pls. 7.

SHARP, L. W.   1911.   The embryo sac of *Physostegia*.   Bot. Gaz. **52**: 218–225.   pls. 6, 7.

SPEK, J.   1918*a*.   Oberfläschenspannungsdifferenzen als eine Ursache der Zellteilung. Arch. Entw. **44**: 1–113.   figs. 25.

   1918*b*.   Die amöboiden Bewegungen und Strömungen in den Eizellen einiger Nematoden während der Vereinigung der Vorkerne.   Arch. Entw. **44**: 217–255. figs. 15.

STRASBURGER, E.   1875.   Ueber Zellbildung und Zelltheilung.   Jena.

   1882.   Ueber den Bau und das Wachstum der Zellhäute.   Jena.

   1884.   Die Controversen der indirekten Kernteilung.   Arch. Mikr. Anat. **23**: 246– 304.   pls. 13, 14.

   1888.   Ueber Kern- und Zellteilung im Pflanzenreich, nebst einem Anhang über Befruchtung.   Hist. Beitr. **1**.   pp. 258.   pls. 3.

   1889.   Ueber das Wachstum vegetabilischer Zellhäute.   Ibid. **2**.

1892.   Schwärmsporen, Gameten, pflanzliche Spermatozoiden und das Wesen der Befruchtung.   Ibid. **4**: 49–158.   pl. 3.

1897.   Ueber Cytoplasmastrukturen, Kern- und Zellteilung.   Jahrb. Wiss. Bot. **30**: 375–405.   figs. 2.

1898.   Die pflanzlichen Zellhäute.   Ibid. **31**: 534–598.   pls. 15, 16.

1907.   Apogamie bei *Marsilia*.   Flora **97**: 123–191.   pls. 3–8.

SWINGLE, D. B.   1903.   Formation of the spores in the sporangia of *Rhizopus nigricans* and of *Phycomyces nitens*.   U. S. Dept. Agric. Bur. Plt. Indus. Bull 37.

SWINGLE, W. T.   1897.   Zur Kenntniss der Kern- und Zelltheilung bei den Sphacelariaceen.   Jahrb. Wiss. Bot. **30**: 296–350.   pls. 15, 16.

TAYLOR, W. R.   1920.   A morphological and cytological study of reproduction in the genus *Acer*. Contrib. Bot. Lab. U. of Pa. **5**: pp. 30.   pls. 6–11.

TIMBERLAKE, H. G. 1900.   The development and function of the cell plate in higher plants.   Bot. Gaz. **30**: 73–99.   154–170.

1902.   Development and structure of the swarm spores of *Hydrodictyon*.   Trans. Wis. Acad. Sci. **13**: 486–522.   pls. 29, 30.

TISCHLER, G.   Zellstudien an sterilen Bastardpflanzen.   Arch. Zellf. **1**: 33–151.

TREUB, M.   1878.   Quelques recherches sur le rôle du noyau dans la division des cellules végétales.   Amsterdam.

WATASE, S.   1894.   Origin of the centrosome.   Biol. Lectures, Woods Hole.

WIESNER, J.   1864.   Untersuchungen über das Auftreten von Pectinkörper in dem Geweben der Runkelrübe.   Sitzber. k. Akad. Wiss.   Wien, Math.-Naturw.   Cl. II **50**: 442–453.

1886.   Untersuchungen über die Organization der vegetabilischen Zellhaut.   Ibid. I **93**: 17–80.   figs. 5.

WIGAND, A.   1856.   Ueber die feinste Struktur der Zellenmembran.   Schriften d. Ges. z. Beförd. d. ges. Naturwiss. zu Würzburg.

WILLIAMS, C. L.   1899. ' The origin of the karyokinetic spindle in *Passiflora cærulea*. Proc. Calif. Acad. Sci. III Bot. **1**: 189–206.   pls. 33–40.

WILSON, E. B.   1899.   On protoplasmic structure in the eggs of echinoderms and some other animals.   Jour. Morph. **15**: Suppl.   1–23.

1900.   The Cell in Development and Inheritance.

1901.   Experimental studies in cytology.   I. A cytological study of parthenogenesis in sea urchin eggs.   Arch. Entw. **12**: 529–596.   pls. 11–17.   figs. 12.

ZACHARIAS, E.   1888.   Ueber Kern und Zellteilung.   Bot. Zeit. **46**: 33–40, 51–62. pl. 2.

1891.   Ueber das Wachstum der Zellhaut bei Wurzelhaaren.   Flora **74**: 466–491. pls. 16, 17.

ZIMMERMANN, A.   1887.   Die Pflanzenzelle.

1893.   Sammel-Referate.   6.   Beih. Bot. Centr. **3**: 342–354

CHAPTER X

OTHER MODES OF NUCLEAR DIVISION

In accordance with the well established principle which states that
only through the simpler organisms can an adequate understanding of
those higher in the scale of complexity be approached, search has been
made for primitive modes of nuclear division with the hope that light
may thereby be thrown upon the origin and significance of the elaborate
karyokinetic process which is so universally found in the cells of higher
animals and plants. It is to be acknowledged that such a phylogenetic
explanation of mitosis is very far from being reached, but many of the
observations recorded are nevertheless of a very suggestive nature. To
botanists the most interesting of these have been made upon the Cyanophyceæ, which have long been a subject of controversy in this connection.

Cyanophyceæ.—For many years the nature of the "central body"
of the cells of such blue-green algæ as *Oscillatoria* remained very obscure.
Bütschli (1890), Dangeard (1892), Scott (1888), and others believed it
to be a nucleus of a somewhat primitive type, whereas other investigators,
among them Zacharias (1892) and Chodat (1894), denied its nuclear
nature. Zukal (1892) held that the peripheral portion of the cell represents a chromatophore, the central body consisting of cytoplasm with a
number of minute nuclei imbedded in it.

One of the first critical accounts based partly on the study of sections
was that of Fischer in 1897. Fischer concluded that the central body,
in which he found no chromatin, is the main portion of the cytoplasm,
and not to be regarded as the forerunner of the nucleus or indeed as an
independent organ at all. He also investigated the nature of the peripheral portion of the protoplast. By treating the plants with 10 per
cent hydrofluoric acid he dissolved away the other parts of the cell, leaving this portion intact; and as the result of comparative studies on other
plants he concluded, in harmony with Zukal, that it is a single large
chromatophore.

Since Fischer's work the most important contributions are those of
Hegler, Kohl, Olive, Phillips, Gardner, and Miss Acton. Contrary to
the view of Fischer, all of these cytologists interpret the central body
as a nucleus, and the first three regard its division as essentially mitotic.
The opinions of these workers with respect to the organization of the cell
of the Cyanophyceæ and the behavior of its nucleus are summarized
below.

According to Hegler (1901) the nucleus contains granules of chromatin but no nucleolus or nuclear membrane, division occurring by a simple form of mitosis. The coloring matter exists in the form of minute granules or *cyanoplasts*. Two other kinds of bodies are also present: albuminous *slime globules* and albuminous crystals (*cyanophycin granules*) representing reserve food.

Kohl's (1904) description of the cell of *Tolypothrix* (Fig. 70) is one of the most detailed which has been given in this group of researches. Kohl shows that the nucleus of this form has extensions reaching outward

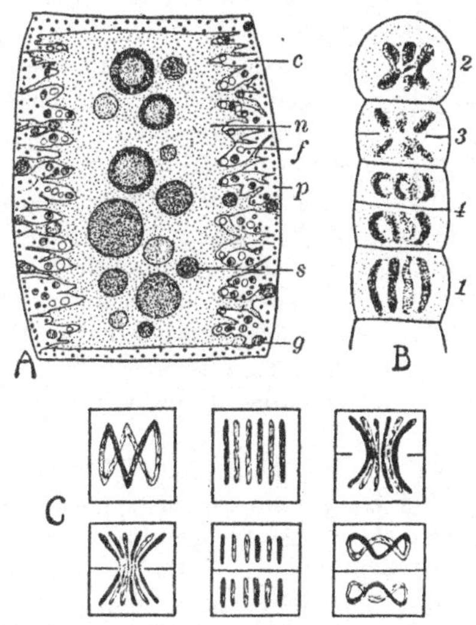

FIG. 70.—Structure and division of the cell of *Tolypothrix lanata.*

A, cell in the vegetative state: *c*, cytoplasm; *n*, nucleus; *f*, fat droplets; *p*, phycocyanin and chlorophyll granules; *s*, slime globules; *g*, granules of cyanophycin. *B*, four stages of cell-division in *Tolypothrix*, showing transverse division of chromosomes. *C*, diagram showing 6 stages of cell-division. (*After Kohl*, 1903.)

toward the cell wall, and that they are withdrawn at the time of nuclear division. The nucleus, which contains chromatin, also includes a number of large *Zentralkörner*, or slime globules, while in the cytoplasm are fat droplets, cyanophycin granules of reserve albumen, and granules of chlorophyll and phycocyanin. The nucleus, which is very rarely in the resting state, divides as follows: the chromatic material forms a spireme which segments into a definite number of chromosomes; these lie in the direction of the long axis of the cell and break transversely as the separating wall grows inward from the periphery. Their halves are thus included in the two daughter cells, where they form daughter nuclei without membranes.

Olive (1904)[1] finds that the nucleus of *Oscillatoria* (Fig. 71) consists of a fibrous achromatic framework with a number of very small chromatin granules, and is nearly always in some stage of division. A spireme is formed carrying 16 chromatin granules (8 in *Glœocapsa* and 32 in one species of *Oscillatoria*), each representing a chromosome. The spireme and its chromatin granules are split longitudinally, and the daughter spiremes with the daughter granules separate, a distinct central spindle extending between them. The dividing wall is formed as a centripetally growing partition. In *Glœocapsa* the cell divides by simple constriction. The vegetative nuclei of *Oscillatoria* very rarely approach the resting condition, but in spores and heterocysts they soon pass into this state, a nuclear membrane and vacuole being developed. In the heterocyst the protoplast disorganizes. Olive regards the central body of the Cyanophyceæ as not essentially different from the nucleus of the higher plants, although it is relatively primitive in several features. In the cytoplasm he finds both cyanophycin granules and slime globules, but no cyanoplasts, the coloring matters being diffused in the peripheral portion of the protoplast.

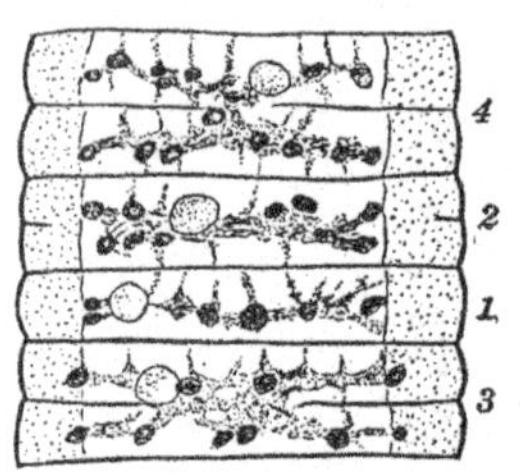

Fig. 71.—Nuclear division in *Oscillatoria Froelichia*. 1, 2, 3, 4, four successive stages. (*After Olive*, 1904.)

Fischer (1905), in reply to the claims of Kohl and Olive, reasserted his view that the central body is not a nucleus, but rather an accumulation of carbohydrate materials. The glycogen formed as a result of assimilatory activity gathers in the central body where it is transformed into another carbohydrate, *anabœnin*, which assumes the form of sausage-shaped structures. At the time of cell-division these masses of reserve material are distributed by a process of "pseudomitosis" to the daughter cells. Fischer therefore regards the mitotic figures observed by others as significant in connection with nutrition rather than with the functions usually attributed to nuclei.

Gardner (1906), investigating a number of species, found nuclei of three kinds, which he called the diffuse type, the net karyosome type, and the primitive mitosis type respectively. The "diffuse type" of nucleus, which has no very definite delimitation from the peripheral portion of the protoplast, contains an indefinite number of chromatin masses. As the cell divides this central aggregation of chromatic material divides into approximately equal portions. In the "net karyosome" type, found in *Dermocarpa*, the distinction between the nucleus and the surrounding cytoplasm is much clearer. The nucleus has an achromatic

---

[1] A very convenient tabulation of the results of researches on cell structure in the Cyanophyceæ up to 1904 is given by Olive.

network with chromatin granules at its nodes, and constricts simultaneously into a large number of daughter nuclei which pass to the conidia. In *Synochocystis aquatilis* occurs the "primitive mitosis" type: here Gardner found the only case of anything approaching mitotic behavior. A spireme develops and segments into three pieces which arrange themselves parallel to the long axis of the cell and divide transversely; the daughter pieces then separate and a centripetally growing cell wall completes the division of the cell. Gardner thus finds in the Cyanophyceæ "a series of nuclear structures, beginning with a very simple

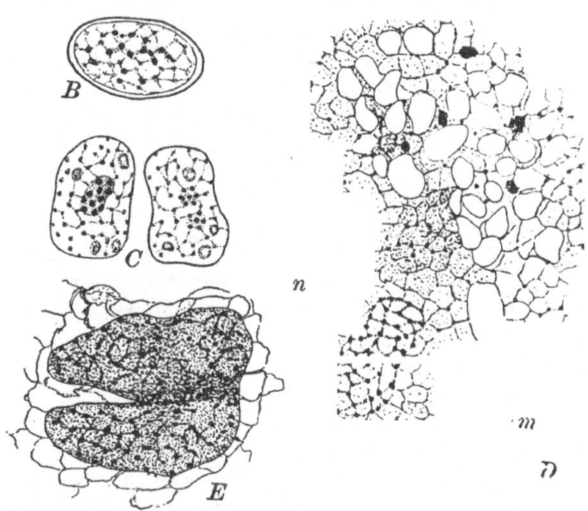

—FIG. 72. The nuclei of various members of the Chroococcaceæ.

A, cell of *Chroococcus turgidus* with scattered metachromatin granules (*m*) and plasmatic microsomes (*p*); division beginning. × 2500. B, cell of *Glococapsa* with chromatic granules. C, *Merismopedia elegans*, showing two stages of nuclear division. × 1500. D. *Chroococcus macrococcus*: *n*, nucleus; *m*, metachromatin; *v*, vacuole. × 2500. E, dividing nucleus of *Chroococcus macrococcus*. × 2500. (*After Acton, 1914.*)

form of nucleus scarcely differentiated from the surrounding cytoplasm and dividing by simple direct division" and passing "by very gradual steps to a highly differentiated form of nucleus which in dividing shows a primitive type of mitosis, and in structure approximates the nucleus of the Chlorophyceæ and the higher plants."

In a more recent investigation of the Chroococcaceæ Miss Acton (1914) finds that the nucleus is in general much simpler than that of the higher plants. Like Gardner, however, she points out a series beginning with a form in which definite organization is almost entirely lacking and ending with one in which the structure of the higher plant nucleus is closely approached (Fig. 72). In *Chroococcus turgidus* the protoplast is made up of a ground substance with a reticulum bearing bodies of two sorts: granules of *metachromatin* closely similar to chromatin in reaction, and cyanophycin granules, or *plasmatic microsomes*. Although there is no definitely delimited central region in the cell the metachromatin is found mostly at the center and the cyanophycin mostly nearer the periphery. When the metachromatin granules become numerous division sets in, a centripetally growing wall cleaving the protoplast into two daughter cells. In *Glœocapsa* the central region is somewhat more definite and may often show a spireme-like appearance such as Olive describes; but this may possibly be an artifact. In *Merismopedia elegans* there is a definitely delimited nucleus, not like that of the higher plants but merely an accumulation of chromatin or chromatin-like material which divides just before the cell constricts into two portions. In *Chroococcus macrococcus*, finally, the nucleus and cytoplasm are sharply distinct, the former having a reticulum with chromatin granules at its nodes and dividing by a sort of constriction at the time of cell-division.

As a result of these observations Miss Acton advances a theory of the evolution of nucleus and cytoplasm, which is briefly as follows. The excess food elaborated by the protoplast with its pigments was first stored as plasmatic microsomes composed of a carbohydrate, cyanophycin. As the reserve material became more complex in nature the nucleo-protein metachromatin was elaborated; this became aggregated at the center of the cell, insuring its equal distribution in cell-division, as in *Merismopedia*. There thus arose in the cell a physiological and morphological differentiation, the nucleo-protein with its portion of the supporting reticulum becoming a stable nucleus, as in *Chroococcus macrococcus*, and the ground substance remaining as the cytoplasm.

Summary.—In the Cyanophyceæ, therefore, although these forms in all probability had nothing directly to do with the evolution of the higher plants, we see a series of stages such as may well have occurred in the evolution of the nucleus and its complicated mitotic division. In the simplest forms the material concerned with those cell activities which in higher organisms are associated with the nucleus, is scattered throughout the cell without the morphological distinctness characteristic of an organ in the strict sense. It is passively distributed to the daughter cells when the cleavage wall is formed at the time of cell-division. In other cases this material reacts more strongly like true chromatin and may form a more or less definite aggregation separating into two masses as the cell divides. This metachromatin, which is a nucleic acid compound, has also been observed in other algæ, in Protozoa, and in fungi, including the yeasts. It appears to represent a reserve material, though it may also have other functions. Finally, definite and well organized nuclei are present in certain of the forms described in the foregoing pages, and although these nuclei may lack some of the features exhibited

by the nuclei of higher organisms, they show in the division and distribution of their chromatic elements many of the characteristics of true
mitosis.   In the evolution of the nucleus through such a series of stages
we have an illustration of "the conception of cell structure which im-

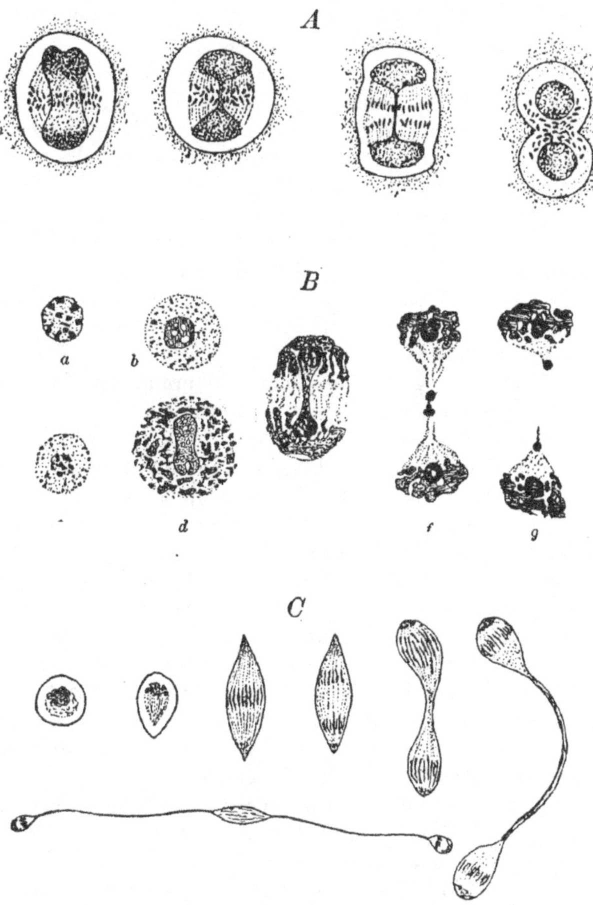

Fig. 73.—Nuclear division in Protozoa.

A, one form of mitosis in *Amœba diplomitotica*. (*From Minchin, after Aragao.*) *B*,
Nuclear division in *Coccidium schubergi*.   × 2250· (*From Minchin, after Schaudinn.*)
C, mitotic division of micronucleus of *Paramœcium* (horizontal figure on smaller scale
than others.) (*From Minchin, after Hertwig*)

plies differentiated regions of a colloidal system in which special processes
have become localized and tend to remain fixed" (Harper 1919).

Protozoa.—Such an apparent derivation of mitosis from a simpler,
more indefinite division of a less sharply delimited mass of special

chromatic substance is seen also in the protozoa. Here the chromatic material in a number of species is held together in loose granular aggregations, and the achromatic figure is seen in curious, relatively simple manifestations. In many cases, however, even among the Rhizopoda, there occurs a very advanced type of mitosis, with spindle, centrosomes, asters, and a definite number of chromosomes.

One of the simplest types of nuclear division found among the protozoa is known as "chromidial fragmentation." Here the nucleus is resolved into a large number of chromatin granules, or chromidia, which reassemble in two or more groups and form new nuclei. In nuclei of the "vesicular type," found commonly among the Microsporidia, the chromatin is concentrated in a single large body, or karyosome, which becomes dumbbell-shaped and divides, the rest of the nucleus then dividing also. In other forms, such as *Coccidium* (Fig. 73, *B*), the nucleus contains in addition a second kind of chromatin which is approximately halved in nuclear division. It is but a short step from such non-mitotic division as this to the simplest types of mitosis ("promitosis") seen in many protozoa. Within the group are found all gradations in complexity from such primitive modes of division up to the advanced types showing a complete achromatic figure with chromosomes which are regular in form, number, arrangement, and division, just as in the higher animals.[1]

Both Metcalf and Kofoid (1915) have emphasized the fundamental similarity of protozoan and metazoan nuclei. The process of mitosis has the same succession of phases in the two cases, though many minor variations occur. In some representatives of all the main groups of protozoa are found elongated chromosomes which Metcalf regards as linear aggregates of chromatin granules, and which split longitudinally, giving exact equivalence to the daughter-nuclei. Although the cell mechanism of Mendelian inheritance is thus held to be present in members of each great protozoan group and to operate as in the metazoa at the sexual stages, Metcalf believes this mechanism is not kept intact through the vegetative phases as it is in the higher groups.

Other Cases in Plants.—With regard to the myxomycetes, the researches of Strasburger (1884), Harper (1900), Jahn (1904, 1911), Olive (1907), and Winge (1912) have shown that nuclear division is essentially mitotic, and that in some cases the chromosomes are not only definite in number but undergo a reduction prior to spore formation. As an example of an exceptional condition may be taken *Sorodiscus* (Fig. 74, *A*), in which Winge describes two sorts of chromatin: vegetative *trophochromatin* and generative *idiochromatin*, the two forming a single mass at the center of the nucleus. As nuclear division begins this mass takes the form of three or four bodies very similar to chromosomes.

[1] For a description of mitotic phenomena in protozoa see Minchin (1912, Chapter VII).

The two kinds of chromatin now separate, the "trophochromatin placing itself in the center and the generative or idiochromatin lying like a thin equatorial plate around it." As the nucleus elongates the trophochromatin body becomes dumbbell-shaped and breaks into two, while the idiochromatin plate splits into daughter plates which apparently move to the poles and coöperate with the trophochromatin in the formation of the daughter nuclei.

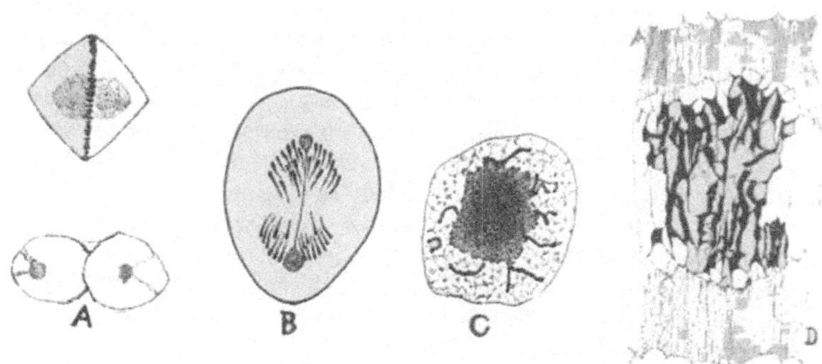

Fig. 74.

A, two stages of mitosis in *Sorodiscus*. (*After Winge*, 1912.) *B*, anaphase of nuclear division in *Euglena*. Chromosomes grouped about dividing "nucleolo-centrosome." (*After Keuten*, 1895.) *C*, chromosomes developing from nucleolus in *Spirogyra*. × 1335. (*After Berghs*, 1906.) *D*, Mitosis in *Spirogyra crassa*. (*After Merriman*, 1913.)

A process with much the same appearance at certain stages is seen in the flagellate, *Euglena* (Keuten 1895) (Fig. 74, *B*). Here the chromosomes group themselves about the large nucleolus which soon takes the form of a dumbbell-shaped "central spindle" or "centrodesmose." The nucleolus completes its division, the chromosomes meanwhile separating into two groups which pass to the poles and reorganize the daughter nuclei. In certain other flagellates Kofoid (1915) reports a split spireme and a definite number of chromosomes which differ markedly in size and shape.

In *Cladophora* (Carter 1919) nearly all the chromatin is contained in one or more large chromatin nucleoli, or karyosomes. After the numerous chromosomes have arrived at the two poles at the close of the anaphase the spindle connecting the two groups constricts and completes the division of the nucleus.

Another unusual condition is found in *Spirogyra* (Fig. 74, *C*, *D*). In this form nearly all of the chromatic material is lodged in the large nucleolus, the nuclear reticulum being very delicate and almost invisible in many preparations. According to Berghs (1906), Karsten (1908), and Tröndle (1912) all the chromosomes which appear in the prophase and split as usual are derived from this nucleolus, most of its material

14

being used in their formation. In the opinion of Miss Merriman (1913) the chromatic bodies observed by the above workers are not true chromosomes, but are rather more indefinite chromatic aggregations which are variable in number and appearance, and which are irregularly pulled apart as mitosis proceeds. She finds here "no evidence throughout the karyokinesis of an equational division of autonomous bodies."

In *Zygnema* both Escoyez (1907) and van Wisselingh (1914) find that the reticulum, and not the nucleolus, gives rise to all the chromosomes. Although the nucleolus furnishes no morphological element, chromatic material may flow from it to the chromosomes as they develop from the reticulum. Much the same condition is found in *Marsilia* (Strasburger 1907; Berghs 1909). Strasburger points out that in the somatic nuclei (in the cells of the root and the young prothallium) most of the chromatic substance is held in the nucleolus during the resting stages (Fig. 17, *E*), and that the material of the reticular framework, which is very delicate, is to be regarded as the substance of importance in heredity. Berghs shows that the nucleolus consists of an achromatic substratum which appears independently of the reticulum in the telophase and soon becomes impregnated with chromatic material transferred to it from the chromosomes. In the next prophase the chromatic material flows back to the delicate reticulum, from which the chromosomes gradually develop. As the chromosomes increase in distinctness the nucleolus becomes paler, and when the nuclear membrane breaks down the nucleolus dissolves in the protoplasmic liquid. It is therefore clear that in *Marsilia* the nucleolus is not a mere aggregation of the chromosomes of the telophase, as might at first be supposed. The chromosomes arise from the reticulum as usual, and not from the nucleolus as reported for *Spirogyra*. In these observations we have additional evidence favoring the view of Haecker, Boveri, Maréchal, and others (see Chapter VIII) that it is the achromatic substratum of the chromosome, and not the chromatic substance which it carries, that should be regarded as the persistent structural unity representing the basis of inheritance.

Amitosis.—In amitotic or direct nuclear division the nucleus simply constricts and separates into two portions while in the "resting" condition, no condensed chromosomes, centrosomes, spindle, or asters being formed. As a general rule such a division of the nucleus is not followed by a division of the cell; cells with two or more nuclei therefore commonly result. As examples may be cited the tapetal cells in the anthers of angiosperms, the internodal cells of *Chara* (Fig. 75) (Johow 1881), and certain glandular cells of animals. The presence of more than one nucleus cannot by itself be regarded as evidence that amitosis has occurred, however. Amitosis appears to be of rather frequent occurrence among the lower organisms, some of which show other methods of division also. For example, amitosis occurs regularly in budding yeasts, though the divisions giving rise to the ascospore nuclei have been shown to be mitotic in certain cases. (See Guilliermond 1920.) Amitosis was once believed to be the normal mode of nuclear division, mitosis being looked upon as very exceptional. The true condition, so far as higher organisms are concerned, has turned out to be quite the reverse: it is evident that amitosis occurs frequently in certain kinds of cells, but the mitotic method of division has been found to be almost universal.

What the physiological significance of amitosis may be is not well known. It was once suggested (Chun 1890) that it aids the processes of metabolism by increasing the nuclear surface in the cell, since it is of such frequent occurrence in cells with a distinctively nutritive function. This view has recently been restated by Nakahara (1917) as a result of his work on the larva of *Pieris*.[1] The most generally held opinion regarding amitosis in the higher organisms was for many years that expressed by Flemming (1891), namely, that it represents a degeneration phenomenon or aberration of some kind, which would explain why it is so often found in degenerating and pathological tissues. In the words of vom Rath (1891), "when once a cell has undergone amitotic

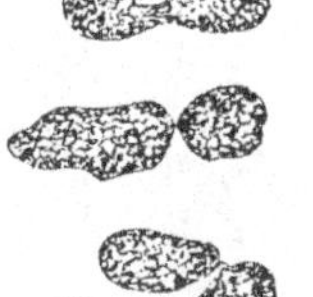

Fig. 75.—Amitosis in internodal cell of *Chara.* × 413.

division it has received its death-warrant; it may indeed continue to divide for a time by amitosis, but inevitably perishes in the end."

That the view of vom Rath must be modified has been indicated by the results of a number of investigations. For instance, Pfeffer (1899) and Nathansohn (1900) found that if *Spirogyra* filaments are placed in a ½ to 1 per cent solution of ether the nuclei divide by amitosis only, and that when the filaments are returned to pure water the mitotic method of division is resumed, with no evidence of degeneration. Haecker, however, working on the eggs of *Cyclops*, came to view such artificially induced behavior not as true amitosis but rather as a much modified mitotic division, which he termed "pseudoamitosis." Other cytologists observed nuclear divisions that seemed intermediate in character between mitosis and amitosis (Dixon in the endosperm of *Fritillaria*, 1895; Sargant in the embryo sac of *Lilium*, 1896; R. Hertwig in *Actinosphærium*, 1898; Buscalioni in the endosperm of *Corydalis*, 1898; and Wasielewski in the roots of *Vicia faba*, 1902, 1903). Hertwig accordingly concluded that mitosis and amitosis are separated by no sharp boundary line, but are connected by an unbroken series of transition stages.

[1] In a second paper (1918) Nakahara gives a convenient review of the literature of the subject.

As a result of his recent researches on chloralized cells (Fig. 76) Sakamura (1920) interprets all such unusual types of nuclear division as those described by Hertwig and Wasielewski as the effect of disturbed mitotic division, but denies the claim of those authors that such types of division represent actual transition stages between amitosis and mitosis. True amitosis he regards as a fundamentally different process. and as essentially a degeneration phenomenon.

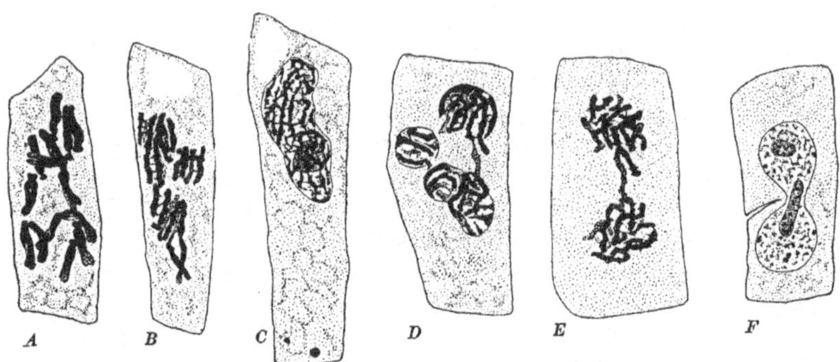

Fig. 76.—Abnormal mitosis in chloralized root cells of *Vicia*.

A, chromosomes distributed irregularly in cell. *B*, scattered chromosomes beginning to assume nuclear form. *C*, nucleus reconstructed by scattered chromosomes. *D*, scattered chromosomes reconstructing 3 separate nuclei. *E*, chromosomes reconstructing 2 nuclei connected by bridge. *F*, amitosis-like appearance resulting from condition shown in *E*. (*After Sakamura*, 1920.)

On the contrary, Des Cilleuls (1914) reports that in the rabbit periods of amitosis and mitosis succeed each other regularly in the same cell lineage without affecting the vitality of the cells. In his opinion, therefore, amitosis does not necessarily place the stigma of senescence upon the cell. A similar conclusion is reached by Arber (1914), who finds amitosis supplementing mitosis in the early growth stages of the leaves and adventitious roots of *Stratiotes aloides;* and by McLean (1914), who asserts that it is the sole method of nuclear division in the cortical parenchyma of several aquatic angiosperms. Saguchi (1917) likewise states that the nuclei in the ciliated cells of vertebrates divide by amitosis only.

Amitosis and Heredity.—One of the most important theoretical questions raised by the phenomenon of amitosis is that of the effect which the process may have upon the hereditary mechanism of the cell. According to the chromosome theory of heredity and development in its usual form it has been thought that, although amitosis may occur in connection with an altered metabolism in cells not to undergo further differentiation, mitosis must occur exclusively in the germ cell lineage, in order that the chromosomes and the hereditary elements they contain shall be properly distributed to the reproductive cells; and also in developing tissues and organs, so that differentiation may proceed normally. On the other hand, several workers (Meves; Flemming in his later papers) admit that amitosis may not affect any hereditary powers which the nuclei concerned may possess. Child (1907, 1911), who reports amitosis in both the somatic and germ cells of certain animals, where it appears to play an important rôle in the developmental cycle, strongly urges that such facts render the hypothesis of chromosome individuality highly improbable, and that our conceptions of the rôle of the cell organs in heredity must be greatly altered.

The hopelessly unsettled state of opinion on this question may be illustrated by the list of authors and their views cited by Conklin (1917). That amitosis frequently occurs in the process of normal cell differentiation, and therefore constitutes evidence against the chromosome theory, has been held by Nathansohn (1900), Wasielewski (1902, 1903), Gurwitsch (1905), Hargitt (1904, 1911), Child (1907, 1911), Patterson (1908), Glaser (1908), Jordan (1908), Jörgensen (1908), Maximow (1908), Moroff (1909), Knoche (1910), Nowikoff (1910), and Foot and Strobell (1911). Several of these investigators, together with R. Hertwig (1898), Lang (1901), Calkins (1901), Herbst (1909), Godlewski (1909), and Konopacki (1911), see no principal distinction between amitosis and mitosis, believing that both may occur without interfering with normal differentiation.

Haecker (1900), Němec (1903), and Schiller (1909) dissented from the above view, which was also strongly contested by Boveri (1907) and Strasburger (1908). Richards (1909, 1911) and Harman (1913) failed to confirm the results of Child on amitosis in cestodes, but Child (1911) reasserted his view, which was supported by Young (1913). Schürhoff (1919), working on *Podocarpus*, emphatically states that a nucleus which has once undergone true amitosis is incapable of dividing mitotically. Sakamura (1920) is of the same opinion.

In a careful study of maturation and cleavage in *Crepidula plana* Conklin (1917) finds that the nuclei divide only by mitosis. There are many apparent cases of amitosis, but upon careful examination they all prove to be only various modifications of the regular mitotic process. Such modifications are these: the scattering of the chromosomes and their failure to unite into a single nucleus; mitosis without cytokinesis, giving cells with two or more nuclei; the failure of certain daughter chromosomes to pull apart, leaving a chromatic bridge between the daughter nuclei; the persistence of the nuclear membrane, with a division of the chromosomes by mitosis and of the nuclear vesicle by constriction. Conklin concludes as a result of his many observations and an examination of the evidence offered by others, that there is not known a single conclusive case of true amitosis in a normally differentiating cell, and that all attacks upon the chromosome theory on the ground of amitosis have signally failed. The results obtained by Sakamura (1920) in his study of modified mitosis in chloralized plant cells are strikingly similar to those of Conklin, and his conclusions regarding the chromosome theory are essentially the same.

From the foregoing it is evident that the problem of the effect of amitosis upon the differentiation of the tissues in which it occurs and upon the hereditary powers of the nucleus is by no means easy of solution, and that much care must be used in interpreting supposed amitotic phenomena in fixed preparations. The work of Conklin and Sakamura has shown clearly that many of the phenomena reported as amitosis are in reality aberrations of the mitotic process, and that the opinions of many writers are undoubtedly due to a failure to recognize this fact. Should it be proved, however, that true amitosis may occur in the lineage of normally functioning germ cells a serious obstacle would be placed in the way of the chromosome theory of inheritance in its current form, for this theory requires that, no matter what happens in cells not in the direct line of the germ cells, nuclear division in this line must be exclusively mitotic in order that the hereditary mechanism in the nucleus shall be preserved. This mechanism, as we shall see in later chapters, is supposed to be of such a nature that amitosis would seriously derange its organization. In each daughter nucleus of an amitotic division some of the elements necessary for normal functional activity would presumably be lacking, owing to the simple mass division of the chromatin. With reference to this point it has been contended by Child that the nucleus is a dynamic system capable of regenerating its lost parts and "producing a whole" after amitosis. But it is a well established fact that when chromosomes are lost in abnormal mitotic division they are not regenerated by the daughter nuclei (non-disjunction; Chapter *XVII*).

In this connection an experiment performed by Chambers (1917) is of interest. This investigator succeeded in pinching the nucleus of an animal egg into two pieces. The two "amitotic" nuclei so produced reunited upon touching, after which the egg was fertilized and passed through the early cleavage stages in the normal manner. It is known that the character of these early stages is largely independent of the nuclei present, being the outgrowth of an organization already present in the egg cytoplasm. (See Chapter XIV.) The later stages, in which the effects of the hereditary constitution of the nucleus appear, were not reached in the present experiment. Moreover, the entire chromatin outfit was present in the reunited nucleus, which is not supposed to be true of a daughter nucleus of an amitotic division. From this experiment, therefore, it can only be concluded that whatever disturbance of the spatial arrangement of the nuclear elements may have been caused by the temporary separation of the nucleus into two parts, it had no serious effect on the nutritive functions performed by the nucleus during the early cleavage stages. Development did not proceed far enough to warrant any conclusion regarding the effect upon the rôle of the nucleus in differentiation and inheritance.

Although what probably represents amitosis has been observed in young germ cells, it has not been shown with certainty in any case that descendants of these amitotically dividing nuclei become the nuclei of normally functioning gametes. To gain conclusive evidence for such an occurrence it would be necessary to trace the descendants of the amitotically dividing nuclei through to particular gametes or spores and then to note the effect upon the individuals produced by them. This would be a matter of extreme experimental difficulty, and not at all possible in most organisms. If it were successfully accomplished and the individuals were found to be normal in every respect, not only in the cleavage stages but throughout development, the revision of the chromosome theory which various workers have advised would at once become necessary.

Bibliography 10

Other modes of nuclear division

ACTON, E. 1914. Observations on the cytology of the Chroococcaceæ. Ann. Bot. **28**: 433-454. pls. 23, 24.

ARBER, A. 1914. On root development in *Stratiotes atoides* L., with special reference to the occurrence of amitosis in an embryonic tissue. Froc. Camb. Phil. Soc. **17**: 369-379. pls. 2.

BERGHS, J. 1906. Le noyau et la cinèse chez le *Spirogyra*. La Cellule **23**: 53-86. pls. 3.

1909. Les cinèses somatiques dans le *Marsilia*. Ibid. **25**: 73-84. 1 pl.

BOVERI, TH. 1907. Zellen-Studien VI. Die Entwicklung dispermer Seeigeleier. Jena.

BUSCALIONI, L. 1898. Osservazioni richerche sulla cellula vegetale. Ann. Inst. Bot. Roma **7**.

BÜTSCHLI, O. 1890. Ueber den Bau der Bakterien und verwandter Organismen. Leipzig.

1896. Weitere Ausführungen über den Bau der Cyanophyceen und Bakterien. Leipzig.

1902. Bemerkungen über Cyanophyceen und Bakterien. Arch. Protist. **1**.

CALKINS, G. N. 1901. The Protozoa. New York.

CARTER, N. 1919. The cytology of the Cladophoraceæ. Ann. Bot. **33**: 467-478. pl. 27. figs. 2.

CHAMBERS, R. 1917. Microdissection studies. I. The visible structure of the cell protoplasm and death changes. Am. Jour. Physiol. **43**: 1-12. figs. 2.

CHILD, C. M. 1907a. Amitosis as a factor in normal and regulatory growth. Anat. Anz. **30**: 271-297. figs. 12.

1907b. Studies on the relation between amitosis and mitosis. III-VI. Biol. Bull **13**: 138-160, 165-184. pls. 2-10.

1911. The method of cell division in *Monczia*. Ibid. **21**: 280-296. figs. 16.

CHODAT, R. 1894. Contenue cellulaire des Cyanophycées. Arch. Sci. Phys. Math Geneve III **32**: 637-641.

CHUN, C. 1890. Ueber die Bedeutung der direkten Zelltheilung. Sitzber. Schr. Phys.-Oekon. Ges. Königsberg.

CONKLIN, E. G. 1903. Amitosis in the egg follicle cells of the cricket. Am. Nat. **37**: 667–675. figs. 8.

1912. Experimental studies on nuclear and cell division in the eggs of *Crepidula*. Jour. Acad. Nat. Sci. Phila. **15**: 503–591. pls. 43–49.

1917. Mitosis and amitosis. Biol. Bull. **33**: 396–436. pls. 10.

DANGEARD, P. 1892. Le noyau d'une Cyanophycée. Le Botaniste **3**: 28–31. pl. 2.

DES CILLEULS, J. 1914. Recherches sur la signification physiologique de l'amitose. Arch. d'Anat. Micr. **16**: 132–148. pls. 7, 8.

DIXON, H. H. 1895. Note on the nuclei of the endosperm of *Fritillaria imperialis*. Proc. Roy. Irish Acad. III **3**: 721–726. pl. 24.

ESCOYEZ, E. 1907. Le noyau et la caryocinèse chez le *Zygnema*. La Cellule **24**: 355–366. 1 pl.

FISCHER, A. 1897. Untersuchungen über den Bau der Cyanophyceen und Bakterien. Jena.

1905. Die Zelle der Cyanophyceen. Bot. Zeit. **63**: 51–129. pls. 4, 5.

FLEMMING, W. 1890. Amitotische Kerntheilung im Blasenepithel des Salamanders. Arch. Mikr. Anat. **34**: 437–451. pl. 27.

1891. Neue Beiträge zur Kenntniss der Zelle. II. Arch. Mikr. Anat. **37**: 685–751. pls. 38–40.

FOOT, K. and STROBELL, E. C. 1911. Amitosis in the ovary of *Protenor belfragi* and a study of the chromatin nucleolus. Arch. Zellf. **7**: 190–230. pls. 12–20.

GARDNER, N. L. 1906. Cytological studies in Cyanophyceæ. Univ. Calif. Publ. Bot. **2**: 237–296. pls. 21–26.

GLASER, O. 1908. A statistical study of mitosis and amitosis in the entoderm of *Fasciolaria tulipa* var. *distans*. Biol. Bull. **14**: 219–248.

GODLEWSKI, E. 1909. Das Vererbungsproblem, usw. Vorträge u. Aufsätze Entw. Mech. **9**.

GRIFFITHS, B. M. 1915. On *Glaucocystis Nostochinearum*, Itz. Ann. Bot. **29**: 423–432. pl. 19.

GRIGGS, R. F. 1909. Some aspects of amitosis in *Synchytrium*. Bot. Gaz. **47**: 127–138. pls. 3, 4.

GUILLIERMOND, A. 1906. Contribution a l'étude cytologique des Cyanophycées. Rev. Gén. Bot. **18**: 392–408, 447–465.

1920. The Yeasts. (Engl. transl. by F. W. Turner.) N. Y.

GURWITSCH, A. 1904. Morphologie und Biologie der Zelle. Jena.

HAECKER, V. 1900. Mitosen im Gefolge amitosenähnlicher Vorgänge. Anat. Anz. **17**: 9–20. figs. 16.

HARGITT, C. W. 1904. The early development of *Eudendrium*. Zool. Jahrb. **20**: 257–276. pls. 14–16.

1911. Some problems of Coelenterate ontogeny. Jour. Morph. **22**: 493–550. pls. 3.

HARMAN, M. T. 1913. Method of cell division in the sex cells of *Tænia teniæformis*. Jour. Morph. **24**: 205–244. pls. 8.

HARPER, R. A. 1900. Cell and nuclear division in *Fuligo varians*. Bot. Gaz. **30**: 217–251. pl. 14.

1914. Cleavage in *Didymium melanosporum* (Pers.) Macbr. Am. Jour. Bot. **1**: 127–144. pls. 11, 12.

1919. The structure of protoplasm. Am. Jour. Bot. **6**: 273–300.

HEGLER, R. 1901. Untersuchungen über die Organization der Phycochromzelle. Jahrb. Wiss. Bot. **36**: 229–354. pls. 5, 6. figs. 5.

HERBST, C. 1909. Vererbungsstudien VI. Arch. Entw. **27**: 266–308. pls. 7–10.

HERTWIG, R. 1898. Ueber Kerntheilung, Richtungskörperbildung und Befruchtung von *Actinosphærium eichornii*. Abh. Bäyer. Akad. Wiss. **19**.

1908. Ueber neue Probleme der Zellenlehre. Arch. Zellf. **1**: 1–32. figs. 9.

JAHN, E.   1904.   Myxomycetenstudien.   3. Kernteilung und Geisselbildung bei den Schwärmern von *Stemonitis flaccida* Lister.   Ber. Deu. Bot. Ges. **22**: 84–92. pl. 6.

1908.   Myxomycetenstudien.   7. *Ceratiomyxa*.   Ibid. **26a**: 342–352.

1911.   Myxomycetenstudien.   8. Der Sexualakt. Ibid. **29**: 231–247.   pl. 11.

JOHOW, F.   1881.   Die Zellkerne von *Chara fœtida*.   Bot. Zeit. **39**: 729–743, 745–753. pl. 7.

JORDAN, H. E.   1908.   The accessory chromosome in *Aplopus mayeri*.   Anat. Anz. **32**: 284–295.   figs. 48.

JÖRGENSEN, M.   1908.   Untersuchungen über die Eibildung bei *Nephilis*, usw. Arch. Zellf. **2**: 279–347.   pls. 20–23.   figs. 4.

KARSTEN, G.   1908.   Die Entwicklung der Zygoten von *Spirogyra jugalis* Ktzg. Flora **99**: 1–11.   pl. 1.

KEUTEN, J.   1895.   Die Kerntheilung von *Euglena viridis* Ehrenberg.   Zeit. Wiss. Zool. **60**: 215–235.   pl. 11.

VON KNOCHE, E. 1910.   Experimentelle und andere Studien am Insektenovarium. Zool. Anz. **35**: 261–265.   figs. 3.

KOFOID, C. A.   1915.   The evolution of the protozoan nucleus and its extranuclear connections.   Science **42**: 658.

KOHL, F. G.   1903.   Ueber die Organization und Physiologie der Cyanophyceenzelle, und die mitotische Teilung ihres Kerns.   Jena.

KONOPACKI, M.   1911.   Ueber den Einfluss hypotonischen Lösungen auf befruchtete Echinideneier.   Arch. Zellf. **7**: 139–184.   pls. 9–11.

LANG, A.   1901.   Lehrbuch der vergleichenden Anatomie.   II Aufl.   Jena.

MACKLIN, C. C.   1916.   Amitosis in cells growing *in vitro*.   Biol. Bull. **30**: 445–466.

MAXIMOW, A.   1908.   Ueber Amitose in den embryonalen Geweben bei Säugtieren. Anat. Anz. **33**: 89–98.   figs. 11.

McLEAN, R. C.   1914.   Amitosis in parenchyma of water plants.   Proc. Camb. Phil. Soc. **17**: 380–382.   1 fig.

MERRIMAN, M. L.   1913.   Nuclear division in *Spirogyra crassa*.   Bot. Gaz. **56**: 319–330.   pls. 11, 12.

METCALF, M. M.   1915.   Chromosomes in protozoa.   Science **42**: 658.

MINCHIN, E. A.   1912.   An Introduction to the Study of the Protozoa.   London.

MOROFF, TH.   1909.   Oögenetische Studien.   I. Copepoden.   Arch. Zellf. **2**: 432–494.   pls. 34–36.

NAKAHARA, W.   1917.   Preliminary note on the nuclear division in adipose cells of insects.   Anat. Record **13**: 81–86.   figs. 11.

1918.   Studies in amitosis: Its physiological relations in the adipose cells of insects, and its probable significance.   Jour. Morph. **30**: 483–526.   pls. 6.   figs. 36.

NATHANSOHN, A.   1900.   Physiologische Untersuchungen über amitotische Kernteilung.   Jahrb. Wiss. Bot. **35**: 48–78.   pls. 2, 3.

NĚMEC, B.   1903.   Ueber die Einwirkung des Chloralhydrats auf die Kern und Zellteilung.   Jahrb. Wiss. Bot. **39**: 645–730.   figs. 157.

NOWIKOFF, M.   1910.   Zur Frage nach der Bedeutung der Amitose.   Arch. Zellf. **5**: 365–374.   figs. 2.

OLIVE, E. W.   1904.   Mitotic division of the nuclei of the Cyanophyceæ.   Beih. Bot. Centr. **18**: 9–44.   pls. 1, 2.

1907.   Cytological studies on *Ceratiomyxa*.   Trans. Wis. Acad. Sci. **15**: 754–773. pl. 47.

PATTERSON, J. T.   1908.   Amitosis in the pigeon's egg.   Anat. Anz. **32**: 117–125. figs. 24.

PFEFFER, W.   1899.   Bericht über amitotische Kerntheilung.   Ber. d. Math.-Phys. Kl. d. Kgl. Sächs. Ges. Wiss.

PHILLIPS, O. P. 1904. A comparative study of the cytology and movements of the Cyanophyceæ. Contrib. Bot. Lab. Univ. Pa. **2**: No. 3.

VOM RATH, O. 1891. Ueber die Bedeutung der amitotischen Kerntheilung im Hoden. Zool. Anz. **14**: 331, 342, 355. figs. 3.

RICHARDS, A. 1909. On the method of cell division in *Tænia*. Biol. Bull. **17**: 309–326.

1911. The method of cell division in the development of the female sex organs of *Monezia*. Ibid. **20**: 123–178. 8pls.

SAGUCHI, S. 1917. Studies on ciliated cells. Jour. Morph. **29**: 217–279. pls. 1- 4.

SAKAMURA, T. 1920. Experimentelle Studien über die Zell- und Kernteilung mit besonderer Rücksicht auf Form, Grosse und Zahl der Chromosomen. Jour. Coll. Sci. Imp. Univ. Tokyo **39**: pp. 221. pls. 7.

SARGANT, E. 1896. Direct nuclear division in the embryosac of *Lilium martagon*. Ann. Bot. **10**: 107–108.

SCHILLER, I. 1909. Ueber kunstliche Erzeugung "primitiver" Kernteilungsformen bei *Cyclops*. Arch. Entw. **27**: 560–609. figs. 62.

SCHÜRHOFF, P. N. 1915. Amitosen von Riesenkernen im Endosperm von *Ranunculus acer*. Jahrb. Wiss. Bot. **55**: 499–519. pls. 3, 4.

1919. Das Verhalten des Kerns in den Knöllchenzellen von *Podocarpus*. Ber. Deu. Bot. Ges. **37**: 373–379.

SCOTT, D. H. 1888. On nuclei in *Oscillatoria* and *Tolypothrix*. Jour. Linn. Soc. Bot. **24**: 188–192. pl. 5. figs. 1–4.

STRASBURGER, E. 1884. Zur Entwicklungsgeschichte der Sporangien von *Trichia fallax*. Bot. Zeit. **42**: 305–316, 321–326. pl. 3.

1907. Apogamie bei *Marsilia*. Flora **97**: 123–191. pls. 3–8.

1908. Chromosomenzahlen, Plasmastrukturen, Vererbungsträger und Reduktionsteilung. Jahrb. Wiss. Bot. **45**: 479–568. pls. 1–3.

TISCHLER, G. 1900. Verh. Naturhist. Med. Ver. Heidelberg **6**.

TRÖNDLE, A. 1912. Der Nucleolus von *Spirogyra* und die Chromosomen höherer Pflanzen. Zeit. f. Bot. **4**: 721–747. pl. 9.

WAGER, H. 1903. The cell structure of the Cyanophyceæ. (Prelim. note.) Proc Roy. Soc. London **72**: 401–408. figs. 3.

VON WASIELEWSKI, W., 1902, 1903. Theoretische u. Experimentelle Beiträge zur Kenntniss der Amitose, I, II. Jahrb. W. Bot. **38**: 377–420, pl. 7; **39**: 581, 606. figs. 10.

WINGE, O. 1912. Cytological studies in the Plasmodiophoraceæ. Arkiv. för Botanik **12**: 1–39. pls. 3.

VAN WISSELINGH, C., 1914. On the nucleolus and karyokinesis in *Zygnema*. Rec. Trav. Bot. Neer. **11**: 1–13.

YOUNG, R. T. 1913. The histogenesis of the reproductive organs of *Tænia pisiformis*. Zool. Jahrb. **35**: 355–418. pls. 18–21.

ZACHARIAS, E. 1887. Beiträge zur Kenntniss des Zellkerns und der Sexualzellen. Bot. Zeit. **45**: 297–304. pl. 4.

1890. Ueber die Zellen der Cyanophyceen. Ibid. **48**: 1–. pl. 1.

1892. Ueber die Zellen der Cyanophyceen. Ibid. **50**: 617–624.

1903. Jahrb. Hamb. Wiss. Anst. **21**.

1907. Ueber die neuere Cyanophyceen-Literatur. Bot. Zeit. **65**: 264–287.

ZUKAL, H. 1892. Ueber den Zellinhalt der Schizophyten. (Vorl. Mitt.) Ber. Deu. Bot. Ges. **10**: 51–55.

CHAPTER XI

THE REDUCTION OF THE CHROMOSOMES

The subject of chromosome reduction is one of the most important to be met with in the study of cytology. Many of the problems, both theoretical and practical, upon which biological investigators are expending their most intense efforts seem to be bound up directly or indirectly with the reduction of the chromosomes. The essential feature of reduction is relatively simple in nature, and must be thoroughly grasped in order that the discussions in the following chapters may be intelligible. The entire process by which reduction is accomplished, on the other hand, is very complicated and extremely difficult to observe and interpret with any degree of confidence. In spite of the enormous amount of work already done there still exists much difference of opinion regarding some of the significant steps in the series of changes undergone by the nuclear material. In the present chapter a number of these opinions will be reviewed, but our main purpose will be to make clear the fundamental feature of chromosome reduction.

We have seen that all the cells of the body in a given species are characterized by the presence of a certain number of chromosomes in their nuclei, and that this number is held constant throughout development by an equational division of every chromosome at every somatic mitosis. When we speak of "reduction" we ordinarily refer to the fact that at a certain stage in the life history of the organism the number of chromosomes is reduced one-half. This mere change in the number of chromosomes, though very important, is not in itself the essential feature of the reducing process, as will be seen further on. The whole number is restored at the time of fertilization, when two nuclei, each with the reduced number, unite. In all organisms reproducing sexually reduction and fertilization thus represent the two most critical stages in the life cycle so far as the chromosomes are concerned; hence the exhaustive researches on these two processes.

Discovery.—The discovery of reduction was made by van Beneden, who in 1883 announced that the nuclei of the egg and spermatozoön of *Ascaris* each contain one-half the number of chromosomes found in the body cells. Although van Beneden and other early workers believed that the change in number was brought about by the simple casting out of half the chromosomes during the growth of the germ cells, it was soon shown that this view was incorrect, and that *"reduction is effected by a rearrange-*

ment and redistribution of the nuclear substance without loss of any of its essential constituents" (Wilson 1900, p. 233).

In plants the discovery of reduction came somewhat later. Strasburger in 1888 showed that in angiosperms the number of chromosomes in the egg and male nuclei is fixed by a reduction occurring in the mother-cells of the embryo sac and pollen respectively. This was at once confirmed by Guignard (1889, 1891). E. Overton (1893) found that the female gametophyte cells in the cycad, *Ceratozamia*, have half the number of chromosomes found in the cells of the sporophyte. He further suggested that reduction probably occurs in the sporocytes in mosses and ferns. In the liverwort, *Pallavicinia*, Farmer (1894) found the gametophyte cells to have four chromosomes and the sporophyte cells eight. That Overton's theory of a reduction in the sporocytes of bryophytes and pteridophytes was correct was demonstrated by Strasburger (1894), who postulated the occurrence of a periodic reduction of the chromosomes in all organisms reproducing sexually.

The Stage in the Life Cycle at which Reduction Occurs.—The reduction of the chromosomes is accomplished during the course of two nuclear divisions which, since in animals they have to do with the maturing of the gametes, early came to be known as the *maturation divisions*. Because of its peculiar character the first of these divisions was termed the *heterotypic* by Flemming (1887), while the second, which is essentially like a somatic division, was called the *homœotypic* (sometimes written *homotypic*). Although the essential act of reduction usually occurs at the first division, the entire process, to which the name *meiosis* has been applied, is of such a nature that the second division is normally necessary for its completion. As a result of the two divisions the "reduced" nuclei or cells are formed in groups of four, or *tetrads*, though all members of a tetrad may not function. The point in the life cycle at which these divisions take place in various organisms will now be noted.

In animals, almost without exception, reduction occurs at gametogenesis (Fig. 77). In the male those cells (*spermatogonia*) in the testes whose ultimate descendants are to become spermatozoa multiply by divisions of the ordinary equational type until a certain number are produced. These cells, now called *primary spermatocytes*, enlarge a little and quickly undergo two successive divisions: the first division in each is heterotypic and results in two cells called *secondary spermatocytes;* the second is homœotypic and divides the two secondary spermatocytes into four *spermatids*, each of which becomes transformed into a spermatozoön. The four spermatozoa are therefore the immediate result of the two maturation divisions. In the female the situation is somewhat different: here nearly all of the differentiation of the gamete is accomplished before the nuclear divisions bringing about reduction actually occur. The *primary oöcytes* (*ovocytes*) are the descendants of a number of generations of *oögonia* (*ovogonia*).   The oöcyte, usually while its nucleus is in the prophases of the first maturation division, enlarges greatly ("growth period"), becomes filled with stored food, and develops the general features characterizing the egg.   The oöcyte is now called the "ovarian egg," and it actually is an egg in all respects save one of much importance: its nucleus still has the full number of chromosomes.   At a comparatively late stage, in many cases even after the spermatozoön has entered the egg at fertilization, the oöcyte nucleus (*germinal vesicle*), having passed through some of the prophasic changes characteristic of the heterotypic

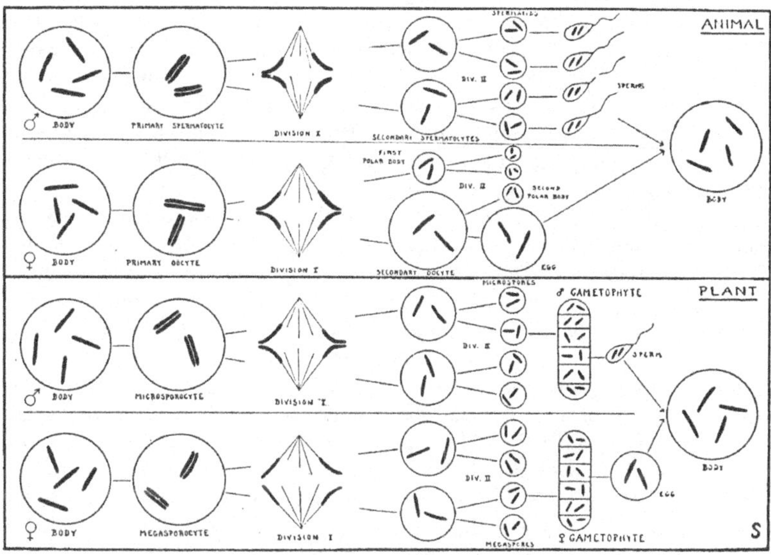

Fig. 77.—Diagram showing the history of the chromosomes in the ordinary life cycles of animals and plants.

mitosis before and during the growth period, gives rise to a mitotic figure which is often surprisingly small for the volume of the nucleus.   The spindle takes up a position perpendicular to the surface of the cell, and at telophase the chromosomes passing to the outer pole are included in the *first polar body*, a small cell budded off at this point.   (See Fig. 106.)   A second spindle is rapidly formed about the chromosomes remaining in the egg (called at this stage the *secondary oöcyte*) and the second maturation mitosis occurs, one daughter nucleus being included in the *second polar body*.   In the course of these two divisions chromosome reduction is accomplished.   The first polar body may divide to form two, thus completing the tetrad of cells corresponding to the tetrad of spermatozoa in the male.   Although the polar bodies are normally functionless they are generally looked upon as eggs historically: the maturation divisions probably resulted formerly in a tetrad of eggs, whereas now only one relatively large and highly differentiated egg is produced at the expense of the other three cells, which remain small and functionless.

Among the protozoa (see Minchin 1912) it has been found that in those forms which appear to have their chromatin aggregated into no definite number of chromosomes, there often occur two successive nuclear divisions suggestive in certain respects of maturation divisions, a part of the products then degenerating. Some have regarded this as a "casting out of effete vegetative chromatin," an interpretation which was at one time placed upon the maturation process generally. In many cases this "reduction" of the chromatin occurs immediately prior to syngamy (sexual union),[1] and so agrees with reduction in higher forms in taking place at gametogenesis; but in other cases it immediately follows syngamy, as in certain algæ mentioned below. Other protozoa have been shown to have a definite chromosome number which is regularly reduced in a manner essentially comparable to that in the metazoa.

In plants it is among the members of the lower groups (thallophytes) that a striking diversity is shown in the stage of the life cycle at which reduction takes place: in the groups above the thallophytes it is regularly accomplished at sporogenesis. In the myxomycete, *Ceratiomyxa*, it has been shown by Olive (1907) and Jahn (1908) that spore formation is accompanied by a chromosome reduction. In the green algæ it is in the first two divisions of the zygote (either a zygospore or a fertilized egg) that reduction occurs: this has been definitely established in *Spirogyra* (Karsten 1908; Tröndle 1911), *Zygnema* (Kurssanow 1911), *Coleochæte* (Allen 1905c), and *Chara* (Oelkers 1916). In a number of other forms, such as *Ulothrix*, *Œdogonium*, *Sphæroplea*, and *Closterium*, in which the chromosomes are not well known, it is probable that the same condition holds, since the zygote upon germination gives rise with considerable regularity to four cells; in some cases (*Œdogonium*) these four cells are zoöspores.

In the BROWN ALGÆ *Cutleria* (Yamanouchi 1912), *Zanardinia* (Yamanouchi), and *Ectocarpus* (Kylin 1918a) reduction occurs in connection with zoöspore formation. In *Fucus*, however, an exceptional condition is found: here reduction takes place in the antheridium and oögonium initials, in the first two divisions following the one delimiting the stalk cell (Fig. 78, *B*). Since there are only three divisions in the oögonium, which thus produces eight eggs, the eggs are but one division removed from the four products of the maturation mitoses, a condition closely approaching that in animals. That reduction in *Fucus* is associated with gametogenesis was inferred by Strasburger (1897) and Farmer and Williams (1898) and demonstrated by Yamanouchi (1909).

[1] See the cases of *Actinophrys sol* and *Amœba albida*, Chapter XII.

In the RED ALGÆ reduction occurs in the two divisions differentiating the nuclei of the tetraspores when the latter are present in the life history. Such is the case in *Polysiphonia* (Yamanouchi 1906) (Fig. 78, *A*), *Griffithsia* (Lewis 1909), and *Corallina* (Yamanouchi). The brown algæ *Dictyota* (Williams 1904) and *Padina* (Wolfe 1918) also conform to this scheme. In *Nemalion*, which has no tetraspores, it was long supposed (Wolfe 1904) that reduction occurs in connection with carpospore formation, but Cleland (1919) has recently shown that it takes place at the time the zygote germinates, as in so many green algæ.[1]

In the ASCOMYCETES reduction occurs in the course of the first two of the three mitoses initiated by the primary ascus nucleus (Figs. 22, 61) and resulting in the eight ascospore nuclei. It was for a long time generally thought that there were two nuclear fusions in the life history—one in the archicarp and one in the ascus (see p. 290), and the three divisions in the ascus were accordingly regarded as a process whose function was to reduce the "quadrivalent" chromosomes to the univalent condition (Harper 1905; Overton 1906). Such a double reduction was described by Miss Fraser (1907, 1908) for *Humaria rutilans:* the first mitosis she found to be heterotypic, the second homœotypic, and the third "brachymeiotic," the last bringing about a further reduction by the separation of the chromosomes into two smaller groups. This was also reported for *Otidea aurantia* and *Peziza vesiculosa* (Fraser and Welsford 1908), *Lachnea stercorea, Ascobolus furfuraceus,* and *Humaria granulata* (Fraser and Brooks 1909), and *Helvella crispa* (Carruthers 1911). Harper (1900, 1905), although he thought two fusions occurred, found no double reduction, holding rather that the fusion of the two ascus nuclei and their chromsomes is so complete as to render the quadrivalent character of the latter entirely invisible. Other investigators also find no double reduction in the ascus. They show rather that the first two mitoses correspond to the heterotypic and homœotypic mitoses of other organisms, and that the third division is purely vegetative or equational in character. As instances may be cited the work of Faull (1905, 1912)

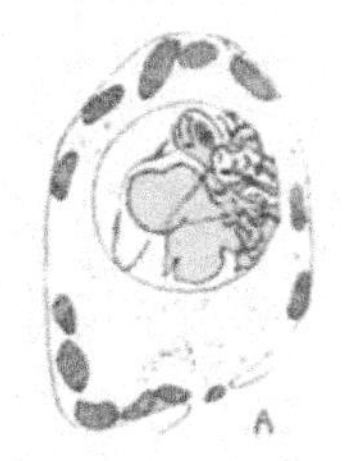

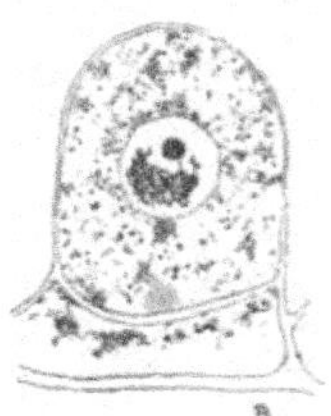

Fig. 78.

A, prophase of heterotypic division in the tetrasporocyte of *Polysiphonia.* (*After Yamanouchi,* 1906.) *B,* prophase of heterotypic mitosis in oögonium of *Fucus.* (*After Yamanouchi,* 1909.)

---

[1] For a review of sexual reproduction and alternation of generations in the algæ see Bonnet (1914). Davis (1916) gives a convenient summary of the life histories of the red algæ. Dodge (1914) summarizes and compares the life histories of red algæ and ascomycetes. See Atkinson (1915) for a complete review of researches on ascomycetes. For the cytology of the yeasts see Guilliermond 1920.

on *Hydnobolites, Neotiella,* and *Laboulbenia,* and that of Claussen (1912) on *Pyronema.* Furthermore, it is becoming increasingly apparent (see p. 291) that there is but one fusion in the life cycle—that in the ascus, so that the necessity for a second reduction is removed.

In the BASIDIOMYCETES it has been shown by the researches of Juel (1898), Maire (1905), Guilliermond (1910), Kniep (1911, 1913), Levine (1913), and others on the hymenomycetes, and by those of V. H. Blackman (1904), Dietel (1911), Fitzpatrick (1918), and others on the rusts,[1] that reduction occurs in the two mitoses giving rise to the four basidio-

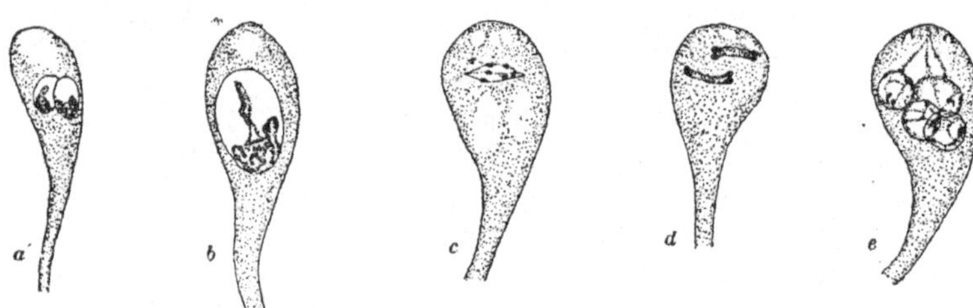

FIG. 79.—Sexual fusion and maturation divisions in the basidium of
Nidularia pisiformis.

a, two sexual nuclei about to unite. *b,* prophase of heterotypic division in fusion nucleus. *c,* heterotypic mitosis. *d,* homœotypic mitosis. *e,* the four basidiospore nuclei. × 1800. (*After Fries,* 1911.)

spore nuclei (Fig. 79). As in the ascomycetes, it thus follows immediately upon the nuclear fusion: in the basidium in hymenomycetes and in the teleutospore in rusts. An exception is reported in the case of *Hygrophorus conicus,* in which Fries (1911) finds in the basidium neither a nuclear fusion nor a reduction.

In the BRYOPHYTES reduction, so far as known, is universally brought about by the two mitoses which differentiate the four nuclei of each spore tetrad. It was at one time reported (van Leeuwen-Reijnvaan 1907) that in *Polytrichum* there is a second reduction at spermatogenesis and oögenesis: the sporophyte was said to have 12 chromosomes, the spore and gametophyte six, and the gametes three. This double reduction was thought to be compensated for by the fusion of the ventral canal cell with the egg, raising the number in the latter to six, in combination with the entrance of two sperms into the egg at fertilization, making the sporophytic number 12. This interpretation has been shown to be false by both Vandendries (1913) and Walker (1913), who find the life cycle normal in every respect: a reduction from 12 to 6 occurs at sporogenesis but no second reduction follows at gametogenesis.

[1] A summary of researches on rusts is given by Maire (1911). A list of numbers of nuclei in the cells of basidiomycetes is given by Levine (1913).

In VASCULAR PLANTS reduction in all normal life cycles in both homosporous and heterosporous forms occurs uniformly in the divisions differentiating the spore tetrads (Fig. 77). The sporocytes, particularly the microsporocytes ("pollen mother-cells"), of the higher plants have long been favorite objects for the study of reduction. Since the gametophyte generation in the higher plants is so abbreviated, reduction closely precedes fertilization in these forms. In the ordinary angiosperm embryo sac in which the eight nuclei are derived from a single megaspore of the tetrad, the egg nucleus is removed from the product of reduction (megaspore nucleus) by only three mitoses. In some cases, of which *Lilium* is the best known example, walls fail to form between the four megaspore nuclei (Fig. 80, *B*), leaving them in a common cavity (embryo sac) where they undergo but one further division to produce the eight nuclei of the female gametophyte. The egg here is consequently removed from the product of reduction by a single mitosis. In one known case, *Plumbagella* (Dahlgren 1915), the four reduced nuclei, formed as in *Lilium,* divide no further, one of them functioning directly as the egg nucleus. Here, therefore, the condition characteristic of animals has been reached: the gamete nucleus is itself the direct product of reduction, and the haploid generation usually produced by the spore is eliminated. The male gametophyte also has undergone much abbreviation in higher

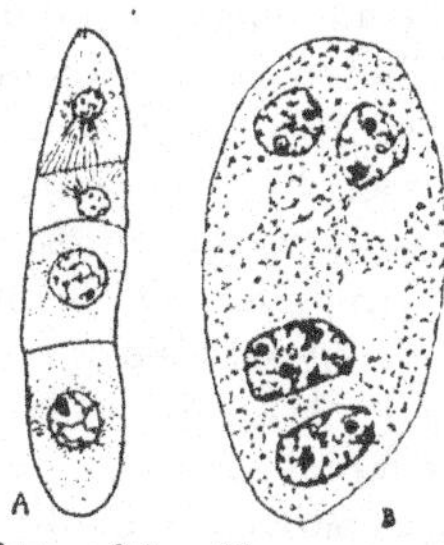

FIG. 80.—Megaspore tetrads in Angiosperms.

A, tetrad of walled cells in *Physostegia virginiana;* formation of two upper ones just being completed. × 462. (*After Sharp,* 1911.) *B,* tetrad of megaspore nuclei in *Lilium canadense.*

plants, but the male nucleus is still removed from the reduction product (microspore nucleus) by two mitoses. In no known case does the microspore nucleus function directly as a gamete nucleus.

The term *gonotokont* was introduced by Lotsy (1904) to designate any cell, whatever its origin or position in the life cycle, in which the reduction process is initiated. In animals the gonotokonts are therefore the primary spermatocyte and the primary oöcyte. In most green algæ the gonotokont is the zygote; in the red algæ it is usually the tetrasporocyte; in the ascomycetes it is the ascus; in the basidiomycetes it is the basidium; and in the bryophytes and vascular plants it is the sporocyte— the microsporocyte and megasporocyte in the case of heterosporous forms.

The Meaning of Reduction.—In order that the true meaning of reduction may be appreciated it will be necessary to indicate the main points of a theory first suggested by Roux (1883) and later developed particularly by Weismann (1887, 1891, 1892). It had been believed by the earlier workers that reduction was merely a process whose function was "to prevent a summation through fertilization of the nuclear mass and of

15

the chromatic elements" (Hertwig 1890). But the chromatic mass is actually quartered at reduction, whereas the number of chromosomes is halved. Moreover, great changes in nuclear volume occur with no change in the number of chromosomes. This careful guarding, so to speak, of the chromosome number was siezed upon as a most significant fact by Roux, who "argued that the facts of mitosis are only explicable under the assumption that the chromatin is not a homogeneous substance, but differs qualitatively in different regions of the nucleus; that the collection of the chromatin into a thread and its accurate division into two halves is meaningless unless the chromatin in different regions of the thread represents different *qualities* which are to be divided and distributed to the daughter cells according to some definite law. He urged that if the chromatin were qualitatively the same throughout the nucleus, direct division would be as efficacious as indirect, and the complicated apparatus of mitosis would be superfluous."[1] Upon this conception Weismann based his remarkable theory, the starting point of which was "the hypothesis of De Vries that the chromatin is a congeries or colony of invisible self-propagating vital units or *biophores*, somewhat like Darwin's 'gemmules,' each of which has the power of determining the development of a particular quality. Weismann conceives these units as aggregated to form units of a higher order known as 'determinants,' which in turn are grouped to form 'ids,' each of which . . . is assumed to possess the complete architecture of the germ-plasm characteristic of the species. The 'ids' finally, which are identified with the visible chromatin-granules, are arranged in linear series to form 'idants' or chromosomes. It is assumed further that the 'ids' differ slightly in a manner corresponding with the individual variations of the species, each chromosome therefore being a particular group of slightly different germ-plasms and differing qualitatively from all the others.

"We come now to the essence of Weismann's interpretation. The end of fertilization is to produce new combinations of variations by the mixture of different ids. Since, however, their number, like that of the chromosomes which they form, is doubled by the union of two germ-nuclei, an infinite complexity of the chromatin would soon arise did not a periodic reduction occur. Assuming, then, that the 'ancestral germ-plasms' (ids) are arranged in a linear series in the spireme thread or the chromosomes derived from it, Weismann ventured the prediction (1887) that two kinds of mitosis would be found to occur. The first of these is characterized by a longitudinal splitting of the thread, as in ordinary cell-division, 'by means of which all the ancestral germ-plasms are equally distributed in each of the daughter-nuclei after having been divided into halves.' This form of division, which he called *equal division* (Aequationstheilung), was then a known fact. The second

---

[1] This and the following quotations are from Wilson (1900, pp. 245–246).

form, at that time a purely theoretical postulate, he assumed to be of such a character that each daughter-nucleus should receive only half the number of ancestral germ-plasms possessed by the mother-nucleus. This he termed a *reducing division* (Reduktionstheilung), and suggested that this might be effected either by a *transverse* division of the chromosomes, or by the elimination of entire chromosomes without division. By either method the number of 'ids' would be reduced; and Weismann argued that such reducing divisions must be involved in the formation of the polar bodies, and in the parallel phenomena of spermatogenesis."

Reduction in Weismann's sense, then, is a reduction of the number of kinds of germ-plasm or ancestral hereditary qualities present, this reduction being brought about by means of a redistribution, half of the qualities to one daughter nucleus and the remainder to the other daughter nucleus. The change in the number of chromosomes is a consequence of the manner in which this redistribution is accomplished, as we shall see.

Interpretations Based on Weismann's Theory.—As would be expected, there were announced certain interpretations of chromosome behavior based on Weismann's idea. Several cytologists thought that they found the chromsomes actually dividing transversely at one or the other of the two maturation mitoses. This interpretation, however, proved to be incorrect. Much light was thrown on the problem when Henking (1891), Rückert (1891, etc.), Haecker (1890–9), vom Rath (1892–3), and others showed that the double chromosomes appearing in the reduced number on the spindle at the first maturation mitosis are not split chromosomes like those seen in somatic divisions, but are pairs of chromosomes, or *bivalent chromosomes*, each arising by an end-to-end conjugation (*synapsis*) of two somatic chromosomes. The two parts of each bivalent then separate at the first or second maturation division, the entire chromosomes thus being segregated into two groups, each with the reduced number. Thus it appeared unnecessary that a single chromosome, representing a linear series of different qualities, should be transversely divided in order for Weismannian reduction to occur: it was only necessary to assume that the whole chromosomes differ qualitatively from one another, so that when the two members of a bivalent pair separate there would be a segregation of different qualities. It is in the light of this bivalent chromosome conception that we are to interpret the many early reports of a transverse division of the chromosome during maturation. What was called a transverse division was merely the separation of two entire chromsomes placed end-to-end.

A number of workers soon found that in many cases there is nothing even simulating a transverse division, either of single chromosomes or of bivalent pairs, but that both maturation divisions are apparently longitudinal (Flemming, Brauer 1893, Moore 1896, Meves 1896, Grégoire, etc.). How, then, is there any Weismannian reduction if there is neither a transverse division of the chromosome nor a transverse separation of bivalents? Montgomery (1901), von Winiwarter (1900), Sutton (1902), Boveri (1904), and a number of others showed that here the chromosomes conjugate side-by-side rather than end-to-end. Thus when they separate there is an appearance of a longitudinal division, but reduction is nevertheless accomplished, since entire somatic chromosomes supposedly qualitatively different, and not the longitudinal halves of split chromosomes, are separating. The appearance of a longitudinal division may also be present after an end-to-end conjugation, for the two members may bend around to a side-by-side position before finally separating.

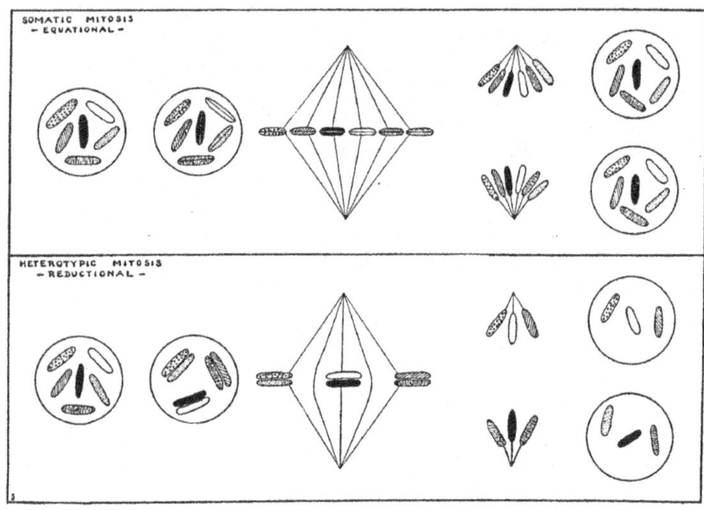

Fig. 81.—Diagram showing essential difference between somatic and heterotypic mitoses.

As a matter of fact, the maturation divisions in nearly all cases, especially those studied by botanists, are both longitudinal in appearance. End-to-end conjugation later came to be called *telosynapsis* or *métasyndèse*, and side-by-side conjugation *parasynapsis* or *parasyndèse*.

Somatic and Heterotypic Mitoses Compared.—Before taking up a more detailed account of the process of reduction as it has been described by various investigators it is of the utmost importance to fix clearly in mind the essential difference between somatic and heterotypic mitosis in order to realize what constitutes the cardinal feature of reduction, and thereby to detect the significant points of the various theories. This essential difference is illustrated in Figs. 81 and 82. *In a somatic or vegetative mitosis every chromosome is split into two exactly similar longitudinal halves which are distributed to the two daughter nuclei. The daughter nuclei are therefore like each other and like the mother nucleus in the quality*

*of their substance. In the heterotypic mitosis, the first of the two matura-
tion mitoses, the chromosomes conjugate two by two during the prophase to
form the reduced number of bivalent chromosomes, which take their place
on the spindle. The members of each pair, which are supposed to differ
qualitatively from each other, separate and pass to the two daughter nuclei.
These nuclei are therefore qualitatively unlike each other, having different
members of the full chromosome group; and also unlike the mother nucleus,
since each of them has only half as many chromosomes as the latter.* The
second maturation mitosis (not shown in the diagrams) is essentially a
vegetative mitosis in most cases: each chromosome splits longitudinally

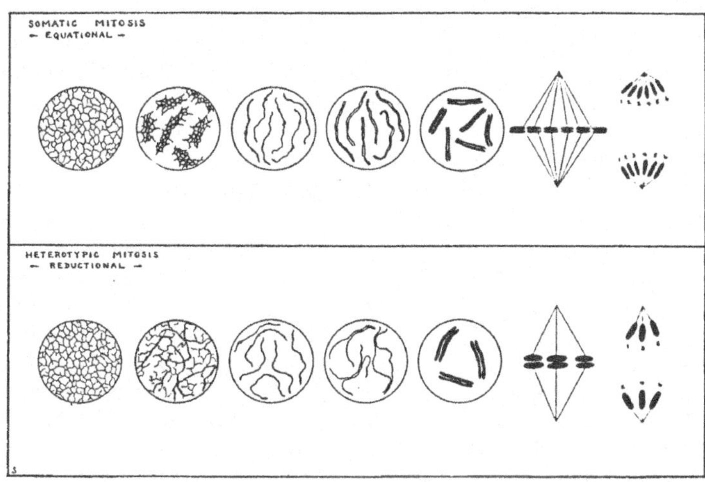

Fig. 82.—Diagram showing essential difference between somatic and heterotypic mitoses.

and the halves are distributed to the daughter nuclei. The four nuclei,
and consequently the four cells, resulting from the two maturation divisions are therefore of two kinds: two of them have half of the chromosomes
of the original nucleus and the other two have the remaining ones.

Assuming, then, that the chromatin of the nucleus represents the
principal physical basis of inheritance (see Chapter XIV), *reduction is
essentially this: a reduction in the number of kinds of hereditary units by
the separation and distribution of qualitatively different masses of chromatin
to different cells and eventually into different hereditary lines, rather than an
equational division and distribution of all the qualities as in somatic mitosis.*
As has already been stated, the change in the number of chromosomes
("numerical reduction") is a consequence of the method by which this
qualitative reduction is brought about, this method being the distribution
of entire chromosomes, each representing one or more particular qualities,
to different cells.

Another important feature of the reduction process should be noted before proceeding further. In many cases, chiefly among animals, the chromosomes appearing on the spindle of the first maturation mitosis are not merely double, but quadruple. This is due to the fact that each of the conjugating chromosomes is already longitudinally split, giving the bivalent chromosome the form of a *chromosome tetrad* (not to be confused with tetrads of cells or nuclei). The four constituents of the chromosome tetrad are known as *chromatids,* and are distributed by the two maturation mitoses to the four resulting cells. It is thus seen that in the case of chromosome tetrads the lines of separation for both maturation mitoses are marked out in the prophase of the first. It should be borne in mind that one of these lines represents a plane of chromosome conjugation, and the other a plane of true longitudinal splitting. When the chromatids separate along the conjugation plane reduction occurs, whether this be at the first or second mitosis, and when separation along the plane of splitting occurs the mitosis is equational, as in somatic division.

MODES OF CHROMOSOME REDUCTION

In all cytology there is scarcely a subject upon which there has been entertained so great a variety of opinion as upon the question of the exact behavior of the chromosomes during the meiotic phases. Entirely aside from the theoretical interpretations placed upon the process of maturation, cytologists have yet failed to arrive at any universally accepted conclusion regarding all the structural changes which occur. This diversity of opinion is due in part to the complexity of the process and the difficulty of interpreting its various stages, some of which fail to stand out clearly in preparations made by our available methods. On the other hand, a great variety of organisms have been studied, and these undoubtedly differ considerably in the details of the reduction process, so that agreement in all particulars is not to be expected. The attempt has too often been made to apply universally an interpretation founded upon a study of one or two organisms. Certain essential features of meiosis may be expected to show close agreement in all organisms reproducing sexually, as Strasburger pointed out, but it is evident that there is no full correspondence as regards the exact manner in which the essential changes are accomplished. In the following pages are given brief descriptions of a few representative interpretations advanced by various cytologists.[1]

The two interpretations of reduction which have been most conspicuous in the literature of recent years are diagrammed in Figs. 83 and 89.

[1] No attempt can be made in a work of this scope to give a complete summary and classification of all the interpretations that have been put upon the maturation phenomena. Only enough will be presented to afford a starting point for a study of this complex subject. For a review and criticism of all views expressed up to 1910 see Grégoire's two invaluable works (1905, 1910). A useful list of works on somatic and heterotypic mitosis in angiosperms is given by Picard (1913).

Nearly all of the accounts of reduction now appearing, especially those given by botanists, conform in general to one or the other of these two schemes, though they vary greatly in detail. Both theories have been upheld by competent observers, and it may be possible that both modes of reduction actually occur; but the same objects have been so differently described by the two opposing schools that it seems very probable that interpretation is chiefly responsible for the persistent diversity of opinion. For convenience the two theories will be referred to as *Scheme A* and *Scheme B*.

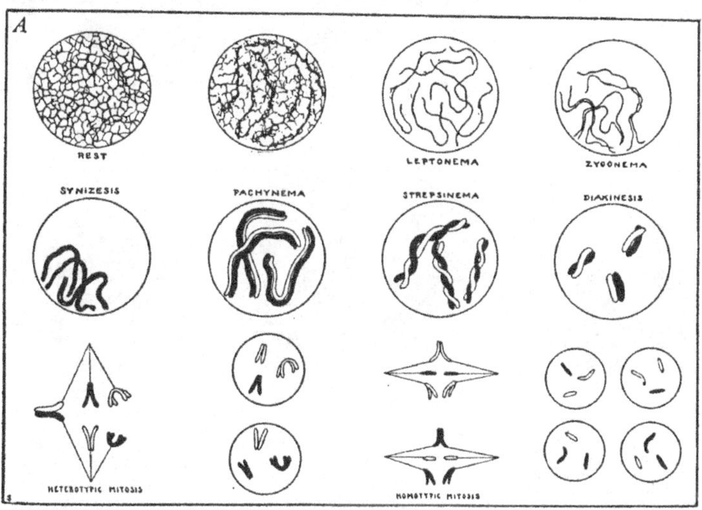

Fig. 83.—The method of chromosome reduction according to *Scheme A*. Explanation in text.

Scheme A.—The first of the two main interpretations of reduction came into prominence in 1900 and shortly after, when von Winiwarter (1900), Grégoire (1904, 1907, 1909), A. and K. E. Schreiner (1904–1908), and Berghs (1904, 1905) applied it to the phenomena observed by them in several animals and plants. Its essential points are as follows (Figs. 83–88):

At the beginning of the heterotypic prophase the nuclear reticulum, without breaking down into such distinct elementary nets or alveolar units as are seen in the somatic prophase, takes the form of long slender threads (*leptotène* or *leptonema* stage).[1] During the very early prophase

---

[1] The terms *leptotène, synaptène, pachytène,* and *diplotène* were proposed by von Winiwarter (1900); *leptonema, zygotène, pachynema,* and *strepsinema* by Grégoire (1907); *amphitène* by Janssens (1905); *strepsitène* by Dixon (1900); *diakinesis* by Haecker (1897); *synapsis* by Moore (1896); *synizesis* by McClung (1905); and *meiosis* by Farmer and Moore (1905). The terms ending in *-tène* are ordinarily used as adjectives.

these threads conjugate in pairs side-by-side (*parasynaptically; parasyndetically*). The association does not take place at all regions of the threads at once: it begins at one or two points, commonly at one end,

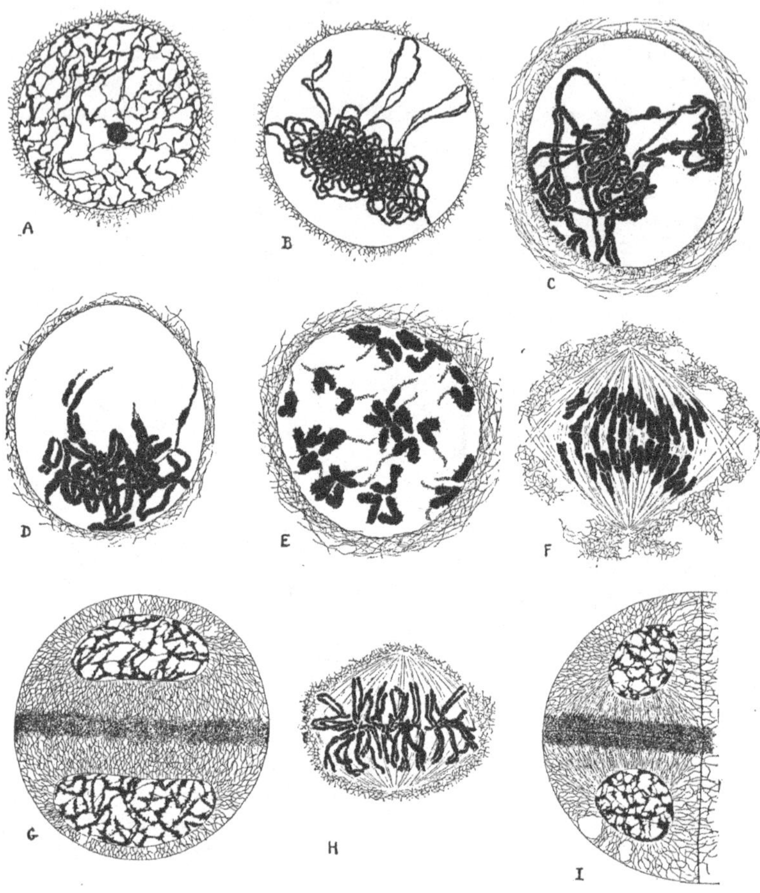

Fig. 84.—Reduction in sporocyte of *Nephrodium*.

A, leptonema. *B*, recovery from synizesis; parallel conjugation consummated during synizesis. *C*, pachynema. *D*, second contraction of bivalent spireme. *E*, diakinesis. *F*, anaphase of heterotypic mitosis. *G*, interkinesis. *H*, homœotypic mitosis. *I*, two of the four spore cells. (*After Yamanouchi, 1908.*)

and gradually involves all portions, so that at stages when the process is yet incomplete the two threads may be closely paired at some points and widely divergent at others, giving an appearance very unlike that of the halves of a longitudinally split chromosome in a somatic cell. This is known as the *zygotène, zygonema, synaptène,* or *amphitène* stage. Usually before the union is complete the nucleus enlarges somewhat and the threads contract, forming a tight knot at one side of the nucleus. This stage, formerly called synapsis (see p. 255), is now more properly known as *synizesis.* The pairing threads come into very close association during synizesis, which, though variable in time, usually ensues at about this stage. When the closely paired threads recover from the synizesis contraction they extend more uniformly throughout the nucleus ("open spireme"), and are now seen to be much thicker (*pachytène; pachynema*).

In many cases they may again contract into a loose knot with loops extending from it ("second contraction"). As they continue to decrease in length and increase in diameter the members of each pair twist more or less tightly about each other for a short time (*strepsitène; strepsinema; diplotène*). Eventually they become very short and thick, and the various pairs (*gemini; bivalent chromosomes*), present in the haploid or reduced number, lie scattered throughout the nucleus (*diakinesis*). The two components of each geminus may now separate slightly at one or both ends or at the middle, which gives them the form of **Y**s, **V**s, **X**s, and **O**s. The bivalent chromosomes are now fully formed and ready to take their places on the spindle, which soon forms.

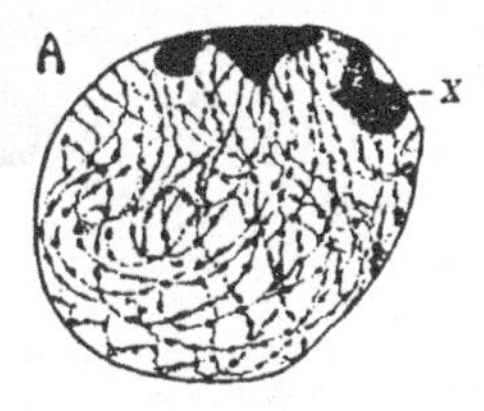

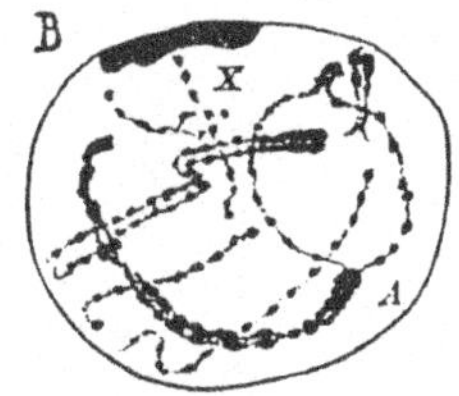

Fig. 85.—Parasynapsis in *Phrynotettix magnus.*

A, leptonema; *x,* sex-chromosome. *B,* conjugation of "chromosome *A.*" Portions of other unconjugated threads and one other bivalent also present in section. × 2000. (*After Wenrich,* 1916.)

In the case of the animal egg the "growth period" introduces a complication. In the sporocytes of plants and the spermatocytes of animals there is some enlargement of the cell and nucleus during the stages just described, but the chromosomes pass directly from the strepsinema stage to diakinesis. During the relatively enormous growth of the oöcyte on the other hand, the chromosomes, which have usually reached the strepsinema stage when the enlargement begins, become greatly modified in form. Their achromatic framework takes the form of fine threads extending out in all directions, giving the chromosome an irregular brush-like form (Fig. 86, *C, D*), while the chromatic substance either may flow into the nucleolus, leaving the chromosome framework uncolored and very difficult to observe, or by loss of its staining capacity through chemical change it may disappear from view completely. As the growth period comes to an end, however, the original staining capacity returns and the chromosomes again assume the compact form and pass into the diakinesis stage.

In the case of most animals, and apparently in certain plants also, the split which is to function in the homœotypic mitosis may develop during diakinesis or even much earlier, the result being the formation of chromosome tetrads. This introduces another element of complication which will be touched upon later (p. 243).

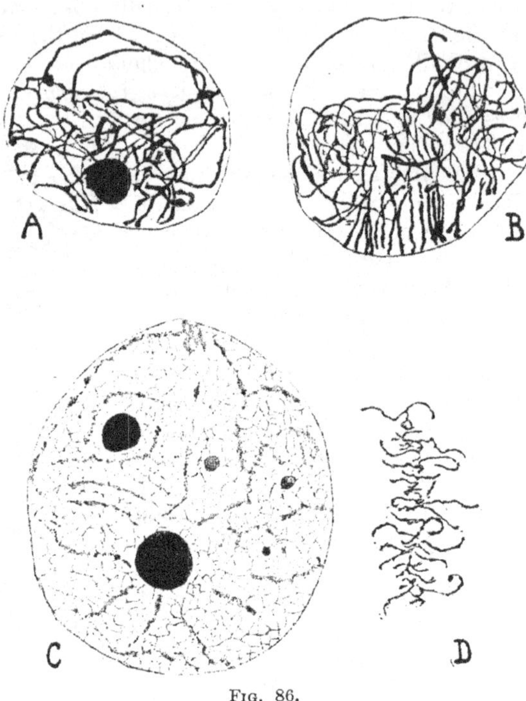

Fig. 86.

A, parasynapsis in *Allium fistulosum.*   *B*, parasynapsis in *Osmunda regalis.*   *C*, nucleus of oöcyte of *Scyllium canicula* (Selachian) in "growth stage."   *D*, single chromosome in growth stage, showing the fine subdivision of its substance.   (*A and B after Grégoire*, 1907, *C and D after Maréchal* 1907.)

The diakinesis stage is terminated by the dissolution of the nuclear membrane and the formation of the spindle, upon which the bivalent chromosomes, whether secondarily split or not, now become arranged. Because of the peculiar form and consistency of the heterotype chromosomes the mitotic figure presents a striking contrast in appearance to the ordinary figure of somatic cells. This is especially true as the chromosomes are drawn into various curious shapes as their anaphasic separation begins. The two univalent components of each bivalent chromosome eventually become free from each other and pass to the two daughter nuclei, bringing about reduction. During the anaphase the separating

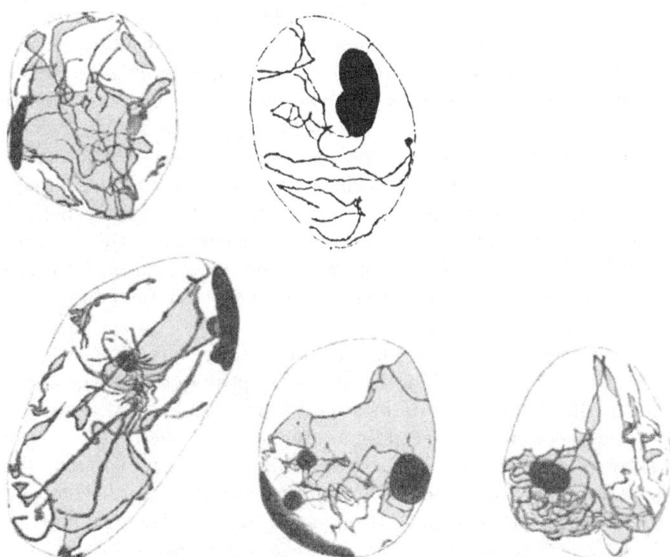

FIG. 87.—Nuclei from microsporocytes of *Vicia faba*, showing parasynapsis. Synigesis beginning in No. 6.   × 1900.

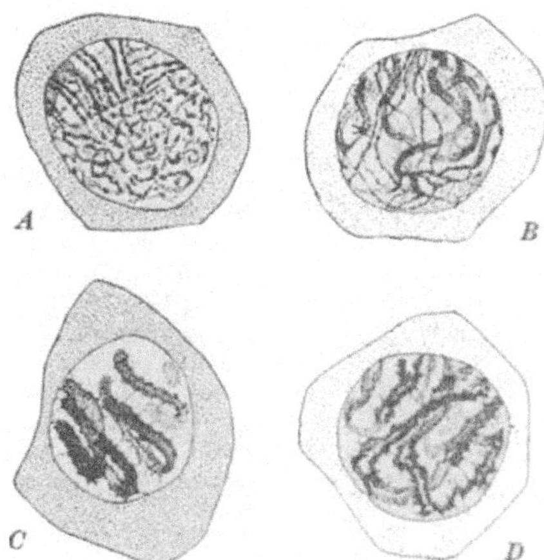

FIG. 88.—Heterotypic prophases in spermatocyte of *Tomopteris onisciformis*.

A, pairing of leptotène threads beginning. B, pairing complete in some threads and only beginning in others. C, conjugation complete; pachynema stage. D, resplitting of pachytène threads (separation of conjugated chromosomes.) (*After A. and K. E. Schreiner*, 1905.)

univalents, if not already double, rapidly develop a longitudinal split, in some cases even before they are entirely free from each other. The resulting halves tend to open out along this split; chromosomes being drawn endwise to the poles thus take the form of simple **V**s, while those to which the fibers are attached at the middle appear as double **V**s. After reaching the poles the split chromosomes begin the reconstruction of the daughter nuclei. As a rule this does not proceed very far, since the homœotypic mitosis follows very quickly upon the heterotypic. Well organized daughter nuclei are often formed, whereas in the animal egg there may be no reconstruction whatever, the daughter chromosomes of the first mitosis at once taking their places on a newly formed spindle for the second mitosis.

In the homœotypic mitosis the chromosomes, if there has been an intervening interkinesis of any length, usually appear much longer and thinner than in the heterotypic mitosis, and separate along the longitudinal line of fission seen in the preceding anaphase. The homœotypic mitosis is therefore equational in character, and differs from an ordinary somatic mitosis only in the number of its chromosomes and the precocity of their splitting. In each of the four nuclei resulting from the two maturation mitoses there is now the haploid number of univalent chromosomes, and meiosis is complete.

The foregoing interpretation of reduction has been widely accepted from the first by both botanists and zoölogists. The following is a partial list of works in which it has been described.

PLANTS

Grégoire	1904, '07	*Lilium, Allium, Osmunda*
Berghs	1904, '05	*Allium, Drosera, Helleborus,* etc.
Rosenberg	1905, '07, '08, '09	*Drosera,* Compositæ
Allen	1905*bc*	*Lilium, Coleochæte*
J. B. Overton	1905, '09	*Thalictrum, Calycanthus, Richardia*
Strasburger	1905, '07, '08, '09	*Lilium, Galtonia,* etc.
Miyake	1905	*Lilium, Funkia, Iris, Allium, Tradescantia, Galtonia*
Tischler	1906	*Ribes*
Cardiff	1906	*Acer, Salomonia, Botrychium, Ginkgo*
Lagerberg	1906, '09	*Adoxa*
Yamanouchi	1906, '08, '10	*Polysiphonia, Nephrodium, Osmunda*
Martins Mano	1909	*Funkia*
Lundegårdh	1909, '14	*Trollius*
Frisendahl	1912	*Myricaria*
McAllister	1913	*Smilacina*
Schneider	1913, '14	*Thelygonium*
Weinzieher	1914	*Xyris*
Sakamura	1914	*Vicia*
de Litardière	1917	*Polypodium*

ANIMALS

von Winiwarter	1900	Rabbit, Man
Maréchal	1904, '05, '07	Tunicates, Selachians, Teleosts, *Amphioxus*
A. and K. E. Schreiner	1906, '07, '08	*Tomopteris, Ophryotrocha, Zoögonus, Enteroxenos, Myxine, Salamandra, Spinax*
Lerat	1905	*Cyclops*
Deton	1908	*Thysanozoön*
Grégoire	1909	*Zoögonus*
Janssens	1905, '09	*Batracoseps*
Janssens et Willems	1909	*Alytes*
Schleip	1906, '07	*Planaria*
Debaisieux	1909	*Dytiscus*
Montgomery	1911	*Euschistus*
Kornhauser	1914, '15	*Hersilia, Enchenopa*
Wenrich	1916, '17	*Phrynotettix, Chorthippus*
Fasten	1914, '18	*Cambarus, Cancer*
Malone	1918	*Canis*
Pratt and Long	1917	*Mus*
Robertson	1916	Insects

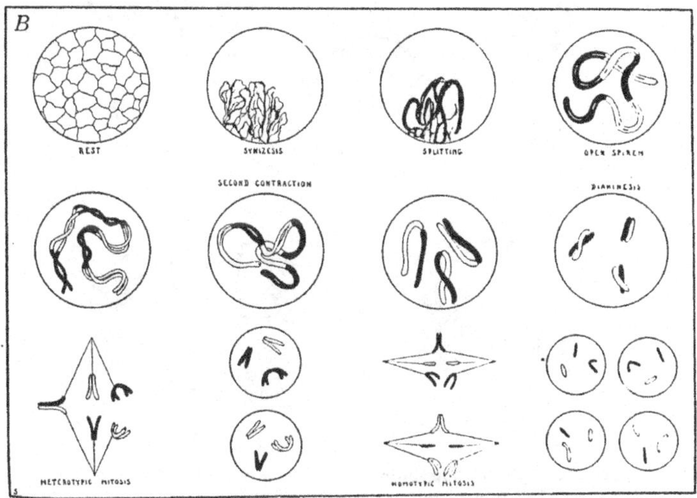

FIG. 89.—The method of chromosome reduction according to *Scheme B.* Explanation in text.

Scheme B.—The second of the two conspicuous interpretations was advanced by Farmer and Moore (1903, 1905), and is essentially as follows (Figs. 89–92): In the early heterotypic prophase the reticulum becomes more thready in structure and contracts into a tight knot (*synizesis*). When this knot loosens up the chromatic material has assumed the form of a continuous spireme which is double.   This doubleness is believed to represent a true longitudinal split, and although it usually disappears from view during the later prophases it is thought to

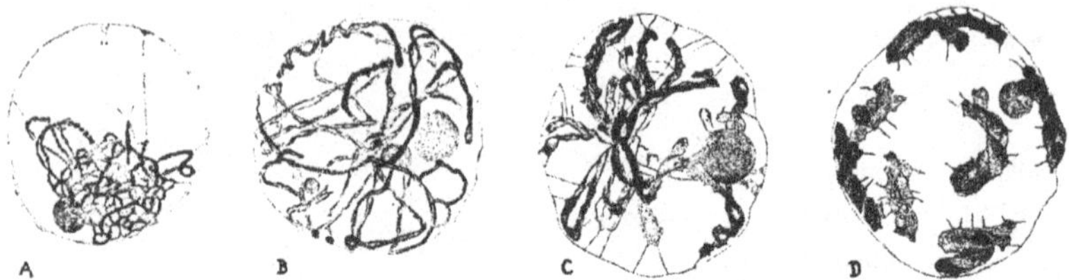

Fig. 90.—The heterotypic prophases in *Lilium*, according to Mottier (1907.)

A, synizesis knot loosening up; threads splitting; note chromomeres.   *B*, hollow spireme. *C*, second contraction.   *D*, diakinesis.   × 900.

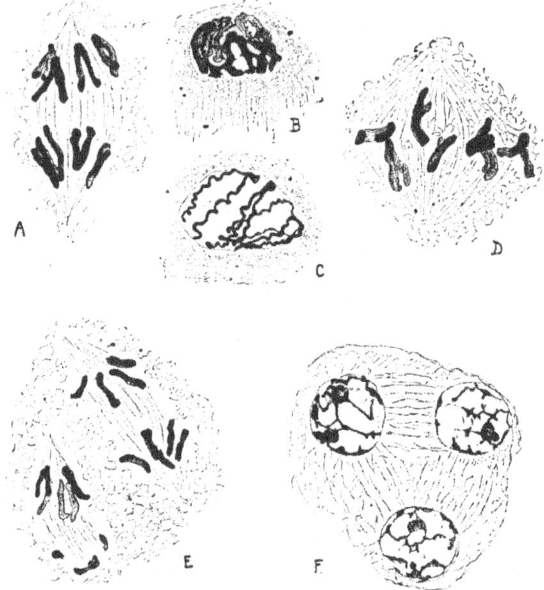

Fig. 91.—Maturation mitoses in microsporocyte of *Vicia faba.*

A, anaphase of heterotypic mitosis; split for second mitosis evident in separating daughter chromosomes.   *B*, one daughter nucleus in early telophase of heterotypic mitosis. *C*, later telophase.   *D*, metaphase of homœotypic mitosis.   *E*, anaphase of same, showing portions of both spindles.   *F*, three of the four microspore nuclei.   × 1335.   (*After Fraser*, 1914.)

persist and reappear at a much later stage.   After extending loosely throughout the nucleus ("open spireme"), the double spireme, now considerably thickened and twisted (*strepsinema*), contracts again and is thrown into loops ("second contraction"). These loops then break apart from one another through a segmentation of the spireme; each of them is composed of two split chromosomes arranged end-to-end. Chromosome conjugation has thus occurred *telosynaptically* (*metasyndetically*) either while the spireme was being formed or when the daughter spiremes were formed in the preceding telophase. The two members of each pair are brought around to a side-by-side position by the looping at the second contraction, usually but not always remaining closely connected at the original point of conjugation. The resulting bivalent chromosomes, with their split obscured, become much shortened and thickened (*diakinesis*) and take up their positions on the first maturation spindle. In case the original split, instead of being wholly obscured, is visible at this time or earlier, chromosome tetrads are evident. In the heterotypic anaphase the bivalents are separated into their component univalents, bringing about reduction. During the anaphase the univalents often widen out along the line of fission which had been temporarily obscured, giving them the form of simple or double Vs as described for Scheme *A*. They remain through interkinesis in the double condition, and in the homœotypic mitosis separate along this line of fission.

The following is a list of the principal works in which this theory of reduction has been advocated.

PLANTS

Farmer and Moore	1903, '05	*Lilium, Osmunda, Psilotum, Aneura*
Farmer and Digby	1910	*Galtonia*
Farmer and Shove	1905	*Tradescantia*
Mottier	1907, '09, '14	*Lilium, Acer, Allium, Podophyllum, Tradescantia, Staphylea*
Gregory	1904	Ferns
Lewis	1908	*Pinus, Thuja*
Schaffner	1906, '09	*Agave*
Digby	1910, '12, '14, '19	*Galtonia, Primula, Crepis, Osmunda*
Fraser	1914	*Vicia*
Lawson	1912	*Smilacina*
McAvoy	1912	*Fuchsia*
Beer	1912, '13	*Equisetum, Crepis, Tragopogon*
Woolery	1915	*Smilacina*
Nothnagel	1916	*Allium*

ANIMALS

Farmer and Moore	1905	*Periplaneta*, Elasmobranchs
Montgomery	1903, '04, '05, '06, '10	Hemiptera, Amphibia
Moore and Embleton	1906	Amphibia
Griggs	1906	*Ascaris*
Zweiger	1907	*Forficula*
H. S. Davis	1908	Insects
Nakahara	1920	*Perla*

Some of the above named investigators, notably Miss Digby (1910, 1912, 1914, 1919), Miss Fraser (1914), and Miss Nothnagel (1916), have laid emphasis upon the view that the split seen in the early heterotypic prophase has its origin in the telophase of the last premeiotic division, each chromosome persisting through the intervening resting stage in the double condition. It is consequently held, as fully stated by Miss Digby (1919) in her account of the archesporial and meiotic phases of *Osmunda* (see Fig. 92), that the lateral pairing of thin threads in the

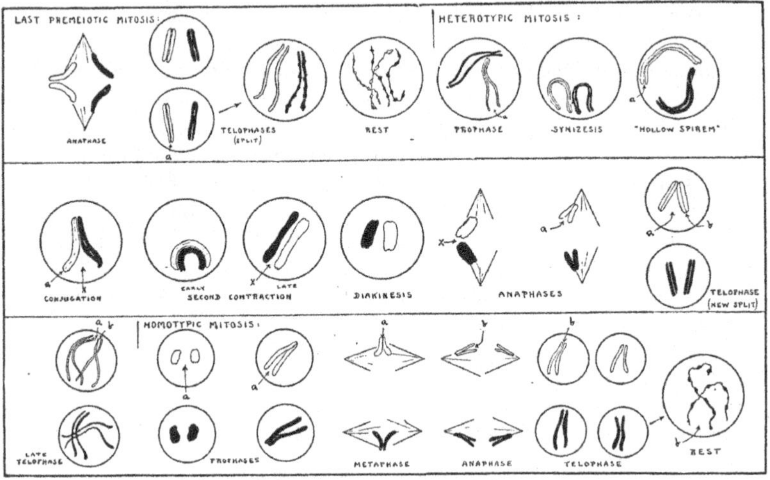

Fig. 92.—Diagram showing behavior of chromosomes in premeiotic and meiotic phases in *Osmunda*, according to Digby (1919).

a, split which originates in telophase of premeiotic mitosis, persists (though obscured at times) through heterotypic prophases, reappears in heterotypic anaphase, and becomes effective in homœotypic mitosis. *b*, split which originates in heterotypic telophase, persists obscured through homœotypic prophases, reappears in homœotypic anaphase, and becomes effective in post-homœotypic division. *x*, plane of conjugation.

heterotypic prophase which the advocates of Scheme *A* have regarded as a conjugation of entire chromosomes is in reality only the reassociation of the two halves of one chromosome which had been split in the preceding telophase. Such a reassociation is thought to occur in every prophase, somatic and heterotypic, since these workers regard chromosome splitting as regularly a thlophasic phenomenon. The split which forms in the last premeiotic telophase functions in the homœotypic mitosis: the homœotypic division is therefore looked upon as the continuation of the premeiotic division, the heterotypic mitosis being an interpolated process bringing about numerical reduction. Not only does this premeiotic split reappear in the anaphase of the heterotypic mitosis to function in the homœotypic, but a new split develops in the heterotypic telophase, and after being temporarily obscured functions in the post-homœotypic division.[1]

A variation of Scheme *B* has been observed in *Œnothera* (Gates 1908, 1909, 1911; Geerts 1908; B. M. Davis 1909, 1910, 1911); in *Fucus* and *Cutleria* (Yamanouchi 1909, 1912); in *Bufo* (King 1907); and in a few other forms. Here the spireme in the heterotypic prophase does not become double, the split for the second division appearing first in the heterotypic anaphase.

Comparison of Schemes A and B.—According to both of the foregoing prominent theories of reduction the conjugated chromosomes separate at the first maturation mitosis, thus causing reduction, and

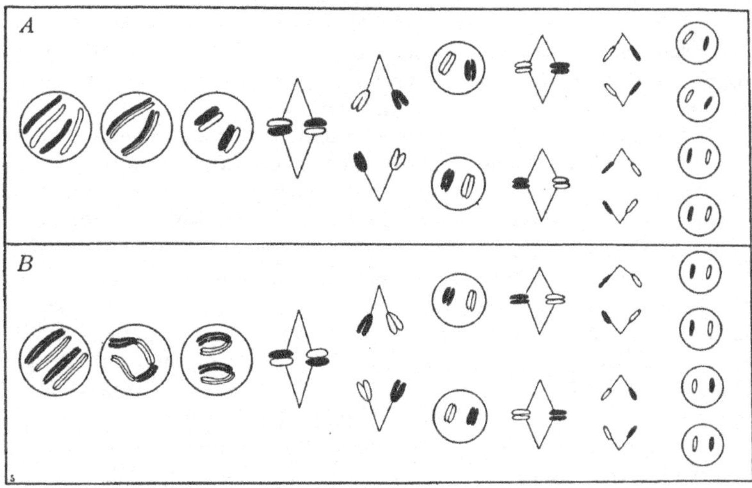

FIG. 93.—Diagram showing distinction between *Schemes A* and *B*. See text.

divide longitudinally (equationally) at the second mitosis, so that the final result is essentially the same: two of the resulting four nuclei differ qualitatively from the other two in their chromatin content (Fig. 93). The distinction between the two interpretations is nevertheless an important one, and may be emphasized in the following summary.

According to Scheme *A* the double character of the chromatin spiremes of the early heterotypic prophase is due to a lateral pairing of simple threads each representing an entire somatic chromosome, the second contraction not being significant as regards pairing. The bivalent chromosomes so formed, after much shortening and thickening, are separated in .the heterotypic mitosis, during the anaphase of which (or earlier in the

1 A more detailed summary of this view may be found in a review of Miss Digby's paper on *Osmunda* by the present author (1920*a*).

16

case of chromosome tetrads) the split that is to function in the homœotypic mitosis makes its appearance. The doubleness in the heterotypic prophase is therefore not homologous with that in the somatic prophase: in the former it is due to a conjugation and in the latter to a split.

According to Scheme *B* the doubleness of the heterotypic prophase is due to a true splitting as in the case of somatic division. In both cases, moreover, the split may have its origin in the preceding telophase. The bivalent chromosome is formed by the association in pairs (often at first end-to-end in the spireme but later side-by-side) of segments of this split spireme at the time of the second contraction. The two split univalents composing the bivalent are separated in the heterotypic mitosis, while in the homœotypic mitosis the separation is along the line of the split originating in the last premeiotic telophase and seen in the spireme of the early heterotypic prophase. The doubleness of the early heterotypic prophase is therefore regarded as homologous with that of the somatic prophase: in both cases it represents a true split.

It cannot yet be said what the outcome of this controversy is to be. The advocates of Scheme *A* believe that those of Scheme *B* have misinterpreted the changes occurring in the early heterotypic prophase and in all telophases, while the latter charge the former with a neglect of the second contraction stage. Scheme *B* as fully elaborated by Miss Digby has certain advantages: it allows one interpretation to be placed upon the double spireme in both somatic and heterotypic prophases, irrespective of the exact time at which the split originates, and it also helps to explain the sudden appearance of the split for the second maturation mitosis in the anaphase of the first. Scheme *A*, on the other hand, is preferred by geneticists because of the earlier and much longer continued association of the conjugating chromosomes, which allows a greater opportunity for "crossing-over" to occur. The significance of this point will be brought out in Chapter *XVII*.

This question, however, must be settled primarily by direct evidence. It is obvious that its solution depends upon the exact manner in which the telophasic transformation of the chromosomes and the derivation of the latter from the reticulum in the prophase are accomplished. It is granted by both schools that the alveolar or reticulate condition in which the chromosomes are found in late telophase is continuous with the similar condition seen in the succeeding prophase. If, then, it is true (1) that the telophasic transformation (alveolation) represents a true splitting, and (2) that the early prophasic reticulate condition passes directly into the double spireme, it follows that this doubleness in every prophase is due to the split originating in the preceding telophase. But workers on mitosis are not at all agreed that the evolution of the chromosomes is that stated in (1) and (2). It has been shown in *Vicia faba* (Sharp 1913), *Tradescantia* (Sharp 1920*b*), and a number of other instances (see Chapter

VIII) not only that the telophasic alveolation is too irregular to be regarded as a splitting, but also that the reticulate condition of the pro_ phase, instead of developing directly into the definitive split, gives rise to simple thin threads in which a new split is developed. From this it cannot be concluded that in no form does the split develop directly from the early reticulate condition, or that the telophasic alveolation, though irregular, may not later become so equalized as to constitute the first stages of the split; but it does follow that it is quite unsafe to use the principle of telophasic splitting as a premise from which to draw the conclusion that the approximation of thin threads in the early heterotypic prophase represents the reassociation of the halves of a single split chromosome. It is well to emphasize the possible importance of the premeiotic telophase, but any ultimate solution of this perplexing prob- lem must be reached mainly through a more refined analysis of those

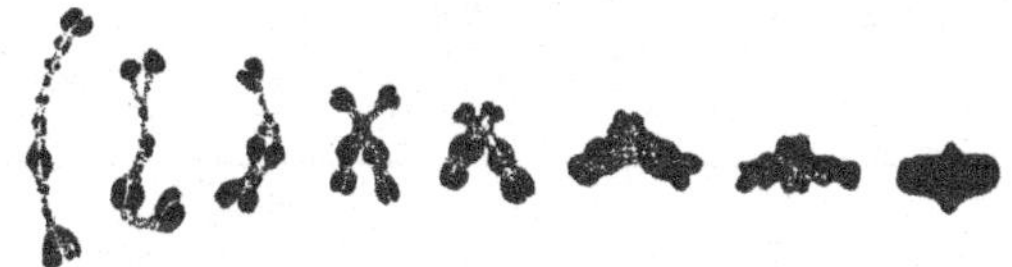

prophasic changes which have led a long list of investigators to the con- clusion that the early heterotypic association of slender threads represents a conjugation of entire chromosomes which separate in the first matura- tion mitosis.

One of the most convincing pieces of direct evidence favoring Scheme *A* is found in Wenrich's recent work on *Phrynotettix* (1916). Wenrich is able to trace a single pair of chromosomes, distinguishable by their peculiar form and the arrangement of their chromatic accumulations or chromomeres, through every stage from the spermatogonia to the sperma- tids. During the heterotypic prophase the two members of the pair conjugate parasynaptically while in the form of slender filaments. Simi- larly strong arguments are advanced by Robertson (1916) as the result of his detailed analysis of the chromosome groups in other Tettigidæ and Acrididæ, in which the homologous members can be followed with much certainty because of their frequent inequality in size.

Reduction With Chromosome Tetrads.—As already pointed out, the marking out of the lines of separation for both maturation divisions during the heterotypic prophase, with the resulting formation of chromo- some tetrads, increases in no inconsiderable manner the difficulty of interpreting the essential changes at these stages. The four chromatids composing the tetrad represent two conjugated chromosomes each of which is longitudinally split. Because of the variety of ways in which these may arrange themselves with reference to one another—in the form of simple or compound rods, crosses, and rings—their distribution to the daughter nuclei, as well as the manner of their origin, is very difficult to follow with certainty. The accompanying diagrams will serve to illustrate the more common modes of behavior described for chromosome tetrads, which are found chiefly in the cells of animals.

Figure 95, *D* represents an exceptional method of tetrad formation described by Henking (1891) for *Pyrrochoris* and by Korschelt (1895) for *Ophryotrocha*. The continuous spireme segments to form the diploid number of chromosomes,[1] which then split longitudinally and shorten.

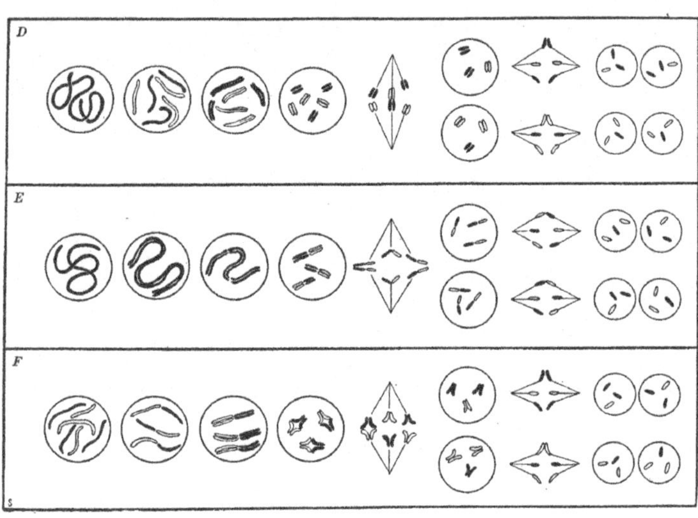

FIG. 95.—Reduction with chromosome tetrads.

D, in *Pyrrochoris* (Henking) and *Ophryotrocha* (Korschelt.) *E*, in certain copepods (Rückert, Haecker, and vom Rath.) *F*, in *Anasa* and *Allolobophora* (Paulmier; Foot and Strobell).

No conjugation occurs until the metaphase, when the split chromosomes come together end-to-end, forming tetrads. They at once separate in the anaphase, bringing about reduction. In the second mitosis they divide along the original split, so that each of the four resulting nuclei receives the haploid number of chromosomes, two of the nuclei thus differing from the other two as the result of the separation of entire (though secondarily split) chromosomes at the first mitosis. According to Goldschmidt (1905), the chromosomes of *Zoögonus mirus*, after thus undergoing no prophasic conjugation, divide longitudinally at the first

---

[1] For the sake of uniformity and clearness the diploid number is represented as 6 in all of these diagrams.

mitosis and separate into two haploid groups at the second. To this simple form of reduction Goldschmidt applied the term "Primärtypus." Grégoire (1909a), on the contrary, found parasynapsis and the usual mode of reduction in *Zoögonus*.

The interpretation at one time given by Rückert (1893, 1894), Haecker (1895), and vom Rath (1895) for certain copepods is shown in Fig. 95, *E*. The continuous spireme splits throughout its length and then breaks into the haploid number of segments. These again break transversely, forming chromosome tetrads, each composed of two split chromosomes arranged end-to-end. In some species the chromatids open out to form four-parted rings, whereas in others they maintain the rod form. A separation occurs along the line of the original split at the first mitosis, which is therefore equational, and along the plane of conjugation at the second mitosis, which is therefore reductional. In *Dicrocœlium* Goldschmidt (1908) reported that such tetrads divide reductionally at the first mitosis. Lerat (1905), moreover, has found that in *Cyclops strenuus*, one of the forms used by the earlier workers, the tetrads arise by a parallel conjugation of thin threads which later split.

A third mode of tetrad behavior is that reported by Paulmier (1899) for *Anasa tristis* and by Foot and Strobell (1905, 1907) for *Anasa* and *Allolobophora fœtida* (Fig. 95, *F*). Here the chromosomes conjugate end-to-end, the bivalents so formed then splitting longitudinally, giving tetrads which take on a cross or ring form. At the first mitosis the separation is along the plane of conjugation, effecting reduction, and at the second it is along the plane of splitting. According to McClung (1902), Sutton (1902, 1905), Robertson (1908), and others, such tetrads separate reductionally at the second mitosis (postreduction) rather than at the first (prereduction) in certain orthopterans studied by them.

Figure 96 illustrates the origin of chromosome tetrads of five characteristic types by the two prominent modes of reduction described in detail in foregoing pages. According to Scheme A (A_1–C_1), two chromosomes conjugate parasynaptically while in the form of slender threads. Instead of remaining unsplit as in most plants, each member then splits longitudinally in a plane at right angles to the conjugation plane, thus giving a tetrad composed of four parallel strands (chromatids) (D). According to Scheme B (A_2–C_2), the two chromosomes are at first arranged telosynaptically in the spireme and the latter splits throughout its length. The two conjugating members then take up a side-by-side position, and their split, instead of becoming obscured as usually occurs in plants, remains open, giving the tetrad of parallel strands (D).

The tetrad, by whichever method it has arisen, may now undergo a variety of alterations, some of which are shown at E and F. The chromatids may simply shorten and thicken, the tetrad at diakinesis maintaining the form of parallel rods (E_1, F_1). They may open out along the plane of

conjugation (E_2) and take the form of rod tetrads (F_2) like those described
by Rückert and Haecker.   While opening out in this manner the longi-

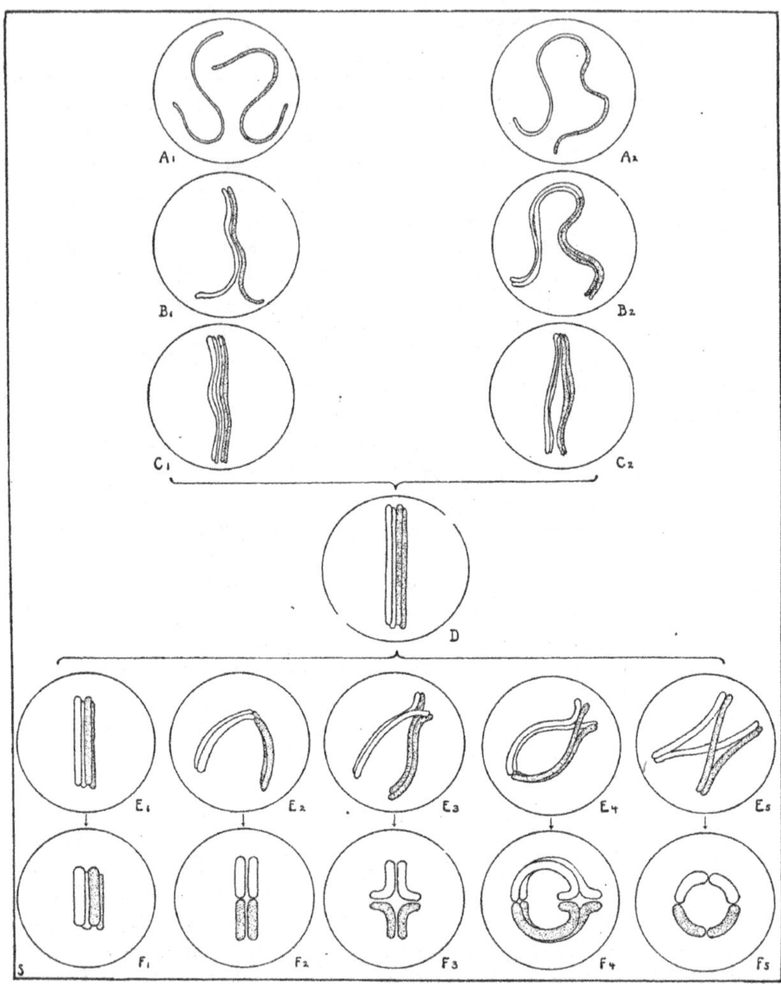

Fig. 96.—Diagram showing the origin of the tetrad of chromatids (*D*) according to *Scheme A* (A_1-C_1) and *Scheme B* (A_2-C_2), and the further transformation of this tetrad into tetrads of five types (F_1-F_5).

tudinal halves of each chromosome may diverge where the two chromosomes remain in contact (E_3), the tetrad eventually taking the form of a
cross (F_3) as in the cases described by Paulmier and by Foot and Strobell.

If the conjugated chromosomes remain in contact at both ends (E_4) a complete ring results (F_4). In certain orthopterans the four chromatids open out along the conjugation plane in some regions and along the plane of splitting in other regions; this results in the curious compound rings (Fig. 156) found in the cells of these insects. Finally, the chromatids may open out from one end along the conjugation plane and from the other end along the splitting plane (E_5), the tetrad then assuming the form of a ring composed of four parts (F_5). In all cases the tetrads usually condense into compact quadruple bodies by the time they take their places on the spindle of the heterotypic mitosis.

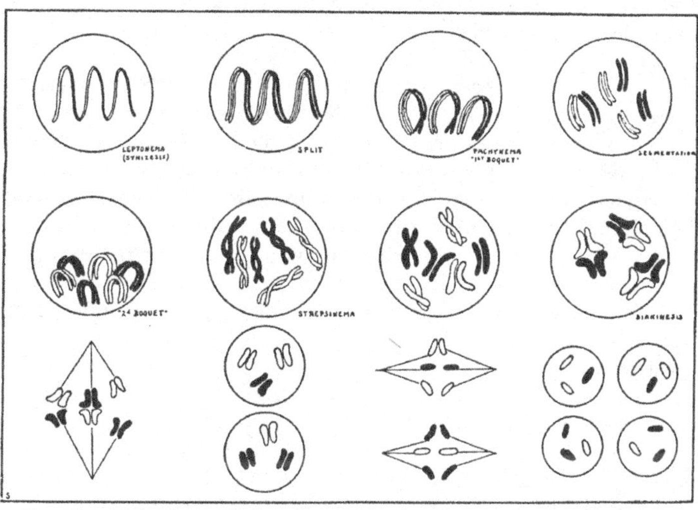

Fig. 97.—Reduction with chromosome tetrads in *Fasciola hepatica*, according to Schellenberg (1911). Explanation in text.

The four chromatids composing the completed tetrad are in most cases exactly similar in appearance, so that it is a matter of much difficulty to determine along which plane they are separated at the first maturation mitosis. According to the two theories of tetrad origin illustrated in the foregoing diagram, however, the chromatids are supposed in almost all cases to separate along the plane of conjugation at the first mitosis, and this conclusion is supported by the behavior of those bivalent chromosomes which are not divided into tetrads of chromatids.

A further interpretation of reduction involving chromosome tetrads has been given by Schellenberg (1911) for the parasitic flatworm, *Fasciola hepatica* (Fig. 97). The chromatin in the heterotypic prophase takes the form of a long slender filament which splits longitudinally soon after synizesis. This double thread then segments into the haploid number of pieces, each representing two chromosomes end-to-end; these have the form of loops with a definite orientation ("first boquet stage"). Each segments again, giving the diploid number of split chromosomes, which again assume the form of oriented loops ("second boquet stage"). The halves twist tightly about each other, shorten to form the double bodies seen at diakinesis in the diploid rather than the haploid number, and then conjugate to form the haploid number of chromosome tetrads. The conjugating members (each split) separate at the first mitosis, bringing about reduction; at the second mitosis the separation is along the line of the original split. According to this interpretation, therefore, the doubleness of the early heterotypic prophase is due to a split, as in Scheme *B*, but the chromosomes arranged end-to-end in the spireme soon become separated and do not conjugate again until diakinesis.

For a number of years it was thought (Carnoy 1886; Boveri 1887; Hertwig 1890; Brauer 1893) that the chromosome tetrad in *Ascaris megalocephala* was exceptional in being formed by two longitudinal fissions of a primary chromatin rod, there being as a consequence no qualitative reduction in the two maturation divisions unless the organization of the chromatin were different from that of other organisms. But it has since been shown that they arise as in other organisms by the conjugation of two split chromosomes (Sabaschnikoff 1897; Tretjakoff 1904; Griggs 1906). In the oögenesis Griggs reports telosynapsis with prereduction, whereas in the spermatogenesis Tretjakoff describes parasynapsis followed by postreduction. In *Ascaris canis* (Marcus 1906; Walton 1918) the four chromatids each show a transverse constriction, the chromosomes on the first maturation spindle having the form of octads.

Although the formation of well differentiated chromosome tetrads occurs very commonly in animals, it appears to be very rare in plants. Farmer (1895) described tetrads in *Fossombronia*, and they have since been reported in at least three other bryophytes: *Pallavicinia* (Moore 1905), *Sphagnum* (Melin 1915), and *Chiloscyphus* (Florin 1918). They have also been described in a few vascular plants: *Equisetum* (Osterhout 1897), *Pteris* (Calkins 1897), *Arisæma* (Atkinson 1899), *Tricyrtis* (Ikeda 1902), *Thalictrum, Calycanthus*, and *Richardia* (Overton 1909) (Fig. 98), *Spinacia* (Stomps 1911), *Primula* (Digby 1912), and *Lopezia* (Täckholm 1914).

According to Grégoire (1905) such structures in plants are not true tetrads, but resemble them because the chromosomes are often bent and have their material accumulated largely at their ends. Sakamura (1920) interprets them as conjugated *constricted* chromosomes, and denies that the quadripartite condition has anything to do with reduction in such cases. He likewise accounts for the metasynaptic rod tetrads (Fig. 95, *E*) described by several investigators of maturation in animals, holding that they represent two constricted chromosomes conjugated parasynaptically rather than two split ones placed end-to-end. In support of this contention he cites the following observations: such "tetrads" are seen not only in the oöcytes and spermatocytes but also in oögonia, spermatogonia, and somatic cells; the supposed telosynaptically conjugated members are often very unequal in size; such tetrads are sometimes divided in the transverse plane at neither maturation mitosis; not only tetrads, but also octads and hexads are often observed, even in

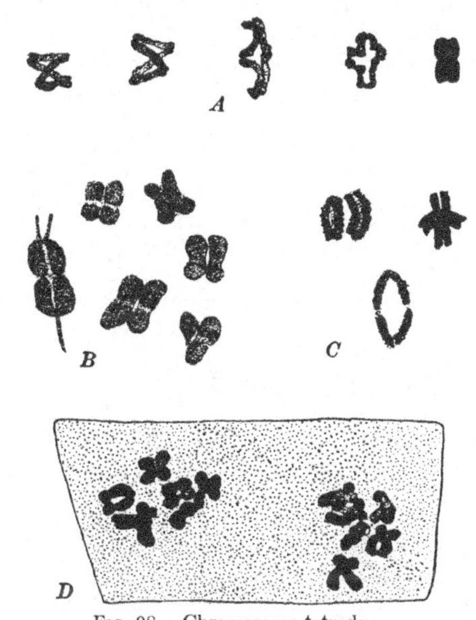

FIG. 98.—Chromosome tetrads.

A, five stages in the development of the tetrad in the spermatocyte of *Anasa tristis.* × 3000.   From a preparation by Dr. H. E. Stork.   *B*, tetrads in sporocyte of *Chiloscyphus.* Enlarged;   × 2800.   (*After Florin*, 1918.)   *C*, tetrads in *Richardia africana.* (*After Overton*, 1909.)   *D*, false tetrads in somatic cells of *Pisum* due to action of chloral hydrate on constricted chromosomes.   (*After Sakamura*, 1920.)

the same cell, and these are plainly due to the presence of additional accentuated constrictions.   Robertson (1916) also interprets such telosynaptic rod tetrads as those observed by Haecker in the copepods as constricted chromosomes.   The constrictions, according to this writer, represent points of temporary union between non-homologous elements. From these considerations it is evident that constrictions have much to do with the appearances assumed by chromosomes, and that they should be taken into account in interpreting the chromosome tetrad.

Numerical Reduction Without Qualitative Reduction.—Figure 99 illustrates the behavior of the chromosomes in maturation according to three not widely accepted views.   A few workers, including Fick (1907,

1908), Meves (1907, 1908, 1911), Giglio-Tos (1908), and Granata (1910), reject the theory of chromosome individuality and specificity, and therefore do not regard the chromosomes which are distributed to the four cells at maturation as at all identical with those of the divisions immediately preceding, except in so far as they are composed of the same nuclear material. Accordingly they recognize no qualitative reduction, but only a numerical one. This reduction in number results from the fact that the spireme formed in the heterotypic prophase (Fig. 99, *A*) segments into the haploid number of pieces instead of the diploid number, these pieces being simply divided longitudinally at both maturation divisions, and the four resulting nuclei being qualitatively similar.

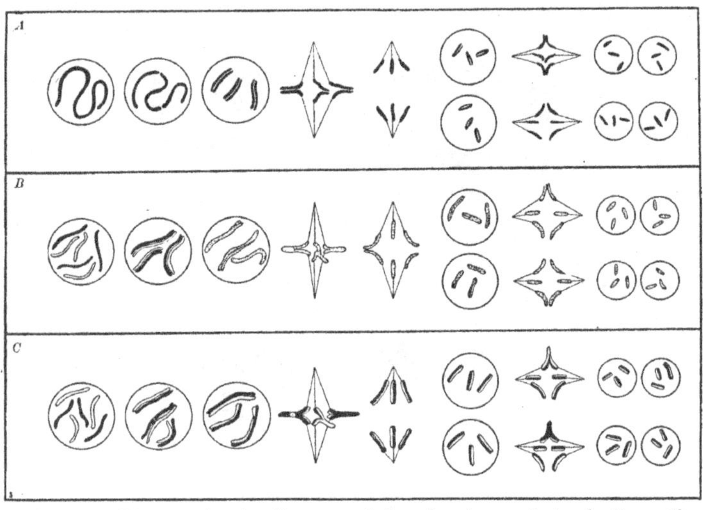

Fɪɢ. 99.—Diagram showing three reported modes of numerical reduction without qualitative reduction.

A, according to Fick *et al*. *B*, according to Vejdowský; complete fusion of conjugating members. *C*, according to Bonnevie; bivalents arranged on spindles in juxtaposition; fusion of conjugating members eventually becomes complete.

According to Vejdowský (1907) (Fig. 99, *B*) the chromosomes appear in diploid number in the heterotypic prophase and conjugate parasynaptically. The members of the pair fuse completely and lose their individual identity, so that the chromosomes appearing on the first maturation spindle in haploid number are new entities, and not merely temporary pairs of somatic chromosome individuals. At both divisions these bodies split longitudinally, giving equivalence to the four resulting nuclei. Here, as in the foregoing example, there is no definite qualitative reduction in Weismann's sense, though a numerical reduction is brought about by means of a complete fusion at the time of chromosome conjugation.

An interpretation put forward by Bonnevie (1906, 1908) is shown in Fig. 99, *C*. Here the chromosomes conjugate parasynaptically and come into very intimate union: although they appear to undergo a real fusion their identity is maintained for a time. Owing to the fact that these bivalent chromosomes are inserted upon the spindle with their halves in juxtaposition (side-by-side with respect to the poles) rather than in superposition (one toward each pole), the members of a conjugated pair separate neither at the first division nor at the second. As a result each of the four cells receives the haploid number of chromosomes, all of which are bivalent, and no qualitative reduction occurs. Bonnevie believes that the conjugating members of each pair finally fuse completely in the subsequent stages. In this case, therefore, as in the preceding one, numerical reduction is supposed to result from a complete fusion of the chromosomes in pairs.

Whether any confidence is to be placed in such interpretations or not —and according to most cytologists none should be—they at least serve to show how it is possible that numerical reduction may occur without effecting any qualitative reduction, and that the essential feature of the reduction of the chromosomes is something other than the mere change in their number, as pointed out at the beginning of the chapter.

SYNAPSIS, OR CHROMOSOME CONJUGATION

The phenomenon of chromosome conjugation, or *synapsis*, which we have seen above is such an important feature of the reduction process, must now be somewhat more closely examined. Attention will be directed to three points: the relationship of the conjugating members (the "synaptic mates"), the stage at which the synaptic union takes place, and the exact nature of this union.

Relationship of the Synaptic Mates.—We may first inquire into the relationship which may exist between the two chromosomes pairing to form a given bivalent chromosome: is any chromosome of the duplex group (the two intermingled parental chromosome sets in the individual's nuclei) present in the gonotokont free to pair with any other chromosome, or does the pairing take place according to more restricting rules?

It was suggested by Henking (1891) that the two synaptic mates are ultimately derived from the two parents at the previous fertilization, one from the father and the other from the mother: the chromosomes of one parental set pair with those of the other parental set to form the haploid number of bivalent chromosomes appearing on the first maturation spindle. This idea was later emphasized and developed by Montgomery (1900–4), Sutton (1902), Boveri (1901), and others, who found for it much supporting evidence in organisms with chromosomes differing in size and shape. An observation made by Rosenberg (1909) on *Drosera* hybrids is significant in this connection. When *Drosera rotundi-*

folia (20 chromosomes) is crossed with *D. longifolia* (40 chromosomes) there results a hybrid with 30 chromosomes, of which 10 are contributed by *rotundifolia* and 20 by *longifolia*. When synapsis occurs preparatory to reduction in this hybrid only 10 bivalents are formed, 10 chromosomes remaining unpaired. This was taken by Rosenberg to mean that the 10 *rotundifolia* chromosomes pair with 10 of the *longifolia* ones, leaving the other 10 of *longifolia* without synaptic mates. Had any chromosome of the duplex group of 30 been free to pair with any other, 15 bivalents would have been produced.

Other instances of this phenomenon may be mentioned. By crossing *Œnothera Lamarckiana* (seven chromosomes in gamete) with *Œ. gigas* (14 in gamete) individuals with 21 chromosomes are obtained. Geerts (1911) found that, preparatory to reduction, the seven *Lamarckiana* chromosomes pair with seven of the *gigas* chromosomes, leaving the other seven of *gigas* unpaired. On the contrary, however, Gates (1909) found that the 21 chromosomes in a *lata-gigas* hybrid simply separate into two approximately equal groups, usually of 10 and 11 chromosomes respectively. Kihara (1919) reports that in some 35-chromosome wheat hybrids formed by crossing *Triticum polonicum* (14 chromosomes in gamete) with *T. spelta* (21 in gamete) there are present in the heterotypic prophase 14 bivalents (*polonicum* conjugated with *spelta*) and seven univalents (*spelta*). The 14 bivalents are arranged on the spindle and separate as usual, whereas the seven unpaired *spelta* chromosomes split longitudinally at the first mitosis and distribute themselves irregularly at the second (Fig. 100). An analogous condition is found in *Pigœra* hybrids by Federley (1913).

A very significant additional suggestion with respect to synapsis was made by McClung (1900) and Sutton (1902): not only are the two chromosomes which conjugate derived from the two parents, but they are homologous—each chromosome of one parental set pairs with a particular chromosome of the other parental set, the two members of the resulting bivalent being presumably of corresponding hereditary value, as will be shown in Chapter XV. The evidence for this important hypothesis was found chiefly in *Brachystola* (Fig. 101) and a number of other insects having chromosome complements made up of members with constant characteristic differences in size and shape. Many such cases have been subsequently discovered, especially by McClung and his coworkers in their extensive researches on insect spermatocytes. As examples among plants may be cited *Crepis virens*, *Najas major*, *N. marina*, and *Vicia faba*.

Crepis virens (Rosenberg 1909) (Fig. 102) has six chromosomes: two long, two medium sized, and two short. When synapsis occurs the like chromosomes pair, forming bivalents of three sizes. The members of each pair separate and pass to the daughter cells at the first maturation mitosis, each microspore (after the second mitosis) having as a result three chromosomes: one long, one medium sized, and one short. Since the gamete receives such a simplex group of three chromosomes, and the

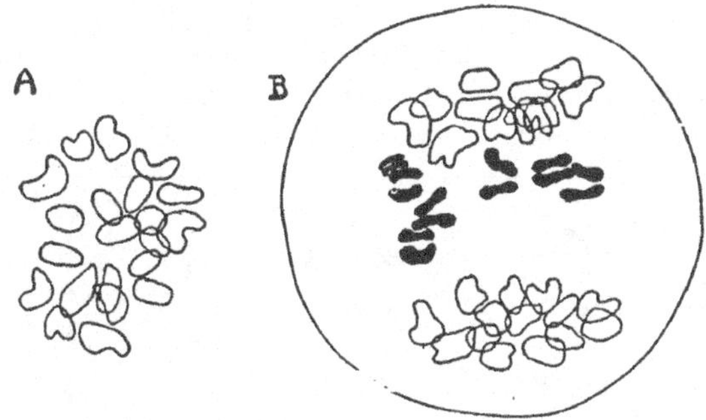

FIG. 100.—Heterotypic mitosis in *Triticum polonicum* × *T. spelta*.
A, the 21 chromosomes (polar view). *B*, 14 bivalents separated into component univalents; 7 unpaired *spelta* chromosomes have split and are about to be distributed. (*After Kihara*, 1919.)

somatic cells of the new individual show six (two of each length), it is evident that the other gamete furnishes a similar simplex group of three.

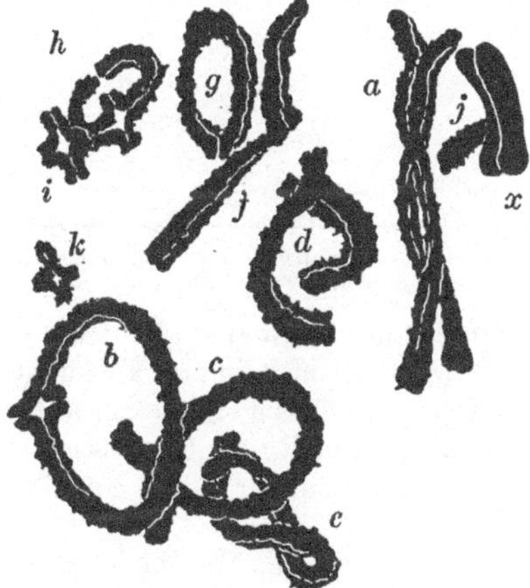

FIG. 101.—The chromosome complement in the spermatocyte of *Brachystola magna*. (*After Sutton*, 1902.)

In *Najas marina* and *Najas major* (Müller 1911; Tschernoyarow 1914) the duplex group of 14 chromosomes is made up of seven visibly different pairs (Fig. 56 *bis*). In the heterotypic prophase these conjugate selectively to form seven bivalents, the reduced nuclei therefore receiving a set of seven visibly different chromosomes. Sakamura (1920) holds the number here to be six rather than seven. (See p. 160.)

In *Vicia faba* (Sharp 1914; Sakamura 1915, 1920) there are in the somatic cells 12 chromosomes, two of them being about twice as long as the other 10 (Figs. 56 and 102). At synapsis in the microsporocyte there are formed six bivalents, one of them having about twice the length of the other five. Hence it is clear that the two long chromosomes pair with each other. In the heterotypic mitosis the synaptic mates separate

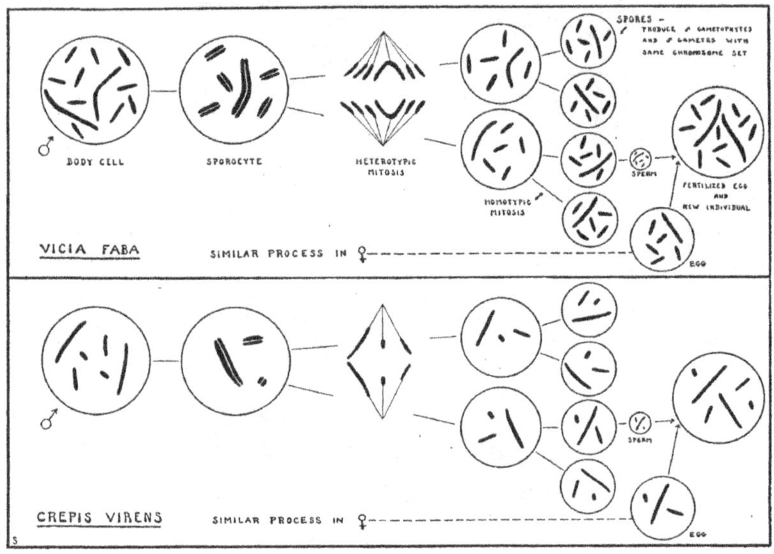

Fig. 102.—Chromosome cycles in *Vicia faba* and *Crepis virens*, showing homologous pairing.

and pass to the daughter nuclei, bringing about reduction. At the close of the homœotypic mitosis the microspore, and hence the male gamete to which it later gives rise, receives a simplex group of six chromosomes: one long and five short. Since the somatic cells contain each of these in duplicate it is evident that a similar set is contributed by the female gamete.

Summing up, we may draw from the above facts certain very important conclusions: (1) *Each parent furnishes the offspring with a set of chromosomes, the members of the two sets being intermingled in all the nuclei of the new individual.* This point will receive further attention in the following chapter on fertilization. (2) *The two members of each bivalent chromosome formed at synapsis are derived one from each parental set.* (3) *Each chromosome of the paternal set conjugates with a particular chromosome of the maternal set: the two are in some sense homologous.*

It should be pointed out that cytologists and geneticists have generally assumed that each synaptic pair is independent of all the others as regards the manner in which it is oriented on the heterotypic spindle. In some pairs the paternal members are directed toward one pole and in other pairs toward the other pole. It is conceivable that in some cases all the paternal members might go to one pole and all the maternal members to the other. Direct evidence that the assortment of the various chromosome pairs is in this respect a random one as originally assumed has been furnished by Miss Carothers (1913, 1917). In the grasshopper, *Trimerotropis,* she finds that the components of some of the bivalents are visibly different in size, in the mode of attachment to the spindle fibers, and in the presence of constrictions; and that these differences make it possible to show beyond question that the several pairs are entirely independent of one another as regards their orientation on the spindle and their consequent distribution to the daughter cells.

From the precise manner in which the distribution of chromosomes at the time of reduction and at other stages of the life cycle parallels the distribution of the hereditary characters it is inferred that such homologous chromosome pairs represent the material basis for the allelomorphic pairs of Mendelian characters exhibited by the organism. This subject is to be taken up in Chapter XV.

The Stage at Which Conjugation Occurs.—In the great majority of observed cases chromosome conjugation occurs during the prophase of the first maturation division. Since the chromatin threads at some time during these prophases usually take the form of a tightly contracted knot out of which they emerge in an obviously double condition, it was suggested (Moore 1896) that the contraction is an important factor in bringing about the conjugation, and the contraction itself came to be called "synapsis." But an examination of the various modes of reduction shows that the conjugation may begin very early, before the contraction (Fig. 83) or, on the other hand, not until the spindle is established (Fig. 95, *D*). The conjugation of the chromosomes is therefore to be distinguished from the contraction. It has now become customary to refer to the former, at whatever stage it occurs, as *synapsis,* and to the latter as *synizesis.*

In an increasing number of reported cases the paired association apparently begins even before the heterotypic prophase. The chromosomes have been observed in several instances to undergo pairing during the anaphase and telophase of the last premeiotic division. Such is the condition in certain Hemiptera (Montgomery 1900, 1901), *Oniscus* (Nichols 1902), *Brachystola* (Sutton 1902), *Scolopendra* (Blackman 1903, 1905), *Pedicellina* (Dublin 1905), and a number of more recent cases. Furthermore, the pairing has been stated to begin in the spermatogonia several cell generations before maturation in certain Hemiptera and *Ascaris* (Montgomery 1904, 1905, 1908, 1910), *Alytes* (Janssens and Willems 1909), *Helix* and *Sagitta* (Stevens 1903; Ancel 1903), certain Diptera (Stevens 1908, 1911), and *Pediculus* (Doncaster 1920).

More recently it has been shown that the homologous chromosomes may begin to show a paired arrangement even earlier in the cycle, in some cases directly after the parental groups are brought together at fertilization.   In the Diptera, for example, Metz (1916*a*) has shown that the association, which at certain stages is so close as to constitute a synapsis, begins before the cleavage of the fertilized egg, and that the paired condition is maintained in all cells, somatic and germinal, throughout the life cycle.   Metz examined 80 species and in all of them found such a somatic pairing.   In *Culex* (Stevens 1910, 1911; Taylor 1914, 1917), which has six chromosomes, the association can be seen in the nuclei of the segmenting egg, and in the early larval stages there follows an actual parasynaptic fusion, so that the somatic cells thereafter show three bivalent chromosomes rather than six univalents.   In the maturation divisions the members of each pair separate, the gametes receiving three chromosomes each, just as they would had conjugation begun in the heterotypic prophase as usual.

A loosely paired arrangement of the chromosomes in the somatic cells of plants has been reported by Strasburger (1905, 1907, 1910) for *Galtonia candicans, Funkia Sieboldiana, Pisum sativum, Melandrium, Mercurialis,* and *Cannabis;* by Sykes (1908) for *Hydrocharis, Lychnis, Begonia, Funkia,* and *Pisum;* by Overton (1909) for *Calycanthus;* by Müller (1909, 1911) for *Yucca* and other forms; by Stomps (1910, 1911) for *Spinacia;* by Kuwada (1910) for *Oryza;* by Tahara (1910) for *Morus;* and by Ishikawa (1911) for *Dahlia.*   This is another matter that will be considered further in Chapter XII.

The Nature of the Synaptic Union.—Because of the manner in which chromosome behavior is at present being applied to the solution of the problems of inheritance, no question concerning chromosome conjugation is more important than that of the exact nature of the synaptic union. In reviewing some of the opinions of this subject it will be convenient to list separately the views of the telosynaptists and the parasynaptists.

In such cases as those described by Henking and by Goldschmidt Fig. 95, *D*) there is only a momentary end-to-end association of the fully formed chromosomes on the spindle of the heterotypic mitosis, there being no real fusion and almost no opportunity for an "interchange of influences."   In the other tetrad chromosomes formed by telosynapsis (Fig. 95, *E* and *F*; Fig. 90) there is only slightly greater opportunity for such interchange.   According to Scheme *B* (Fig. 89) the synaptic mates are at first arranged end-to-end, and only later, when partially condensed, do they take up a side-by-side position, allowing a more intimate and extensive union for a short time.

Generally speaking, the parasynaptists have given more attention to the details of the synaptic union than have the telosynaptists. Although cases are on record in which there is only a momentary parasynaptic association of fully formed chromosomes (von Voss 1914), the association usually extends over considerable time. Most parasynaptists hold that the conjugation begins with the association of the leptotène threads before or during synizesis, continues through the remainder of the prophase, and ends with the anaphasic separation (Scheme *A*). The association of the synaptic mates is thus long and intimate. Concerning the closeness of the union, however, opinions differ widely.

A few investigators (Vejdowský 1907; Bonnevie 1906, 1908, 1911; Winiwarter and Sainmont 1909; Schneider 1914) have thought that the conjugating members fuse completely and lose their individual identity, the "mixochromosome" so formed then undergoing two true longitudinal splits along new planes at the two maturation divisions. In some cases (Bonnevie; see p. 251) the fusion may not be fully consummated until during the post-meiotic divisions. Others believe the split for the heterotypic mitosis to be along the plane of conjugation (Cardiff 1906, Fasten 1914, and others). Probably the most widely advocated view is that there is no actual fusion of the synaptic threads, the latter maintaining their identity completely. Although their association may at times be so intimate that they seem to constitute a single thick thread, the doubleness, if thus lost to view, reappears during later stages (Berghs 1904, 1905; A. and K. E. Schreiner 1905, 1906; Maréchal 1907; Overton 1905, 1909; Robertson 1915, 1916) (Fig. 88). Several careful observers have reported that the doubleness can be seen at all stages (Grégoire 1907, 1910; Schleip 1906, 1907; Montgomery 1911; Kornhauser 1914, 1915; Wenrich 1915, 1917). Grégoire, who has argued strongly for this interpretation, has emphasized the ease with which the closely appressed threads may be mistaken for a single thick structure.

One of the most important suggestions which has been made concerning chromosome conjugation is embodied in the "Chiasmatype Hypothesis" of Janssens (1909). According to Janssens, the pairing threads, though remaining separate throughout the greater part of their length, fuse at one or more points as they twist about each other. When they again separate a break occurs at each of these fusion points, but along a new plane, so that each of the two resulting chromosomes is composed of portions of both conjugating members (Fig. 149). This interpretation, which has been admitted as possible by several of the investigators named in the preceding paragraph, is significant in that it shows how an orderly evolution of chromosomes with new constitutions may occur, a point of great importance in connection with current conceptions of the

17

physical basis of heredity. Special attention will be devoted to this question in Chapter *XVII*.

Chromomeres.—An important rôle has been attributed to the chromomeres by many students of synapsis. Allen (1905), for example, maintained that the fusion of the leptotène threads in *Lilium* involves a fusion of their chromomeres, the subsequent division of the fused chromomeres initiating the resplitting of the pachytène thread. Allen found the chromomeres to be composed of still smaller chromatic elements, and offered various suggestions concerning the manner in which the resplitting of the pachyténe thread might be supposed to effect a redistribution of the "idioplasms." That chromosome conjugation is primarily a conjugation of small chromatic elements within the chromosome was held by Strasburger (1904, 1905) and the Schreiners (1906). The visible chromatin granules, or "pangenosomes," were conceived by Strasburger to represent complexes of "pangens" such as were postulated by de Vries, conjugation involving an interchange of these latter units.

The chromomere interpretation has been adversely criticised by Grégoire (1907, 1910) on the basis of further evidence obtained from a study of the chromatic structures themselves. This author points out several serious objections to the view that the chromomeres are autonomous bodies, and concludes that they are rather to be regarded simply as swellings or thicker portions of the chromatin thread, such thick portions remaining as the thread undergoes a stretching which is not uniformly resisted at all points. Their frequently striking correspondence or paired arrangement in the synaptic threads is explained as the result of the response of the two closely associated threads to the same stretching force. This interpretation is also shown to account for the variability in the dimensions of the chromomeres, their tapering form, the often reported absence of correspondence between the chromomeres of the two threads, and various other aspects. Wenrich (1916, 1917), on the other hand, has found that in *Phrynotettix* (Fig. 155) the chromomeres show a remarkable individual constancy in size and position in a given member of the chromosome complement, not only in the various cells of a given individual, but also in those of different individuals. These facts strongly suggest an autonomy of the bodies in question.

Because of their great theoretical importance (see Chapter *XVII*) it is to be regretted that after such a large amount of research so many points regarding the process of synapsis should remain in such an unsettled state. It is hoped that further refinements in microtechnique may remove some of the obscurity which at present surrounds them.

OTHER OPINIONS ON THE HETEROTYPIC PROPHASE

Although the phenomena of the heterotypic prophase, particularly synizesis and synapsis, are generally looked upon as normal occurrences of considerable significance, not all investigators concur in this opinion. That synizesis is an artifact due to faulty fixation is an interpretation which, though it may be justified for certain cases in which the contraction may be very slight or absent, is not of general application. Fixation often serves to accentuate the appearance of contraction, but the characteristic synizesis figure has been observed widely enough in faithfully preserved, and even in living material to make it evident that we are dealing here with a normal feature of the heterotypic prophase. It may not, however, be of universal occurrence.

R. Hertwig (1908) came to the conclusion, as a deduction from his theory of the nucleoplasmic relation, that the phenomena of the heterotypic prophase represent an abortive mitosis: the disturbed nucleoplasmic balance is restored to the normal by a multiplication of chromatin without an actual mitosis, the process taking the form of the changes peculiar to the heterotypic prophase. This view of Hertwig, which was denied by Grégoire (1909*b*), is supported by Kingsbury and Hirsch (1912), who state:

"According to this view, on the one hand, synizesis represents 'an attempt on the part of' the spermatogonia to divide again—which fails; while on the other hand, the reputed conjugation of chromosomes occurring at about this time is but the imperfect fission and subsequent fusion of daughter chromosomes of such abortive division."

The above quoted authors regard synizesis and synapsis as indications of the onset of degeneration. In this conclusion they are supported by Kingery (1917), who, in his investigation of the white mouse, finds synizesis in the primitive germ cells which degenerate, but not in the definitive germ cells. Observations of a similar nature were made by Wodsedalek (1916) in the mule. If, as Kingery (1917) and Popoff (1908) point out, the "heterotypic" changes are due to degeneration, they should be found in abnormal somatic cells. Marcus (1907), in fact, had observed a contraction similar to that of synizesis preceding degeneration in the cells of the thymus gland. Němec (1903) and Kemp (1910) also found that in the cells of roots treated with chloral hydrate the nuclei come to have an abnormally high number of chromosomes ("syndiploid nuclei"), this number, according to Němec, being gradually restored to the normal during the subsequent mitoses, which show phenomena of a heterotypic nature. Strasburger (1911), while agreeing with Němec that the syndiploid condition gradually disappears, denied that any truly heterotypic phenomena are concerned. The "heterotypic" changes observed by Němec he held to be only peculiar vegetative mitoses with a superficial resemblance to genuine reduction divisions.

Němec's conclusion regarding a reduction in chloralized vegetative cells is also contradicted by Sakamura (1920), who has made a particularly exhaustive study of these phenomena. Sakamura finds that a variety of agencies, including chloral hydrate, benzene vapor, ether, chloroform, and the gall-producing secretions of *Heterodera*, may be employed to bring about aberrations of the mitotic process.   After the chromosomes are divided and partially distributed they may be reorganized in a single "didiploid" nucleus.   In other cases the chromosomes may reorganize as two or more nuclei with various chromosome numbers, and these may often fuse to form "syndiploid" nuclei.   To all the kinds of stimuli applied the chromosomes react by becoming shorter and thicker, and thus appear like heterotypic chromosomes.   Furthermore, latent or obscure constrictions are rendered more conspicuous, so that some of the split chromosomes appear like chromosome tetrads.   Sakamura shows that these false tetrads do not represent heterotypic phenomena in any true sense: they are merely the result of the response of split and constricted chromosomes to the abnormal conditions induced in the cell, and have nothing to do with any reducing process.   No such autoregulative reduction occurs in these didiploid cells, their gradual decrease in relative number being due to their lowered capacity for, and rate of, division.

Child (1915) emphasizes the physiological significance of maturation, and shows that the heterotypic phenomena are associated with a low metabolic rate in the cells, that they may occur occasionally in other cells having a low rate, and that they can be induced artificially with narcotics as Němec stated.

All of these observations are interesting in that they indicate the nature of some of the physiological changes occurring at the time of maturation.   The description of the heterotypic phenomena upon which will be based our ultimate interpretation of its significance, will not be complete until the physiological as well as the morphological changes have been exhaustively examined and correlated.   But because it has been found that the onset of the meiotic process is associated with a lowering of the rate of metabolism which, if continued, may result in degeneration; or because appearances similar to those of the heterotypic prophase may occur in other cells with disturbed metabolism; it does not at all follow that the heterotypic phenomena are at bottom phases of a degeneration process, or that they have no other significance in the normal life cycle. These phenomena occur almost universally throughout the whole world of living organisms at a very critical stage in the life cycle and lead to significant results with a high degree of regularity.   The lowered rate of metabolism accompanying them offers in the vast majority of cases no check to the normal functioning of the products of the maturation divisions.   It therefore seems more reasonable to regard the observed degeneration as a secondary effect that may occasionally set in during the normal heterotypic prophases because the metabolic rate is already at a relatively low level at that time, than to look upon the heterotypic changes as a part of a degeneration process which is only exceptionally completed—unless, indeed, all changes in the organism which are accompanied by a fall in the metabolic rate be regarded as degenerative in character.

Bibliography 11

Chromosome Reduction

ALLEN, C. E. 1904. Chromosome reduction in *Lilium canadense*. Bot. Gaz. **37**: 464–469.

  1905a. Nuclear division in the pollen mother cells of *Lilium canadense*. Ann. Bot. **19**: 189–258. pls. 6–9.

  1905b. Das Verhalten der Kernsubstanzen der Synapsis in den Pollenmutterzellen von *Lilium canadense*. Jahrb. Wiss. Bot. **42**: 72–82. pl. 2.

  1905c. Die Keimung der Zygoten bei *Coleochæte*. Ber. Deu. Bot. Ges. **23**: 285–292. pl. 13.

ANCEL, P. 1903. Histogénèse et structure de la glande hermaphrodite d'*Helix pomatia* (Linn.). Arch. de Biol. **19**: 389–652. pls. 12–18.

ATKINSON, G. F. 1899. Studies on reduction in plants. Bot. Gaz. **28**: 1–26. pls. 1–6.

  1915. Phylogeny and relationships in the ascomycetes. Ann. Mo. Bot. Garden **2**: 315–376. figs. 10.

BEER, R. 1912. Studies in spore development. II. On the structure and division of the nuclei in the Compositæ. Ann. Bot. **26**: 705–726. pls. 66, 67.

  1913. Studies in spore development. III. The premeiotic and meiotic nuclear divisions of *Equisetum arvense*. Ibid. **27**: 643–661. pls. 51–53.

VAN BENEDEN, E. 1883. Recherches sur la maturation de l'oeuf, la fécondation et la division cellulaire. Arch. de Biol. **4**.

BERGHS, J. 1904. La formation des chromosomes hétérotypiques dans la sporogénèse végétale. II. La Cellule **21**. 383–394. 1 pl.

  1905. La formation des chromosomes, etc. IV. Ibid. **22**: 141–160. 2 pls.

BLACKMAN, M. W. 1903. Spermatogenesis of the myriopods. II. On the chromatin in the spermatocytes of *Scolopendra heros*. Biol. Bull. **5**: 187–217. figs. 22.

  1905. Spermatogenesis of the myriopods. III. The spermatogenesis of *Scolopendra heros*. Bull. Mus. Comp. Zool. Harvard **48**: 1–137. pls. 9.

BLACKMAN, V. H. 1904. On the fertilization, alternation of generations, and general cytology of the Uredineæ. Ann. Bot. **18**: 323–369. pls. 21–24.

BONNET, J. 1914. Reproduction sexuée et alternance des générations chez les algues. Prog. Rei Bot. **5**: 1–126. figs. 65.

BONNEVIE, K. 1906. Untersuchungen über Keimzellen: I. Beobachtungen an den Keimzellen von *Enteroxenos östergreni*. Jen. Zeitschr. **41**: 229–428. pls. 16–23. 10 figs.

  1908. Chromosomenstudien. II. Arch. Zellf. **2**: 201–278. pls. 13–19. figs. 23.

  1911. Chromosomenstudien. III. Chromatinreifung in *Allium cepa*. Ibid. **6**: 190–253. pls. 10–13.

BOVERI, TH. 1887. Zellen-Studien. I. Die Bildung der Richtungskörper bei *Ascaris megalocephala* und *Ascaris lumbricoides*. Jen. Zeitsch. **21**: 423–515. pls. 25–28.

  1890. Zellen-Studien. III. Ueber das Verhalten der chromatischen Kernsubstanz bei der Bildung der Richtungskörper und bei der Befruchtung. Ibid. **24**: 314–401. pls. 11–13.

  1891. Befruchtung. Ergeb. Anat. u. Entw. **1**: 386–485. figs. 15.

  1904. Ergebnisse über die Konstitution der chromatischen Kernsubstanz. Jena.

Brauer, A. 1893. Zur Kenntniss der Spermatogenese von *Ascaris megalocephala.* Arch. Mikr. Anat. **42**: 153–212. pls. 11–13.

Calkins, G. N. 1897. Chromatin-reduction and tetrad-formation in Pteridophytes. Bull. Torr. Bot. Club **24**: 101–115. pls. 295, 296.

Carothers, E. E. 1917. The segregation and recombination of homologous chromosomes as found in two genera of Acrididæ (Orthoptera). Jour. Morph. **28**: 445–521. pls. 14. figs. 5.

Carruthers, D. 1911. Contributions to the cytology of *Helvella crispa* Fries. Ann. Bot. **25**: 243–252. pls. 18, 19.

Child, C. M. 1915. Senescence and Rejuvenescence. Chicago.

Claussen, P. 1912. Zur Entwicklungsgeschichte der Ascomyceten. *Pyronema confluens* Zeitschr. f. Botanik **4**: 1–64. pls. 1–6. figs. 13.

Cleland, R. E. 1919. The cytology and life history of *Nemalion multifidum*, Ag. Ann. Bot. **33**: 323–352. pls. 22–24. figs. 3.

Dahlgren, K. V. O. 1915. Der Embryosack von *Plumbagella*, ein neuer Typus. Ark. f. Botanik **14**: 1–10. figs. 5.

Davis, B. M. 1909. Pollen development of *Œnothera grandiflora.* Ann. Bot. **23**: 551–571. pls. 41, 42.

    1910. The reduction division of *Œnothera biennis.* Ibid. **24**: 631–651. pls. 52, 53

    1911. Cytological studies on *Œnothera.* III. A comparison of the reduction divisions of *Œnothera Lamarckiana* and. *Œ Gigas.* Ibid. **25**: 941–974. pls. 71–73.

    1916. Life histories in the red algæ. Am. Nat. **50**: 502–512.

Davis, H. S. 1908. Spermatogenesis in Acrididæ and Locustidæ. Bull. Mus. Comp. Zool. Harvard Coll. **53**: 159–212. pls. 9.

Debaisieux, P. 1909. Les débuts de l'ovogénèse dans le *Dytiscus marginalis.* La Cellule **25**: 207–236. pls. 2.

Deton, W. 1908. L' "étape synaptique" dans le *Thysanozoon Brochii.* La Cellule **25**: 133–146. 1 pl.

Dietel, P. 1911. Versuche über die Keimungsbedingungen der Teleutosporen einiger Uredineen. Centr. Bakt. II **31**: 95–106.

Digby, L. 1910. The somatic, premeiotic, and meiotic nuclear divisions of *Galtonia candicans.* Ann. Bot. **24**: 727–757. pls. 59–63.

    1912. The cytology of *Primula Kewensis* and of other related *Primula* hybrids. Ibid. **26**: 357–388. pls. 41–44. figs. 2.

    1914. A critical study of the cytology of *Crepis virens.* Arch. Zellf. **12**: 97–146. pls. 8–10.

    1919. On the archesporial and meiotic mitoses of *Osmunda.* Ann. Bot. **33**: 135–172. pls. 8–12. 1 fig.

Dodge, B. O. 1914. The morphological relationships of the Florideæ and the ascomycetes. Bull. Torr. Bot. Club **41**: 157–202. figs. 13.

Doncaster, L. 1920. On the spermatogenesis of the louse (*Pediculus corporis* and *P. capitis*), with some observations on the maturation of the egg. Quar. Jour. Micr. Sci. **64**: 303–328. pl. 16. 1 fig.

Dublin, L. I. 1905. The history of the germ cells in *Pedicellina americana* (Leidy). Ann. N. Y. Acad. Sci. **16**: 1–64. pls. 1–3.

Duesberg, J. 1908. La spermatogénèse chez le rat (*Mus decumanus* Pall., variété *albinos*). Arch. f. Zellf. **1**: 399–449. pl. 10. 1 fig.

    1909. Note complementaire sur la spermatogénèse du rat. Ibid. **3**: 553–562.

Farmer, J. B. 1894. Studies in the Hepaticæ: On *Pallavicinia decipiens.* Ann. Bot. **9**: 35–52. pls. 6, 7.

    1895. On spore formation and nuclear division in the Hepaticæ. Ibid. **9**: 469–523. pls. 16–18.

FARMER, J. B. and WILLIAMS, J. L. 1898. Contributions to our knowledge of the Fucaceæ; their life history and cytology. Phil. Trans. Roy. Soc. London B **190**: 623-645. pls. 19-24.

FARMER, J. B. and MOORE, J. E. S. 1903. New investigations in the reduction phenomena of animals and plants. Proc. Roy. Soc. London **72**: 104-108. 6 figs.

— 1905. On the meiotic phase (reduction divisions) in animals and plants. Quar. Jour. Micr. Sci. **48**: IV 489-557. pls. 34-41.

FARMER, J. B. and SHOVE, D. 1905. On the structure and development of the somatic and heterotype chromosomes of *Tradescantia virginica*. Quar. Jour. Micr. Sci. **48**: 559-569. pls. 42, 43.

FARMER, J. B. and DIGBY, L. 1910. On the somatic and heterotype mitoses in *Galtonia candicans*. Rep. Brit. Assn. Sheffield. p. 778.

FARMER, J. B. 1912. Telosynapsis and parasynapsis. Ann. Bot. **26**: 623-624.

FASTEN, N. 1914. Spermatogenesis in the American crayfish, *Cambarus virilis* and *Cambarus immunis* (?) with special reference to synapsis and the chromatoid bodies. Jour. Morph. **25**: 587-649. pls. 1-10. 1 fig.

— 1918. Spermatogenesis of the Pacific coast edible crab, *Cancer magister*, Dana. Biol. Bull. **34**: 277-306. pls. 1-4.

FAULL, J. H. 1905. Development of ascus and spore formation in ascomycetes. Proc. Boston Soc. Nat. Hist. **32**: 77-113. pls. 7-11.

— 1912. The cytology of *Laboulbenia chætophora* and *L. Gyrinidarum*. Ann. Bot. **26**: 325-355. pls. 37-40.

FEDERLEY, H. 1913. Das Verhalten der Chromosomen bei der Spermatogenese der Schmetterlinge *Pygaera anachoreta, curtula* und *pigra* sowie einiger ihrer Bastarde. Zeitschr. Ind. Abst. Vererb. **9**: 1-110. pls. 1-4. figs. 5.

FICK, R. 1905. Betrachtungen über die Chromosomen, ihre Individualität, Reduktion und Vererbung. Arch. Anat. u. Physiol.. Anat. Abt., Suppl. 179-228.

— 1907. Vererbungsfragen, Reduktions- und Chromosomenhypothesen, Bastardregeln. Ergeb. Anat. u. Entw. **16**: 1-140. (Review.)

— 1908. Zur Konjugation der Chromosomen. Arch. Zellf. **1**: 604-611.

FITZPATRICK, H. M. 1918. The cytology of *Eucronartium muscicola*. Am. Jour. Bot. **5**: 397-419. pls. 30-32.

FLEMMING, W. 1887. Neue Beiträge zur Kenntniss der Zelle. Arch. Mikr. Anat. **29**: 389-463. pls. 23-26.

FLORIN, R. 1918. Cytologische Bryophytenstudien. I. Ueber Sporenbildung bei *Chiloscyphus polyanthus* (L) Corda. Ark. f. Bot. **15**: 1-10. 1 pl.

FOOT, K. and STROBELL, E. C. 1905. Prophases and metaphase of the first maturation spindle of *Allolobophora fœtida*. Am. Jour. Anat. **4**: 199-243. pls. 1-9.

— 1909. The nucleoli in the spermatocytes and germinal vesicles of *Euschistus variolarius*. Biol. Bull. **16**: 215-238. 1 pl.

— 1910. Pseudo-reduction in the oögenesis of *Allolobophora fœtida*. Arch. Zellf. **5**: 149-165. pls. 11, 12. 1 fig.

— 1912. A study of chromosomes and chromatin nucleoli in *Euschistus crassus*. Ibid. **9**: 47-62. pls. 2-4.

FRASER, H. C. I. 1907. On the sexuality and development of the ascocarp of *Lachnea stercorea* Pers. Ann. Bot. **21**: 349-360. pls. 29, 30

— 1908. Contributions to the cytology of *Humaria rutilans*. Ibid. **22**: 35-55. pls. 4, 5.

— 1914. The behavior of the chromatin in the meiotic divisions of *Vicia Faba*. Ibid. **28**: 633-642. pls. 43, 44.

FRASER, H. C. I. and WELSFORD, E. J. 1908. Further contributions to the cytology of the ascomycetes. Ann. Bot. **22**: 465-477. pls. 26, 27.

FRASER, H. C. I. and BROOKS, W. E. 1909. Further studies in the cytology of the ascus. Ann. Bot. **23**: 538–549.

FRASER, H. C. I. and SNELL, J. 1911. The vegetative divisions in *Vicia faba*. Ann. Bot. **25**: 845–855. pls. 62, 63.

FRIES, R. E. 1911. Zur Kenntniss der Cytologie von *Hygrophorus conicus*. Svensk. Bot. Tids. **5**: 241–251. pl. 1.

FRISENDAHL, A. 1912. Cytologische und entwicklungsgeschichtliche Studien an *Myricaria germanica* Dear. Kgl. Svensk. Vet. Handl. **48**.

GATES, R. R. 1908. A study of reduction in *Œnothera rubrinervis*. Bot. Gaz. **48**: 1–34. pls. 1–3.

— 1909. The behavior of chromosomes in *Œnothera lata* × *O. gigas*. Ibid. **48**: 179–199. pls. 12–14.

— 1911. The mode of chromosome reduction. Ibid. **51**: 321–344.

GIGLIO-TOS, E. e GRANATA, L. 1908. I mitocondrii nelle cellule seminali di *Pamphagus marmoratus*, Burm. Biologica **2**: No. 4.

GOLDSCHMIDT, R. 1905. Eireifung, Befruchtung und Embryonalentwicklung des *Zoögonus mirus* Lss. Zool. Jahrb. **21**: 607–654. pls. 36–38.

— 1908c. Die Chromatinreifung der Geschlechtszellen des *Zoögonus mirus* Lss. und die Primärtypus der Reduktion. Arch. Zellf. **2**: 348–370. pls. 24, 25. figs. 6.

— 1908a. Ueber das Verhalten des Chromatins bei der Eireifung und Befruchtung des *Dicrocælium lanceolatum* Stil. et Has. (*Distomum lanceolatum*.) Ibid. **1**: 232–244. pl. 7.

— 1908b. Ist eine parallele Chromosomenkonjugation bewiesen? Ibid. **1**: 620–622.

GOLDSMITH, W. M. 1919. A comparative study of the chromosomes of tiger beetles (Cicindelidæ). Jour. Morph. **32**: 437–487. pls. 1–10.

GRANATA, L. 1910. Le cinesi spermatogenetische di *Pamphagus marmoratus*, Burm. Arch. Zellf. **5**.

GRÉGOIRE, V. 1899. Les cinèses polliniques chez les Liliacées. La Cellule **16**: 235–297. pls. 2.

— 1904. La reduction numerique des chromosomes et les cinèses de maturation. Ibid. **21**: 297–314.

— 1905. Les resultats acquis sur les cinèses de maturation dans les deux regnes. Ibid. **22**: 221–376. (Review.)

— 1907. La formation des gemini hétérotypiques dans les végétaux. Ibid. **24**: 369–420. pls. 2.

— 1909a. La reduction dans le *Zoögonus mirus* Lss. et le "Primärtypus." Ibid. **25**: 245–285. pls. 2.

— 1909b. Les phénomènes de l'étape synaptique represent-ils une caryocinèse avortée? Ibid. **25**: 87–99.

— 1910. Les cinèses de maturation dans les deux regnes. L'unite essentielle du processus meiotique. Ibid. **26**: 223–422. figs. 145.

— 1912. La verité du schema hétérohomeotypique. Compt. Rend. Acad. Sci. Paris **155**: 1098–1100.

GREGORY, R. P. 1904. Spore formation in leptosporangiate ferns. Ann. Bot. **18**: 445–458. pl. 31. 1 fig.

GRIGGS, R. F. 1906. A reduction division in *Ascaris*. Ohio Nat. **6**: 519–528. pl. 33.

GUIGNARD, L. 1891. Nouvelles études sur la fécondation. Ann. Sci. Nat. Bot. VII **14**: 163–296. pls. 9–18.

— 1899. Le developpement du pollen et la reduction dans le *Naias minor*. Arch. d'Anat. Micr. **2**: 455–509. pls. 19, 20.

GUILLIERMOND, A. 1910. La sexualité chez les champignons. Bull. Sci. France et Belg. **44**: 109–196.

Haecker, V. 1890. Ueber die Reifungsvorgänge bei *Cyclops*. Zool. Anz. **13**: 551–558. 1 fig.

1892. Die Eibildung bei *Cyclops* und *Canthocamptus*. Zool. Jahrb. **5**: 211–248. pl. 19.

1895a. The reduction of the chromosomes in the sexual cells. Ann. Bot. **9**: 95–102.

1895b. Die Vorstadien der Eireifung. Arch. Mikr. Anat. **45**: 200–272. pls. 14–17.

1897. Weitere Uebereinstimmungen zwischen den Fortpflanzungsvorgänge der Thiere und Pflanzen. Biol. Centralbl. **17**: 689–705, 721–745. figs. 36.

1899. Die Reifungserscheinungen. Ergebn. Anat. u. Entw. **8**: 847–922. figs. 22.

Harman, M. T. 1920. Chromosome studies in Tettigidæ. II. Biol. Bull. **38**: 213–230.

Harper, R. A. 1900. Sexual reproduction in *Pyronema confluens* and the morphology of the ascocarp. Ann. Bot. **14**: 321–396. pls. 19–21.

1905. Sexual reproduction and the organization of the nucleus in certain mildews. Carnegie Inst. Wash. Publ. **37**.

von Henking, H. 1891. Untersuchungen über der ersten Entwicklungsvorgänge in den Eiern der Insekten. Zeit. Wiss. Zool. **51**: 685–736. pls. 35–37.

1892. Same title. III. Ibid. **54**: 1–274. pls. 1–12. figs. 12.

Hertwig, R. 1908. Ueber neue Probleme der Zellenlehre. Arch. Zellf. **1**: 1–32. figs. 9.

Heuser, F. 1884. Beobachtungen über Zellkerntheilung. Bot. Centralbl. **17**: 27, 57, 85, 117, 154. pls. 1, 2.

Ikeda, T. 1902. Studies on the physiological functions of antipodals and related phenomena of fertilization in Liliaceæ. 1. *Tricyrtis hirta*. Bull. Coll. Sci. Agr. Tokyo **5**.

Ishikawa, M. 1911. Cytologische Studien über Dahlien. Bot. Mag. Tokyo **25**: 1–8. 1 pl.

Jahn, E. 1908. Myxomycetenstudien. 7. *Ceratiomyxa*. Ber. Deu. Bot. Ges. **26a**: 342–352.

Janssens, F. A. 1905. Spermatogénèse dans les batrachiens. III. Evolution des auxocytes mâles du *Batracoseps attenuatus*. La Cellule **22**: 379–425. pls. 7.

1909. La theorie de la chiasmatypie. Ibid. **25**: 389–411. pls. 2.

Janssens, F. A. et Willems, J. 1909. Spermatogénèse dans les batraciens. IV. Ibid. **25**: 151–178. pls. 2.

Juel, H. O. 1898. Die Kerntheilungen in den Basidien und die Phylogenie der Basidiomyceten. Jahrb. Wiss. Bot. **32**: 361–388. pl. 4.

Karsten, G. 1908. Die Entwicklung der Zygoten von *Spirogyra jugalis*, Ktzg Flora **99**: 1–11. pl. 1.

Kemp, H. P. 1910. On the question of the occurrence of "heterotypical reduction" in somatic cells. Ann. Bot. **24**: 775–803. pls. 66, 67.

Kihara, H. 1919. Ueber cytologische Studien bei einigen Getreidearten. Bot. Mag. Tokyo **32**: 17–38. figs. 21.

King, H. D. 1907. The spermatogenesis of *Bufo lentiginosus*. Am. Jour. Anat. **7**: 345–388. pls. 3.

1908. The oögenesis of *Bufo lentiginosus*. Jour. Morph. **19**: 369–438. pls. 4.

Kingery, H. M. 1917. Oögenesis in the white mouse. Ibid. **30**: 261–316. 5 pls.

Kingsbury, B. F. and Hirsch, P. E. 1912. The degeneration of the secondary spermatogonia of *Desmognathus fusca*. Jour. Morph. **23**: 231–253. pls. 3.

Kniep, H. 1911. Ueber das Auftreten von Basidien im einkernigen Mycel von *Armillaria mellea* Fl. Don. Zeit. f. Bot. **3**: 529–553. pls. 3, 4.

KOERNICKE, M. 1904, 1905. Die neueren Arbeiten über die Chromosomenreduktion im Pflanzenreich und daran anschliessende karyokinetische Probleme. Bot. Zeit. **62**: II, 305–314; **63**: II, 289–307.

KORNHAUSER, S. I. 1914. A comparative study of the chromosomes in the spermatogenesis of *Enchenopa binotata* (Say) and *Enchenopa* (*Campyleuchia* Stäl) *curvata* (Fabr.). Arch. Zellf. **12**: 241–298. pls. 18–22.

1915. A cytological study of the semi-parasitic copepod, *Hersilia apodiformis* (Phil.), with some general considerations of copepod chromosomes. Ibid. **13**: 399–445. pls. 17–19. figs. 9.

KORSCHELT, E. 1895. Ueber Kerntheilung, Eireifung, und Befruchtung bei *Ophryotrocha puerilis*. Zeitschr. Wiss. Zool. **60**: 543–688. pls. 28–34.

KURSSANOW, L. 1911. Ueber Befruchtung, Reifung, und Keimung bei *Zygnema*. Flora **104**: 65–84. pls. 1–4.

KUWADA, Y. 1910. A cytological study of *Oryza sativa* L. Bot. Mag. Tokyo **24**: 267–281. pl. 8.

1919. Die Chromosomenzahl von *Zea Mays* L. Jour. Coll. Sci. Tokyo **39**: 1–248. pls. 2. figs. 4.

KYLIN, H. 1918. Studien über die Entwicklungsgeschichte der Phaeophyceen. Svensk. Bot. Tidskr. **12**: 1–64.

LAGERBERG, T. 1906. Ueber die präsynaptische und synaptische Entwicklung der Kerne in den Embryosackmutterzellen von *Adoxa moschatellina*. Bot. Stud. Kjelmann, Uppsala.

1909. Studien über die Entwicklungsgeschichte und systematische Stellung von *Adoxa moschatellina*, L. Kgl. Svensk. Vet. Akad. Handl. **44**.

LAWSON, A. A. 1911. The phase of the nucleus known as synapsis. Trans. Roy. Soc. Edinburgh **47**: III, 591–604. pls. 1, 2.

1912. A study in chromosome reduction. Ibid. **48**: 601–627. pls. 3.

LERAT, P. 1905. Les phénomènes de maturation dans l'ovogénèse et la spermatogénèse du *Cyclops strenuus*. La Cellule **22**: 163–198. pls. 4.

LEVINE, M. 1913. The cytology of Hymenomycetes, especially the Boleti. Bull. Torr. Bot. Club **40**: 137–181. pls. 4–8.

LEVY, F. 1914–1915. Studien zur Zeugungslehre. IV. Ueber die Chromatinverhältnisse in der Spermatogenese von *Rana esculenta*. Arch. Mikr. Anat. **86**: 85–177. pls. 3–5. figs. 15. (Bibliography.)

LEWIS, I. F. 1909. The life history of *Griffithsia bornetiana*. Ann. Bot. **23**: 639–689.

LEWIS, I. M. 1908. The behavior of the chromosomes in *Pinus* and *Thuja*. Ibid. **22**: 529–556. pls. 29, 30.

DE LITARDIÈRE, R. 1912. Formation des chromosomes hétérotypiques chez le *Polypodium vulgare* L. Compt. Rend. Acad. Sci. Paris **155**: 1023–1026.

LOTSY, J. P. 1904. Die Wendung der Dyaden beim Reifung der Tiereier als Stütze für die Bivalenz der Chromosomen nach der numerische Reduktion. Flora **93**: 65–86. figs. 19.

LUNDEGÅRDH, H. 1909. Ueber Reduktionsteilung in den Pollenmutterzellen einiger dikotylen Pflanzen. Svensk. Bot. Tidskr. **3**: 78–124. pls. 2, 3.

1914. Zur Kenntniss der heterotypischen Kernteilung. Arch. Zellf. **13**: 145–157. pl. 4.

MAIRE, R. 1905. La mitose hétérotypique et la signification des protochromosomes chez les Basidiomycétes. Compt. Rend. Soc. Biol. Paris **57**[1]: 726–728.

1911. La biologie des Uredinales. Prog. Rei. Bot. **4**: 110–162.

MALONE, J. Y. 1918. Spermatogenesis of the dog. Trans. Am. Micr. Soc. **37**: 97–110. pls. 9, 10.

MARCUS, H. 1906. Ei und Samenreife bei *Ascaris canis* (Werner) (*Ascaris mystax*). Arch. Mikr. Anat. **68**: 441–490. pls. 29, 30. figs. 10.

1907. Ueber die Thymus. Verh. Anat. Gesell. Würzburg **21**.

Maréchal, J. 1904. Ueber die morphologische Entwicklung der Chromosomen im Keimbläschen des Selachiereies. Anat. Anz. **25**: 383–398. figs. 25.

1905. Ueber die morphologische Entwicklung der Chromosomen im Teleostierei, usw. Ibid. **26**: 641–652. figs. 27.

1907. Sur l'ovogénèse des Selachiens. La Cellule **24**: 1–239. pls. 10.

Martins Mano, T. 1909. La microsporogénèse dans le *Funkia ovata*. Brotéria, ser. bot., **8**.

McAllister, F. 1913. On the cytology and embryology of *Smilacina racemosa* Trans. Wis. Acad. Sci. **17**[1]: 599–660. pls. 56–58.

McAvoy, B. 1912. The reduction divisions in *Fuchsia*. Ohio Nat. **13**: 1–18. pls. 1, 2.

McClung, C. E. 1900. The spermatocyte divisions of the Acrididæ. Kans. Univ. Quarterly **9**: 73–100. pls. 15–17.

Melin, E. 1915. Die Sporogenese von *Sphagnum squarrosum* Pers., nebst einigen Bemerkungen über das Antheridium von *Sphagnum acutifolium* Ehrh. Svensk. Bot. Tidskr. **9**: 261–293.

Metz, C. W. 1914. Chromosome studies in the Diptera. I. A preliminary survey of five different types of chromosome groups in the genus *Drosophila*. Jour. Exp. Zool. **17**: 45–56. 1 pl.

1916a. Chromosome studies in the Diptera. II. The paired association of the chromosomes in the Diptera, and its significance. Ibid. **21**: 213–280. pls. 1–8.

1916b. Chromosome studies in the Diptera. III. Additional types of chromosome groups in the Drosophilidæ. Am. Nat. **50**: 587–599. pl. 1. fig. 1.

Meves, Fr. 1896. Ueber die Entwicklung der männlichen Gesechlechtszellen von *Salamandra maculosa*. Arch. Mikr. Anat. **48**: 1–83. pls. 1–5.

1907. Die Spermatocytenteilungen bei der Honigbiene (*Apis mellifica*) nebst Bemerkungen über Chromatinreduktion. Ibid. **70**: 414-491. pls. 27, 28. figs. 4.

1908. Es gibt keine parallele Konjugation der Chromosomen. Arch. Zellf. **1**: 612–619. 1 fig.

1911. Chromosomenlängen bei *Salamandra*, nebst Bemerkungen zur Individualitätstheorie der Chromosomen. Arch. Mikr. Anat. **77**: II, 273–300. pls. 11, 12.

Minchin, E. A. 1912. An Introduction to the Study of the Protozoa. London.

Miyake, K. 1905. Ueber Reduktionsteilung in den Pollenmutterzellen einigen Monokotylen. Jahrb. Wiss. Bot. **42**: 83–120. pls. 3–5.

Montgomery, T. H. 1901a. A study of the germ cells of Metazoa. Trans. Am. Phil. Soc. **20**: 154–236. pls. 4–8.

1901b. Further studies on the chromosomes of the Hemiptera heteroptera. Proc. Acad. Nat. Sci. Phila. **53**: 261–271. pl. 10.

1903. The heterotype maturation mitoses in Amphibia and its general significance. Biol. Bull. **4**: 259–269. figs. 8.

1904. Some observations and considerations upon the maturation phenomena of the germ cells. Ibid. **6**: 137–158. pls. 3.

1905. The spermatogenesis of *Syrbula* and *Lycosa* with general considerations upon chromosome reduction and the heterochromosomes. Proc. Acad. Nat. Sci. Phila. **57**: 162–205. pls. 9, 10.

1906. Chromosomes in the spermatogenesis of the Hemiptera heteroptera Trans. Am. Phil. Soc. **21**: 97–174. pls. 9–13.

1908. On the morphological difference of the chromosomes in *Ascaris megalocephala*. Arch. Zellf. **2**: 66–75. pls. 6, 7.

1910. On the dimegalous sperm and chromosomal variation in *Euschistus*, with reference to chromosomal continuity. Ibid. **5**: 121–145. pls. 9, 10. 1 fig.

1911. The spermatogenesis of an Hemipteran, *Euschistus*. Jour. Morph. **22**: 731–798. pls. 1–5.

MOORE, A. C. 1905. Sporogenesis in *Pallavicinia*. Bot. Gaz. **40**: 81–96. pls. 3, 4.

MOORE, J. E. S. 1896. On the structural changes in the reproductive cells during the spermatogenesis of the Elasmobranchs. Quar. Jour. Micr. Sci. **38**: 275–314. pls. 13–16.

MOORE, J. E. S. and EMBLETON, A. L. 1906. On the synapsis in Amphibia. Proc. Roy. Soc. London **77**: 555–562. pls. 20–23.

MOTTIER, D. M. 1897. Beiträge zur Kenntniss der Kerntheilung in den Pollen-mutterzellen einiger Dikotylen und Monokotylen. Jahrb. Wiss. Bot. **30**: 169–204. pls. 3–5.

—— 1903. The behavior of the chromosomes in the spore mother-cells of higher plants and the homology of the pollen and embryo-sac mother-cells. Bot. Gaz. **35**: 250–292. pls. 11–14.

—— 1905. The development of the heterotypic chromosomes in pollen mother-cells. Bot. Gaz. **40**: 171–177.

—— 1907. The development of the heterotypic chromosomes in pollen mother-cells. Ann. Bot. **21**: 309–347. pls. 27, 28.

—— 1909. Prophases of heterotypic mitosis in the embryo-sac mother-cells of *Lilium*. Ibid. **23**: 343–352. pl. 23.

—— 1914. Mitosis in the pollen mother-cells of *Acer negundo*, L., and *Staphylea trifolia*, L. Ibid. **28**: 115–133. pls. 9, 10.

MOTTIER, D. M. and NOTHNAGEL, M. 1913. The development and behavior of the chromosomes in the first or heterotypic mitosis of the pollen mother-cells of *Allium cernuum* Roth. Bull. Torr. Bot. Club **40**: 555–565. pls. 23, 24.

MÜLLER, H. A. CL. 1909. Ueber karyokinetische Bilder in den Wurzelspitzen von *Yucca*. Jahrb. Wiss. Bot. **47**: 99–117. pls. 1–3.

—— 1911. Kernstudien an Pflanzen. I u. II. Arch. Zellf. **8**: 1–51. pls. 1, 2.

NAKAHARA, W. 1919. A study of the chromosomes in the spermatogenesis of the stonefly, *Perla immarginata* Say, with special reference to the question of synapsis. Jour. Morph. **32**: 509–530. pls. 3.

—— 1920. Side-to-side versus end-to-end conjugation of chromosomes in relation to crossing over. Science **52**: 82–84.

NĚMEC, B. 1903. Ueber die Einwirkung des Chloralhydrats auf die Kern- und Zell-teilung. Jahrb. Wiss. Bot. **39**: 645–730. figs. 157.

NICHOLS, G. E. 1908. The development of the pollen of *Sarracenia*. Bot. Gaz. **45**: 31–37. pl. 5.

NOTHNAGEL, M. 1916. Reduction divisions in the pollen mother-cells of *Allium tricoccum*. Bot. Gaz. **61**: 453–476. pls. 28–30. fig. 1.

OEHLKERS, F. 1916. Beitrag zur Kenntniss der Kernteilung bei den Characeen. Ber. Deu. Bot. Ges. **34**: 223–227.

OLIVE, E. W. 1907. Cytological studies on *Ceratiomyxa*. Trans. Wis. Acad. Sci. **15**: 754–773. pl. 47.

OSTERHOUT, W. J. V. 1897. Ueber Entstehung der karyokinetischen Spindel bei *Equisetum*. Jahrb. Wiss. Bot. **30**: 159–168. pls. 1, 2.

OVERTON, E. 1893. On the reduction of the chromosomes in the nuclei of plants. Ann. Bot. **7**: 139–143.

OVERTON, J. B. 1905. Ueber Reduktionsteilung in den Pollenmutterzellen einiger Dikotylen. Jahrb. Wiss. Bot. **42**: 121–153. pls. 6, 7.

—— 1906. The morphology of the ascocarp and spore formation in the many-spored asci of *Theocatheus pelletieri*. Bot. Gaz. **42**: 450–492. pls. 29, 30.

—— 1909. On the organization of the nuclei in the pollen mother cells of certain plants, with special reference to the permanence of the chromosomes. Ann. Bot. **23**: 19–61. pls. 1–3.

Paulmier, F. C. 1898. Chromosome reduction in the Hemiptera. Anat. Anz. **14**: 514–520. figs. 19.

1899. The spermatogenesis of *Anasa tristis*. Jour. Morph. **15**: Suppl. 223–272. pls. 13, 14.

Picard, M. 1913. A bibliography of works on meiosis and somatic mitosis in the angiosperms. Bull. Torr. Bot. Club **40**: 575–590.

Popoff, M. 1908. Experimentelle Zellstudien. Arch. Zellf. **1**: 245–380. figs. 18.

Pratt, B. H. and Long, J. A. 1917. The period of synapsis in the egg of the white rat, *Mus norvegicus albinus*. Jour. Morph. **29**: 441–458. pl. 1. figs. 2.

vom Rath, O. 1892. Zur Kenntniss der Spermatogenese von *Gryllotalpa vulgaris* Latr. Arch. Mikr. Anat. **40**: 102–132. pl. 5.

1893. Beiträge zur Kenntniss der Spermatogenese von *Salamandra maculosa*. Zeitschr. Wiss. Zool. **57**: 97–185. pls. 7–9.

1895. Neue Beiträge zur Frage der Chromatinreduktion in der Samen- und Eireife. Arch. Mikr. Anat. **46**: 168–238. pls. 6–8.

Robertson, W. R. B. 1916. Chromosome studies. I. Jour. Morph. **27**: 179–332. pls. 1–26.

Rosenberg, O. 1907. Zur Kenntniss der praesynaptischen Entwicklungsphasen der Reduktionsteilung. Svensk. Bot. Tidskr. **1**.

1909a. Zur Kenntniss von den Tetradenteilungen der Compositen. Ibid. **3**: 64–77. pl. 1.

1909b. Cytologische und morphologische Studien an *Drosera longifolia* × *rotundifolia*. Kgl. Svensk. Vet. Handl. **43**: 3–64. pls. 4. figs. 33.

1918. Chromosomenzahlen und Chromosomendimensionen in der Gattung *Crepis*. Ark. f. Bot. **15**: 1–16. figs. 6.

Roux, W. 1883. Ueber die Bedeutung der Kerntheilungsfiguren. Leipzig.

Rückert, J. 1892. Zur Entwicklungsgeschichte des Ovarialeies bei Selachiern. Anat. Anz. **7**: 107–158. figs. 6.

1893. Die Chromatinreduktion bei der Reifung der Sexualzellen. Ergeh. d. Anat. u. Entw. **3**: 517–583. (Review.)

1894. Zur Eireifung der Copepoden. Ibid., Anat. Hefte. **4**: 261–352. pls. 21–25.

Sabaschnikoff, M. 1897. Beiträge zur Kenntnis der Chromatinreduktion in der Ovogenesis von *Ascaris*. Bull. Soc. Nat. Moscow **1**.

Sakamura, T. 1914. Ueber die Kernteilung bei *Vicia cracca*. Bot. Mag. Tokyo **28**: 131–147. pl. 2.

1915. Ueber die Einschnürung der Chromosomen bei *Vicia Faba* L. Ibid. **29**: 287–300. pl. 13. figs. 12.

1920. Experimentelle Studien über die Zell- und Kernteilung mit besonderer Rücksicht auf Form, Grosse und Zahl der Chromosomen. Jour. Coll. Sci. Imp. Univ. Tokyo **39**: pp. 221. pls. 7.

Schaffner, J. H. 1909. The reduction division in the microsporocytes of *Agave virginica*. Bot. Gaz. **47**: 198–214. pls. 12–14.

Schleip, W. 1906. Die Entwicklung der Chromosomen im Ei von *Planaria gonocephala* Dug. Zool. Jahrb. (Anat. Abt.) **23**: 357–380. pls. 23, 24.

1907. Die Samenreifung bei Planarien. Ibid. **24**: 129–174. pls. 14, 15.

Schellenberg, A. 1911. Ovogenese, Eireifung und Befruchtung von *Fasciola hepatica*, L. Arch. Zellf. **6**: 443–484. pls. 24–26.

Schneider, H. 1914. Ueber die Prophasen der ersten Reifeteilung in Pollenmutterzellen, inbesondere bei *Thelygonium Cynocrambe* L. Arch. Zellf. **12**: 359–371. pl. 28.

Schreiner, A. u. K. E. 1904. Die Reifungsteilungen bei den Wirbeltieren. Anat Anz. **24**: 561–578. figs. 24.

1905. Ueber die Entwicklung der männlichen Geschlechtszellen von *Myxine glutinosa* (L.). Arch. d. Biol. **21**: 183–355. pls. 5–14.

1906. Neue Studien über die Chromatinreifung der Geschlechtszellen. Arch. d. Biol. **22**: 1–70, pls. 1–3; 419–492, pls. 23–26; Anat. Anz. **29**: 465–479. figs. 17.

SHARP, L. W. 1913. Somatic chromosomes in *Vicia*. La Cellule **29**: 297–331. pls. 2.

1914. Maturation in *Vicia*. (Prelim. Note). Bot. Gaz. **57**: 531.

1920a. Mitosis in *Osmunda*. (Review). Ibid. **69**: 88–91.

1920b. Somatic chromosomes in *Tradescantia*. Am. Jour. Bot. **7**: 341–354. pls. 22, 23.

STEVENS, N. M. 1903. On the ovogenesis and spermatogenesis of *Sagitta bipunctata*. Zool. Jahrb. **18**: 227–240. pls. 20, 21.

1908. A study of the germ cells of certain Diptera, with reference to the heterochromosomes and the phenomena of synapsis. Jour. Exp. Zool. **5**: 359–374. pls. 4.

1911. Further studies on the heterochromosomes of the mosquitoes. Biol. Bull. **20**: 109–120. figs. 38.

STOMPS, T. J. 1910. Kerndeeling en synapsis bij *Spinacia oleracea*. pp. 162. pls. 2. See Biol. Centr. **31**: 257–320. pls. 1–3. 1911.

STRASBURGER, E. 1888. Ueber Kern- und Zellteilung im Pflanzenreich, nebst einem Anhang über Befruchtung. Hist. Beitr. **1**. pp. 258. pls. 3.

1894. The periodic reduction of chromosomes in living organisms. Ann. Bot. **8**: 281–316.

1897. Kerntheilung und Befruchtung bei *Fucus*. Jahrb. Wiss. Bot. **30**: 351–374. pls. 27, 28.

1900. Ueber Reduktionsteilung, Spindelbildung, Centrosomen und Cilienbildner im Pflanzenreich. Histol. Beitr. **6**. pp. 224. pls. 4.

1904a. Anlage des Embryosacks und Prothalliumbildung bei der Eibe nebst anschliessenden Erörterungen. Festschr. f. Hæckel. Jena.

1904b. Ueber Reduktionsteilung. Sitzber. Berlin. Acad. Wiss., phys.-math.Kl. **18**: 587–614. figs. 9.

1905. Typische und allotypische Kernteilung. Jahrb. Wiss. Bot. **42**: 1–71. pl. 1.

1907. Ueber die Individualität der Chromosomen und die Propfhybriden-Frage. Ibid. **44**: 472–555. pls. 5–7.

1908. Chromosomenzahlen, Plasmastrukturen, Vererbungsträger und Reduktionsteilung. Ibid. **45**: 479–568. pls. 1–3.

1910. Ueber geschlechtsbestimmende Ursachen. Ibid. **47**: 427–520. pls. 9, 10.

1911. Kernteilungsbilder bei der Erbse. Flora **102**: 1–23. pl. 1.

SUTTON, W. S. 1902. On the morphology of the chromosome group in *Brachystola magna*. Biol. Bull. **4**: 24–39. figs. 11.

SVEDELIUS, N. 1914a. Ueber die Tetradenteilung in den vielkernigen Tetrasporangiumanlagen bei *Nitophyllum punctatum*. Ber. Deu. Bot. Ges. **32**: 48–57. pl. 1. 1 fig.

1914b. Ueber Sporen an Geschlechtspflanzen von *Nitophyllum punctatum*, usw. Ibid. **32**: 106–116. pl. 2. 1 fig.

SYKES, M. G. 1908. Nuclear division in *Funkia*. Arch. Zellf. **1**: 381–398. pls. 8, 9. 1 fig.

TÄCKHOLM, G. 1914. Zur Kenntniss der Embryosackentwicklung von *Lopezia coronata* Andr. Svensk. Bot. Tidskr. **8**.

TAHARA, M. 1910. Ueber die Kernteilung bei *Morus*. Bot. Mag. Tokyo **24**: 281–289. pl. 9.

TAYLOR, M. 1914, 1917. The chromosome complex of *Culex pipiens*. Quar. Jour. Micr. Sci. **60**: 377–398. pls. 27, 28; **62**: 287–302. pl. 20.

Tischler, G. 1906. Ueber die Entwicklung des Pollens und der Tapetenzellen bei *Ribes*-Hybriden. Jahrb. Wiss. Bot. **42**: 545–578. pl. 15.

1916. Chromosomenzahl, -Form und -Individualität in Pflanzenreich. Prog. Rei. Bot. **5**: 164–284. (Bibliography).

Tretjakoff, D. 1904. Die Spermatogenese bei *Ascaris megalocephala*. Arch. Mikr. Anat. **65**: 383–438. pls. 22–24. 1 fig.

Tröndle, A. 1911. Ueber die Reduktionsteilung in den Zygoten von *Spirogyra* und über die Bedeutung der Synapsis. Zeitschr. f. Bot. **3**: 593–619. 1 pl. 20 figs.

Tschernoyarow, M. 1914. Ueber die Chromosomenzahl und besonders beschaffene. Chromosomen im Zellkerne von *Najas major*. Ber. Deu. Bot. Ges. **32**: 411–416. pl. 10.

Vandendries, R. 1913. Le nombre des Chromosomes dans la spermatogénèse des *Polytrichum*. La Cellule **28**: 257–261. figs. 11.

van Leeuwen-Reijnvaan, D. 1907. Ueber eine zweifache Reduktion bei einigen *Polytrichum*-Arten. Rec. Trav. Bot. Neerl. **4**.

1908. Ueber die Spermatogenese der Moose. Ber. Deu. Bot. Ges. **26a**: 301–308. pl. 5.

Vejdowský F. 1907. Neue Untersuchungen über Reifung und Befruchtung. Kgl. Böhm. Ges. Wiss. Prag.

1912. Zum Problem der Vererbungsträger. pp. 184. pls. 12. Prag. 1911–2.

von Voss, H. 1914. Cytologische Studien an *Mesostoma Ehrenbergi*. Arch. Zellf. **12**: 159–194. pls. 12–14. figs. 5.

deVries, H. 1889. Intracelluläre Pangenesis. Jena.

Walker, N. 1913. On abnormal cell-fusion in the archegonium; and on spermatogenesis in *Polytrichum*. Ann. Bot. **27**: 115–132. pls. 13, 14.

Walton, A. C. 1918. The oögenesis and early embryology of *Ascaris canis* Werner. Jour. Morph. **30**: 527–604. pls. 9. 1 fig.

Weinzieher, S. 1914. Beiträge zur Entwicklungsgeschichte von *Xyris indica* L. Flora **106**: 393–432. pls. 6, 7. figs. 10.

Weismann, A. 1887–1902. Ueber die Zahl der Richtungskörper und über ihre Bedeutung für die Vererbung. Jena 1887. Essays upon Heredity. Oxford 1891–2. Das Keimplasma. 1892. (Engl. 1893.) The Evolution Theory. 1902.

Wenrich, D. H. 1915. Synapsis and the individuality of the chromosomes. Science **41**: 440.

1916. The spermatogenesis of *Phrynotettix magnus*, with special reference to synapsis and the individuality of the chromosomes. Bull. Mus. Comp. Zool. Harvard Coll. **60**: 55–136. 10 pls.

1917. Synapsis and chromosome organization in *Chorthippus* (*Stenobothrus*) *curtipennis* and *Trimerotropis suffusa* (Orthoptera). Jour. Morph. **29**: 471–516. pls. 1–3.

Williams, J. L. 1904. Studies in the Dictyotaceæ. II. The cytology of the gametophyte generation. Ann. Bot. **18**: 183–204. pls. 12–14.

Wilson, E. B. 1900. The Cell in Development and Inheritance. 2d ed.

von Winiwarter, H. 1900. Recherches sur l'ovogénèse et l'organogénèse de l'ovaire des mammifères (Lapin et Homme). Arch. d. Biol. **17**: 33–199. pls. 3–8.

von Winiwarter, H. et Sainmont. 1909. Nouvelles recherches sur l'ovogénèse et l'organogénèse de l'ovaire des Mammifères (Chat). Chapter IV. Ibid. **24**: 1–142, 165–276, 373–432, 627–652. pls. 11.

Wodsedalek, J. E. 1916. Causes of sterility in the mule. Biol. Bull. **30**: 1–56.

Wolfe, J. J. 1904. Cytological studies on *Nemalion*. Ann. Bot. **18**: 607–630. pls. 40, 41. 1 fig.

1918.   Alternation and parthenogenesis in *Padina*.   Jour. Elisha Mitchell Sci. Soc.
    34: 78–109.
WOOLERY, R.   1915.   Meiotic divisions in the microspore mother-cells of *Smilacina
    racemosa* (L) Desf.   Ann. Bot. **29**: 471–482.   pl. 22.   1 fig.
YAMANOUCHI S.   1906.   The life history of *Polysiphonia violacea*.   Bot. Gaz. **42**:
    401–449.   pls. 19–28.
    1908.   Sporogenesis in *Nephrodium*.   Ibid. **45**: 1–30.   pls. 1–4.
    1909.   Mitosis in *Fucus*.   Ibid. **47**: 173–197.   pls. 8–11.
    1910.   Chromosomes in *Osmunda*.   Ibid. **49**: 1–12.   pl. 1.
    1912.   The life history of *Cutleria*.   Ibid. **54**: 441–502.   pls. 26–35.   figs. 15.
ZWEIGER.   1907.   Die Spermatogenese von *Forficula auricularia*.   Jen Zeitschr. **42**:
    143–172.   pls. 11–14.

CHAPTER XII

FERTILIZATION

We have already pointed out that reduction and fertilization constitute the two principal cytological crises in the life cycles of all organisms reproducing sexually. Although the first of these processes was not discovered until 1883, some of the grosser features of fertilization had been made out many years previously (Chapter I). But the central feature of this process—the union of the two parental nuclei—was not known until 1875, when O. Hertwig discovered it in animals, Strasburger's parallel discoveries in plants following in 1877 (*Spirogyra*) and 1884 (angiosperms). As the finer details of fertilization and the significance of its results become better understood, the aptness of Huxley's (1878) often quoted simile, in which he compares the organism to "a web of which the warp is derived from the female and the woof from the male," becomes increasingly striking.

We shall first describe the morphology of the fertilization process as it is typically shown in many animals, after which attention will be given to some of its physiological aspects. The second half of the chapter will be devoted to a review of fertilization in the various groups of the plant kingdom.

FERTILIZATION IN ANIMALS[1]

The Gametes.—The spermatozoa of different animals exhibit a surprising variety of form and structure (Fig. 103). What may be referred to as the "typical" spermatozoön consists of three fairly distinct parts: *head, middle piece,* and *tail* or *flagellum* (Fig. 104). The *head* represents the nuclear portion of the sperm cell: it consists almost wholly of an extremely compact mass of chromatin. It has an envelope of cytoplasm which in few forms is very conspicuous and in many cases is scarcely distinguishable. Anterior to the nucleus there may be an *acrosome,* and the end of the head often has the form of a sharp point, the *perferatorium.* Posterior to the head is the *middle piece;* this is made up of cytoplasm in which are located the centrosomal structures, together with chondriosomes and other inclusions, such as the "Golgi bodies." The *flagellum,* or *tail,* consists of a slender *axial filament,*

---

[1] In the preparation of this portion of the chapter the author has drawn very freely upon Professor F. R. Lillie's admirable and concise presentation of the subject, *Problems of Fertilization* (1919).

which grows out from the centrosome in the middle piece or in some cases apparently from the base of the nucleus, and a cytoplasmic *sheath* which usually extends not quite to its end. The sheath sometimes has the form of an undulating membrane. The spermatozoa of crustaceans and nematodes are non-flagellate, and in other groups various departures from the "typical" form and structure are known. A few of the many known types are shown in Fig. 103.

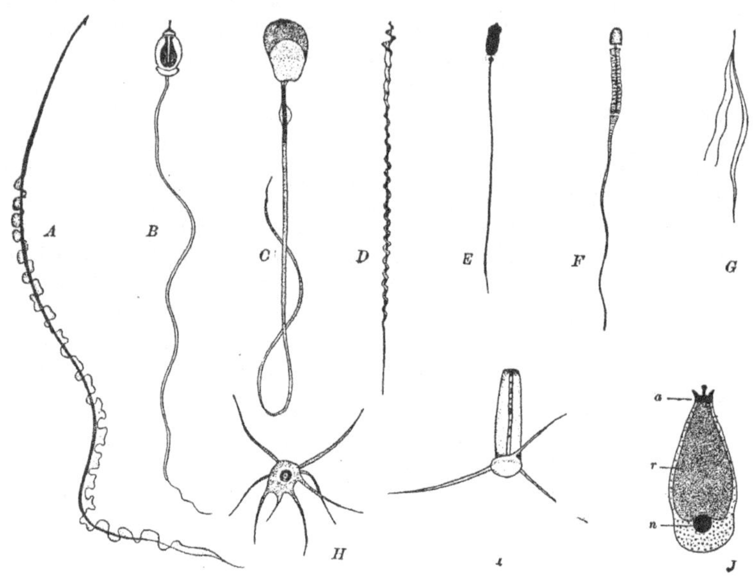

FIG. 103.—Various types of spermatozoa.

A, Triton (salamander). (*After Ballowitz.*) *B, Nereis* (annelid). (*After Lillie,* 1912.) *C,* guinea pig. (*After Meves.*) *D, Phyllopneuste* (bird). (*After Ballowitz.*) *E,* sturgeon (*After Ballowitz.*) *F, Vesperugo* (bat). (*After Ballowitz.*) *G, Castrada hofmanni* (turbellarian). (*After Luther.*) *H, Pinnotheres veterum* (crustacean). (*After Koltzoff.*) *I, Homarus* (lobster). (*After Herrick*). *J, Ascaris* (nematode); *a,* apical body; *n,* nucleus; *r,* "refractive body." (*After Scheben.*)

The ovum undergoes nearly or quite all of its elaborate differentiation before the maturation divisions occur. Certain cells in the ovary gradually become greatly enlarged (Fig. 105), and during this "growth period" the cytoplasm may not only differentiate into visibly distinct regions but may also become stored with energy-containing materials ("food"), which in the case of some animals, such as birds, is present in relatively enormous amounts. The "ovarian egg" or *primary oöcyte,* as the egg cell is called before the maturation mitoses take place, may have a definite limiting membrane at its surface, but in many forms this cannot be demonstrated. The nucleus of the primary oöcyte is known as the *germinal vesicle:* it is very large and contains in addition to its chromatin a considerable amount of material which appears to take no part in the formation of the chromosomes when division ensues. After the cytoplasmic differentiation is complete and the oöcyte has reached its full size—even after the spermatozoön has entered in many cases—the oöcyte nucleus undergoes two divisions in rapid succession at the periphery of the egg, which at this point buds off two small nucleated cells, the *polar bodies* (Fig. 106). The first polar body may or may not divide again. The details of chromosome behavior in these two mitoses have been described in the preceding chapter. The reduced or haploid number of chromosomes left in the egg organize the egg nucleus ("female pronucleus"), rendering the egg ready for the sexual fusion.

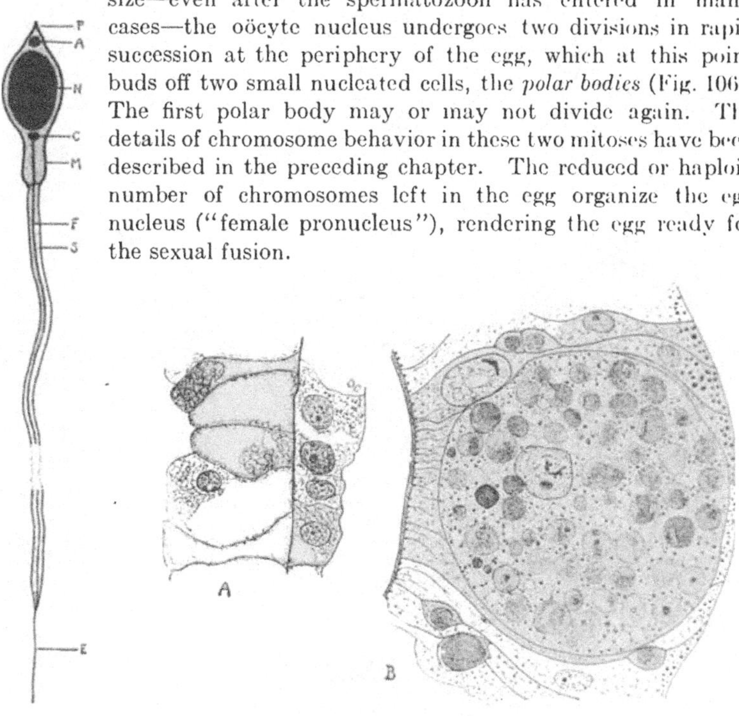

Fig. 104.
Fig. 105.

Fig. 104.—Diagram of typical flagellate spermatozoön.

P, perferatorium; *A*, acrosome; *N*, nucleus; *M*, middle piece; *F*, axial filament; *S*, cytoplasmic sheath; *E*, end piece. (*After Wilson.*)

Fig. 105.—The differentiation of the oöcyte in *Hydra*.

A, very young oöcyte lying between ectodermal cells at right. *B*, oöcyte after growth period, with yolk globules, × 500. (*After Downing,* 1909.)

The time relation of the maturation of the egg and the entrance of the spermatozoön varies considerably in different animals. In echinoderms and some other forms maturation is completed before the spermatozoön penetrates. In some other animals it proceeds as far as the metaphase of the heterotypic mitosis (*Chætopterus, Cerebratulus*) or the prophase of the homœotypic mitosis (many vertebrates), but does not go further unless penetration occurs. In the marine annelid, *Nereis*, finally, the germinal vesicle undergoes no change unless the spermatozoön has entered the egg.

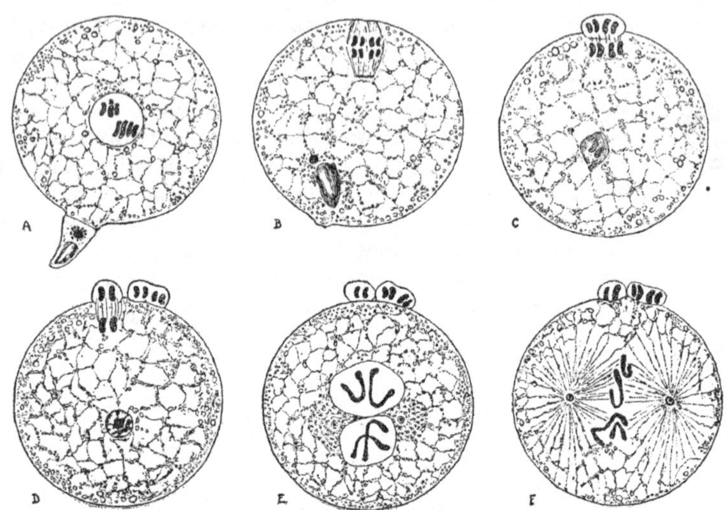

Fig. 106.—Maturation and fertilization in *Ascaris*.

A, spermatozoön about to enter egg. *B*, spermatozoön inside; first maturation mitosis in progress. *C*, first maturation mitosis completed; first polar body budded off. *D*, second maturation mitosis, forming second polar body; sperm nucleus below. *E*, male and female pronuclei, each with 2 chromosomes, meeting. *F*, first cleavage mitosis, showing 2 paternal and 2 maternal chromosomes. (*After Hertwig.*)

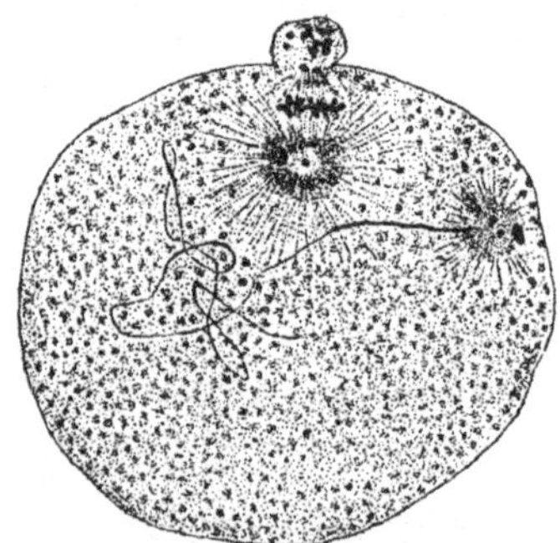

Fig. 107.—Fertilization in *Physa* (snail.) Sperm head and amphiaster at right, with long flagellum extending toward left. Second maturation mitosis in progress. (*After Kostanecki and Wierzyski*, 1896.)

The Fusion of the Gametes.—In most cases the whole spermatozoön enters the egg (Fig. 107). In some sea urchins only the head and middle piece enter, while in *Nereis* the head alone passes in, the middle piece

and tail being left on the egg surface. The process in *Nereis* as described by Lillie (1912, 1919) is as follows. The egg of this worm has a
tough vitelline membrane, an alveolar cortical layer, many yolk and oil
droplets, and a large central germinal vesicle (nucleus). If many
spermatozoa are present in the vicinity a large number attach themselves
to the egg, but usually all but one are carried away by an outflow of jelly
from the alveolæ of the cortical layer. This layer now takes the form of
a zone traversed by radial protoplasmic plates representing the walls of
the alveolæ. A transparent "fertilization cone" extends from the inner
part of the egg across this zone and touches the membrane at the point
where the spermatozoön is beginning to penetrate. The perforatorium
pierces the egg membrane and becomes attached to the transparent cone.
The latter is now withdrawn, carrying the head of the spermatozoön
into the egg with it. Thus it appears that the initiative for the final act
of penetration lies with the egg rather than with the spermatozoön.
Since only the head enters the egg in *Nereis* it seems clear that the only
necessary portion of the spermatozoön in the actual union is the nucleus:
the middle piece and tail are accessory and function only as locomotor
organs.

The immediate visible effects of the entrance of the sperm are seen
chiefly in changes in the appearance of the cortical region of the egg. If
a vitelline membrane is present, as in vertebrates, a "perivitelline
space" usually appears between the membrane and the egg; and this
space may in some cases (frog) be great enough to permit the rotation
of the egg within the membrane. In the sea urchin a *fertilization membrane* is formed as the result of fertilization: it first appears at the point
where the spermatozoön is attached and spreads over the egg with great
rapidity. It seems probable that a delicate membrane already present is
raised and thus made more conspicuous. In *Ascaris*, which is parasitic
in the intestine of the horse, this membrane becomes very thick and later
acts as a protection against the digestive juices of the host. These
cortical changes do not depend upon the actual entrance of the spermatozoön into the egg: in *Nereis* they occur before the slow penetration
can be completed, or even if the spermatozoön is shaken loose shortly
after penetration has begun.

In describing the remarkable transformation undergone by the
spermatozoön within the egg the behavior of its different organs will for
the sake of clearness be considered separately.

The Nucleus.—Immediately after gaining entrance to the egg (Fig.
108) the sperm head begins to enlarge and assumes the usual form and
structure of a nucleus. Meanwhile it advances toward the egg nucleus.
As Lillie points out, both male and female pronuclei pass toward a position of equilibrium in a cell preparing to divide and consequently meet:
the assumption of an attractive force between them is unnecessary. By

the time they meet the male pronucleus has usually, but not always, become equal in size and appearance to the female pronucleus. The union of the two pronuclei to form a fusion nucleus, or *synkaryon*, usually

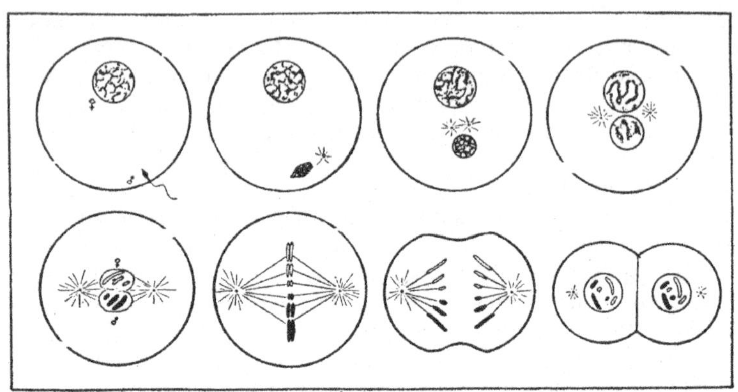

Fig. 108.—Diagram of fertilization and cleavage in an animal. It is assumed that in this case the egg has undergone maturation before the penetration of the spermatozoön.

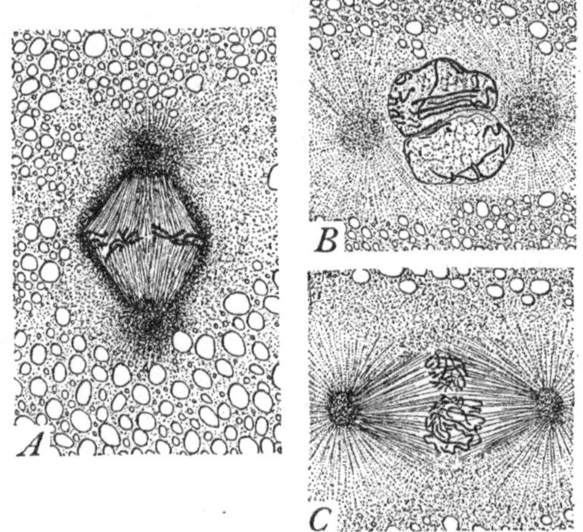

Fig. 109.—Independence of parental chromatin contributions in the cleavage of the egg of *Cryptobranchus*.

A, first cleavage mitosis. B, C, prophase and metaphase of fourth cleavage mitosis (*After Smith*, 1919.)

occurs at once after they meet. In a great many cases there may be no actual fusion of the pronuclei at all: as they come close to one another each passes through the prophase stages and gives rise independently to its group of chromosomes, the two groups arranging themselves on a common spindle which organizes when the nuclear membranes dissolve. The first cleavage mitosis (first embryonal division) then ensues, and the two daughter nuclei receive longitudinal halves of each and every chromosome.   Thus in the act of fertilization, in both animals and plants, *each parent furnishes the offspring with a haploid set of chromosomes, the two intermingled sets constituting the diploid set of the new individual. Since every chromosome divides equationally at every subsequent somatic mitosis, every cell of the body receives half of its chromosome complement from each parent.*   The cardinal importance of this fact in connection with current theories of heredity will be apparent in subsequent chapters.

The two groups of chromoosmes, paternal and maternal, can often be distinguished not only on the spindle of the first cleavage division, but in several divisions thereafter.   As examples may be cited *Cyclops* (Rückert 1895; Hæcker 1895), *Crepidula* (Conklin 1901), and *Crypto-branchus* (Smith 1919) (Fig. 109).   This phenomenon is especially evident in hybrids (p. 160).   There is much reason to believe that the chromatins of the two parents, although intermingled in the nuclei of the offspring, never actually fuse, unless it is at the time of synapsis in the next maturation; and it has already been pointed out (Chapter XI) that they may not fuse even then.   This fact also has an important bearing on the chromosome theory of heredity.

The Centrosome.—(See Wilson 1900, pp. 208 ff.)   Shortly after the entrance of the spermatozoön into the egg (Figs. 106–108) an aster develops at the base of the sperm head, and in the aster a centrosome appears. Since the centrosome thus arises in the position of the middle piece, and since the centrosome of the spermatid is included in the middle piece during spermatogenesis, a widely accepted theory has been that the newly appearing centrosome is in reality that of the spermatid.   Whatever its origin, it soon divides to form the two which function in the first cleavage mitosis.   These facts had much to do with the formulation of a theory of fertilization set forth by Boveri (1887, 1891), who was much impressed by the conspicuous part played by the centrosomes in cell-division.   According to Boveri's theory the egg is not able to undergo division because of the lack of any centrosome to initiate the process, while the spermatozoön has a centrosome but not sufficient cytoplasm in which to act.   Through the union of the gametes all the organs necessary for division are brought together and cleavage proceeds.   This theory has recently been recalled by Walton (1918) in his work on *Ascaris canis.*

Another early view of the origin of the cleavage centrosomes was that of van Beneden (1887) and Wheeler (1895, 1897), who believed them to be the centrosomes of the egg cell.

The theory that the cleavage centrosomes arise from both egg and spermatozoön is of some historic interest.   It was suggested by Rabl

(1889) that if the centrosome is a permanent cell organ the conjugation of the gametes must involve not only a union of nuclei but also a union of centrosomes (Wilson, p. 210). Fol (1891), in his work on echinoderm eggs, thought that he observed just such a process, which he termed "The Quadrille of the Centers." The egg centrosome and the centrosome brought in by the spermatozoön were both supposed to divide, the products then fusing in pairs to form the two cleavage centrosomes. A similar thing was reported by certain other investigators, but none of the cases stood the test of later work. Another theory now abandoned was that advanced by Carnoy and Lebrun (1897), who also attempted to derive one centrosome from each gamete. The cleavage centrosomes were thought to arise *de novo* and separately, one inside each pronucleus, to migrate thence into the cytoplasm.

Much less confidence is now placed in the persistence of the spermatid or egg centrosomes through the fertilization stages. Since the middle piece, which is thought to contain the centrosome, does not enter the egg at all in *Nereis*, it seems probable that the male nucleus in some way induces the formation of asters and centrosomes by the egg cytoplasm. Lillie found that even a portion of the sperm head will bring about this effect. In *Unio* (Lillie 1897, 1898) and *Crepidula* (Conklin 1897) it seems not unlikely that each pronucleus causes the formation of one cleavage centrosome. In the sea urchin Wilson (1901) concluded that the cleavage centrosomes in all probability arise by the division of one which originates *de novo* at the nuclear membrane. In almost every case there are gaps in the known history of the centrosome in fertilization, and it seems very doubtful whether the cleavage centrosomes are continuous with those of either gamete. This conclusion is supported by the fact that the formation of asters with centrosomes in the egg cytoplasm can be artificially induced by treating the eggs with certain chemicals, such as weak $MgCl_2$. It is possible that the spermatozoön carries a substance which brings about centrosome formation in a similar way. However this may be, the importance of the centrosome undoubtedly lies in its relation to cleavage rather than to fertilization.

Cytoplasm and Chondriosomes.—In some cases (*Nereis*) no cytoplasm can be shown to enter the egg with the spermatozoön, whereas in others (*Ascaris*) a relatively large amount is brought in. Its great indefiniteness in behavior makes it seem probable that it has no special significance in the fertilization process.

The importance of the chondriosomes in fertilization has been emphasized by Meves (1911, 1915), who finds that many of these bodies are present in the large cytoplasmic mass accompanying the sperm nucleus in *Ascaris*, and that they mingle with the chondriosomes of the egg. Meves (1908, 1915, 1918), together with other writers, accordingly thinks that they are concerned in the transmission of certain hereditary characters. Several observations, however, fail to harmonize with this view. In some echinoderms the middle piece, which contains the chondriosomal material, is not distributed to the daughter cells during the cleavage divisions, but remains in one of them. It is unlikely that hereditary material would behave in this manner. Furthermore, in the worm, *Nereis*, the middle piece does not even enter the egg. In *Peripatus* (Montgomery 1912) the chondriosomal mass is thrown out of the spermatid. Wildman (1913) points out that in *Ascaris* also the chondriosomes may be largely lost during spermatogenesis. Although their almost constant presence in the spermatozoön indicates that they have a special physiological rôle comparable to that in other cells, there is as yet no adequate reason to regard them as important in the transmission of the factors of inheritance.

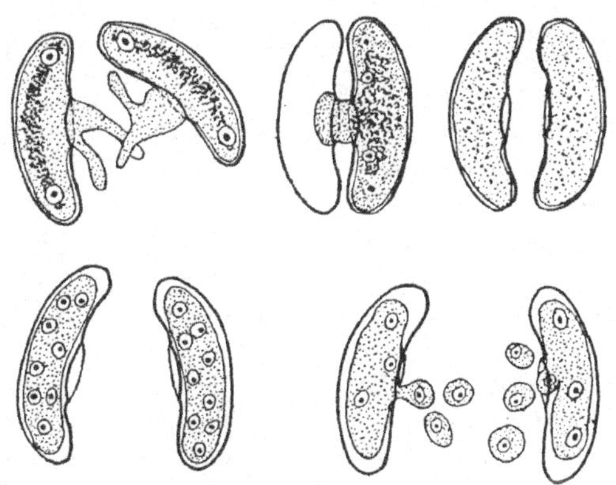

Fig. 110.—Chromidiogamy in *Arcella*. (*From Minchin, after Swarczewsky.*) Somewhat modified.

Fertilization in Protozoa.—Among the protozoa there are found several modes of sexual union very different from that described above for the higher animals. Four illustrative cases may be cited (see Minchin 1912).

In *Arcella*, a rhizopod with a hemispherical shell, the protoplast has two "primary nuclei" and many scattered chromidia. Two individuals come together with their shell openings opposite one another (Fig. 110), and the protoplasm of one flows almost entirely into the other shell, where it mingles with that of the other individual. The primary nuclei degenerate, while the chromidia become divided up into fine granules. The protoplasm with the fine chromidia now becomes evenly distributed in the two shells, after which the latter separate. In each individual the chromidia now form "secondary nuclei;" around these are organized uninucleate amœbulæ which escape from the shell and give rise to new *Arcella* individuals. This process, in which the chromatin chiefly concerned is that of the chromidia and not that of the larger nuclei, is quite rare and is known as *chromidiogamy*.   *Arcella* also has the other form of sexual union, *karyogamy*.

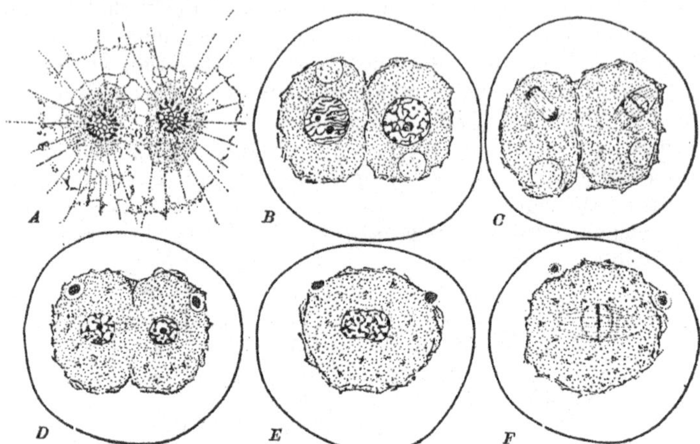

FIG. 111.—Copulation in *Actinophrys sol.*   × 850.   (*From Minchin, after Schaudinn.*)

In *Actinophrys sol* (Schaudinn 1896) (Fig. 111) two individuals, each with a single nucleus, approach each other and become enclosed in a common cyst.   In each of them the nucleus now undergoes two preliminary mitotic divisions, at each of which a small "reduction nucleus" is expelled from the body in a manner very similar to the expulsion of chromatin into the polar bodies of higher animals.   The two individuals,

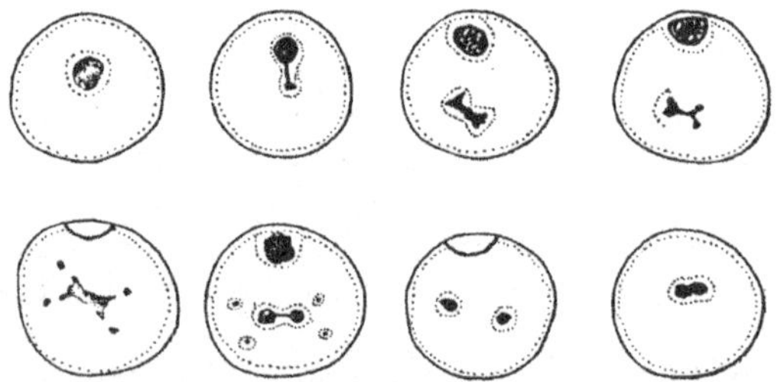

FIG. 112.—Autogamy in *Amœba albida*.   (*From Minchin, after Nägler.*)

or gametes, as they may now be called, fuse completely, their nuclei uniting to form a synkaryon.   Soon the synkaryon divides mitotically, this being followed by the division of the cell to form two individuals which escape from the cyst and resume the vegetative state.

In *Amœba albida* (Nägler 1909) (Fig. 112) a peculiar process known as *autogamy* occurs while the organism is in the encysted state.   The single nucleus divides, forming a large vegetative nucleus and a smaller generative nucleus. The former moves to the surface of the cell and degenerates, while the latter gives rise to the gamete nuclei in the following manner. After becoming elongated and incompletely divided it buds off four "reduction nuclei"—two from one end and then two from the

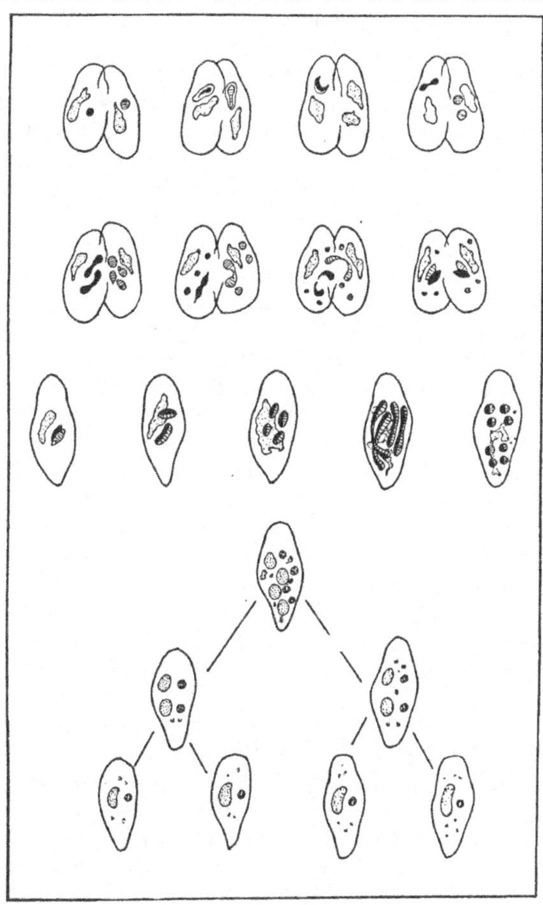

Fig. 113. Conjugation in *Paramœcium*. (*After Morgan, 1913.*)

other. It then completes its division to form the two gamete nuclei, or pronuclei, while the four reduction nuclei degenerate. After lying apart for some time the two pronuclei approach each other and fuse; sexual union thus takes place between sister nuclei.

The complicated nuclear behavior occurring at the time of conjugation in *Paramœcium caudatum* (Fig. 113) may best be described in the words of Wilson (1900, pp. 224–226). In "*Paramœcium caudatum,* which possesses a single macronucleus and micronucleus, . . . conjugation is temporary and fertilization mutual. The two animals become united by their ventral sides and the macronucleus of each begins to degenerate, while the micronucleus divides twice to form four spindle-shaped bodies. Three of these degenerate, forming the 'corpuscles de rebut,' which play no further part. The fourth divides into two, one of which, the 'female pronucleus,' remains in the body, while the other, or 'male pronucleus,' passes into the other animal and fuses with the female pronucleus. Each animal now contains a cleavage-nucleus equally derived from both the conjugating animals, and the latter soon separate. The cleavage-nucleus in each divides three times successively, and of the eight resulting bodies four become macronuclei and four micronuclei. By two succeeding fissions the four macronuclei are then distributed, one to each of the four resulting individuals. In some other species the micronuclei are equally distributed in like manner, but in *P. caudatum* the process is more complicated, since three of them degenerate, and the fourth divides twice to produce four new micronuclei. In either case at the close of the process each of the conjugating individuals has given rise to four descendants, each containing a macronucleus and a micronucleus derived from the cleavage-nucleus. From this time forward fission follows fission in the usual manner, both nuclei dividing at each fission, until, after many generations, conjugation recurs."

The Physiology of Fertilization.—The principal results of fertilization are two: the activation of the egg, and, in dioecious organisms, biparental heredity. Both of these have their physiological as well as their morphological aspects, and in the present section the first of them will be considered with special reference to its physiology. What is the nature of the physiological reactions through which the development of the egg is initiated? In the terms used by Child (1915), how does fertilization bring about the rejuvenation of the egg, which is a physiologically old cell? The attack upon this problem has been carried out along two main lines: by a study of artificial parthenogenesis and by a direct analysis of the chemical constitution of the gametes at these stages.

Artificial Parthenogenesis.—This line of attack has been followed particularly by Loeb, who has found a number of methods by which the parthenogenetic development of unfertilized eggs may be artificially induced.[1] As stated in the foregoing pages, the first externally visible effect of fertilization is in many animals the formation of a fertilization membrane. The formation of this membrane, which seems to be a condition necessary to the future development of the egg, Loeb was able to induce in the California sea urchin by placing the eggs for 2 minutes in a solution made up of 50 c.c. of sea water and 3 c.c. of a tenth-normal

[1] For a convenient summary of such methods see Harvey (1910).

fatty acid (butyric, propionic, or valerianic), and then back into pure sea water: the membrane then forms by a cytolysis of the cortical layer of the egg. Although in some forms (starfish) this one treatment is sufficient to bring about successful development, in most cases (sea urchin) the eggs become sickly and die. Loeb found that this sickliness may be prevented, allowing normal development, by either of two second treatments. If, after membrane formation, the eggs are placed for 20 minutes in hypertonic sea water or other solution with an osmotic pressure 50 per cent above that of ordinary sea water, they will develop normally when returned to pure sea water. The same effect may be brought about, though not always so successfully, by placing the eggs for 3 hours in sea water free from oxygen, or into sea water with a trace of KCN. It is therefore concluded by Loeb that the stimulus to such parthenogenetic development has two phases: the inducement of membrane formation by cytolysis, and the subsequent effect of the hypertonic solution. In rare cases the first treatment alone is sufficient for normal development, but in all cases it at least starts the egg into activity. As a result of these experiments Loeb has interpreted the action of the spermatozoön in normal fertilization on the assumption that it carries two substances: first, a lysin which brings about membrane formation by cytolysing the cortical layer of the egg, and which can act even if the spermatozoön does not enter the egg; and second, a substance which produces an effect similar to that of the hypertonic sea water employed in the experiments. The quite different explanation offered by Lillie will be mentioned further on.

How it is that cytolysis of the cortical layer of the egg brings about activation Loeb attempts to explain in the following manner. A calcium lipoid compound forms a continuous layer just beneath the surface of the egg, and the solution of this layer would probably result in the destruction of the cortical emulsion. It is assumed that in this cortical region there is a catalytic agent which increases the metabolism (rate of oxidation, etc.) of the egg. Following Warburg (1914) Loeb suggests that the cytolysis releases the catalyzer by breaking down the cortical emulsion; this results in an increase in the rate of oxidation and other reactions, and development proceeds.

That the process of activation is bound up primarily with reactions occurring in the cortical region of the egg is shown further by the experiments of Guyer (1907), Herlant (1913, 1917), McClendon (1912), Loeb and Bancroft (1913), and particularly Bataillon (1910), who have shown that the egg of the frog may be made to develop by pricking it with a needle, especially if some blood enters the egg with it; and also by the researches of R. S. Lillie (1908, 1915), who finds that starfish eggs may be made to develop parthenogenetically by exposing them to high temperatures for definite periods. (See F. R. Lillie, 1919, Chapter VII.)

Heilbrunn (1920) shows that the egg of *Cumingia* can be induced to undergo maturation by agencies which release the fluid cytoplasm from the restraint of the tough vitelline membrane. If the membrane be swollen, elevated above the egg surface, ruptured, or removed the maturation changes begin at once.

The sickliness and death of those eggs given only the first treatment Loeb thought to be due to the continued action of the cytolytic agent. Against this conception it is urged by F. R. Lillie that since any activated egg not developing normally cytolyzes sooner or later from internal causes, it is more probable that the sickliness and death are due to some internal cause resulting from activation, and points out that such a conclusion is supported by the cytological phenomena in eggs activated by Loeb's method. To these phenomena we may turn for a moment.

Eggs which have been given the first treatment alone do not begin to disorganize for many (12 to 24) hours. During this period Herlant (1917) has observed the following events. After the formation of the membrane and a hyaline zone, alterations cease, and the nucleus becomes the seat of a series of conspicuous changes. The nuclear membrane dissolves, and around the chromosomes there is formed a monaster (one-poled group of achromatic fibers), but no amphiaster develops. The chromosomes divide but do not separate, and although the cytoplasm becomes active no cytokinesis ensues. The chromosomes then return to the resting condition. This process is repeated several times, the nucleus increasing in bulk each time, but it soon becomes very irregular and the egg ultimately breaks down by general cytolysis. The second treatment (Loeb's method) in some way gives the egg the capacity to divide regularly. Morgan (1899) and Wilson had long before shown that such treatment with hypertonic sea water causes aster formation in the unfertilized sea urchin egg. Herlant shows that one of these asters and a second aster formed in connection with the egg nucleus together form an amphiaster, normal division then ensuing.

In the light of these facts it seems evident that the death of the egg after the first treatment alone is not due to the continued action of the cytolytic agent employed, but rather to irregularities in the activation processes aroused by the cortical changes in the absence of a proper coördination of nuclear and cell division. The second treatment produces a regulatory effect, partly through aster formation, resulting in normal development. This recalls Boveri's morphological theory of normal fertilization.

Direct Analysis of the Fertilization Process.—In contrast to the theory that the spermatozoön contributes organs (Boveri) or substances (Loeb) necessary for the activation, Lillie (1919, Chapter VII) regards the egg itself as an "independently activable system." "The egg possesses all substances needed for activation; the spermatozoön is an inciting cause of those reactions within the egg system upon which development depends." As a result of his direct analysis of the gametes during the fertilization period Lillie has identified a substance in the egg which he calls *fertilizin*. This substance is present in the egg for a short time only; its formation usually begins at about the time the germinal vesicle begins to break down, and immediately after fertilization its production ceases, possibly through the neutralizing action of a second substance, called "anti-fertilizin." As a rule it is only during the period at which fertilizin is present that spermatozoa will enter the egg; the egg remains fertilizable for but a short time. Hence it seems clear that it is not the fertilization membrane that prevents the entrance of other spermatozoa, as Fol thought, but rather the physiological state of the egg. That the protection is thus a physiological rather than a mechanical one is indicated by the fact that membraneless egg fragments without fertilizin are not entered by spermatozoa.

Fertilizin has two effects: it first acts by causing an agglutination of the spermatozoa at the surface of the egg, and later causes the activation of the egg. It may thus be said to stand between the spermatozoön and the activation reactions in the egg. Being present in the egg secretion at a certain period it binds the spermatozoön to the surface of the egg, and the spermatozoön, without necessarily penetrating the egg at all, by means of a substance which it bears releases the activity of the fertilizin within the egg, which results in development. In brief, the activating substance is already present in the egg and is not brought to it by the spermatozoön. It may be incited to activity by the spermatozoön, but by other agencies as well.

In concluding this sketch of the physiological features of fertilization we may state briefly the immediate physiological consequences of the process as summarized by Lillie (1919, Chapter V). The rate of oxidation increases in most cases in which it has been investigated. In the sea urchin egg (Warburg 1908–1914) this rate increases as much as six- or seven-fold; in *Strongylocentrotus*, four- or five-fold (Loeb and Wasteneys, 1912, 1913); in the starfish, apparently not at all. The egg membrane becomes more permeable to oxygen, CO_2, pigment, water, alkalis, intravitam stains, and a number of other substances. The protoplasm becomes less fluid after fertilization (Heilbrunn 1915). This gelation effect Chambers (1917) believes to center upon the sperm aster. The volume of the egg decreases and its electrical conductivity rises. The most conspicuous chemical change is seen in the loss of the fertilizin, and with it the loss of capacity for further fertilization reaction.

FERTILIZATION IN PLANTS

Although the central act of the process of fertilization is regularly the union of two sexually differentiated nuclei, the morphological features associated with this fusion are more varied in plants than in animals. This is especially true of the algæ and fungi.

Algæ.—In *Ulothrix* fertilization consists in the complete union of two morphologically similar, motile biciliate gametes (Fig. 114, *A*). In *Fucus* the two gametes are very dissimilar: the male (spermatozoid) is small, laterally biciliate, and actively motile (Fig. 114, *B*), while the female (egg), though discharged from the oögonium, is large and passive, as in all higher plants and animals. In *Œdogonium* (Fig. 114, *D, E*) the

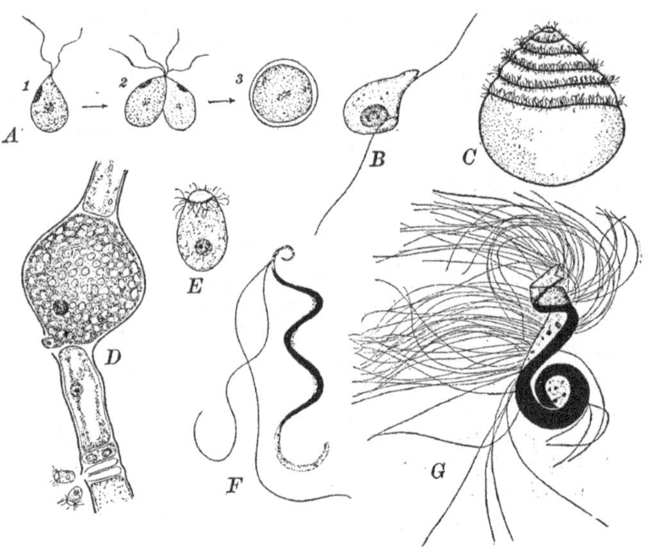

Fig. 114.—Spermatozoids of plants.

A, Ulothrix: 1, gamete; 2, gametes fusing (isogamy); 3, zygospore. *B, Fucus.* (*After Guignard.*) *C, Zamia.* (*After Webber.*) *D,* bit of filament of *Œdogonium;* spermatozoids escaping from antheridial cells below; spermatozoid about to enter egg above. (*After Coulter.*) *E,* spermatozoid of *Œdogonium. F, Chara.* (*After Belajeff.*) *G, Onoclea.* (*After Steil.*) For figures of spermatozoids of *Blasia, Polytrichum, Equisetum,* and *Marsilia,* see Figs. 28, 29, 30, and 32.

egg is not shed from the cell which produces it, but is fertilized *in situ,* a condition which is retained in all the higher plant groups. The spermatozoid in this genus has a crown-like ring of cilia. In *Spirogyra* (and other Conjugatæ) certain vegetative cells, without further morphological differentiation, function as gametes. The entire contents of such a cell pass through a conjugation tube to a similar cell in an adjacent filament, where the two unite to form the zygospore. The two nuclei fuse, but the chloroplasts furnished by the contributing ("male") gamete may eventually degenerate (*Zygnema*). In *Polysiphonia* a non-motile male gamete (spermatium) comes in contact with a prolongation (trichogyne) of the female sex organ (carpogonium).   Solution of the intervening walls allows the nucleus of the spermatium to pass into the trichogyne and down to the female nucleus in the base of the carpogonium.   In *Polysiphonia* we have one of the few cases among lower plants in which the fusion of the sexual nuclei has been minutely described.   According to Yamanouchi (1906) the male nucleus, by the time it has reached the female nucleus, has resolved itself into a group of 20 chromosomes (Fig. 115, *A*).   In this

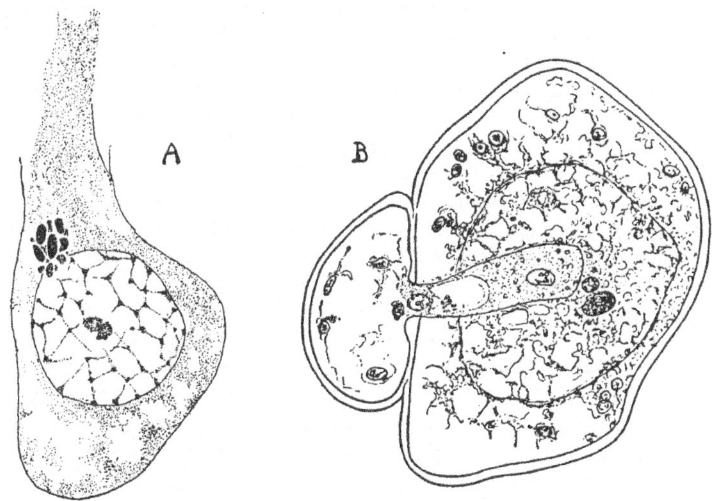

Fig. 115.

A, fertilization in *Polysiphonia*.   Group of male chromosomes about to enter female nucleus.   (*After Yamanouchi*, 1906.)   *B*, fertilization in *Albugo candida*.   Female nucleus lying in center of oöplasm near the "cœnocentrum" (larger dark body.)   Antheridial tube about to discharge a male nucleus; another male nucleus in neck of tube.   Additional nuclei in periplasm surrounding the oöplasm.   (*After Davis*, 1900.)

condition it enters the female nucleus while the latter is yet in the reticulate state.   Soon the female reticulum becomes transformed into 20 chromosomes, which arrange themselves with the 20 paternal chromosomes upon the spindle as the fusion nucleus divides.

Fungi.—In the Phycomycetes sexual reproduction occurs in two principal forms, which serve to divide the group into two main divisions: Oömycetes and Zygomycetes.

In the Oömycetes the cytological phenomena are best known in the Peronosporales and Saprolegniales.   In the former there is differentiated in the oögonium a single large egg into which the contents of an antheridium are discharged through a penetrating tube.   In *Albugo bliti* and *A. portulaccæ* (Stevens 1899, 1901) the egg has a large number of nuclei,

19

and after the entrance of the antheridial nuclei about 100 sexual fusions occur. In *Albugo candida* (*Cystopus candidus*) (Wager 1896; Davis 1900), *Peronospora parasitica* (Wager 1900), *Albugo tragopogonis* and *A. ipomœae* (Stevens 1901) the mature egg has but one nucleus, which fuses with a single male nucleus discharged into the egg by an antheridium (Fig. 115, *B*). In all cases an oöspore results.

In the Saprolegniales, as shown by the researches of Davis (1903, 1905), Miyake (1901), Trow (1895–1905), and Claussen (1908), there are two general conditions. In *Saprolegnia* (Trow, Davis, Claussen) from 10 to 15 uninucleate eggs are formed within an oögonium. One or more antheridia send in conjugating tubes and deliver a male nucleus to each egg, in which a single sexual fusion then occurs. In *Pythium* (Trow 1901; Miyake 1901) a single uninucleate egg is produced, the fertilization process closely resembling that in *Albugo candida*.

In the Zygomycetes, represented chiefly by the Mucoraceæ, the sexual process consists in the union of the contents of two similar (except oc casionally in size) multinucleate gametangia, the result of the fusion being a zygospore. As shown by Blakeslee (1904) these two gametangia are borne on the same mycelium in some species ("homothallic" species), whereas in other species ("heterothallic" species) they are regularly borne on different mycelia, no zygospores being formed in the latter species on a mycelium arising from a single spore. Owing to the extremely minute size of the nuclei their behavior at these stages is not well known. By some investigators (Macormick on *Rhizopus nigricans*, 1912) it is held that only one fusion occurs, the remaining nuclei degenerating. Others (Keene on *Sporodinia grandis*, 1914) think it probable that although some degeneration occurs, the nuclei nevertheless fuse in pairs in considerable numbers. Until further researches have been carried out very little of a definite nature can be said concerning the nuclear history of the Zygomycetes.

In the Ascomycetes (see Atkinson 1915) the fusion of two nuclei in the ascus was first described for several species by Dangeard (1894) (Fig. 116, *A*), who regarded it as a sexual fusion and the ascus as an oögonium. The matter soon became complicated when a number of cytologists, beginning with Harper (1895 etc.), found what they believed to be a nuclear fusion at an earlier stage in the life cycle. This fusion was described as occurring (*a*) in the archicarp when fertilized by the contents of an antheridium (Harper on *Sphœrotheca castagnei*, 1895, 1896, *Erisiphe* 1896, *Pyronema confluens* 1900, and *Phyllactinia* 1905; Blackman and Fraser on *Sphœrotheca* 1905; Claussen on *Boudiera* 1905); (*b*) in the archicarp when the antheridium is functionless or absent (Blackman and Fraser on *Humaria granulata* 1906; Fraser on *Lachnea stercorea* 1907; Welsford on *Ascobolus furfuraceus* 1907; Dale on *Aspergillus repens* 1909); or (*c*) in the vegetative cells when the archicarp is functionless or absent (Fraser on *Humaria rutilans* 1907, 1908; Carruthers on *Helvella crispa* 1911; Blackman and Welsford on *Polystigma rubrum* 1912). By the above investigators this early fusion was regarded as a sexual one, that in the ascus being vegetative in nature; and some described a "double reduction" in the ascus to compensate for the two nuclear fusions. (See p. 223.)

In a series of somewhat later researches another group of observers found the evidence for an early fusion to be very unsatisfactory, and concluded that the only nuclear union in the life cycle is that occurring in the ascus: with Dangeard they saw in this union the sexual act. Furthermore, no "double reduction" was found in the ascus. Among the researches supporting this view, which now appears to be the more probable, may be cited the following: Claussen on *Pyronema confluens*

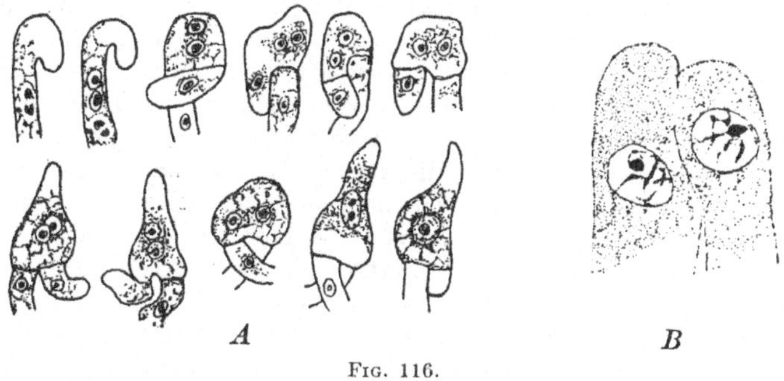

A   B

Fig. 116.

A, nuclear fusion in the ascus of *Peziza vesiculosa*. (*After Dangeard, 1894.*) *B*, cell fusion in æciospore sorus of *Phragmidium speciosum*. *After Christman, 1905.*

1907, 1912; Schikorra on *Monascus* 1909; W. H. Brown on *Pyronema confluens* 1909, *Lachnea scutellata* 1911, and *Leotia* 1910; Faull on *Laboulbenia* 1911, 1912; Blackman on *Collema pulposum* 1913; Nienburg on *Polystigma rubrum* 1914; Ramlow on *Ascophanus carneus* and *Ascobolus immersus* 1914; Brooks on *Gnomonia erythrostoma* 1910; McCubbin on *Helvella elastica* 1910; H. B. Brown on *Xylaria tentaculata* 1913; and Fitzpatrick on *Rhizina undulata* 1918a.

As the two nuclei fuse in the young ascus Harper (1905) observed in the case of *Phyllactinia corylea* that not only the chromatin systems but also the nucleoli and "central bodies" (centrosomes), upon which the chromatin strands converge, unite. In the Ascomycetes generally the fusion nucleus, or "primary ascus nucleus," undergoes three successive mitoses to form the eight ascospore nuclei, the spore walls in each case being formed in association with the curving astral rays which focus upon the centrosome. (See p. 80.)

In certain yeasts it has been shown (see Guilliermond 1920) that the production of ascospores is preceded by a copulation of two cells with a fusion of their nuclei, the fusion nucleus dividing to form the spore nuclei. A somewhat similar copulation of the ascospores themselves has also been observed in a few cases.

Among the BASIDIOMYCETES the nuclear phenomena are best known in the case of the rusts, owing to the researches of Blackman (1904), Christman (1905), and a number of later writers. In the typical rust life cycle there is a fusion of uninucleate cells at the base of the aecial sorus (Fig. 116, *B*). The binucleate cells thus arising produce the binucleate aeciospores; and these upon germination form a mycelium with binucleate cells, the two nuclei dividing in unison ("conjugately") at each cell-division. After producing a series of crops of binucleate uredospores this mycelium eventually bears teliospores which may consist of one or more cells. In each cell of the teliospore the two nuclei delivered to it as the result of the conjugate divisions throughout the binucleate mycelium finally unite, initiating the uninucleate phase of the life cycle. Here the fusion of sexual cells and the fusion of their nuclei—two events which in most organisms occur very near each other in time—are widely separated in the cycle. The two nuclei dividing conjugately constitute together a synkaryon in many respects equivaleng to a diploid nucleus. Since there is as yet no evidence to show in what degree the two effects of fertilization (the stimulus to development and the mixing of hereditary lines) are brought about in the rusts by the fusion of the sexual cells on the one hand and by the final union of their nuclei on the other, it seems best to regard the two fusions as two phases of the fertilization process in spite of their wide separation in the life history.

In the Hymenomycetes it has been known for some time that a fusion of two nuclei occurs in the basidium, itself the terminal cell of a binucleate hypha, prior to the formation of the four basidiospore nuclei (Fig. 79). The origin of the binucleate condition in the mycelium which has apparently arisen from a uninucleate spore has long been an obscure point. It has recently been shown by Miss Bensaude (1918) in the case of *Coprinus fimetarius* that the binucleate hyphæ arise as the result of cell fusions ("plasmogamy;" "pseudogamy") between uninucleate hyphæ arising from different spores, and that no carpophores are produced upon a uninucleate mycelium arising from a single spore. Thus it appears that in at least some hymenomycetes the sexual process is initiated by a fusion of two cells of different strains ("plus" and "minus"), as in the heterothallic molds.

Bryophytes and Pteridophytes.—In bryophytes and pteridophytes the details of the union of the motile spermatozoid with the egg in the archegonium have been described in very few cases. In the former group may be cited the works of Garber (1904) and Black (1913) on

Riccia, Meyer (1911) on *Corsinia,* Graham (1918) on *Preissia,* and
Woodburn (1920) on *Reboulia.* It appears that in bryophytes the body
of the biciliate spermatozoid, which consists mainly of nuclear material,
undergoes in the egg cytoplasm a transformation into a reticulate nucleus
before fusing with the egg nucleus (Fig. 117). The fate of the non-
nuclear structures (cytoplasm, blepharoplast, and cilia) is not known
with certainty, but it is probable that they are absorbed in the egg cytoplasm. In the liverwort, *Preissia
quadrata,* Miss Graham has found
two centrosomes with weakly developed asters in the cytoplasm of
the egg at the time the two pronuclei
are about to fuse (Fig. 23, *A*). It is
not known what relation their appearance may have to the entrance
of the spermatozoid.

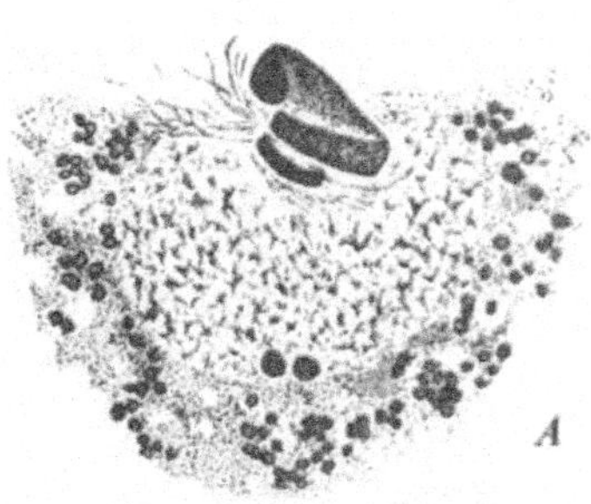

The most detailed account of fertilization in a pteridophyte is that
given by Yamanouchi (1908) for

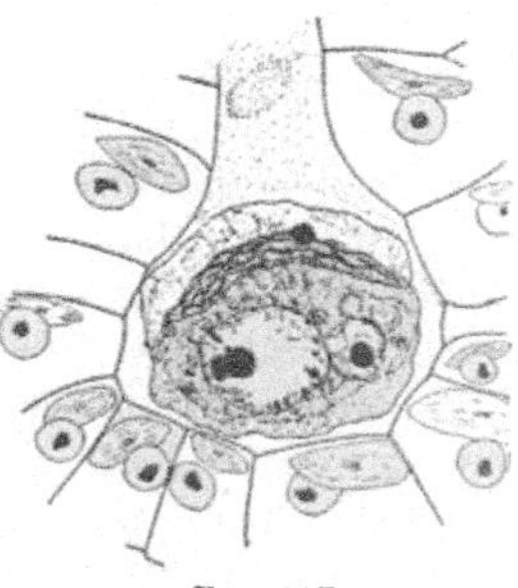

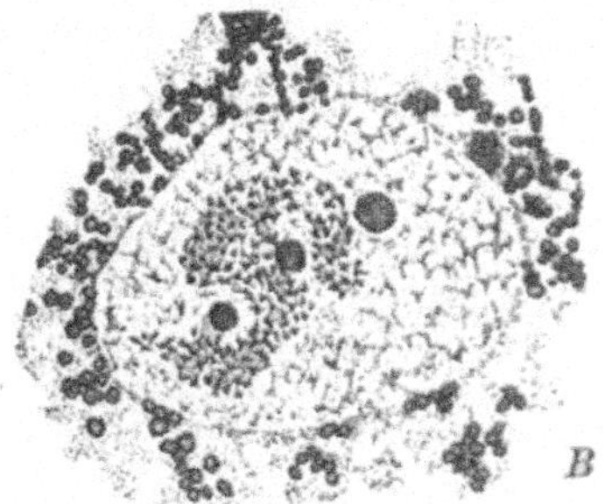

Fig. 117.

Fig. 118.

Fig. 117.—Fertilization in *Anthoceros.* Male and female pronuclei about to fuse in
lower part of egg in venter of archegonium; elongated plastid above them. Gametophyte
cells show one nucleus and one plastid each. × 1050.

Fig. 118.—Fertilization in *Nephrodium.*

A, spermatozoid entering egg nucleus. *B,* spermatozoid becoming reticulate in midst
of female reticulum. (*After Yamanouchi,* 1908.)

Nephrodium (Fig. 118). In *Nephrodium* the multiciliate spermatozoid
enters bodily into the egg nucleus with no previous alteration into the
reticulate state. Here it gradually becomes reticulate and irregular in
shape, until finally its limits are indistinguishable, the chromatic material
contributed by the two gametes apparently forming a single fine-meshed
network.

Gymnosperms.—Among living gymnosperms the Cycadales and
Ginkgoales are characterized by the possession of motile spermatozoids.

These spermatozoids are very much alike in structure and behavior in the two groups, and are unusually large, being easily visible to the naked eye. The body is made up of a large nucleus surrounded by a thin cytoplasmic layer in which is imbedded a long, spirally coiled blepharoplast bearing many cilia (Fig. 114, *C*). The behavior of the spermatozoid in fertilization has been studied in *Ginkgo* by Hirasé (1895, 1918) and Ikeno (1901); in *Cycas revoluta* by Ikeno (1898); in *Zamia floridana* by Webber (1901); and in *Dioon edule, Ceratozamia mexicana,* and *Stangeria*

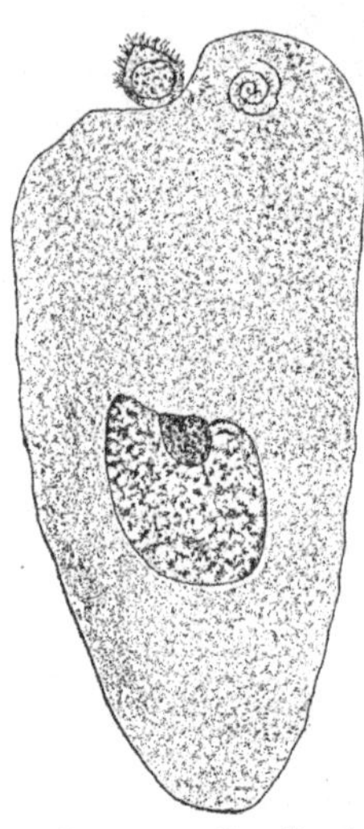

paradoxa by Chamberlain (1910, 1912, 1916). In all cases the entire spermatozoid penetrates into the egg cytoplasm, where the nucleus frees itself from the cytoplasmic sheath with its blepharoplast and cilia and advances alone to the egg nucleus, with which it fuses (Fig. 119). The behavior of the chromatin during the fusion is not well known in either *Ginkgo* or the cycads.

In the Coniferales and Gnetales the male cells have no motile apparatus. Each consists of a nucleus surrounded by a more or less sharply delimited mass of cytoplasm. In most cases this cytoplasm remains intact until after the male cell has entered the egg, but in other forms, such as *Pinus,* it mingles with the cytoplasm of the pollen tube, so that only male nuclei, rather than completely organized male cells, are delivered to the egg. All the nuclei present in the pollen tube—stalk nucleus, tube nucleus, the two male nuclei, and in certain species free prothallial nuclei—may be discharged into the egg. All but the functioning male nucleus usually degenerate at once, but in some cases they have been observed to undergo division.

When a complete male cell enters the egg the cytoplasm of the former shows two general modes of behavior. In some species it may be left behind in the peripheral region of the egg as the male nucleus frees itself and advances alone to the female nucleus. This type of behavior has been reported in *Pinus* (Ferguson 1901, 1904), *Thuja* (Land 1902), *Juniperus* (Norén 1904), *Cryptomeria* (Lawson 1904), and *Libocedrus* (Lawson 1907). In *Sequoia* (Lawson 1904) the male nuclei escape from their cytoplasm before their discharge from the pollen tube, and enter the egg alone. In a second group of species the male cytoplasm remains intact and invests the fusing sexual nuclei, being clearly distinguishable from the cytoplasm of the egg. The pollen tube cytoplasm often plays a conspicuous part in the formation of this "mantle." This phenomenon, the significance of which can only be conjectured, is found in *Taxodium* (Coker 1903), *Torreya californica* (Robertson 1904), *Torreya taxifolia* (Coulter and Land 1905), *Cephalotaxus Fortunei* (Coker 1907), *Ephedra* (Berridge and Sanday 1907; Land 1907), *Phyllocladus* (Kildahl 1908), *Juniperus* (Nichols 1910), *Agathis* (Eames 1913), and *Taxus* (Dupler 1917).

Chromosome Behavior.—The behavior of the chromosomes during the fusion of the sexual nuclei and the first embryonal division has been described in a number of conifers. As a general rule, to judge from the data at hand, the chromatin contributions of the two pronuclei do not become intimately associated in the fusion nucleus, but remain distinguishable until the first embryonal mitosis occurs. Each of the pronuclei then gives rise to its complement of chromosomes which

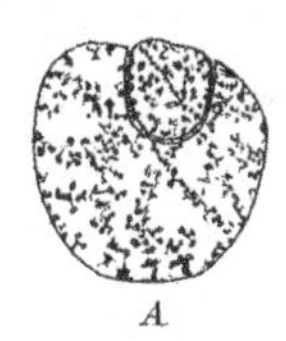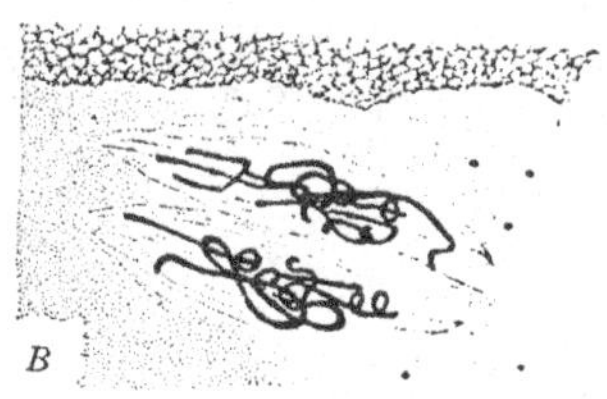

FIG. 120—Fertilization in *Pinus*.

A, male nucleus pressing into female nucleus. × 140. *B*, first embryonal mitosis, showing separate paternal and maternal chromosome groups. × 472. (*After Ferguson. 1904.*)

become arranged, often as two separate groups, upon a common spindle. Such an independent formation of the male and female chromosome groups has been observed in *Pinus* (Blackman 1898; Chamberlain 1899; Ferguson 1909, 1904) (Fig. 120), *Larix* (Woyciki 1899), *Tsuga candensis* (Murrill 1900), *Juniperus communis* (Norén 1907), *Cunninghamia* (Miyake 1910), and *Abies* (Hutchinson 1915). In *Sequoia*, on the other hand, Lawson (1904) reports that the two nuclei form a common reticulum in which the male and female constituents cannot be distinguished. With regard to the first embryonal mitosis the general opinion has been that all the chromosomes, paternal and maternal, split longitudinally, the daughter chromosomes being distributed to the daughter nuclei as in any other somatic mitosis. This type of behavior was described for the chromosomes of *Pinus* by Miss Ferguson (1904) and at once came to be regarded as general for conifers, as it had been for other organisms.

A new interpretation differing in certain fundamental points from the above has been more recently suggested by Hutchinson (1915), as a result of his work on *Abies balsamea*. According to Hutchinson (Fig. 121) there appear in the fusion nucleus two groups of chromosomes, each containing the haploid number (16). A spindle is differentiated about each group; and the two spindles soon unite to form one, thus bringing the two chromosome groups, representing the two parental contributions, into closer association. The chromosomes now approximate two by two to form 16 pairs. The members of each pair twist about each other and become looped; each of them becomes transversely segmented at the apex of the loop, forming 32 (2x) pairs of segments; these pairs separate to form 64 (4x) chromosomes; a new spindle is formed and 32 (2x) chromosomes pass to each pole.

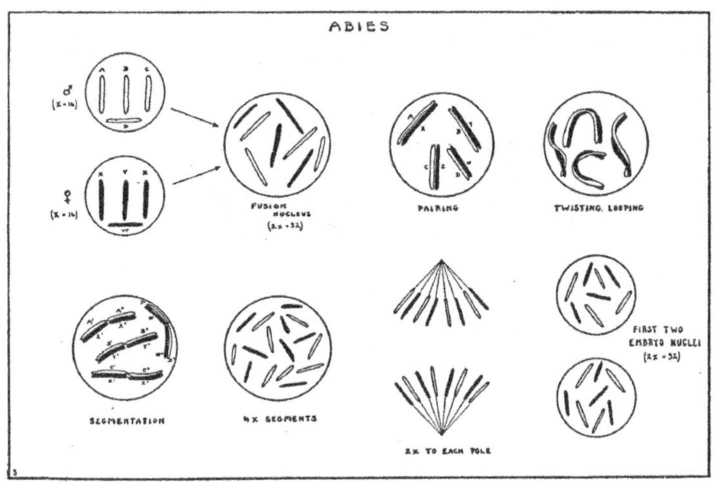

Fig. 121.—The behavior of the chromosomes in fertilization and the first embryonal mitosis in *Abies*, according to Hutchinson. (1915.)

This interpretation of chromosome behavior at fertilization is remarkable not only because it indicates features resembling those of the heterotypic prophase, but chiefly because it actually calls for a qualitative reduction of the chromatin at the first embryonal mitosis if the chromatin is not qualitatively the same throughout the nucleus. This implication has not been discussed by the advocates of the new theory. The chromosomes pair and twist about one another in a way that parallels closely their behavior during the prophase of a reduction division. That the doubleness seen is due to a pairing and not to a splitting as has heretofore been held is supported by the assertion that the pairs are present in the haploid number, rather than in the diploid number as would be the case were a splitting of all the chromosomes occurring. If the two members of each pair were to separate at the first embryonal mitosis, a reduction, qualitative as well as numerical, in all respects similar to that accomplished in the regular heterotypic mitosis, would be brought about if the pairing members are qualitatively different. But instead of such a separation, each member of each pair segments transversely, giving 4x segments which are equally distributed to the two daughter nuclei, each of the latter receiving the diploid number. Since the 4x segments become more or less intermingled before their distribution it is probably impossible to determine just which ones pass to each pole. If both halves of one transversely divided chromosome pass to one pole (see Fig. 122), that daughter nucleus only, and not the other, will receive the kind of chromatin carried by that chromosome, so that

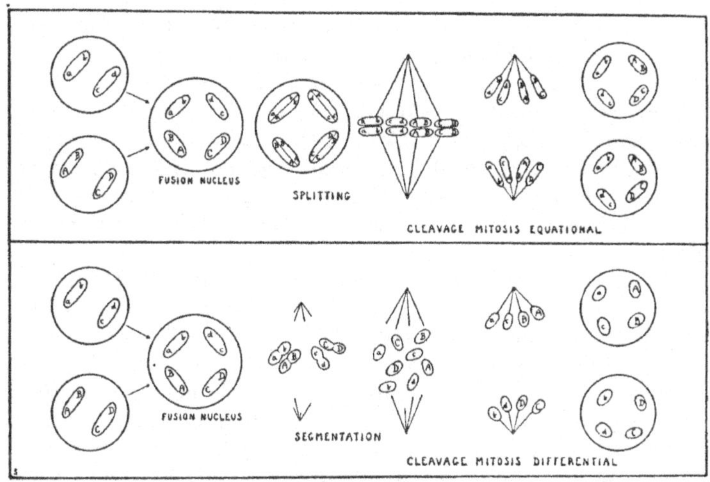

FIG. 122.—Diagram showing the behavior of the chromosomes in fertilization and the first embryonal mitosis as usually interpreted (upper part) and according to Hutchinson's interpretation (lower part).

the two nuclei will be qualitatively different. A qualitative reduction will have occurred, but without a change in the number of chromosomes, since each old chromosome has become two new ones. If, on the other hand, the two halves of the transversely segmented chromosome regularly pass to opposite poles, each daughter nucleus will receive a half of each and every parental chromosome: thus if there are just as many kinds of chromatin as there are chromosomes, these nuclei will be qualitatively alike, just as they would be had the division been longitudinal instead of transverse. But, as has been stated in the chapter on reduction and will be developed at greater length in Chapter XVII, there is a considerable body of evidence which indicates that each chromosome is not only qualitatively different from its fellows, but possesses a linear differentiation of some sort; so that the separation of the two halves of a transversely divided chromosme would constitute a qualitative reduction. If such actually is the condition of the chromatin, and if the chromosomes do behave as Hutchinson supposes, a qualitative reduction must immediately follow each fertilization, and half of the resulting body cells must have a constitution differing from that of the other half. Since there are known no chromosome fusions in which a restoration in the number of qualities is known to occur, the number of these qualities in a single chromosome would in a few generations be reduced to one: in view of the large number of past generations this must have already occurred.

This new interpretation of chromosome behavior at fertilization and the ensuing mitosis is thus seen to offer a direct challenge to those theories of heredity that are based upon the idea of chromosomes carrying linear series of differentiated units. It has now been put forward by Hutchinson (1915) for *Abies balsamea*, by Chamberlain (1916) for *Stangeria paradoxa*, and by Miss Weniger (1918) for *Lilium philadelphicum* and *L.. longiflorum*. Consequently several investigators have renewed the study of fertilization, and evidence contradictory to the new theory has been found by Miss Nothnagel (1918) and Sax (1918), whose researches are summarized in the following section on the angiosperms.

Angiosperms.—The angiosperms are characterized by the occurrence of "double fertilization," a phenomenon discovered independently by Nawaschin (1898) and Guignard (1899). One of the two male nuclei formed by the male gametophyte and brought into the embryo sac by the pollen tube, enters the egg and fuses with its nucleus, thus forming the primary nucleus of the embryo, while the other male nucleus fuses with the two polar nuclei to form the primary endosperm nucleus (Fig. 123, *B*). As the male nuclei pass down the pollen tube they are usually unaccompanied by any specially differentiated cytoplasm: the male gametes are naked nuclei and not complete cells. In some cases, however, male cells have been reported (Fig. 123, *A*). When they are liberated in the embryo sac by the rupture of the end of the pollen tube any such cytoplasm is indistinguishable from that of the sac and that discharged from the pollen tube. The male nuclei may appear in all respects similar to other nuclei, or they may be distinctly vermiform, as was observed by Mottier (1898) and later by many other workers (Fig. 123, *E*). That such vermiform nuclei have the power of independent movement has been held by Nawaschin (1899, 1900, 1909, 1910) for *Lilium* and *Fritillaria*, by Guignard (1900) for *Tulipa*, and by Blackman and Welsford (1913) and Miss Welsford (1914) for *Lilium Martagon* and *L. auratum*. The vermiform condition may persist until the time of fusion, but in other cases, such as *Fritillaria* (Sax 1916), it gives way to the ordinary shape. This change may occur more rapidly

in one male nucleus than in the other, so that the two may appear quite
unlike during the later stages.   Miss Welsford also sees in the male
cytoplasm certain granules which she thinks may represent the vestiges
of blepharoplasts.

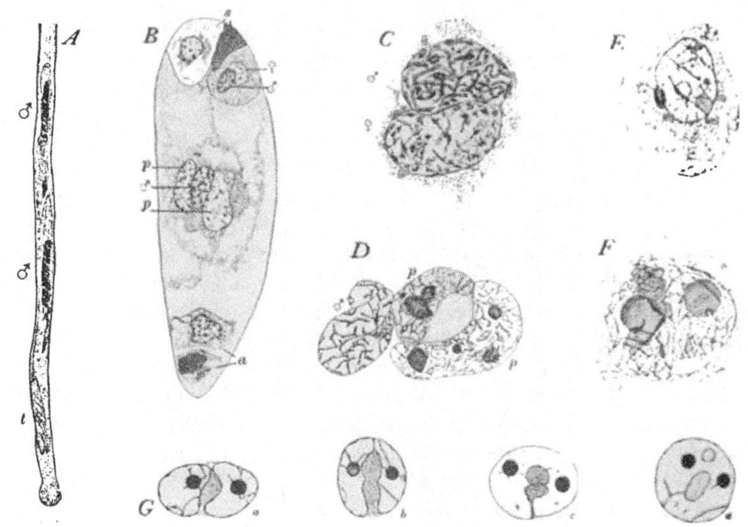

Fig. 123.—Fertilization in angiosperms.

A, end of pollen tube from basal portion of style of *Lilium auratum*, showing two male
cells and tube nucleus.   × 250.   (*After Welsford*, 1914.)   *B*, double fertilization in *Lilium
canadense*: male and female nuclei about to fuse in egg; second male and two polar nuclei
fusing at center of embryo sac; *s*, synergids, one degenerated; *a*, antipodals   × 250.
C, fusion of sexual nuclei in egg of *Lilium philadelphicum*.   × 1000.   (*After Weniger*,
1918.)   *D*, the second male and two polar nuclei in *Lilium Martagon*.   × 750.   (*After
Nothnagel*, 1918.)   *E*, vermiform male nucleus in contact with egg nucleus in *Triticum
durum*.   × 600.   (*After Sax*, 1918.)   *F*, spireme stage of triple fusion nucleus in *Triticum
durum*, showing distinctness of three chromatin contributions.   × 750.   (*After Sax*, 1918.)
G, inclusion of cytoplasm in fusing sexual nuclei of *Peperomia sintenesii*.   (*After Brown*,
1910.)

Fusion in Egg.—As already stated, one male nucleus passes into
the egg and fuses with the egg nucleus.   So far as observations enable
one to say, only the male nucleus, and no cytoplasm, enters the egg, a
point of much importance in connection with the transmission of hereditary characters from the male parent.   It would be a matter of extreme
difficulty, however, to demonstrate conclusively that in passing through
the egg membrane the male nucleus is absolutely freed of all adhering
cytoplasm or chondriosomes; and it must be admitted that such a
demonstration has not yet been given in any case.   The fusion of the two
sexual nuclei probably occurs in most cases very soon after they come
in contact, though in certain forms the actual fusion is known to be

considerably delayed. The chromatin of the two nuclei at the time these unite may be in the reticulate (resting) condition, the male and female chromatins being indistinguishable in the fusion nucleus. This situation was described by most of the earlier workers, including Strasburger (1900, 1901), Mottier (1904), Nawaschin (1898, 1899), and Ernst (1902). It has also been reported by Sax (1916) in his recent work on *Fritillaria*. In other cases, as early reported by Guignard (1891), the chromatin has already reached the spireme stage characteristic of the prophase, the male and female elements being distinguishable on the spindle in the ensuing division of the fertilized egg. Such is the condition, for instance, in *Calopogon* (Pace 1909), *Trillium* (Nothnagel 1918), and *Lilium* (Weniger 1918). That the same species may show considerable variation in this respect is indicated by the situation in *Fritillaria*, in which Sax (1916, 1918) finds that fusion, though it usually occurs in the resting stage, sometimes takes place after the spiremes have been developed. Miss Weniger (1918) reports that in *Lilium philadelphicum* and *L. longiflorum* the egg nucleus is in the resting condition and the male nucleus in the spireme stage at the time of union.

Chromosome Behavior.—With regard to the behavior of the two parental groups of chromosomes, it has been generally held that, whatever their condition at the time of the nuclear union, all of them, both paternal and maternal, split longitudinally at the first division of the fertilized egg, the daughter chromosomes so formed being distributed to the two resulting nuclei, just as in all the subsequent somatic divisions. Recently, however, Miss Weniger (1918) has reported a condition in *Lilium* similar to that described by Hutchinson (1915) for *Abies:* the maternal and paternal chromosomes form pairs and divide transversely into daughter segments which pass to the poles. (See p. 296.) With this conclusion other recent investigations of fertilization in angiosperms are not in agreement. Miss Nothnagel (1918) finds in *Trillium* no such pairing and cross segmentation as Hutchinson and Miss Weniger describe, and states that each chromosome splits longitudinally as held by Miss Ferguson for *Pinus* and by cytologists in general. Sax (1918) also shows that each paternal and maternal chromosome in *Fritillaria* divides longitudinally, the diploid number (24) passing to each pole. In *Trillium* he reports an essentially similar state of affairs, the diploid number here being about 28. He therefore holds that the first mitosis in the fertilized egg is like any other somatic mitosis, and that no mechanism for the segregation of factors of inheritance, such as occurs at reduction, exists here. The outcome of this controversy is awaited with much interest because of its great theoretical importance.

Endosperm Fusion.—The fusion of the second male nucleus with the two polar nuclei of the embryo sac to form the primary endosperm nucleus may be carried out in a variety of ways. The most commonly reported method is that by which the two polars fuse to form an "embryo sac nucleus" before the entrance of the pollen tube, the male nucleus later being added. Ernst (1902), for example, found this to be the method in *Paris quadrifolia*. Less frequently the male nucleus meets and fuses with the polar nucleus of the micropylar end of the sac, the other polar then fusing with the product. This is the method described by Nawaschin (1898, 1899) in his account of the discovery of double fertilization in *Lilium Martagon* and *Fritillaria tenella*. The simultaneous fusion of all three nuclei appears to be a common occurrence; it has recently been described in some detail by Miss Nothnagel (1918) for *Trillium* and *Lilium*. Just as in the case of the union of the first male nucleus with the egg nucleus, the chromatin of the second male and two polar nuclei may be either in the reticulate or the spireme condition as they come together. In *Lilium Martagon* (Nothnagel 1918) (Fig. 123, *D*) it is in the form of fine strands, intermediate between the resting and spireme stages. Although the three nuclei often appear exactly alike, it is frequently possible to distinguish the male from the polars, not only by its shape and smaller size, but by the condition of its chromatin: in *Lilium longiflorum* (Weniger 1918), for example, the male nucleus is in the spireme stage while the polar nuclei are still in the resting condition. The membranes of the three nuclei may persist for some time after they come into intimate contact, and even after they have dissolved the chromatic elements of the three constituent nuclei may in many cases be distinguished if the section has been made in a favorable plane. When fusion occurs in the resting stage this is not so apparent, but when it occurs in the spireme stage the three chromatic groups are made out with little difficulty.

Endosperm.—As the division of the endosperm nucleus approaches the spiremes of its three constituent nuclei become increasingly distinct, even if one or more of the nuclei have fused in the resting stage. Nothnagel (1918), Weniger (1918), and Sax (1918) in their recent studies all report this condition (Fig. 123, *F*). As the spiremes are being developed into completed chromosomes all of them (3x in number) split longitudinally, no observer reporting such a pairing as some have thought to occur between the chromosomes of the egg and first male nuclei. Miss Nothnagel describes the formation of a tripolar spindle about the chromosomes; the bipolar condition soon develops from this. How frequently this may occur is not known. Eventually in any case the mitosis proceeds along the usual lines and the two daughter nuclei receive 3x chromosomes each. This number is characteristic of all the cells of the endosperm formed by the repeated division of these nuclei. An exceptional condition has been noted by Sax (1918) in *Fritillaria*. Here the lower polar nucleus, because of an irregularity in the mitosis giving rise to it, has 24 (2x) chromosomes instead of the normal 12 (x). Consequently the female parent contributes 36 chromosomes (24 in one polar nucleus and 12 in the other) to the endosperm, while the male parent contributes only 12; thus the endosperm has 48 (4x) chromosomes instead of the normal 36.

Although in the great majority of known examples endosperm is formed by the repeated division of a triple fusion nucleus, cases are known in which it is produced by the polar fusion nucleus (embryo sac nucleus) alone without the male, or by the fusion product of the male and one polar, or by one polar alone. Combinations of these three methods may be found in the same embryo sac. The development of the endosperm may be initiated by the formation of a number of free nuclei which are parietally placed and in later mitoses become separated by walls, or by the formation of walled cells from the start. (See Coulter and Chamberlain, 1903.)

The term *xenia* was applied by Focke (1881) to the effect of foreign pollen on the endosperm of the resulting seed in angiosperms. Thus if maize of a certain strain which produces seeds with white endosperm when self-pollinated, is pollinated with pollen from a plant whose seeds have red endosperm, the endosperm of the resulting hybrid seeds is red like that of the pollen parent. No satisfactory explanation of this phenomenon was at hand until the discovery of double fertilization by Nawaschin and Guignard in 1898–9. It then became clear that the endosperm, which was formerly supposed to contain only maternal nuclear material, may show endosperm characters of the parent furnishing the pollen for the reason that the latter contributes a nucleus to the primary endosperm nucleus, so that every endosperm cell contains some nuclear material from the pollen parent.

Normally the endosperm cells are all alike in containing two chromosome sets from the female parent and one set from the male, and the normal inheritance of endosperm characters as well as all ordinary cases of xenia can be understood on this basis. Mottled or mosaic effects in the endosperm of maize hybrids were attributed by Webber (1900) to such abnormal modes of endosperm origin as were referred to in a foregoing paragraph: some of the cells may have been formed by polar nuclei of purely maternal constitution while other cells were the result of the independent division of the second male nucleus. Although this explanation may fit some cases, it is becoming apparent from the work of Emerson and others that most of them can be better accounted for on the basis of aberrant chromosome behavior, such behavior having been observed in certain other organisms.

Additional evidence is found in all these phenomena for the theory that the nuclear substance in some way represents the physical basis of inheritance. The second male nucleus not only is concerned in the initiation of the development of the endosperm, but xenia shows that it also transmits parental characters. Here, therefore, as in the sexual fusion in the egg, the two principal effects of fertilization may be recognized.

Bibliography 12

Fertilization

ALLEN, C. E. 1917. The spermatogenesis of *Polytrichum juniperinum*. Ann. Bot. **31**: 269–292. pls. 15, 16.

ATKINSON, G. F. 1915. Phylogeny and relationships in the ascomycetes. Ann. Mo. Bot. Garden **2**: 315–376. figs. 10.

BALLOWITZ, K. (Many papers on spermatozoa.) See Anat. Anz. **1**, 1886; **27**, 1905; **28**, 1906; Arch. Mikr. Anat. **32**, 1888 (birds); **36**, 1890 (fishes, amphibia); **65**, 1904; **70**, 1907; Zeitschr. Wiss. Zool. **50**, 1890 (insects); **52**, 1891 (mammals); **91**, 1908; Arch. f. Ges. Physiol. **46**, 1890.

BATAILLON, E. 1910. L'embryogénèse complete provoquée chez les amphibiens par piqure de l'oeuf vierge, larves parthenogenetiques de *Rana fusca*. Compt. Rend. Acad. Sci. Paris **150**: 996–998.

VAN BENEDEN, E. 1883. Recherches sur la maturation de l'oeuf, la fécondation et la division cellulaire. Arch. de. Biol. **4**.

1887. (van Beneden et Neyt) Nouvelles recherches sur la fécondation et la division mitosique chez l'Ascaride mégalocéphale. Bull. Acad. Roy. Belg. III **14**: 215–295. 1 pl.

BENSAUDE, M. 1918. Recherches sur la cycle evolutif et la sexualité chez les Basisiomycetes. Diss. Nemours, pp. 156. pls. 13. (Review in Bot. Gaz. **68**: 67–68. 1919. Also in Bot. Absts. **3**: 49. 1920.)

BERRIDGE, E. M. and SANDAY, F. 1907. Oögenesis and embryogeny in *Ephedra distachya*. New Phytol. **6**: 128–134, 167–174. pls. 3, 4.

BLACK, C. A. 1913. The morphology of *Riccia Frostii*, Aust. Ann. Bot. **27**: 511–532 pls. 37, 38.

BLACKMAN, V. H. 1898. On the cytological features of fertilization and related features in *Pinus sylvestris* L. Phil. Trans. Roy. Soc. London B **190**: 395–426. pls 12–14.

1904. On the fertilization, alternation of generations, and general cytology of the Uredineæ. Ann. Bot. **18**: 323–369. pls. 21–24.

BLACKMAN, V. H. and FRASER, H. C. I. 1906. Further studies on the sexuality of the Uredineæ. Ibid. **20**: 35–48. pls. 3, 4.

BLACKMAN, V. H. and WELSFORD, E. J. 1912. The development of the perithecium of *Polystigma rubrum*. Ibid. **26**: 761–767. 2 pls.

1913. Fertilization in *Lilium*. Ibid. **27**: 111–114. pl. 12.

BLAKESLEE, A. F. 1904. Sexual reproduction in the Mucorineæ. Proc. Am. Acad. Arts and Sci. **40**: 205–319.

1920. Sexuality in the Mucors. Science **51**: 375–382, 403–409. figs. 4.

BOVERI, T. 1887a. Ueber die Befruchtung der Eier von *Ascaris megalocephala*. Sitzber. Gesell. Morph. Physiol. München **3**.

1887b. Ueber die Anteil des Spermatozoon an der Teilung des Eies. Ibid.

1888. Zellenstudien. II. Jenaische Zeitschr. **22**: 685–882. pls. 19–23.

1890. Zellenstudien. III. Ibid. **24**: 314–401. pls. 11–13.

1891. Befruchtung. (Review.) Ergebn. Anat. Entw. **1**: 386–485.

1902. Das Problem der Befruchtung. pp. 48. figs. 19. Jena.

BROOKS, F. T. 1910. The development of *Gnomonia erythrostomum* Pers. Ann. Bot. **24**: 585–605. pls. 48, 49.

BROWN, H. B.   1913.   Studies in the development of *Xylaria*.   Ann. Mycol. **11:** 1–13. pls. 1, 2.

BROWN, W. H.   1909.   Nuclear phenomena in *Pyronema confluens*.   J. H. U. Circ. **6:** 42–45.

1910*a*.   The development of the ascocarp of *Leotia*.   Bot. Gaz. **50:** 443–459. figs. 47.

1910*b*.   The exchange of material between nucleus and cytoplasm in *Peperomia Sintenesii*.   Ibid. **49:** 189–194. pl. 13.

1911.   The development of the ascocarp of *Lachnea scutellata*.   Ibid. **52:** 275–305. pl. 9. figs. 51.

CARNOY, J. B. et LEBRUN, H.   1897.   La fécondation chez *l'Ascaris mégalocéphala*. La Cellule **13:** 63–195. pls. 2.

CARRUTHERS, D.   1911.   Contributions to the cytology of *Helvella crispa* Fries. Ann. Bot. **25:** 243–252. pls. 18, 19.

CHAMBERLAIN, C. J.   1899.   Oögenesis in *Pinus Laricio*.   Bot. Gaz. **27:** 268–280. pls. 4–6.

1910.   Fertilization and embryogeny in *Dioon edule*.   Ibid. **50:** 415–429. pls. 14–17.

1912.   Morphology of *Ceratozamia*.   Ibid. **53:** 1–19. pl. 1. figs. 7

1916.   *Stangeria paradoxa*.   Ibid. **61:** 353–372. pls. 24–26. 1 fig.

CHAMBERS, R.   1917.   Microdissection studies.   II.   Jour. Exp. Zool. **23:** 483–504.

CHILD, C. M.   1915.   Senescence and Rejuvenescence.   Chicago.

CHRISTMAN, A. H.   1905.   Sexual reproduction in the rusts.   Bot. Gaz. **39:** 267– 274. pl. 8.

1907.   The alternation of generations and the morphology of the spore forms in the rusts.   Ibid. **44:** 81–101. pl. 7.

CLAUSSEN, P.   1907.   Zur Kenntniss der Kernverhältnisse von *Pyronema confluens*. Ber. Deu. Bot. Ges. **25:** 586–590.

1908.   Ueber Entwicklung und Befruchtung bei *Saprolegnia monoica*.   Ibid. **26:** 144–161.

1912.   Zur   Entwicklungsgeschichte   der   Ascomyceten.   *Pyronema   confluens*. Zeit. f. Bot. **4:** 1–64. pls. 1–6. figs. 13.

CLELAND, R. E.   1919.   The cytology and life history of *Nemalion multifidum*, Ag. Ann. Bot. **33:** 323–352. pls. 22–24. figs. 3.

COKER, W. C.   1903.   On the gametophytes and embryo of *Taxodium*.   Bot. Gaz. **36:** 1–27, 114–140. pls. 1–11.

1907.   Fertilization and embryogeny in *Cephalotaxus Fortunei*.   Ibid. **43:** 1–10. pl. 1. figs. 5.

CONKLIN, E. G.   1897.   The embryology of *Crepidula*.   Jour. Morph. **13:** 1–226. pls. 1–9.

1901.   The individuality of the germ-nuclei during the cleavage of *Crepidula*. Biol. Bull. **2:** 257–265.

COULTER, J. M. and CHAMBERLAIN, C. J.   1903.   Morphology of Angiosperms. 1910.   Morphology of Gymnosperms.   Chicago.

COULTER, J. M. and COULTER, M. C.   1918.   Plant Genetics.   Chicago.

COULTER, J. M. and LAND, W. J. G.   1905.   Gametophytes and embryo of *Torreya taxifolia*.   Bot. Gaz. **39:** 161–178. pls. A, 1–3.

CUTTING, S. M.   1909.   On the sexuality and development of the ascocarp in *Ascophanus carneus* Pers.   Ann. Bot. **23:** 399–417. pl. 28.

DALE, E.   1909.   On the morphology and cytology of *Aspergillus repens*.   Ann. Mycol. **7:** 215–225. pls. 2, 3.

DANGEARD, P. A.   1894*a*.   Mémoire sur la reproduction sexuelle des Basidio- mycètes.   Le Botaniste **4:** 119–181. figs. 24.

1894*b*.   La reproduction sexuelle des Ascomycètes.   Ibid. **4**: 21–58.   figs. 10.

1895.   Second memoire sur la reproduction sexuelle des Ascomycètes.   Ibid. **5**: 245–284.   figs 17.

Davis, B. M.   1896.   The fertilization of *Batrachospermum*.   Ann. Bot. **10**: 49–76. pls. 6, 7.

1900.   The fertilization of *Albugo candida*.   Bot. Gaz. **29**: 297–311.   pl. 22.

1903.   Oögenesis in *Saprolegnia*.   Ibid. **35**: 233–249, 320–349.   pls. 9, 10.

1905.   Fertilization in the Saprolegniales.   Ibid. **39**: 61–64.

Downing, E. R.   1909.   The ovogenesis of *Hydra*.   Zool. Jahrb. (Anat. Abt.) **28**: 295–324.   pls. 11, 12.   2 figs.

Dupler, A. W.   1917.   The gametophytes of *Taxus canadensis* Marsh.   Bot. Gaz. **64**: 115–136.   pls. 11–14.

Eames, A. J.   1913.   The morphology of *Agathis australis*.   Ann. Bot. **27**: 1–38. pls. 1–4.   92 figs.

Ernst, A.   1902.   Chromosomenreduktion, Entwicklung des Embryosackes und Befruchtung bei *Paris quadrifolia* L. und *Trillium grandiflorum* Salisb.   Flora **91**. 1–46.   pls. 1–6.

Farmer, J. B. and Williams, J. L.   1898.   Contributions to our knowledge of the Fucaceæ; their life history and cytology.   Phil. Trans. Roy. Soc. London B **190**: 623–645.   pls. 19–24.

Faull, J. H.   1911.   The cytology of the Laboulbeniales.   Ann. Bot. **25**: 649–654.

1912.   The cytology of *Laboulbenia chætophora* and *L. Gyrinidarum*.   Ibid. **26**: 325–355.   pls. 37–40.

Ferguson, M. C.   1901.   The development of the pollen tube and the division of the generative nucleus in certain species of *Pinus*.   Ann. Bot. **15**: 193–223. pls. 12–14.

1904.   Contributions to the knowledge of the life history of *Pinus*.   Proc. Wash. Acad. Sci. **6**: 1–202.   pls. 1–24.

1913.   Included cytoplasm in fertilization.   Bot. Gaz. **56**: 501–502.

Fitzpatrick, H. M.   1918*a*.   Sexuality in *Rhizina undulata* Fries.   Bot. Gaz. **65**: 201–226   pls. 3 4.

1918*b*.   The cytology of *Eucronartium muscicola*.   Am. Jour. Bot. **5**: 397–419. pls. 30–32.

Focke, W. O.   1881.   Die Pflanzen-Mischlinge.   Berlin.

Fol, H.   1891.   Die "Centrenquadrille," ein neue Episode aus der Befruchtungsgeschichte.   Anat. Anz. **6**: 266–274.   figs. 10.

Fraser, H. C. I.   1907.   On the sexuality and development of the ascocarp of *Lachnea stercorea* Pers.   Ann. Bot. **21**: 349–360.   pls. 29, 30.

1908.   Contributions to the cytology of *Humaria rutilans*.   Ibid. **22**: 35–55. pls. 4, 5.

Fries, R. E.   1911.   Zur Kenntniss der Cytologie von *Hygrophorus conicus*.   Svensk. Bot. Tids. **5**: 241–251.   pl. 1.

Garber, J. F.   1904.   The life history of *Ricciocarpus natans*.   Bot. Gaz. **37**: 161–177.   pls. 9, 10.

Graham, M.   1918.   Centrosomes in fertilization stages of *Preissia quadrata* (Scop.) Nees.   Ann. Bot. **32**: 415–420.   pl. 10.

Guignard, L.   1899.   Sur les antherozoids et la double copulation sexuelle chez les végétaux angiospermes.   Comp. Rend. Acad. Sci. Paris **128**: 864–871.   figs. 19.

1900.   L'appareil sexuel et la double fécondation dans les Tulipes.   Ann. Sci. Nat. Bot. VIII **11**: 365–387.   pls. 9–11.

1901.   La double fécondation chez les Renonculacées.   Jour. Botanique **15**: 394–408.   figs. 16.

1902.   La double fécondation chez les Solanées.   Ibid. **16**: 145–167.   figs. 45.

20

GUILLIERMOND, A. 1910. La sexualite chez les champignons. Bull. Sci. France et Belg. **44**: 109–196.

1920. The Yeasts. (Engl. transl. by F. W. Turner.) N. Y.

GUYER, M. F. 1907. The development of unfertilized frogs' eggs injected with blood. Science N. S. **25**: 910–911.

HAECKER, V. 1895. Ueber die Selbständigkeit der väterlichen und mütterlichen Kernbestandtheile während der Embryonalentwicklung von *Cyclops*. Arch. Mikr. Anat. **46**: 579–617. pls. 28–30.

1899. Praxis und Theorie der Zellen- und Befruchtungslehre.

HARPER, R. A. 1895. Beitrag zur Kenntniss der Kerntheilung und Sporenbildung im Ascus. Ber. Deu. Bot. Ges. **13**: (67)–(78). pl. 27.

1896. Ueber das Verhalten der Kerne bei den Fruchter. ntwicklung einiger Ascomyceten. Jahrb. Wiss. Bot. **29**: 655–685. pls. 11, 12.

1900. Sexual reproduction in *Pyronema confluens* and the morphology of the ascocarp. Ann. Bot. **14**: 321–396. pls. 19–21.

1905. Sexual reproduction and the organization of the nucleus in certain mildews. Carnegie Inst. Publ. 37.

HARVEY, E. N. 1910. Methods of artificial parthenogenesis. Biol. Bull. **18**: 269–180.

HEILBRUNN, L. V. 1915. Studies in artificial parthenogenesis. II. Physical changes in the egg of *Arbacia*. Biol. Bull. **29**: 149–203.

1920. Studies in artificial parthenogenesis. III. Cortical changes and the iniation of maturation in the egg of *Cumingia*. Biol. Bull. **38**: 317–339.

HERLANT, M. 1913. Le méchanisme de la parthenogénèse esperimentelle. Bull. Sci. France et Belg. VII **50**: 381–404.

1917. Étude sur les bases cytologiques du mécanisme de la parthenogénèse experimentelle chez les amphibiens. Arch. d. Biol. **28**: 505–608.

HERTWIG, O. 1875. Beiträge zur Kenntniss der Bildung, Befruchtung, und Theilung des tierischen Eies, I. Morph. Jahrb. **1**.

HIRASÉ, S. 1895. Études sur la fécondation et l'embryogenie du *Ginkgo biloba*. Jour. Imp. Coll. Sci. Tokyo **8**: 307–322. pls. 31, 32.

1898. Études sur la fécondation et l'embryogenie du *Ginkgo biloba*. (Second memoire.) Ibid. **12**: 103–149. pls. 7–9.

1918. Further studies on the fertilization and embryogeny in *Ginkgo biloba*. Bot. Mag. Tokyo **32**: No. 378.

HOYT, W. D. 1910. Physiological aspects of fertilization and hybridization in ferns. Bot. Gaz. **49**: 340–370. figs. 12.

HUTCHINSON, A. H. 1915. Fertilization in *Abies balsamea*. Bot. Gaz. **60**: 457–472. pls. 16–20. fig. 1.

HUXLEY, T. H. 1878. Evolution in Biology.

IKENO, S. 1898. Untersuchungen über die Entwicklung der Geschlechtsorgane und der Vorgang der Befruchtung bei *Cycas revoluta*. Jahrb. Wiss. Bot. **32**: 557–602. pls. 8–10.

1901. Contributions a l'étude de la fécondation chez le *Ginkgo biloba*. Ann. Sci. Nat. Bot. VIII **13**: 305–318. pls. 2, 3.

JONES, W. N. 1918. On the nature of fertilization and sex. New Phytol. **17**: 167–188.

KEENE, M. L. 1914. Cytological studies of the zygospores of *Sporodinia grandis*. Ann. Bot. **28**: 455–470. pls. 25, 26.

KILDAHL, N. J. 1908. The morphology of *Phyllocladus alpina*. Bot. Gaz. **46**: 339–348. pls. 20–22.

KOLTZOFF, N. K. 1906. Studien über die Gestalt der Zelle: I, Untersuchungen über die Spermien der Decapoden. Arch. Mikr. Anat. **67**: 364–572. pls. 25–29. figs. 37.

LAND, W. J. G. 1900. Double fertilization in Compositæ. Bot. Gaz. **30**: 252–260. pls. 15, 16.

1902. A morphological study of *Thuja*. Ibid. **34**: 249–259. pls 6–8.

1907. Fertilization and embryogeny of *Ephedra trifurca*. Ibid. **44**: 273–292. pls. 20–22.

LAWSON, A. A. 1904*a*. The gametophyte, archegonia, fertilization and embryo of *Sequoia sempervirens*. Ann. Bot. **18**: 1–28. pls. 1–4.

1904*b*. The gametophyte, fertilization and embryo of *Cryptomeria japonica*. Ibid. **18**: 417–444. pls. 27–30.

1907. The gametophytes, fertilization and embryo of *Cephalotaxus drupacea*. Ibid. **21**: 1–23. pls. 1–4.

LEVINE, M. 1913. The cytology of Hymenomycetes, especially the Boleti. Bull. Torr. Bot. Club **40**: 137–181. pls. 4–8.

LILLIE, F. R. 1901. The organization of the egg in *Unio* based on a study of its maturation, fertilization and cleavage. Jour. Morph. **17**: 227–292. pls. 24–27.

1912. Studies of fertilization in *Nereis*. III. The morphology of normal fertilization in *Nereis*. IV. The fertilizing power of portions of the spermatozoön. Jour. Exp. Zool. **12**· 413–476. pls. 1–11.

1914. Studies of fertilization. VI. The mechanism of fertilization in *Arbacia*. Ibid. **16**: 523–590.

1919. Problems of Fertilization. Chicago.

LILLIE, R. S. 1908. Momentary elevation of temperature as a means of producing artificial parthenogenesis in starfish eggs and the conditions of its action. Jour. Exp. Zool. **5**: 375–428.

1915. On the conditions of activation of unfertilized starfish eggs under the influence of high temperatures and fatty acid solution. Biol. Bull. **28**. 260–303.

LOEB, J. 1910. Die Hemmung verschiedener Giftwirkungen auf das befruchtete Seeigelei durch Hemmung der Oxidationen in demselben. Biochem. Zeitschr. **29**.

1911. Auf welche Weise rettet die Befruchtung das Leben des Eies? Arch. Entw. **31**: 658–668.

1912. The Mechanistic Conception of Life. Chicago.

1913. Artificial Parthenogenesis and Fertilization. Chicago.

LOEB, J. and BANCROFT, F. W. 1913. The sex of a parthenogenetic tadpole and frog. Jour. Exp. Zool. **14**: 275–277.

LOEB, J. and WASTENEYS, H. 1910. Warum hemmt Natrium cyanide die Giftwirkung einer Chlornatriumlosung für das Seeigelei? Biochem. Zeitsch. **2**.

1911. Sind die Oxidationsvorgänge die unabhängige Variable in den Lebenserscheinungen? Ibid. **36**: 345–356.

1912. Die Oxidationsvorgänge im befruchteten und unbefruchteten Seesternei. Arch. Entw. **35**: 555–557.

LUTHER, A. 1904. Die Eumesostominen. Zeit. Wiss. Zool. **77**: 1–273. pls. 9: figs. 16.

MACCURDY, H. M. 1919. Division, nuclear organization and conjugation in *Arcella vulgaris*. Mich. Acad. Sci. Rep. 21: 111–113.

MACORMICK, F. A. 1912. Development of zygospore in *Rhizopus nigricans*. Bot. Gaz. **53**: 67–68.

McCLENDON, J. J. 1912. Dynamics of cell division. Artificial parthenogenesis in Vertebrates. Am. Jour. Physiol. **29**: 298–301.

McCUBBIN, W. A. 1910. Development of the Helvellineæ. Bot. Gaz. **49**: 195–206. pls. 14–16

MEVES, FR. 1899. Ueber Struktur und Histogenese der Samenfaden des Meerschweinchens. Arch. Mikr. Anat. **54**: 329–402. pls. 19–21. figs. 16.

1908.   Die Chondriosomen als Träger erblicher Anlagen. Cytologische Studien am Hühnerembryo. Ibid **72** : 816–867. pls. 39–42.

1915.   Ueber Mitwirkung der Plastosomen bei der Befruchtung des Eies von *Filaria papillosa*. Ibid. **37** : 11 12–46. pls. 1–4.

1918.   Die Plastosomentheorie der Vererbung. Ibid. **92** : II 41–136. figs. 18. (Bibliography).

Meyer, K. 1911. Untersuchungen über die Sporophyt der Lebermoose. 1. Entwicklungsgeschichte des Sporogons der *Corsinia marchantioides*. Bull. Soc. Imp. Moscow 236–286.

Minchin, E. A. 1912. An Introduction to the Study of the Protozoa. London.

Miyake, K. 1901. The fertilization of *Pythium deBaryanum*. Ann. Bot. **15** : 653–667. pl. 36.

1910.   The development of the gametophytes and embryogeny in *Cunninghamia sinensis*. Beih. Bot. Centr. **27** : 1–25. pls. 5.

Morgan, T. H. 1899. The action of salt solutions on the unfertilized and fertilized eggs of *Arbacia*, and of other animals. Arch. Entw. **8** : 448–539. pls. 7–10. figs. 21.

1913.   Heredity and Sex. New York.

Mottier, D. M. 1898. Ueber das Verhalten der Kerne bei der Entwicklung des Embryosacks und die Vorgänge bei der Befruchtung. Jahrb. Wiss. Bot. **31** : 125–158 pls. 2, 3.

1904.   Fecundation in Plants. Carnegie Inst. Publ. 15.

Murrill, W. A. 1900. The development of the archegonium and fertilization in the hemlock spruce (*Tsuga canadensis*, Carr.) Ann. Bot. **14** : 583–607. pls. 31, 32.

Nägler, K. 1909. Entwicklungsgeschichtliche Studien über Amöben. Arch. f. Protist. **15** : 1–

Nawaschin, S. 1899. Neue Beobachtungen über Befruchtung bei *Fritillaria* und *Lilium*. Bot. Centr. **77** : 62. (Russian account in 1898.)

1909.   Ueber das selbständige Bewegungsvermögen der Spermakerne bei einigen Angiospermen. Œsterreich. Bot. Zeitschr. **59** : 457.

1910.   Näheres über die Bildung der Spermakerne bei *Lilium Martagon*. Ann. Jard. Bot. Buit. **12** : Suppl. III, 871–904. pls. 33, 34.

Němec, B. 1910. Das Problem der Befruchtungsvorgänge. Jena.

1912.   Ueber die Befruchtung bei *Gagea*. Bull. Internat. Acad. Sci. Boheme, 1–17. figs. 19.

Nichols, G. E. 1910. A morphological study of *Juniperus communis* var. *depressa*. Beih. Bot. Centr. **25** : 201–241. pls. 8–17. figs. 4.

Nienburg, W. 1914. Zur Entwicklungsgeschichte von *Polystigma rubrum* DC. Zeit. f. Bot. **6** : 369–400. figs. 17.

Nothnagel, M. 1918. Fecundation and formation of the primary endosperm nucleus in certain Liliaceæ. Bot. Gaz. **66** : 143–161. pls. 3–5.

Norén, C. O. 1904. Ueber Befruchtung bei *Juniperus communis*. (Vorlauf. Mitt.) Arkiv. Bot. Svensk. Vet.-Akad. **3** : pp. 11.

1907.   Zur Entwicklungsgeschichte des *Juniperus communis*. Uppsala Univ. Arsskrift, pp. 64. pls. 4.

Osterhout, W. J. V. 1900. Befruchtung bei *Batrachospermum*. Flora **87** : 109–115. pl. 5.

Overton, J. B. 1913. Artificial parthenogenesis in *Fucus*. Science **37** : 841, 844.

Pace, L. 1909. The gametophytes of *Calopogon*. Bot. Gaz. **48** : 126–137. pls. 7–9.

Rabl, C. 1889. Ueber Zelltheilung. Anat. Anz. **4** : 21–30. figs. 2.

RAMLOW, G. 1914. Beiträge zur Entwicklungsgeschichte der Ascoboleen. Mycol. Centralbl. **5**: 177–198. pls. 1, 2. figs. 20.

ROBERTSON, A. 1904. Studies in the morphology of *Torreya californica*. II. The sexual organs and fertilization. New Phytol. **3**: 205–216. pls. 7–9.

ROMIEU, M. 1911. La spermiogénèse chez *l'Ascaris mégalocéphala*. Arch. Zellf. **6**: 254–325. pls. 14–17.

RÜCKERT, J. 1895. Ueber die Selbstandigbleiben der väterlichen und mütterlichen Kernsubstanz während der ersten Entwicklung des befruchteten *Cyclops*-Eies. Arch. Mikr. Anat. **45**: 339–369. pls. 21, 22.

SARGANT, E. 1896. The formation of the sexual nuclei in *Lilium Martagon*. I. Oögenesis. Ann. Bot. **10**: 445–477. pls. 22, 23.

——— 1897. Same title. II. Spermatogenesis. Ibid. **11**: 187–224. pls. 12, 13.

SAWYER, M. L. 1917. Pollen tube and spermatogenesis in *Iris*. Bot. Gaz. **64**: 159–164. figs. 18.

SAX, K. 1916. Fertilization in *Fritillaria pudica*. Bull. Torr. Bot. Club **43**: 505-522. pls. 27–29.

——— 1918. The behavior of the chromosomes in fertilization. Genetics **3**. 309–327. pls. 1, 2.

SCHAUDINN, F. 1896. Ueber die Copulation von *Actinophrys sol*. Sitzber. Acad. Wiss. Berlin.

SCHEBEN, L. 1905. Beiträge zur Kenntniss des Spermatozoons von *Ascaris megalo-cephala*. Zeit. Wiss. Zool. **79**: 396–431. pls. 20, 21. 3 figs.

SCHIKORRA, W. 1910. Ueber die Entwicklungsgeschichte von *Monascus*. Zeitschr. f. Bot. **1**: 379–410.

SHARP, L. W. 1912. Spermatogenesis in *Equisetum*. Bot. Gaz. **54**: 89–119. pls. 7, 8.

——— 1914. Spermatogenesis in *Marsilia*. Ibid. **58**: 419–431. pls. 33, 34.

——— 1920. Spermatogenesis in *Blasia*. Ibid. **69**: 258–268. pl. 15.

SMITH, B. G. 1919. The individuality of the germ-nuclei during the cleavage of the egg of *Cryptobranchus alleghaniensis*. Biol. Bull. **37**. 246–287.

STEIL, W. N. 1918. Method for staining antherozoid of fern. Bot. Gaz. **65** 562–563. 1 fig.

STEVENS, F. L. 1899. The compound oosphere of *Albugo bliti*. Bot. Gaz. **28** 149–176. pls. 11–15.

——— 1901. Gametogenesis and fertilization in *Albugo*. Ibid. **32**: 77–98. pls. 1–4.

STRASBURGER, E. 1877. Ueber Befruchtung und Zelltheilung. Jen. Zeitschr. **11**.

——— 1884. Neue Untersuchungen über die Befruchtungsvorgang bei den Phanero-gamen, als Grundlage für eine Theorie der Zeugung. Jena.

——— 1892. Schwärmsporen, Gameten, pflanzlichen Spermatozoiden, und das Wesen der Befruchtung. Histol. Beitr. **4**: 49–158. pl. 3.

——— 1897. Kerntheilung und Befruchtung bei *Fucus*. Jahrb. Wiss. Bot. **30**: 351–374. pls. 27, 28.

——— 1900. Einige Bemerkungen zur Frage nach der "doppelten Befruchtung" bei den Angiospermen. Bot. Zeit. **58**: 293–316.

——— 1901. Ueber Befruchtung. Ibid. **59**: II, 353–368.

TROW, A. H. 1895. The karyology of *Saprolegnia*. Ann. Bot. **9**: 609–652. pls. 24, 25.

——— 1899. Observations on the biology and cytology of a new variety of *Achlya ameri-cana*. Ibid. **13**: 131–179. pls. 8–10.

——— 1901. Biology and cytology of *Pythium ultimum*, n. sp. Ann. Bot. **15**: 269–312. pls. 15, 16.

——— 1904. On fertilization in the Saprolegniaceae. Ibid. **18**: 541–569. pls. 34–36.

WAGER, H. 1896. On the structure and reproduction of *Cystopus candidus* Lev.
     Ann. Bot. **10**: 295–342. pls. 15,16. See also pp. 89–91.
     1899. The sexuality of fungi. Ibid. **13**: 575–597.
WALDEYER, W. 1888. Ueber Karyokinese und ihre Beziehung zu den Befrucht-
     ungsvorgängen. Arch. Mikr. Anat. **32**: 1–122. figs. 14. (Engl. transl. in
     Quar. Jour. Micr. Sci. **30**: 159–281. pl. 14. 1889.)
WALTON, A. C. 1918. The oögenesis and early embryology of *Ascaris canis* Werner
     Jour. Morph. **30**: 527–604. pls. 9. fig. 1.
WARBURG, O. 1908. Beobachtungen über die Oxidationsprozesse im Seeigelei.
     Zeitschr. Physiol. Chem. **57**.
     1910. Ueber die Oxidationen im lebenden Zellen nach Versuchen am Seeigelei.
     Ibid. **66**.
     1911. Untersuchungen über die Oxidationsprozesse im Zellen. Münchener Med.
     Wochenschr. **57**.
     1914. Beiträge zur Physiologie der Zelle, inbesondere über die Oxidationsgesch-
     windigkeit in Zellen. Ergeb. d. Physiol. **14**.
WEBBER, H. J. 1900. Xenia, or the immediate effect of pollen in maize. U. S.
     Dept. Agr., Div. Veg. Path. and Physiol., Bull. **22**: pls. 4.
     1901. Spermatogenesis and fecundation in *Zamia*. U. S. Dept. Agr. Bur. Plt.
     Ind. Bull. **2**. pp. 100. pls. 7.
WELSFORD, E. J. 1907. Fertilization in *Ascobolus furfuraceus*. New Phytol. **6**:
     156.
     1914. The genesis of the male nuclei in *Lilium*. Ann. Bot. **28**: 265–270. pls. 16,
     17.
WENIGER, W. 1918. Fertilization in *Lilium*. Bot. Gaz. **66**: 259–268. pls. 11–13.
WHEELER, W. M. 1895. The behavior of the centrosomes in the fertilized egg of
     Myzostoma glabrum Leukart. Jour. Morph. **10**: 305–311. figs. 10.
     1897. The maturation, fecundation and early cleavage of *Myzostoma glabrum*
     Leukart. Arch. d. Biol. **15**: 1–77. pls. 1–3.
WILDMAN, E. E. 1913. The spermatogenesis of *Ascaris megalocephala* with special
     reference to the two cytoplasmic inclusions, the refractive body and the "mito-
     chondria": their origin, nature and rôle in fertilization. Jour. Morph. **24**:
     421–457. pls. 3.
WILSON, E. B. 1900. The Cell in Development and Inheritance. 2d ed.
     1901. Experimental studies in cytology. I. A cytological study of artificial
     parthenogenesis in sea urchin eggs. Arch. Entw. **12**: 529–596. pls. 11–17.
     figs. 12.
WINGE, O. 1914. The pollination and fertilization process in *Humulus lupulus* L.
     and *H. Japonicus*. Comp. Rend. Trav. Lab. Carlsberg **11**.
WOODBURN, W. L. 1920. Preliminary notes on the embryology of *Reboulia hemis-
     phærica*. Bull. Torr. Bot. Club. **46**: 461–464. pl. 19.
WOYCICKI, Z. 1899. (On fertilization in Coniferæ.) pp. 57. pls. 2. (Russian.)
YAMANOUCHI, S. 1906. The life history of *Polysiphonia violacea*. Bot. Gaz. **42**:
     401–449. pls. 19–28.
     1908. Spermatogenesis, oögenesis, and fertilization in *Nephrodium*. Ibid. **45**:
     145–175. pls. 6–8.

CHAPTER XIII

APOGAMY, APOSPORY, AND PARTHENOGENESIS

APOGAMY AND APOSPORY

The life cycle in all bryophytes and vascular plants is characterized by a regular alternation of two well marked phases or generations: the *gametophyte*, which arises from the spore and produces gametes; and the *sporophyte*, which arises from the fusion product of two gametes and produces spores. In such a normal life cycle the number of chromosomes in the nuclei is doubled at the union of the gametes and reduced to the original number at sporogenesis; the gametophyte is therefore the haploid generation and the sporophyte the diploid generation, their limits being marked by the two cytological crises, fertilization and reduction. Such an alternation of haploid and diploid phases has been discovered in the life cycles of many algæ and fungi also, so that the general conception of alternation of generations has been extended to these lower groups. This, however, is not the place for a discussion of the homologies implied. It should be added that gametophyte and sporophyte may arise not only from each other, but either generation may also multiply by vegetative means.

Many instances in which the above typical life cycle is departed from, and in which the correlation between the alternation of two generations and periodic changes in chromosome number is broken, are now known, the conspicuous examples being found among the ferns and certain angiosperms. The very convenient classification of such abnormalities drawn up by Vines (1911) is given as the basis for the present portion of the chapter. All dates and the matter included within square brackets have been added by the present author.

"In the first place, the sporophyte may be developed either after an abnormal sexual act, or without any preceding sexual act at all, a condition known as *apogamy*. In the second, the gametophyte may be developed otherwise than from a post-meiotic spore, a condition known as *apospory*.[1]

---

[1] [Apogamy in ferns was discovered by Farlow in 1874. Apospory was discovered in mosses by Pringsheim in 1876 and in ferns by Druery in 1884. General discussions of these phenomena are given by Winkler (1908) and Strasburger (1909b).]

Apogamy.—The cases to be considered under this head may be arranged in two groups:

1. *Pseudapogamy: sexual act abnormal.*—The following abnormalities have been observed:

(*a*) Fusion of two female organs: observed (Christman 1905) in certain Uredineæ (*Cæoma nitens, Phragmidium speciosum, Uromyces Caladii*) where adjacent archicarps fuse: male cells (spermatia) are present but functionless.

(*b*) Fusion between nuclei of the same female organ: observed in the ascogonium of certain ascomycetes, *Humaria granulata* (Black-

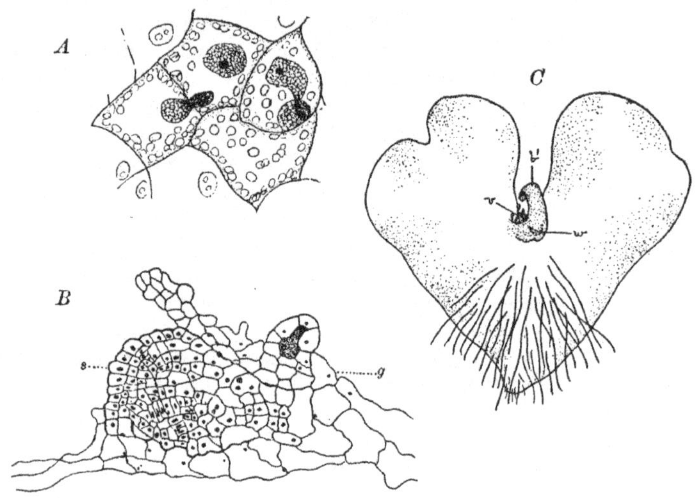

Fig. 124.—Apogamy in ferns.

A, nuclear migration in gametophyte cells of *Lastræa pseudo-mas* var. *polydactyla.* × 500. (*After Farmer and Digby,* 1907.) *B*, section through gametophyte, showing young sporophytic tissue (*s*) "engrafted" into surrounding gametophytic tissue (*g*). (*After Farmer and Digby.*) *C*, sporophyte arising apogamously from gametophyte in *Pteris cretica*: *b*[1], first leaf; *v*, stem apex; *w*, root. (*After de Bary.*)

man 1906), where there is no male organ; *Lachnea stercorea* (Fraser 1907), where the male organ (pollinodium) is present but apparently functionless. [A similar condition has been reported in *Ascobolus furfuraceus* (Welsford 1907), *Aspergillus repens* (Dale 1909), and *Ascophanus carneus* (Cutting 1909).]

(*c*) Fusion of a female organ with an adjacent tissue-cell: observed (Blackman 1904*b*) [Blackman and Fraser 1906] in the archicarp of some Uredineæ (*Phragmidium violaceum, Uromyces Poæ, Puccinia Poarum*): male cells (spermatia) present but functionless.

(*d*) There is no female organ: fusion takes place between two adjacent tissue-cells of the gametophyte; the sporophyte is developed from diploid cells ["grafted tissue"] thus produced, but there is no proper zygote as there is in *a*, *b*, and *c*: observed (Farmer [and Digby] 1907) in the prothallium of certain ferns (*Lastraea pseudo-mas*, var. *polydactyla*) [Fig. 124, *A*]: male organs (and sometimes female) present but functionless. Another such case is that of *Humaria rutilans* (ascomycete), in which nuclear fusion

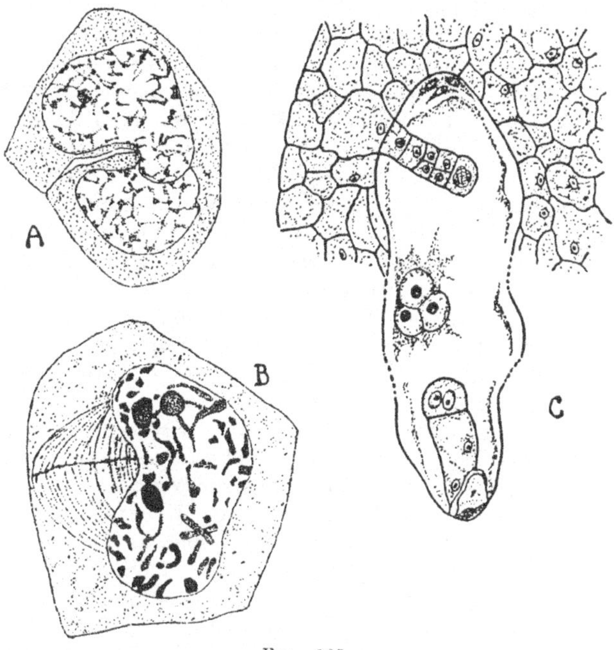

FIG. 125.

A, cell fusion in the sporangium of *Aspidium falcatum*. × 1950. (*After R. F. Allen* 1911.) *B*, incomplete nuclear division in sporangium of *Nephrodium hirtipes*. × 1250. (*After Steil*, 1919.) *C*, apogamy and sporophytic budding in the embryo sac of *Alchemilla pastoralis*: egg developing apogamously below; cell of nucellus forming an embryo above; two polar nuclei and one synergid nucleus at center. (*After Murbeck*, 1902.)

has been observed (Fraser 1908) in hyphæ of the hypothecium: the asci are developed from these hyphæ, and in them meiosis takes place; there are no sexual organs. [A similar condition has been reported in *Helvella crispa* (Carruthers 1911) and *Polystigma rubrum* (Blackman and Welsford 1912). It has already been pointed out (p. 291) that many students of the ascomycetes deny the existence of a nuclear fusion in the archicarp or vegetative cells, holding rather that the only

fusion in the life cycle is that observed in the ascus, and that
this fusion is the real sexual act.]

[(*e*) Fusion of two haploid sporocytes: In *Aspidium falcatum* (R. F.
Allen 1911) a haploid sporophyte arises by vegetative apogamy
from a haploid gametophyte. In the sporangium the 16
haploid sporocytes fuse in pairs, producing eight diploid cells
(Fig. 125, *A*). In these cells reduction occurs, 32 haploid
spores resulting.]

2. *Eu-apogamy:* no kind of sexual act.
   (*a*) The gametophyte is haploid:
      (α) The sporophyte is developed from the unfertilized haploid
oösphere: no such case of *true parthenogenesis* has yet been
observed. [Kusano (1915) has observed the division of
the haploid nucleus of an unfertilized egg in a few exceptional cases in the orchid, *Gastrodia elata*. Parthenogenetic
development proceeds no further. The unfertilized egg of
Fucus has been made to begin development by artificial
means (Overton 1913), but the cytological facts are not
known here. Motile gametes of certain other algæ have
been observed to develop without conjugation, as in
Ectocarpus tomentosus (Kylin 1918).]

      (β) The sporophyte is developed vegetatively from the gameto-
phyte and is haploid: observed in the prothallia of certain
ferns, *Lastræa pseudo-mas*, var. *cristata-apospora* (Farmer
and Digby 1907), and *Nephrodium molle* (Yamanouchi
1908). [In the gametophytes of *Nephrodium molle*, which
has antheridia but no functional archegonia, Yamanouchi
found no nuclear migrations such as Farmer described
in *Lastræa* (see 1*d*); but there was haploid grafted tissue,
from which a haploid sporophyte developed. In *Nephrodium hirtipes* (Steil 1919) a haploid sporophyte arises by
vegetative apogamy from a haploid gametophyte. When
there are eight sporogenous cells in the sporangium there
is an incomplete nuclear and cell division (Fig. 125, *B*),
each nucleus coming to have the diploid number of chromosomes. These eight diploid cells function as sporocytes
and produce 32 haploid spores. Steil at first (1915)
adopted Allen's interpretation (1*e*) for his material, but
later decided that the phenomenon observed was one of incomplete division, and not one of fusion. In this case,
as in *Aspidium falcatum*, apogamy is offset not by apospory
but by an abnormal course of events in the sporangium.
In *Aspidium falcatum* the sporophyte arises as in the
examples mentioned in this paragraph, but because of

the presence of a cell and nuclear fusion it is classified under
1c.]

(*b*) The gametophyte is diploid (see under *Apospory*):

(*α*) The sporophyte is developed from the diploid oösphere:
observed in some Pteridophyta, viz. certain ferns (Farmer
1907), *Athyrium Filix-fœmina*, var. *clarissima*, *Scolopendrium vulgare*, var. *crispum-Drummondæ*, and *Marsilia*
(Strasburger 1907); also in some Phanerogams, viz.,
Compositæ (*Taraxacum*, Murbeck 1904; *Antennaria alpina*,
Juel 1898, 1900; sp. of *Hieracium*, Rosenberg 1906):
Rosaceæ (*Eu-Alchemilla* sp., Murbeck 1901, 1904, Strasburger 1905 [Fig. 125, *C*]): Ranunculaceæ (*Thalictrum
purpurascens*, Overton 1902). [Also in the lily, *Atamosco*
(Pace 1913), and *Burmannia* (Ernst 1909). Besides this
form of apogamy ("oöapogamy" or "generative apogamy") *Antennaria* may also develop embryos from
diploid synergids ("vegetative apogamy") and from cells
of the nucellus ("sporophytic budding"). A similar
variety of embryo origins is found in certain other angiosperms. In many cases the chromosome number in
apogamous species is about twice as large as that of nearly
related forms reproducing sexually (Rosenberg 1909).]

(*β*) The sporophyte is developed vegetatively from the gametophyte: observed (Farmer [and Digby] 1907) in the fern
Athyrium Filix-fœmina, var. *clarissima*.

In all cases enumerated under *Eu-apogamy*, apogamy is
associated with some form of apospory except *Nephrodium
molle*, full details of which have not yet been published.
[It is possible that a behavior like that in *Aspidium
falcatum* (1e) or in *Nephrodium hirtipes* (2aβ) may occur
in *Nephrodium molle*.] Many other ferns are known to be
apogamous, but they are not included here because the
details of their nuclear structure have not been investigated.

Apospory.—The known modes of apospory may be arranged as
follows:

1. *Pseudapospory: a spore is formed but without meiosis, so that it is diploid*
   —observed only in heterosporous plants, viz. certain species
   of *Marsilia* (*e.g. Marsilia Drummondii*) where the megaspore has a
   diploid nucleus (32 chromosomes) and the resulting prothallium and
   female organs are also diploid (Strasburger 1907); and in various
   Phanerogams, some Compositæ (*Taraxacum* and *Antennaria alpina*,
   Juel 1898, 1900, 1904), some Rosaceæ (*Eu-Alchemilla*, Strasburger
   1905), and occasionally in *Thalictrum purpurascens* (Overton 1902),
   where the megaspore ([and] embryo-sac) is diploid; in some species

of *Hieracium* it has been found (Rosenberg 1906) that adventitious diploid embryo-sacs are developed in the nucellus: these plants are also apogamous. [In *Marsilia Drummondii*, which Shaw (1897) and Nathansohn (1909) had shown to be apogamous, Strasburger (1907) found that, although normal reduction occurs in some of the megasporocytes, giving spores with 16 chromosomes, other megasporocytes undergo two divisions neither of which is reductional: the first division is homœotypic in character and the second is an additional vegetative mitosis without a homologue

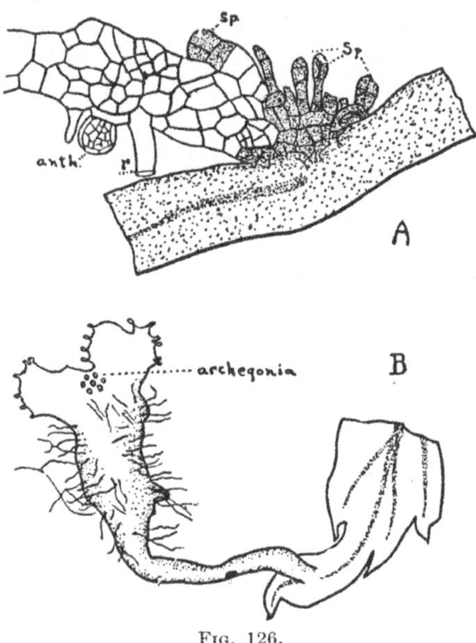

Fig. 126.

A, gametophyte with antheridium (*anth.*) and rhizoids (*r*) arising aposporously from tissue of sorus in *Polystichum angulare* var. *pulcherrimum; sp,* sporangia.   × 70.   (*After Bower.*)   *B*, gametophyte with archegonia arising from tip of pinnule in *Polystichum.* × 10.   (*After Bower.*)

in the normal cases. The resulting spores are therefore diploid, and oöapogamy follows.].

2   *Eu-apospory: no spore is formed*—of this there are two varieties:

   (a) With meiosis: this occurs in some Thallophyta which form no spores; the sporophyte of the Fucaceæ bears no spores, consequently meiosis takes place in the developing sexual organs. The Conjugate Green Algæ also have no spores, meiosis taking place in the germinating zygospore which develops directly into the sexual plant.

(*b*) Without meiosis: the gametophyte is developed upon the sporophyte by budding; that is, spore-reproduction is replaced by a vegetative process: for instance, in mosses it has been found possible to induce the development of protonema, the first stage of the gametophyte, from tissue cells of the sporogonium: [In this way Él. and Ém. Marchal (1909, 1912) were able to produce in *Mnium*, *Bryum*, *Phascum*, and *Amblystegium* diploid gametophytes; these in turn produced tetraploid sporophytes which bore diploid spores. In one case (*Amblystegium*) a tetraploid gametophyte was regenerated from cells of the tetraploid sporophyte.] Similarly, in certain ferns (varieties of *Athyrium Filix-fœmina*, *Scolopendrium vulgare*, *Lastrœa pseudo-mas*, *Polystichum angulare*, and in the species *Pteris aquilina* and *Asplenium dimorphum*), the gametophyte (prothallium) is developed by budding of the leaf of the sporophyte [commonly from the margin of the leaf or from the tissue of the sorus (Fig. 126)], and in some of these cases it has been ascertained that the gametophyte so developed has the same number (2x) of chromosomes in its nuclei as the sporophyte that bears it—that is, it is diploid.

Apospory has been found to be associated frequently with apogamy [in the life cycle]; in fact, in the absence of meiosis, this association would appear to be inevitable."

PARTHENOGENESIS IN ANIMALS[1]

The natural development of an egg without having been fertilized by a male gamete is a phenomenon which is apparently of much more frequent occurrence in animals than in plants. The best known examples are found among the rotifers, crustaceans, and insects, parthenogenesis being the regular mode of reproduction in some species. Other modes also usually occur in such organisms under certain conditions or after a certain number of generations. Parthenogenesis is reported in some protozoa (*Plasmodium vivax*, Schaudinn 1902), where the macrogamete, after certain nuclear changes, continues the life cycle without fusing with a microgamete. Moreover, as has already been described in the preceding chapter, parthenogenesis may be artificially induced in the eggs of other animal groups, notably echinoderms, mollusks, and amphibians, and around this fact centers much of the significant work of modern experimental biology. In commenting upon parthenogenetic development Minchin (1912, p. 137) points out that " . . . the gamete which has this power is always the female; but this limitation receives an explanation from the extreme reduction of the body of the male gamete and its

[1] The cytological results of researches on maturation and development in cases of parthenogenesis have recently been summarized by Paula Hertwig (1920).

feeble trophic powers, rendering it quite unfitted for independent reproduction, rather than from any inherent difference between the two sexes in relation to reproductive activity."

Many normally parthenogenetic animal eggs are known to have the diploid chromosome number as the result of a failure of reduction, a condition paralleling that known as oöapogamy in plants. On the contrary, there are some which, unlike any known vascular plant, are haploid, reduction having taken place in the normal fashion. Parthenogenesis is often associated with certain irregularities in the behavior of the polar bodies, as will be noted in the following descriptions of some well known examples. In the majority of recorded cases the parthenogenetic egg produces but one polar body; in some, however, two are formed as in all zygogenetic eggs (those developing after having been fertilized).

It was long ago noticed by Blochmann (1888; see Wilson 1900, pp. 281–4) that in *Aphis* both zygogenetic and parthenogenetic eggs are produced; the former produce the usual two polar bodies while the latter have but one. It was also seen that the polar bodies are not budded off as separate cells, but remain within the membrane of the egg. Weismann (1886, 1887), working on rotifers, concluded that the second polar body has something to do with parthenogenetic development; and Boveri (1887*d*, 1890), who had seen the chromosomes of the second polar body transform themselves into a nucleus in the egg of *Ascaris*, made the suggestion that this second polar body might unite with the egg nucleus and so initiate development. Brauer (1894) announced that this is precisely what occurs in *Artemia*, a phyllopod crustacean. In this organism two types of parthenogenesis are found. In some cases the nucleus of the second polar body, with 84 chromosomes, actually does unite with the egg nucleus, likewise with 84, causing "fertilization" and the resulting development of an individual with the diploid number (168) of chromosomes. In other cases only one polar body is produced, but reduction is accomplished in the division forming it, and the resulting haploid egg develops parthenogenetically into an individual with only 84 chromosomes.

In *Phylloxera caryæcaulis* (Morgan 1906, 1908, 1909, 1910, 1915) only one polar body appears, but here no reduction occurs: the diploid egg develops parthenogenetically. In *Nematus lacteus* (Doncaster 1906) two polar bodies are produced, but reduction fails and the diploid egg proceeds to develop as in *Phylloxera*.

It has long been known that the eggs of the honey bee, *Apis mellifica*, will develop either zygogenetically into females or parthenogenetically into males. It has been shown in both cases that there are two polar bodies (Blochmann) and that a normal reduction in the number of chromosomes occurs (Nachtsheim 1912, 1913). The fertilized eggs develop into workers or into queens with the diploid number (32) of chromosomes; those not fertilized develop into drones with the haploid number (16). (At the time of spermatogenesis in the drone no further reduction in chromosome number occurs: the spermatozoa retain the number present in the body cells (16).)

In the gall-fly, *Neuroterus lenticularis*, Doncaster (1910–1911) has shown that there are two classes of parthenogenetic females. The egg of the first class gives off no polar bodies, retains the diploid number (20) of chromosomes, and develops parthenogenetically into a sexual female. The egg of the second class gives off two polar bodies, retains the reduced number (10) of chromosomes, and develops parthenogenetically into a male. (The offspring of the sexual females and males constitute the next generation of parthenogenetic females.)

There are thus several organisms in which both zygogenetic and parthenogenetic eggs are produced. In some of them, such as the bee, in which the same egg can develop in either way, the two classes of eggs show no morphological differences. In other forms, such as a species of *Melanoxanthus* (a plant louse) and *Sida crystallina* (crustacean), they may differ considerably. The parthenogenetic egg, for example, may contain much less yolk than the zygogenetic one: it is less highly differentiated, and "still retains the capacity to initiate dedifferentiation and reconstitution independently of union with a male gamete. In this respect it resembles the less highly specialized cells of other tissues rather than the gametes" (Child 1915, p. 408).

It has recently been shown that frogs which have been induced to develop parthenogenetically from punctured eggs (Bataillon's method) are of both sexes (Loeb 1921). The chromosome number in the females has not been determined, but both Parmenter (1920) and Goldschmidt (1920) report the diploid number in males so derived. The origin of this diploid condition has not been satisfactorily explained. Parmenter suggests that it may be due to the retention of one polar body, or to a premature division of the chromosomes without cytokinesis just before the first cleavage. This promises to be an interesting case in connection with the mechanism of sex-determination.

Conclusion.—To review the various theories which have been advanced to account for the origin of parthenogenesis, its relation to other forms of reproduction, and its significance in the life history, is a task which lies beyond the scope of the present work: it has been our purpose only to indicate some of the outstanding cytological facts in certain conspicuous instances of the phenomenon. The cytological features have been accurately ascertained in only a very few cases, and these show little agreement. Furthermore, it is in artificially induced rather than in natural parthenogenesis that the physiological conditions are best known. In view of these facts it appears more than probable that many more cytological and physico-chemical data must be secured before any theory adequately harmonizing all the observed phenomena of parthenogenesis can be formulated.

Bibliography 13

Apogamy; Apospory; Parthenogenesis

(For papers of Bataillon, Harvey, Herlant, McClendon, F. R. Lillie, R. S. Lillie, Loeb, and others on artificial parthenogenesis see Bibl. 12; also for papers of Blackman, Carruthers, Cutting, Dale, Fraser, and Welsford on ascomycetes.)

ALLEN, R. F. 1911. Studies in spermatogenesis and apogamy in ferns. Trans. Wis. Acad. Sci. **17**: 1–56. pls. 1–6.

BLACKMAN, V. H. 1904*a*. On the relation of fertilization to "apogamy" and "parthenogenesis." New Phytol. **3**: 149–158.

1904*b*. On the fertilization, alternation of generations, and general cytology of the Uredineæ. Ann. Bot. **18**: 323–369. pls. 21–24.

1906. On the sexuality and development of the ascocarp of *Humaria granulata* Quel. Proc. Roy. Soc. London Bot. **77**: 354–368. pls. 13–15.

BLACKMAN, V. H. and FRASER, H. C. I. 1906. Further studies on the sexuality of the Uredineæ. Ann. Bot. **20**: 35–48. pls. 3, 4.

BLOCHMANN, F. 1884. Ueber eine Metamorphose der Kerne in den Ovarialeiren und über den Beginn der Blastodermbildung bei den Ameisen. Verh. Nat.-Med. Verein Heidelberg. **3**: 243–246.

1886. Ueber die Eireifung bei Insekten. Biol. Cent. **6**: 554–559.

1887. Ueber die Richtungskörper bei Insekteiren. Morph. Jahrb. **12**: 544–574. pls. 26, 27. See also Biol. Cent. **7**: 108–111.

BOVERI, T. 1887. Zellen-Studien. I. Die Bildung der Richtungskörper bei *Ascaris megalocephala* und *Ascaris lumbricoides*. Jen. Zeitsch. **21**: 423–515. pls. 25–28.

1890. Zellen-Studien. II. Ueber das Verhalten der chromatischen Kernsubstanz bei der Bildung der Richtungskörper und bei der Befruchtung. Ibid. **24**: 314–401. pls. 11–13.

BRAUER, A. 1894. Zur Kenntniss der Reifung des parthenogenetisch sich entwickelenden Eies von *Artemia salina*. Arch. Mikr. Anat. **43**: 162–222. pls. 8–11.

BROWN, E. D. W. 1919. Apogamy in *Camptosorus rhizophyllus*. Bull. Torr. Bot. Club **46**: 27–30. pl. 2.

CHILD, C. M. 1915. Senescence and Rejuvenescence. Chicago.

CHRISTMAN, A. H. 1905. Sexual reproduction in the rusts. Bot. Gaz. **39**: 267–274. pl. 8.

DONCASTER, L. 1907. Gametogenesis and fertilization in *Nematus ribesii*. Quar. Jour. Micr. Sci. **51**: 101–114. pl. 8.

1908. Artificial Parthenogenesis. Sci. Progress.

1910–1911. Gametogenesis of the gall fly, *Neuroterus lenticularis* (*Spathegaster baccarum*). Proc. Roy. Soc. London B **82**: 88–113. pls. 1–3; **83**: 476–489. pl. 17.

DRUERY, C. T. 1884. Observations on a singular mode of development in the lady-fern (*Athyrium Filix-fœmina*). Jour. Linn. Soc. Bot. **21**: 354–357. Further studies on a singular mode of reproduction in *Athyrium Filix-fœmina*. Ibid. 358–360. 2 figs.

ERNST, A. 1909. Apogamie bei *Burmannia cœlistris* Don. Ber. Deu. Bot. Ges. **27**: 157–168. pl. 7.

FARLOW, W. G. 1874. An asexual growth from the prothallium of *Pteris cretica*. Quar. Jour. Micr. Sci. **14**: 266–272. pls. 10, 11. See also Bot. Zeit. **32**: 180–183.

FARMER, J. B., MOORE, J. E. B., and DIGBY, L. 1903. On the cytology of apogamy and apospory. I. Proc. Roy. Soc. London **71**: 453–457. figs. 4.

FARMER, J. B. and DIGBY, L. 1907. Studies in apospory and apogamy in ferns. Ann. Bot. **21**: 161–199. pls. 16–20.

GOLDSCHMIDT, R. 1920. Arch. Zellf. **15**: 283.

HERTWIG, P. 1920. Haploide und diploide Parthenogenese. Biol. Zentralbl. **40**: 145–174.

JUEL, H. O. 1898. Parthenogenesis bei *Antennaria alpina*. (L.) R. Br. Bot. Centr. **74**: 369–372.

1900. Vergleichende Untersuchungen über typische und parthenogenetische Fortpflanzung bei der Gattung *Antennaria*. Handl. Svensk. Vet. Akad. **33**: pp. 59. pls. 6. figs. 5.

1904. Die Tetradenteilung in der Samenanlage von *Taraxacum*. Ark. f. Bot. **2**: 1–9.

1905. Die Tetradenteilung bei *Taraxacum* und anderen Cichoraceen. Kgl. Svensk. Vet. Akad. **39**: 1–20. pls. 1–3.

KUSANO, S. 1915. Experimental studies in the embryonal development in an angiosperm. Jour. Coll. Agr. Tokyo **6**: 7–120. pls. 5–9. figs. 28.

KYLIN, H. 1918. Studien über die Entwicklungsgeschichte der Phæophyceen. Svensk. Bot. Tids. **12**: 1–64.

LOEB, J. 1921, Further observations on the production of parthenogenetic frogs. Jour. Gen. Physiol. **3**: 539–545. figs. 3.

MARCHAL, ÉL. and ÉM. 1909. Aposporie et sexualité chez les mousses. II. Bull. Acad. Roy. Belg. 1249–1288.

1912. Recherches cytologiques sur le genre "*Amblystegium*." Ibid. **51**: 189–203. 1 pl.

MINCHIN, E. A. 1912. An Introduction to the Study of the Protozoa. London.

MORGAN, T. H. 1906. The male and female eggs of phylloxerans of the hickories Biol. Bull. **10**: 201–206. figs. 4.

1908. The production of two kinds of spermatozoa in phylloxerans. Proc. Soc. Exp. Biol. and Med. **5**.

1909*a*. Sex determination and parthenogenesis in phylloxerans and aphids. Science **29**.

1909*b*. A biological and cytological study of sex-determination in phylloxerans and aphids. Jour. Exp. Zool. **7**: 239–352. 1 pl. figs. 23.

1910. The chromosomes in the parthenogenetic and sexual eggs of phylloxerans and aphids. Proc. Soc. Exp. Biol. & Med. **7**.

1913. Heredity and Sex. New York.

MORGAN, T. H., STURTEVANT, A. H., MULLER, H. J., and BRIDGES, C. B. 1915. The Mechanism of Mendelian Heredity. New York.

MOTTIER, D. M. 1915. Beobachtungen über einige Farnprothallien mit bezug auf eingebettete Antheridien und Apogamie. Jahrb. Wiss. Bot. **56**: 65–84.

MURBECK, S. 1901. Parthenogenetische Embryobildung in der Gattung *Alchemilla*. Lunds Arsskr. **36**: pp. 41, pls. 6.

1904. Parthenogenesis bei den Gattungen *Taraxacum* und *Hieracium*. Bot. Not., Lund, 1904. pp. 285–296.

NACHTSHEIM, H. 1912. Parthenogenese, Eireifung und Geschlechtsbestimmung bei der Honigbiene. Sitzber. Ges. Morph. u. Phys., München.

1913. Cytologische Studien über die Geschlechtsbestimmung bei der Honigbiene (*Apis mellifica*). Arch. Zellf. **11**: 169–241. pls. 7–10.

NATHANSOHN, A. 1900. Ueber Parthenogenesis bei *Marsilia* und ihre Abhängigkeit von der Temperatur. Ber. Deu. Bot. Ges. **18**: 99–109. figs. 2.

OSAWA, J. 1913. Studies on the cytology of some species of *Taraxacum*. Arch. f. Zellf. **10**: 450–469. pls. 37, 38.

OVERTON, J. B. 1902. Parthenogenesis in *Thalictrum purpurascens*. Bot. Gaz. **33**: 363–375. pls. 12, 13.

1913. Artificial parthenogenesis in *Fucus*. Science **37**: 841, 844.

PACE, L. 1913. Apogamy in *Atamosco*. Bot. Gaz. **56**: 376–394. pls. 13, 14.

PARMENTER, C. L. 1920. The chromosomes of parthenogenetic frogs. Jour. Gen. Physiol. **2**: 205–206.

PRINGSHEIM, N. 1876. Ueber den Generationswechsel der Thallophyten und seinen Anschluss an den Generationswechsel der Moose. (Vorl. Mitt.) Monatschr. Akad. Wiss. Berlin, 1876. p. 869.

1878. Ueber Sprossung der Moosfrüchte und den Generationswechsel der Thallophyten. Jahrb. Wiss. Bot. **11**: 1–46. pls. 1, 2.

ROSENBERG, O. 1906. Ueber die Embryobildung in der Gattung *Hieracium*. Ber. Deu. Bot. Ges. **24**: 157–161. pl. 11.

1909. Ueber die Chromosomenzahlen bei *Taraxacum* und *Rosa*. Svensk Bot. Tidskr. **3**: 150–162. figs. 7.

SCHAUDINN, F. 1902. Krankheitserregende Protozoen. II. *Plasmodium vivax*. Arb. d. k. Gesundheitsamte, Berlin, **19**: 169.

SHAW, W. R. 1897. Parthenogenesis in *Marsilia*. Bot. Gaz. **24**: 114–117.

STEIL, W. N. 1915. Some new cases of apogamy in ferns. (Prelim. Note.) Science **41**: 293–294.

1918. Studies of some new cases of apogamy in ferns. Bull. Torr. Bot. Club **45**: 93–108. pls. 4, 5.

1919*a*. Apogamy in *Nephrodium hirtipes* Hk. Ann. Bot. **33**: 109–132. pls. 5–7.

1919*b*. Apospory in *Pteris sulcata*, L. Bot. Gaz. **67**: 469–482. pls. 16, 17. figs. 4.

STOKEY, A. G. 1918. Apogamy in the Cyatheaceæ. Ibid. **65**: 97–102. figs. 10.

STORK, H. E. Studies in the genus *Taraxacum*. Bull. Torr. Bot. Club **47**: 199–210.

STRASBURGER, E. 1905. Die Apogamie der Eualchemillen und allgemeine Gesichtspunkte, die aus sich ergeben. Jahrb. Wiss. Bot. **41**: 88–164. pls. 1–4.

1907. Apogamie bei *Marsilia*. Flora **97**: 123–191. pls. 3–8.

1909*a*. Die Chromosomenzahlen der *Wikstroemia indica* (L.) C. A. Mey. Ann. Jard. Bot. Buit. II **3**: suppl. 13–18. figs. 3.

1909*b*. Zeitpunkt der Bestimmung des Geschlechts, Apogamie, Parthenogenesis, und Reduktionsteilung. Hist. Beitr. **7**.

VINES, S. H. 1911. Article on Reproduction in Plants. Encyl. Brit., 11th Ed.

WEISMANN, A. 1886. Richtungskörper bei parthenogenetischen Eiern. Zool. Anz. **9**: 570–573.

1887. Ueber die Zahl der Richtungskörper und über ihre Bedeutung für die Vererbung. Jena. Also in Essays upon Heredity, 1889.

WINKLER, H. 1904. Ueber Parthenogenesis bei *Wikstroemia indica* (L.) C. A. Mey. (Vorl. Mitt.) Ber. Deu. Bot. Ges. **22**: 573–580.

1906. Ueber Parthenogenesis bei *Wikstroemia indica*. Ann. Jard. Bot. Buit. **5**: 208–276. pls. 20–23.

1908. Ueber Parthenogenesis und Apogamie im Pflanzenreiche. Prog. Rei Bot. **2**: 293–454.

WORONIN, HELENE W. 1907. Apogamie und Aposporie bei einigen Farnen. (Vorl. Mitt.) Ber. Deu. Bot. Ges. **25**: 85–86.

1908. Same title. Flora **98**: 101–162. figs. 72.

YAMANOUCHI, S. 1908. Apogamy in *Nephrodium*. Bot. Gaz. **45**: 289–318. pls. 9, 10. (Bibliography.)

CHAPTER XIV

THE RÔLE OF THE CELL ORGANS IN HEREDITY

The chief interest of cytology at the present time probably lies in the relation which it bears to the subject of heredity. From the time when the problems of cell research first began to take definite shape, especially since a connection between the activities of the cell and the phenomena of inheritance was suggested, the efforts of most cytologists have contributed directly or indirectly to the solution of two great and closely interrelated problems of biology: the problem of ontogenetic development and the problem of heredity. The aid which cytology has afforded in these respects has been invaluable. Not only has it been able to discover a large number of the significant facts of individual development, or ontogeny, but it has also thrown a flood of light upon many obscure matters in the field of heredity, and has so come to be an important factor in the study of phylogeny and evolution.

The Law of Genetic Continuity.—"The most fundamental contribution of cell-research to the theory of heredity," says Wilson (1909), "is the law of genetic continuity by cell-division. Cells arise only by the division of preëxisting cells . . . In each generation the germinal stuff runs through the same series of transformations; hence that reappearance of the same traits in successive generations that we call heredity."

It is by the light of the above law that we are enabled to see something of the nature of the material continuity which exists between successive stages of the ontogenetic development, and also between successive generations. It is to be remembered, in the first place, that all the cells of the adult multicellular organism are derived by repeated division from the single cell (ordinarily a zygote or a spore) with which development starts, so that the causes of events occurring at any particular stage are to be sought largely in the reactions of cells at earlier stages; and, in the second place, that the material link connecting two successive generations is a single organized cell, usually a gamete or a spore, which means that the heritage of a long ancestry is in some way represented in this single cell and its capabilities. "The conception that there is an unbroken continuity of germinal substance between all living organisms, and that the egg and the sperm are endowed with an inherited organization of great complexity, has become the basis for all current theories of heredity and development" (Locy, 1915, p. 224).

Cytological studies have therefore centered mainly about the general organization of the egg (chiefly that of animals) as related to the character of the organism arising from it (the problem of development), and about the rôles played by the various cell organs of the gamete in the transmission of heritable characteristics from one generation to the next (the problem of heredity). The character of the principal modern theory of heredity to which these studies have led is due in no small measure to the influence of a number of earlier hypotheses, such as those of Darwin and deVries, and especially that of Weismann. These hypotheses will be reviewed in Chapter *XVIII*, where their relation to the modern cytological interpretation of heredity, set forth in this and the following three chapters, will be discussed.

It is obvious that an account of the physical basis of heredity would require for completeness not only a description of the structural changes by which visible materials are transmitted and distributed during gametogenesis, fertilization, and development; but also a review of many physiological processes which accompany these changes, and through which many characters are brought to expression. In these chapters attention will be limited largely to the structural aspects of the problem. Among the physiological changes those occurring at the time of fertilization are best known, and have already been discussed in Chapter XII.

The Rôle of the Nucleus.—It was Ernst Haeckel (1866) who first advanced the hypothesis that "the nucleus of the cell is the principal organ of inheritance." Cytological evidence in support of this view, announced by Haeckel as a speculation, was brought forward by O. Hertwig (1875 etc.), Strasburger (1878, 1884), van Beneden (1883 etc.), and a number of other investigators, who described the behavior of the nucleus in the various stages of the life cycle, particularly in somatic cell-division, maturation, and fertilization. Two of these workers, O. Hertwig and Strasburger, who had discovered the fusion of the gamete nuclei at the time of fertilization in animals and plants respectively, definitely announced the theory, now supported by a considerable body of observational and experimental evidence, that the nucleus is the "vehicle of heredity." They held that hereditary transmission is through the nuclei of the gametes, and that the chromatin is the special inheritance material, or "idioplasm," about which there had been so much speculation. This view was at once widely adopted by biologists.

The efforts of many cytologists were now directed toward the further elucidation and verification of this nuclear hypothesis of heredity, and many observations and experiments apparently demonstrated its essential correctness. It was noted that, so far as could be discerned, the spermatozoön in many cases brings nothing but nuclear material into the egg, so that hereditary transmission from the male parent must be through the nucleus alone. A similar condition was later reported in plants, Guignard, Nawaschin (1910), and Welsford (1914) pointing out that in *Lilium* only the male nucleus enters the egg, its accompanying cytoplasm being rubbed off and left behind. (See p. 299.) Certain ingenious experiments of Boveri (1889, 1895; also 1909 and 1918) led to the same conclusion regarding the nucleus. Boveri induced the fertilization of enucleated fragments of *Sphærechinus* eggs (a phenomenon

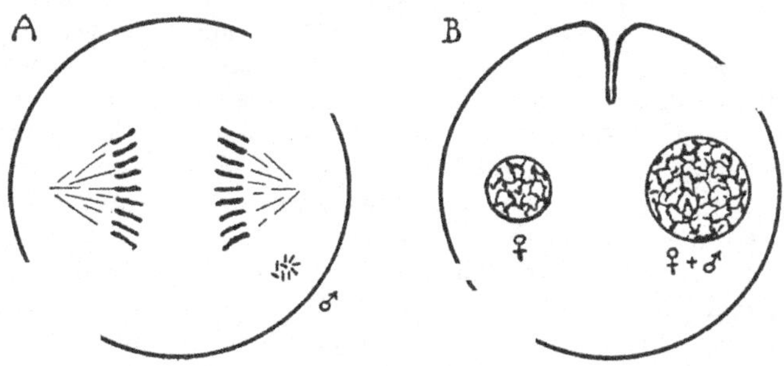

Fig. 127.

A, egg of *Sphærechinus granularis* undergoing artificially induced cleavage mitosis; spermatozoön of *Strongylocentrotus lividus* has entered and taken the form of a chromosome group. B, cytokinesis beginning; one blastomere will have a purely maternal nucleus, and the other a hybrid nucleus. (*Diagrammed after Herbst*, 1909.)

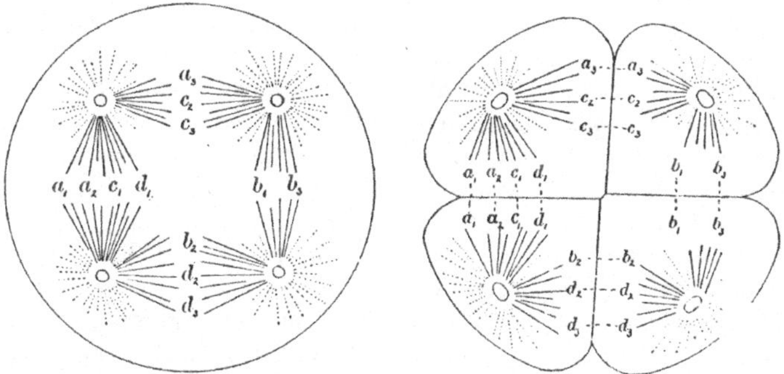

Fig. 127 *bis*.—Diagram showing the irregular distribution of the chromosomes by a quadripolar mitotic figure. (*After Boveri*.)

known as *merogony*) by spermatozoa of *Echinus*, and obtained larvæ which were purely paternal in character. From this it was argued that it is the sperm nucleus alone, and not the egg cytoplasm, that transmits the hereditary characters from one generation to the next in this case. Other experiments of a similar nature, however, turned out differently, as will presently be noted. Certain echinoderm hybrids, furthermore, show paternal larval characters even when the egg nucleus has not been removed.

A strong piece of evidence supporting Boveri's conclusion was furnished by Herbst (1909). By treating eggs of *Sphærechinus* with valerianic acid Herbst caused them to undergo cleavage artificially. While the cleavage mitosis was in progress a spermatozoön of *Strongylocentrotus* was allowed to enter the egg, where it at once gave rise to its group of chromosomes (Fig. 127). These, however, arriving too late to join regularly in the mitosis, were incorporated in neither of the daughter nuclei of the first cleavage: they resumed the form of a nucleus, and this was included in one of the blastomeres. This blastomere therefore contained two nuclei, one maternal and one paternal, which combined during subsequent stages, whereas the other blastomere had a maternal nucleus only. Herbst regarded such nuclear behavior as responsible for the frequently found larvæ which are hybrid in character on one side and purely maternal on the other. This experiment has been held to show not only that it is the nucleus of the spermatozoön which brings in the paternal characters, but also that it is the chromosomes alone that are responsible. It is assumed, though perhaps without sufficient evidence, that the other nuclear materials (karyolymph etc.) which may be present, as well as any cytoplasmic elements, have opportunity to mix generally with the egg cytoplasm, since the membrane of the male nucleus breaks down and leaves the chromosomes lying free before the egg divides into the two blastomeres. The paternal characters, however, appear only where the chromosomes come to be located—that is, in the cells composing one-half of the organism. In his work on multipolar mitoses in dispermic eggs (Fig. 127 *bis*; see also p. 163) Boveri (1902, 1907) was able to show further that abnormal chromosome distribution is associated with abnormalities in development in a very definite way; and that if isolated blastomeres resulting from such abnormal divisions be made to develop independently, completely normal larvæ result only where there is statistical reason to believe that a full complement of the qualitatively different chromosomes is present.

The nuclear theory had its opponents from the beginning. Verworn, Waldeyer, Rauber, and other early investigators held that the cytoplasm as well as the nucleus must be concerned in the hereditary process, since the spermatozoön in many cases does bring cytoplasm into the egg, and also because neither nucleus nor cytoplasm can function independently of the other. This view received support in certain experiments which seemed to discount the power of the nucleus in controlling heredity. Loeb (1903) found that when a sea urchin egg was fertilized by a starfish sperm the resulting larva possessed purely maternal characters, the sperm nucleus exerting no visible hereditary effect. The same thing was noted by Godlewski (1906) in crosses between sea urchins and crinoids. Godlewski made the further significant observation that when enucleated egg fragments of *Parechinus* (sea urchin) were fertilized by sperm of

Antedon (crinoid) the larvæ so produced, contrary to Boveri's results, were maternal in character: they were like the mother, which had presumably contributed cytoplasm only, and not like the father, which had furnished the nucleus. Fertilization by a spermatozoön had here produced a developmental stimulus but no amphimixis (the combining of hereditary lines), so far as could be judged from the appearance of the larvæ. Even if the male cytoplasm were admitted to have no hereditary rôle, it nevertheless seemed that the cytoplasm of the egg was clearly so concerned.

In his work on sea urchin hybrids Baltzer (1910) was able to show why it is that some such larvæ are maternal in character while others have the characters of both parents. When an egg of *Strongylocentrotus* fertilized by a spermatozoön of *Sphærechinus* undergoes its first cleavage division the paternal chromosomes behave irregularly; they fail to become incorporated in the daughter nuclei and are lost. Those individuals which develop far enough show maternal skeletal characters. In the reciprocal cross, on the contrary, all of the chromosomes behave normally and the resulting larvæ are truly hybrid in character. Thus in the first cross, in which the paternal chromosomes are lost, the spermatozoön furnishes only a developmental stimulus and has no appreciable effect on the character of the new individual; whereas in the second cross, in which the paternal chromosomes are included in the blastomere nuclei, the spermatozoön not only furnishes the developmental stimulus but also contributes paternal characters to the new individual. This is particularly convincing evidence in favor of the view that the chromosomes are in some way responsible for the development of parental characters in the offspring.

In a posthumous paper Boveri (1918) reported an additional observation which, he believed, goes far toward explaining the conflicting results of different investigators. He found that egg fragments, and even whole eggs, may often have chromatin in a form that easily escapes observation, but which can exert its usual influence on development. In accordance with his earlier observations, enucleate egg fragments of *Sphærechinus* fertilized by spermatozoa of *Strongylocentrotus* may develop into purely paternal larvæ. Most of them, however, are intermediate in character, resembling the maternal parent also in certain features. Having previously (1895, 1905) demonstrated that the size of the nuclei in merogonic larvæ is proportional to the number of chromosomes they contain (see Chapter IV), Boveri was able to show that the nuclei of the intermediate larvæ are diploid rather than haploid, so that it is clear that the supposedly enucleate fragments in such cases must have contained chromosomes. It is probable, Boveri believed, that the maternal larvæ obtained by Godlewski may be accounted for in a similar fashion.

It was pointed out by Strasburger (1908), who had come to believe in the complete monopoly of the nucleus in the transmission of hereditary characteristics, that the maternal character of Godlewski's larvæ could be explained on the assumption that the early developmental stages do not require the expression of the hereditary capabilities of the nucleus, but are dependent more directly upon mechanical causes. Boveri (1903, 1914), as a result of his hybridization experiments, strongly emphasized the view that the spermatozoön has an influence upon all of the larval characters; but he pointed out that the larval stages by themselves are not sufficient grounds upon which to establish complete conclusions regarding the respective rôles of nucleus and cytoplasm, since the general course of the early developmental stages in such organisms is immediately dependent to a very large extent upon the general organization of the egg.

This brings us to a brief consideration of the "promorphology" of the highly organized animal egg, and of the relation which exists between this organization and the character of the organism developing from it. Of the large amount of work done in this field only a hint can be given here.

The Promorphology of the Ovum.—There arose very early two views regarding the organization of the egg which recall the older theories of preformation and epigenesis (Chapter I). According to one, most fully expressed in W. His's Theory of Germinal Localization (1874; see Wilson 1900, p. 397), the embryo is prelocalized in the general cytoplasm of the egg—not preformed in the old sense of Bonnet, but having its various parts represented by substances with definite relative positions. This view found support in those cases in which a single isolated blastomere of the two-celled stage develops into a half-larva instead of a complete smaller larva (Roux on the frog, 1888; Crampton on the marine gastropod, *Ilyanassa*, 1896); and especially in *Beroë*, a ctenophore, which produces an incomplete larva even if a portion of the unsegmented egg be removed (Driesch and Morgan, 1895).

Opposed to the above view was that which held the egg to be isotropic and without any predetermination of embryonic parts. Certain well known experiments appeared to bear out this conclusion. It was found in the frog (Pflüger 1884; Roux 1885), the sea urchin (Driesch 1892), an annelid (Wilson 1892), and *Ascaris* (Boveri 1910) that very abnormal types of cleavage can be artificially induced, but that normal larvæ nevertheless result. A number of cases were also described in which complete embryos arose from single isolated blastomeres of the two-celled stage (*Fundulus*, Morgan 1895; and other forms), or even from those of the sixteen-celled stage (*Clytia*, Zoja 1895). Had there been any prelocalization of parts in the egg it is difficult to see how normal or complete embryos could have arisen in such abnormal ways as these.

It appears that the eggs of different animal species vary greatly in the degree and fixity of their internal differentiation. In some cases the egg is virtually isotropic, and through several succeeding cell generations the blastomeres are equipotential, that is, equally capable of developing into any part of the body or even into the whole of it. Thus the embryonic parts, and hence many of the individual's characters, are not definitely marked out until a comparatively late stage. On the contrary, there are forms in which the axis of polarity and certain fundamental embryonic parts are roughly delimited in the egg cytoplasm in such a way that an alteration in the relative positions of the egg materials brings about a corresponding alteration in the character of the resulting individual. As an illustration of such internal differentiation may be taken the case of *Styela*, an ascidian, described by Conklin (1915). In the egg of *Styela* there are four or five distinct kinds of plasma arranged in a definite order and distributed in a regular manner as cleavage proceeds, each kind eventually giving rise to a certain portion of the embryo.

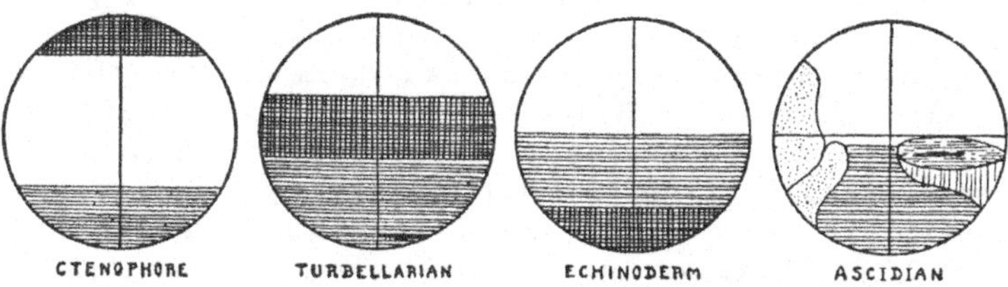

Fig. 128.—Eggs of various animals, showing the patterns assumed by the materials which give rise to the various body regions. In the first three the egg has undergone division, and the plasmas becoming ectoderm, mesoderm, and endoderm are represented in clear white, cross-hatching, and parallel ruling respectively. In the fourth egg two divisions have occurred, and several definitely arranged substances are distinguishable. (*After Conklin*, 1915.)

Substances which are yellow, gray, slate-blue, and colorless give rise respectively to muscle and mesoderm, nervous system and notocord, endoderm, and ectoderm. "Thus within a few minutes after the fertilization of the egg, and before or immediately after the first cleavage, the anterior and posterior, dorsal and ventral, right and left poles are clearly distinguishable, and the substances which will give rise to ectoderm, endoderm, mesoderm, muscles, notocord and nervous system are plainly visible in their characteristic positions" (Conklin 1915, p. 118). If such eggs are placed in a centrifuge the various substances may be made to assume an entirely abnormal stratified arrangement, which in turn "may lead to a marked dislocation of organs; the animal may be turned inside out, having the endoderm on the outside and its skin and ectoderm on the inside, etc." (p. 321). Such a behavior emphasizes the determinative character of the cytoplasmic pattern clearly present in many eggs. It has further been noted that the eggs of various animal phyla are characterized by distinct patterns in the arrangement of their visibly different materials (Fig. 128). "The polarity, symmetry and pattern of a jellyfish, starfish, worm, mollusk, insect or vertebrate are foreshadowed by the characteristic polarity, symmetry and pattern of the cytoplasm of the egg either before or immediately after fertilization" (Conklin, p. 172-5). That the arrangement of the embryonic parts is not solely dependent upon such visible egg substances is shown by the observations of Morgan (1909*c*, 1910*e*) and Boveri (1910*b*) on centrifuged eggs of *Arbacia, Ascaris,* the frog, and other forms. Here it is found that the displacement of the various substances does not necessarily cause a dislocation of the body parts of the embryo: hence the setting apart of the embryonic regions must be dependent upon a polarity in the egg which at least in many cases is not disturbed by the experimental alteration in position of the visible egg substances. But in either case differentiation appears to be related to a cytoplasmic organization.

From this it would appear that the characters which such an organism inherits from the preceding generation do not belong to one category and are not transmitted in the same way. There are first those general characteristics of organization which are the direct outgrowth of a corresponding organization in the egg cytoplasm. Secondly, there are the Mendelian characters which appear later in the ontogeny and which there is every reason to believe are represented in some way in the chromosomes of the gamete nuclei (Chapter XV). Boveri thus distinguished two periods after fertilization: an early one in which the course of development is dependent on the organization of the egg cytoplasm, only general metabolic functions of the chromosomes being active; and a later one in which the specific hereditary powers of the chromosomes are brought to expression, the right chromosomal combination then proving to be necessary for normal development.

The question naturally arises as to how much the cytoplasmic organization may be due in turn to the activity of the nucleus during the differentiation of the egg—as to whether the general characters which are the direct outgrowth of this organization may or may not be ultimately dependent, as are the clearly Mendelian characters, on nuclear factors. Conklin (1915) comments upon this point as follows: "In this differentiation and localization of the egg cytoplasm it is probable that certain influences have come from the nucleus of the egg, and perhaps from the egg chromosomes. There is no doubt that most of the differentiations of the egg cytoplasm have arisen during the ovarian history of the egg, and as a result of the interaction of nucleus and cytoplasm; but the fact remains that at the time of fertilization the hereditary potencies of the two germ cells are not equal, all the early stages of development, including the polarity, symmetry, type of cleavage, and the pattern, or relative positions and proportions of future organs, being foreshadowed in the cytoplasm of the egg cell, while only the differentiations of later development are influenced by the sperm. In short the egg cytoplasm fixes the general type of development and the sperm and egg nuclei supply only the details" (p. 176).

Plastid Inheritance.—Certain cases of "plastid inheritance" have been brought forward to show that the character of an organism may not be entirely due to factors delivered to it by the gamete or spore nuclei. It has been pointed out that two successive generations of cells reproducing by division resemble each other for the obvious reason that the organs of any given cell may actually become the corresponding organs of the daughter cells. Thus in the case of a unicellular green alga the daughter individuals are like the mother individual in being green because the chloroplast of the mother cell is divided and passed on directly to them. In those algæ in which a swarm spore germinates to produce a multicellular individual (*Ulothrix* etc.), or associates with others of its kind to form a colony (*Hydrodictyon, Pediastrum;* Harper 1908, 1918*ab*), the color of the successive colonies or multicellular individuals is a character that is transmitted directly by the repeated division of chloroplasts. Thus, as Harper urges, the nucleus is not required here to account for the resemblance between successive generations of cells or individuals, so far as this character is concerned.

A similar interpretation has been placed by some geneticists upon the inheritance of "chlorophyll characters" in the higher plants, the supposition being that plastids, multiplying only by division, are responsible for the distribution, in the individual plant and through successive generations, of those characters which manifest themselves in these organs. Abnormalities in chlorophyll coloring are accordingly held to be due to an abnormal condition or behavior of the chloroplasts.

Such a case is that of *Mirabilis jalapa albomaculata,* described by Correns (1909). In plants of this race there are some branches with normal green leaves, some with white leaves, and some with "checkered" (green and white) leaves. Flowers are borne on branches of all three types. In all cases crosses between unlikes result in seedlings with the color of the maternal parent: inheritance is strictly maternal. For instance, if a flower on a green branch is pollinated with pollen from a flower on a white branch the offspring are all green. In the reciprocal cross the offspring are all white, and soon die because of the lack of chlorophyll. In neither case does the pollen affect the color of the resulting individual. The explanation offered by Correns for the colorless condition is that it is due to a cytoplasmic disease which destroys the chloroplasts. It is therefore delivered directly to the next generation in the egg cytoplasm, and is not transmitted by the male parent because no male cytoplasm is brought into the egg at fertilization. If it had been due to nuclear factors it would have been transmitted by both parents, since the nuclear contributions of the two are equal. This condition is analogous to that occasionally found in animals, in which bacteria may be carried from one generation to the next in the egg cytoplasm, causing a direct inheritance of the disease. But such pathological cases are not to be confused with, or thought to contradict, normal Mendelian heredity, which, as will be seen in the following chapter, is closely bound up with nuclear phenomena. They are rather to be regarded as examples of repeated reinfection.

Results differing from those of Correns were obtained by Baur (1909) in his researches on *Pelargonium zonale albomarginata*. This form, which is characterized by white-margined leaves, often has pure green and pure white branches, as in *Mirabilis*. Crosses either way between flowers on these two kinds of branches result in every case in mosaic (green and white) offspring: inheritance is here not purely maternal as in *Mirabilis*. Although Baur admits for this case the possibility of a Mendelian interpretation if a segregation of factors for greenness and whiteness in the somatic cells be allowed, he thinks it more probable that inheritance in this instance is not a matter of chromosomes and Mendelism at all, but is rather due to a sorting out of green and colorless plastids, themselves permanent cell organs, in the somatic cells. In order to account for inheritance through both the male and the female, Baur assumes that primordia of plastids are brought in through the male cytoplasm as well as the egg cytoplasm, a conclusion directly contradictory to that of Correns. Ikeno (1917), working on variegated races of *Capsicum annum*, obtained results similar to those of Baur on *Pelargonium*, and concluded that transmission of variegation is not through the nucleus, but through plastids contributed by both parents.

Although the results and interpretations of Correns and Baur are at present irreconcilable except on the basis of assumptions not warranted by known facts, they agree in the conclusion that plastid inheritance is not Mendelian, but is due rather to extra-nuclear factors. Baur reports corroborative evidence in *Antirrhinum* (1918). Opposed to this conclusion is that of Lindstrom (1918), who has clearly shown in the case of certain variegated races of maize that the inheritance of characters due to unusual plastid behavior is strictly Mendelian. This means that the distribution or degree of prominence of the plastids, although these may be organs with their own individuality, depends upon the activity of Mendelian factors in the chromosomes, which represent the only known cell mechanism in which there is at present any hope of finding an explanation for the distribution of Mendelian characters (Chapter XV). In Lindstrom's plants plastid inheritance appears to be as much a nuclear matter as the inheritance of any other character manifested in the extra-nuclear portion of the cell.

On the basis of the data at hand the tentative conclusion seems fully justified that all cases of chlorophyll inheritance do not belong to one category. Some of them are clearly to be accounted for on the same basis with other Mendelian characters, whereas others appear to require an explanation of another kind. The results of work in progress at Cornell University on variegated races of maize points in this direction. One of the most interesting problems in cytology and genetics at present is that concerning the manner in which extra-nuclear bodies, such as plastids and their primordia, may account for certain types of inheritance, and the extent to which their behavior may be influenced by the nucleus.

Aleurone Inheritance.—We may here refer to the attempt which has been made to explain the inheritance of aleurone color in maize endosperm on the basis of a somatic segregation of special cell organs in the form of granular primordia, which multiply by fission and develop into aleurone bodies of various types and colors. But since aleurone and other endosperm characters are inherited in Mendelian fashion, as shown by East and Hayes (1911, 1915), Collins (1911), and Emerson (1918), and since there has been adduced in support of the supposed sorting out of primordia no evidence approaching in cogency that upon which the chromosome theory has been built up, geneticists generally are of the opinion that the chromosomes with their well known mechanism of segregation offer the best promise of an explanation of the inheritance of aleurone characters, though all admit that other organs may play a part in bringing these characters to expression. Furthermore, the case for the self-perpetuity of the aleurone grain is much weakened by the fact of their artificial production by Thompson (1912).

The theory that chondriosomes are concerned in heredity has been discussed in Chapters VI and XII.

General Conclusions.—In conclusion the statement may again be made that as genetical researches multiply it becomes increasingly clear that the characters in which an individual resembles that from which it sprang are not in every case transmitted to it in the same manner. Those characters which are inherited according to Mendelian rules, to anticipate a conclusion based on evidence to be presented in the next chapter, in all probability owe their repeated appearance in successive generations to "factors" of some sort which are transmitted by the chromosomes of the nucleus. This applies also to those characters which, while Mendelian in distribution, depend for their expression upon the presence of other cell organs (plastids) which may have an individuality of their own.

All or nearly all of the hereditary contribution made by the male parent must in most organisms be in the above form, since the male gamete consists almost exclusively of nuclear material. The female gamete, or egg, in addition to the clearly Mendelian characters represented by factors in its nucleus, may at least in the case of many animals contribute certain general characters, such as polarity, symmetry, and general type of early development, which are the direct outgrowth of an elaborate organization present in the egg cytoplasm. It is true that this organization is the result of processes in which the nucleus coöperates during the differentiation of the egg, and those who hold to the universal applicability of the Mendelian interpretation would assume that the type of organization must depend upon Mendelian factors carried in the nucleus. However this may be, the fact remains that the two gametes at the time of fertilization are not equal in hereditary potency, as Conklin states. So far as the clearly Mendelian characters are concerned, however, all evidence goes to show that they are precisely equal.

The direct inheritance of metidentical characters, such as the above mentioned green plastid color in *Pediastrum*, and the indirect inheritance of colony characters in the same form, afford other examples of hereditary transmission otherwise than through the nucleus. With respect to colony characters, Harper has shown in a striking manner, both in *Hydrodictyon* and *Pediastrum*, that the characteristic form and type of organization assumed by the colony are the results of interactions between the form, polarities, adhesiveness, surface tension, etc. of the free-swimming swarm spores which aggregate to build it up. The swarm spore has an individual organization of a particular type, but its capabilities show it to be devoid of any arrangement of its protoplasmic parts corresponding either to its future position in the colony or to the arrangement of the cells in the colony as a whole. The character of the colony thus depends upon the interactions of its component units and is in no way represented in any one of them. Consequently it is held by Harper that no system of spatially arranged factors in a special germ plasm is required to account for the regular reappearance of such cell and colony characters in these organisms, and that such facts must be reckoned with in attempting to explain heredity and development in terms of the cell.

By whatever means they are transmitted, it is evident that most characters must be brought to expression through the activity of the cell system as a whole, the process involving a long series of reactions in which all or nearly all of the cell constituents play their parts. At the present time little or nothing is known of the real nature of the "factor" or of the manner in which it may influence the development of a character. In general, then, we may say that the heritage bequeathed by an individual to its offspring is in most organisms transmitted mainly through the nucleus, since it is very largely upon this organ that the development or non-development of particular characters in the organism depends; but also that the development of the characters in the offspring, however these may be transmitted, involves all of the cell organs as well as a complicated and orderly series of intercellular reactions and responses. These two phases, the transmission of a heritage of factors and the development of the organism's characters as the result of their influence, must both be very much more fully known before either can be adequately understood.

To some of the more cogent evidence upon which these general conclusions are based we shall now turn.

Bibliography at end of Chapter XVIII.

CHAPTER XV

MENDELISM AND MUTATION

MENDELISM

The classic researches carried out by Mendel a half-century ago on the hybridization of garden peas are now so well known that a detailed description of them would be superfluous here. Moreover, since the main principles of Mendelism are illustrated in the results of the simplest of Mendel's experiments, a review of one or two of the latter will for our purposes be sufficient.[1]

A Typical Case of Mendelian Inheritance.—Mendel crossed plants of a pure bred race of tall peas (6 to 7 feet in height) with plants of a pure bred dwarf race ($\frac{3}{4}$ to $1\frac{1}{2}$ feet in height) (Fig. 129). All the plants of the first hybrid generation (F_1) were tall like one of their parents. When these tall hybrids were self-fertilized or bred to one another, it was found that the second hybrid generation (F_2) comprised individuals of the two grandparental types, tall and dwarf, in the relative numerical proportion of 3:1. It was further found that the tall individuals of this generation, though alike in visible characters, were unlike in genetic constitution: one-third of them, if bred for another generation, produced nothing but tall offspring, showing that they were "pure" for the character of tallness; whereas the other two-thirds, if similarly bred, produced again in the next generation both tall and dwarf plants in the proportion of 3 : 1, showing that they were hybrids with respect to tallness and dwarfness. The dwarf plants of the second hybrid generation (F_2) produced nothing but dwarfs when interbred; they were "pure" for dwarfness. From these facts it was evident that the plants of the F_2 generation, although they formed only two visibly distinct classes, were in reality of three kinds: pure tall individuals, tall hybrids, and pure dwarfs, in the relative numerical proportions of 1 : 2 : 1.

The explanation offered by Mendel for these phenomena may be briefly stated as follows (Fig. 129). The germ cells produced by the pure tall plant carry something (now termed a *factor*, represented here

---

[1] Detailed accounts of the many facts of Mendelism may be found in more special works on the subject. See Morgan *et al.* 1915, Chapters 1 and 2; Bateson 1913; Castle, Coulter *et al.* 1912; Castle 1916; Coulter and Coulter 1918; Babcock and Clausen 1918, Chapter 5; Punnet 1919; Darbishire 1911; Morgan 1919*a*; Thomson 1913; East and Jones 1919.

by T) which makes the resulting plant tall. The germ cells of the dwarf plant carry something (t) causing the dwarf condition. In the first

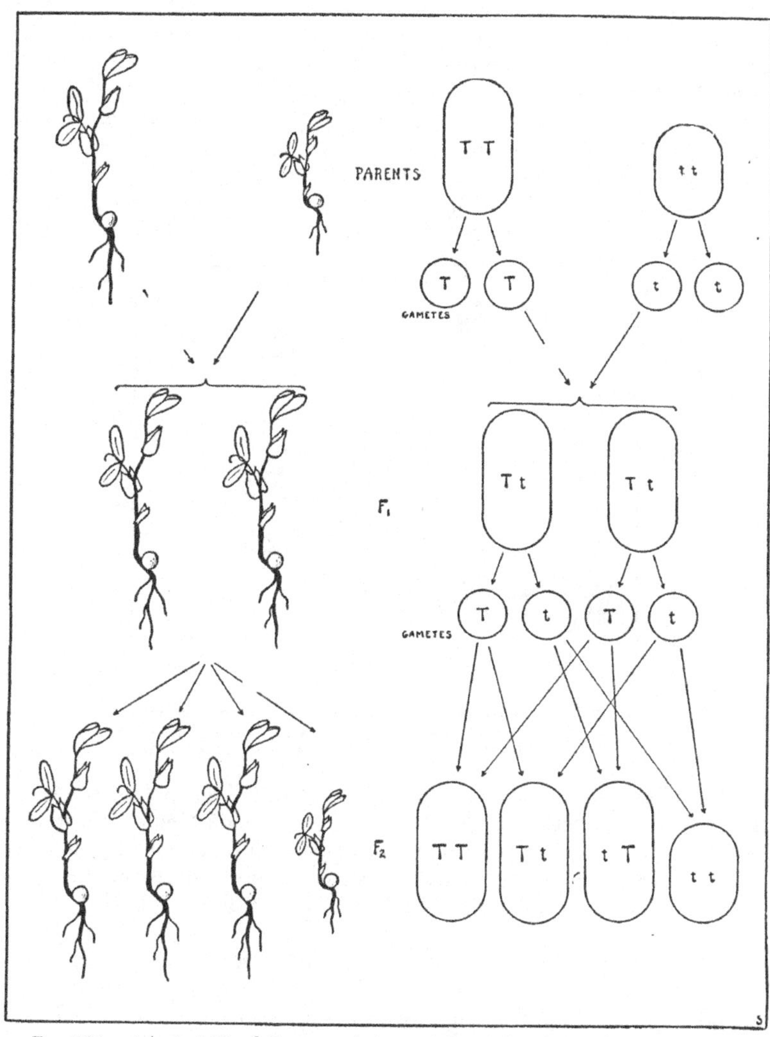

Fig. 129.—A typical Mendelian cross between tall and dwarf peas, showing dominance of the tall over the dwarf condition in the first hybrid generation (F_1), and the 3:1 ratio of tall plants to dwarfs in the second hybrid generation (F_2). At the right is shown the corresponding distribution of the Mendelian factors for tallness (T) and dwarfness (t).

hybrid generation (F_1) both factors are present, T coming from one parent and t from the other, but T "dominates" and prevents the expression of

22

the "recessive" t, so that the plants of this generation are all tall.   When the hybrid (F_1) produces germ cells the two factors for tallness and dwarfness separate, half of the germ cells receiving T and the other half receiving t.   Each gamete therefore carries either one or the other of the two factors in question, but never both: a given gamete is "pure" either for T or for t.   This segregation in the germ cells of factors previously associated in the individual without their having been altered by this association is the central feature of the entire series of Mendelian phenomena, and is often referred to as *Mendel's first law*.   Since, now, the gametes, both male and female, produced by the hybrid plants of the F_1 generation are of two kinds (half of them bearing T and half bearing t)

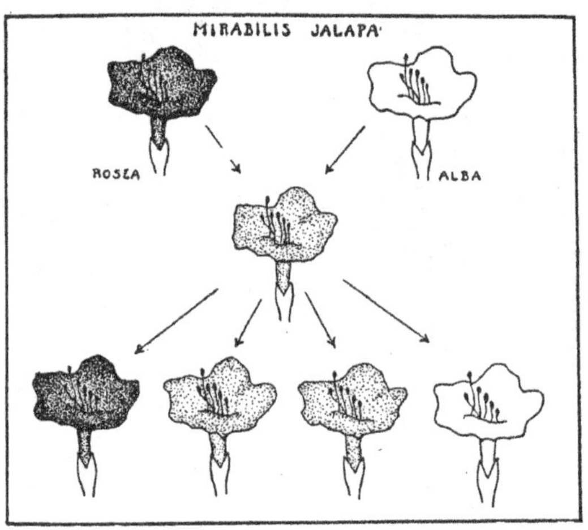

Fig. 130.—Blending inheritance ("incomplete dominance") in *Mirabilis jalapa*, showing 1:2:1 ratio of three genotypes in F_2.   (*Adapted from Correns.*)

four combinations are possible: a T sperm with a T egg, a T sperm with a t egg, a t sperm with a T egg, and a t sperm with a t egg.   These four combinations result respectively in a tall plant (pure dominant, TT), two tall hybrids (Tt and tT), and a dwarf plant (pure recessive, tt).   It is obvious that in the long run these three types will occur in the ratio of $1:2:1$.

Mendel's researches on peas included also a study of six other pairs of heritable characters (now known as *allelomorphic pairs*), the two members of each pair behaving toward each other in a manner similar to that described above for tallness and dwarfness.   He further observed that the seven pairs are entirely independent of each other in inheritance (*Mendel's second law;* now modified; see p. 384).   All these phenomena he interpreted on the basis of the hypothesis that each character is in some way represented by a factor in the cells, new combinations of factors being formed at fertilization and the members of each allelomorphic pair of factors separating when the germ cells are formed.

The Mendelian proportion of pure forms and hybrids is more easily followed in cases of "incomplete dominance," the pure dominants here being visibly distinguishable from the hybrids. Such a case is that of *Mirabilis jalapa*, the four-o'clock (Fig. 130). If plants bearing pure red flowers (var. *rosea*) ·are crossed with those bearing pure white flowers (var. *alba*) the result is an F_1 generation of intermediate pink-flowered plants. When these pink hybrids are bred among themselves the resulting F_2 generation comprises plants of three visibly different types: pure dominants with red flowers, hybrids with pink flowers, and pure recessives with white flowers, in the numerical ratio of $1:2:1$.

Terminology.—We may here introduce certain terms prominent in the literature of genetics. The *genotype* is the entire assemblage of factors which an organism actually possesses in its constitution, irrespective of how many of these may be expressed in externally visible characters. The *phenotype* is the aggregate of externally visible characters, irrespective of any other factors, unexpressed in characters, which may be present in the organism. For illustration: in the case of the tall and dwarf peas there are in the second hybrid generation (F_2) three genotypes (with respect now only to the single character pair discussed): *TT*, *Tt*, and *tt*, represented respectively by pure tall plants, tall hybrids, and dwarfs; but there are only two phenotypes: tall and dwarf, because of the fact that the complete dominance of tallness over dwarfness renders the hybrids externally indistinguishable from the pure tall individuals. Thus one phenotype (tall plants) here includes individuals with two genotypic constitutions, and the two can be distinguished only by a study of their progeny. In *Mirabilis*, however, there are in the F_2 generation not only three genotypes represented, but also three phenotypes, since the incomplete dominance renders the hybrids externally unlike either of the pure forms.

An individual is said to be *homozygous* for a given allelomorphic character pair if it has received the same factor from the two parents—a pea, for example, with the constitution *TT* or *tt*. If it has both members of the pair, such as *Tt*, it is said to be *heterozygous*. It may be homozygous for some allelomorphic pairs and heterozygous for others, or it may conceivably be either homozygous or heterozygous for all of its characters. Thus an organism with the genotypic constitution *AABbcc* is homozygous for the characters represented by *AA* and *cc*, and heterozygous for those represented by *Bb*. It is thus a pure dominant with respect to *A* and *a*, a pure recessive with respect to *C* and *c*, and a hybrid with respect to *B* and *b*. The phenotypic appearance of the organism would be determined by the dominant factors *A* and *B* and by the recessive *c*; a given dominant factor dominates only its recessive allelomorph, and not the recessive factors belonging to other pairs. It is a common practice to represent dominant factors or characters by capital letters and their respective recessive allelomorphs by the corresponding small letters.

The Cytological Basis of Mendelism.—Having before us some of the principal facts of Mendelism and Mendel's interpretation of them, we may now turn to the cytological basis of the Mendelian phenomena, and inquire what visible mechanism there is in the cell which will in any way help us toward an understanding of the striking behavior of the Mendelian characters.

The behavior of the chromosomes at the critical stages of the life cycle as described in the chapters on reduction and fertilization must first be recalled. (See Fig. 131.) It has been shown that their history is as follows. Each parent furnishes the offspring with a set of chromosomes, the two sets (represented in the diagram by *ABCD* and *abcd*) being associated in all the cells of the offspring. When gametes (or spores followed later by gametes in the case of higher plants) are to be

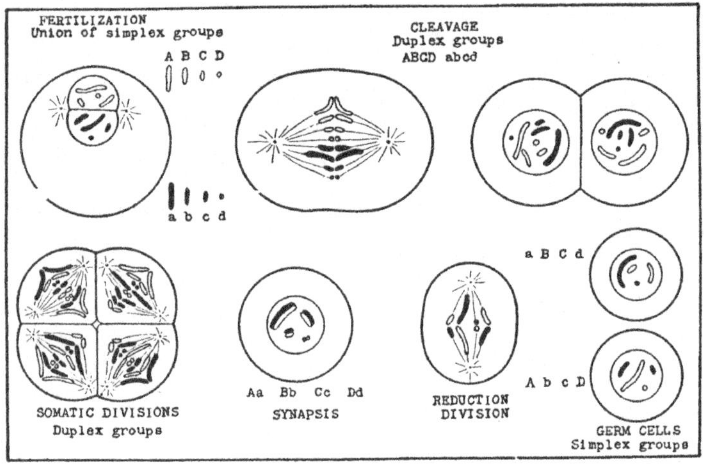

Fig. 131.—Diagram showing the history of the chromosomes in the typical life cycle of animals. (*After Wilson*, 1913.) See also Fig. 77.

formed by the new individual the chromosomes pair two by two (synapsis), the two homologous members of each pair coming from the two parental sets. In the first maturation division (usually) the two members of each pair separate and enter different daughter cells: this is reduction, or the separation of entire chromosomes, presumably qualitatively different, instead of qualitatively similar halves of chromosomes as in somatic division. In the second maturation division all the chromosomes split longitudinally (equationally), so that as the result of the two divisions there are four gametes (or spores), two of them differing from the other two in chromatin content. The somatic chromosomes are therefore segregated into two unlike groups: each gamete (or spore) has a single set of chromosomes, the set being composed of one member of each of the pairs formed at synapsis. This set represents the contribution made to the following generation.

It will be recognized at once that the above is precisely the sort of distribution shown by the characters in Mendel's experiments: two groups

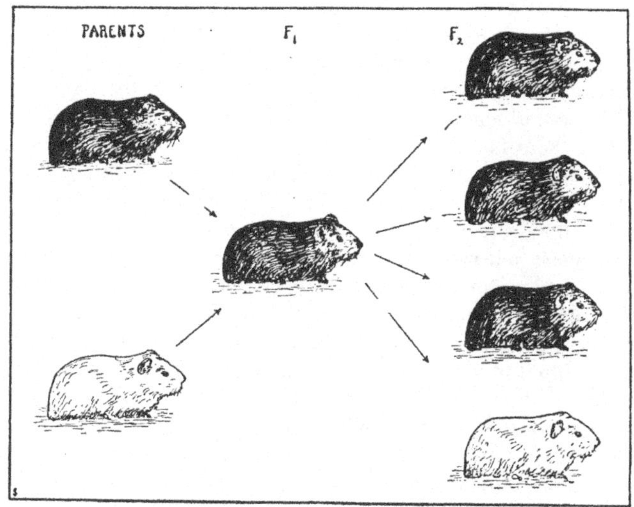

FIG. 132.—Mendelian inheritance in black and albino guinea pigs.

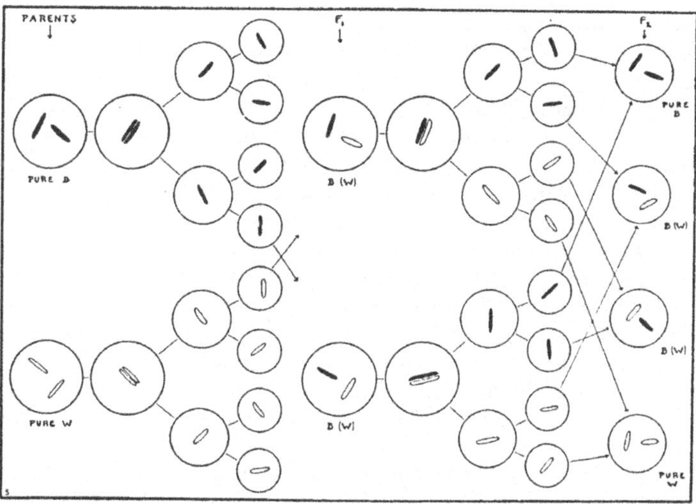

FIG. 133.—Chromosome history in the cross represented in Fig. 132, showing the parallelism between the distribution of a single homologous pair of chromosomes and that of a single allelomorphic pair of Mendelian characters.

of (factors for) characters are brought together at fertilization and are associated in the body of the offspring. When the germ cells are formed the (factors for the) two characters forming each allelomorphic pair separate and pass to different gametes (or spores). Thus the chromosomes and the characters alike form a duplex group in the body cells and a simplex group in the gametes (or spores): the chromosomes, like the characters, form new combinations at fertilization and are segregated when the gametes (or spores) are formed. In the diagram the letters *ABCDabcd* stand equally well either for chromosomes or for characters. In view of these facts it appears extremely probable that chromosomes and Mendelian characters have a definite causal relationship of some kind: it is scarcely conceivable that the exact and striking parallelism that they show can be without significance.

The precise nature of this correspondence between chromosome behavior and character distribution can be even more clearly shown by a consideration of the history of a single homologous pair of chromosomes in a typical Mendelian cross. If a pure white (albino) guinea pig be mated to an individual of a pure black strain the offspring are all black; black is completely dominant over white (Fig. 132). If these black hybrids are bred among themselves they produce in the F_2 generation three black animals to one white, or, more precisely, one pure black to two black hybrids to one pure white. Let us now follow a single pair of chromosomes of each of the original animals through these two generations.

At the left in Fig. 133 are represented the two animals, pure black and pure white, their chromosomes being drawn in solid black and outline respectively. In the black animal the two chromosomes pair at synapsis and separate to the two daughter cells at the first maturation mitosis, and split longitudinally at the second, so that each of the gametes receives a single chromosome representing a longitudinal half of one of the original pair. A similar process occurs in the white individual. Unions between the gametes of the two animals now result in the F_1 hybrids, each of which has one chromosome from its black parent and one from its white parent (not counting the chromosomes of other pairs). When these hybrids form gametes, as is seen at once in the diagram, the paternal and maternal members of the chromosome pair separate, with the result that half the gametes receive one of them and half the other. There are thus two kinds of spermatozoa and two kinds of eggs, one kind carrying the paternal chromosome and the other carrying the maternal one. Chance combinations now result in a generation (F_2) of animals, one-quarter of which have derived both chromosomes of the pair in question from the black grandparent, one-half of which have derived one chromosome of the pair from each grandparent, and one-quarter of which have derived them both from the white grandparent. Moreover, these animals are respectively pure black, hybrid black, and pure white, in the proportion of 1:2:1. Thus it is seen that *there is a direct parallelism, not only between chromosome sets and character groups, but also between the distribution of a given homologous pair of chromosomes and that of a single allelomorphic pair of Mendelian characters.*

This is exactly the condition which would result if two material units, each representing one of the characters of an allelomorphic pair, were located in two homologous chromosomes that pair and separate at reduction. The chromosomes afford precisely the type of mechanism required to account for the distribution of characters if the latter are associated with a definite material basis. It is this parallelism between the behavior of the chromosomes in reduction and that of Mendelian factors in segregation, first emphasized by Boveri and by Sutton, which has led geneticists generally to the view that the characters are actually

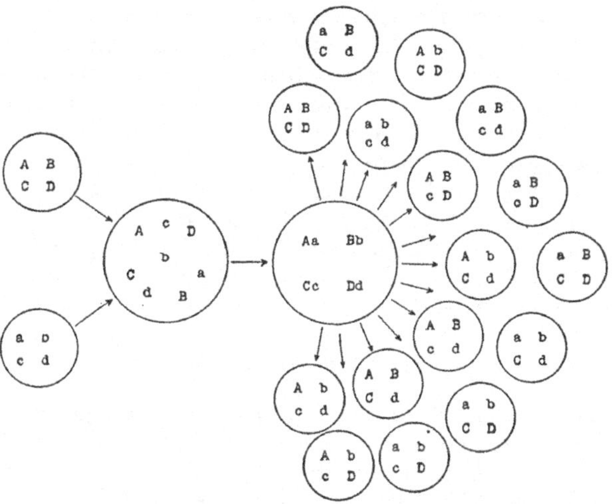

Fig. 134.—Diagram showing the 16 genotypic constitutions which may be present in the gametes of an organism with only 4 pairs of factors. (*After Wilson,* 1913.)

represented in the chromosomes by material factors, or *genes*, which in some unknown manner control the development of the characters in the body.

The earlier view that each character is thus represented by a single material unit or determiner has now given way to the more fully developed Factorial Hypothesis, according to which, on the one hand, a character may be due to the coöperative action of two or more factors ("duplicate" or "cumulative" factors); and, on the other hand, a single factor may have "manifold effects," influencing the development of several characters. The factors, or genes, are thought by some to constitute a complex reaction system, interactions between genes having a marked effect upon their activity in producing characters.[1] "The factorial

[1] A simple and brief explanation of the effects of cumulative factors is given by Coulter and Coulter (1918).

hypothesis does not assume that any one factor produces a particular character directly and by itself, but only that a character in one organism may differ from a character in another because the sets of factors in the two organisms have one difference." "It can not . . . be too strongly insisted upon that the real unit in heredity is the factor, while the character is the product of a number of genetic factors and of environmental conditions" (Morgan *et al.*, 1915, pp. 210, 212).

The abundant opportunity for the formation of new factor combinations should be noted in this connection. An organism with four pairs of chromosomes in its body cells, and only one pair of factors in each chromosome pair, could form, as the result of the independent distribution of the four pairs of chromosomes, gametes with as many as 16 different genotypic constitutions (Fig. 134). Such a diversity being present in the gametes of both sexes, this means that more than 200 different combinations are possible at fertilization. The 12 pairs of chromosomes in man may in the same way form several million such combinations. Since there is good reason to believe that each chromosome carries more than one factor the number of variations actually produced by these means is almost incalculable. This subject will be pursued further in the chapter on Linkage (Chapter XVII), where the evidence for the presence of many factors in a single chromosome will be presented and the consequences of this condition pointed out.

MUTATION

Although opinion is divided over the question of the real nature of the phenomenon of mutation, particularly in *Œnothera Lamarckiana*, one school (deVries *et al.*) holding that it represents the actual origin of new forms, and another (Bateson, Davis, Lotsy) regarding it as the result of segregation in an organism of hybrid constitution, the observed facts in either case are nevertheless very significant with respect to the chromosome theory of heredity. The mutations observed to arise from *Œnothera Lamarckiana* fall into two general classes: first, those accompanied by alterations of the normal chromosome number (seven pairs), and second, those in which the number undergoes no change.[1]

Mutations Accompanied by Changes in Chromosome Number.—It is to be noted first of all that the mutants belonging to this class do not behave in a typically Mendelian fashion when bred to other forms, and that this is correlated with the serious disturbance of the chromosome mechanism. *Œnothera* mutants with many abnormal chromosome numbers have been observed; Gates, for example, found them with 15, 20, 21, 22, 23, 27, 28, 29, and 30 chromosomes.

---

[1] Our knowledge of the cytology of the *Œnotheras* is due mainly to the researches of Gates, Davis, Stomps, and Miss Lutz.

The 28-*chromosome Mutants* (*gigas group*).—In the mutant forms of this group, of which *Œnothera gigas* is a member, the somatic number of chromosomes is 28 rather than 14; the plants are tetraploid (Fig. 135, *B*). How this condition arises is not certainly known. Stomps (1912) believed it to be the result of the union of two unreduced gametes, whereas Gates (1909*b*) suggested that "the doubling in the chromosome number had probably occurred as the result of a suspended mitosis in the fertilized egg or in an early division of the young embryo." Strasburger (1910*b*) also adopted the latter view.

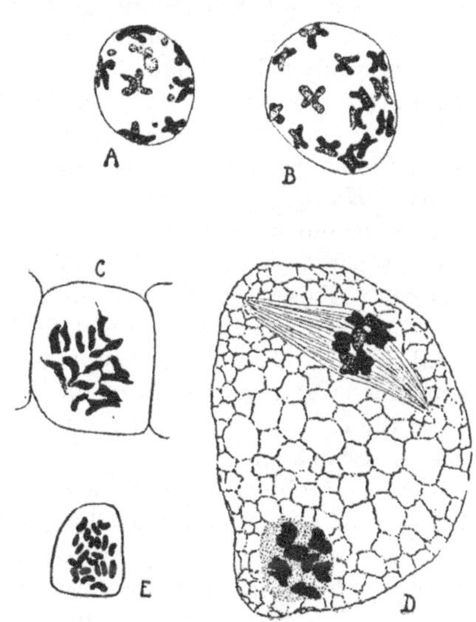

Fig. 135.—Chromosomes in *Œnothera* mutants.

A, interkinesis in *Œ. Lamarckiana;* 7 split chromosomes. *B*, same in *Œ. gigas;* 14 split chromosomes. *C*, somatic cell of *Œ. semilata;* 15 chromosomes. *D*, metaphase of homœotypic mitosis in *Œ. biennis lata*, showing 8 chromosomes on one spindle and 7 on the other. Spores and gametes with these numbers will result. *E*, the 21 chromosomes in a mutant from *Œ. Lamarckiana*. (*A and B after Davis*, 1911; *C and D after Gates and Thomas*, 1914; *E after Lutz*, 1912.)

The mutants of the *gigas* group are characterized chiefly by an unusually large size, not only of the plant as a whole but also of its anatomical constituents. In the tetraploid mutant *Œnothera stenomeres*, for instance, Tupper and Bartlett (1916) found that the change from the diploid to the tetraploid condition is concomitant with a 50 per cent increase in the length of the vessel, a 150 per cent increase in the area of its cross section, a 50 per cent increase in the length and diameter of the tracheids, an increase in the three dimensions of the medullary ray cells, and a breaking up of the tall multiple ray into a number of thin simple rays.

A further significant observation on mutants of this type was that of Gregory (1914) on a tetraploid *Primula*. He showed by breeding experiments that twice the normal number of Mendelian factors are present: thus when the chromosome number is tetraploid the number of factors is also tetraploid; each allelomorphic pair is represented twice.

It should be stated that not all cases of gigantism are accompanied in this manner by an increase in the chromosome number. In *Phragmites communis*, for example, Tischler (1918) finds abnormally large size associated with an increase in the size of the chromosomes, but not in their number. Stomps (1919) points out that among *gigas* mutants of *Œnothera, Narcissus,* and *Primula* there are diploid as well as tetraploid individuals, which must mean that the altered chromosome number is not the sole cause of such mutation but is rather one of the characters of the mutant.

The 15-chromosome Mutants (lata group).—The presence of an extra chromosome in the cells of *Œnothera lata* and other members of this group is due to the fact that the members of one pair of chromosomes in *Œnothera Lamarckiana* fail to separate at the reduction division, both of them going to one daughter cell. This phenomenon is known as *non-disjunction.* As a consequence there are gametes with eight and six chromosomes rather than the normal seven; and a union of an 8-chromosome gamete with a normal 7-chromosome gamete results in an individual with 15 chromosomes instead of the normal 14 (Fig. 135, *C*).

In her study of 15-chromosome mutants Miss Lutz (1917) found 11 or 12 types belonging to this group; only two of them were of the usual *lata* type. This condition may be accounted for on the hypothesis that it is sometimes one pair of chromosomes and sometimes another which fails to separate at the time of reduction, so that the extra chromosome is not in all cases the corresponding one of the complement. If the various chromosomes of the complement differ in hereditary value, as there is much reason to believe, it is evident that this would allow for a great variety of mutants with the same aberrant chromosome number. In *Œnothera scintillans* Hance (1918) has shown by careful measurements that the extra chromosome can be distinguished from the regular 14. Two classes of gametes are formed, some with seven chromosomes and some with eight (Fig. 135, *D*) The union of two 7-chromosome gametes gives *Œnothera Lamarckiana,* the form from which *Œ. scintillans* sprang as a mutant; whereas a union of a 7-chromosome gamete with an 8-chromosome gamete gives *Œ. scintillans.* Hance therefore points out that the *scintillans* characters are plainly associated with the extra chromosome. *Œ. scintillans* was further observed to give rise to a type resembling *Œ. oblonga.* It is possible that this was due to the union of two 8-chromosome gametes.

The 21-chromosome Mutants (semigigas group).—The 21-chromosome condition is brought about by the union of a normal 7-chromosome gamete with a 14-chromosome gamete produced either by a mutant of the *gigas* group or by a normal plant through failure of reduction. These mutants are triploid (Fig. 135, *E*).

As might be expected in forms with aberrant chromosome numbers, especially in those with an odd number like 21, certain irregularities occur in the maturation mitoses, with the result that gametes, and hence progeny, with abnormal chromosome numbers are produced. Accompanying this irregular behavior is an unsettled hereditary condition, the plants showing various unusual character combinations and differing markedly from generation to generation. In the course of such a series of generations there is a gradual settling down to the normal number through the loss of chromosomes in irregular mitoses. When the normal number (14) is finally reached the plants become much more stable in their hereditary behavior: the hereditary mechanism again appears to be in equilibrium. In one such series of mutant forms, which had arisen in the first place from *Œnothera Lamarckiana*, deVries and Stomps found that after the normal chromosome number had thus been settled upon the plants were not of the *Lamarckiana* type. Although the number was that characteristic of the original *Lamarckiana* individual from which the series originated, the assortment was apparently a new one: during the settling down process some chromosome pairs of the complement had been lost completely while others had been duplicated. The plants consequently had certain characters represented in duplicate, while others present in the orignal ancestral plant were entirely lacking. Upon the theory that the chromosomes of the complement differ in hereditary effect, the above facts are readily explained.

Conclusion.—All the evidence goes to show that in the above described mutations the change in chromosome number and the change in the visible characters of the organism occur simultaneously. This fact constitutes a strong support to the theory that in the chromosomes there are factors representing heritable characters, and indicates that mutations, whatever may be their ultimate nature, are causally connected with alterations, often visible, in the cell mechanism.

Bearing on the Origin of Species and Varieties.—The question naturally follows as to what extent the origin of new species and varieties may be bound up with fluctuations in chromosome number. Although the experimental evidence, aside from that of the *Œnothera* mutants which many do not regard as species at all, is as yet quite meager, the following facts are nevertheless very suggestive in this connection.

In published lists of chromosome numbers it is strikingly evident that the numbers shown by the species of a given genus or even of an entire family very commonly form a series of multiples. Also, one or two species of such a group frequently have but one or two chromosome pairs more or less than one of the numbers of the multiple series: From this it is to be inferred, as suggested by McClung (1905, 1907), that there is a relationship of some sort between the constitution of the chromosome complement and the externally visible taxonomic characters. Certain illustrative examples will now be cited.

Plants.—In the Leguminosæ most of the species which have been examined possess either 6, 12, or 24 pairs; *Pisum* has 7. Common numbers in the Rosaceæ are 8, 16, and 32; some species have 6. In a new study of a large number of species of *Rosa* Täckholm (1920) finds the fundamental haploid chromosome number in this genus to be seven rather than eight. The various species of *Chrysanthemum* have 9, 18, 27, 36, and 45 pairs. In both *Triticum* (Sakamura 1918) and *Avena* (Kihara 1919) the pairs number 7, 14, and 28 (Fig. 136). From

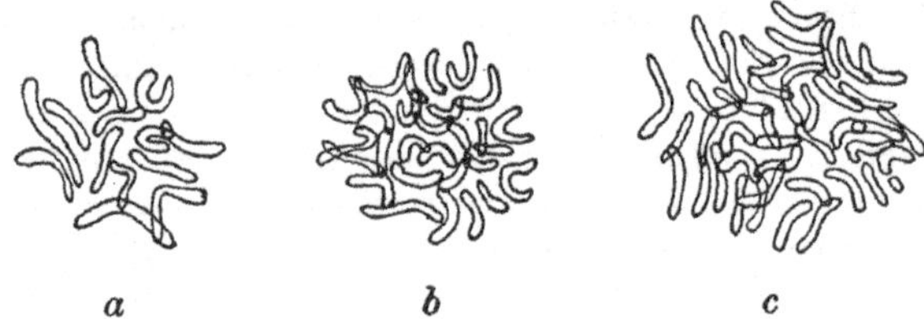

a   *b*   *c*

Fig. 136.—The chromosomes of 3 species of *Avena.* *a, A. strigosa;* 14 chromosomes. *b, A. barbata;* 28 chromosomes. *c, A. sterilis;* 42 chromosomes. (*After Kihara,* 1919.)

these and many other similar cases it is inferred that tetraploid species have been derived from diploid species in much the same way that *Œnothera gigas* with its 14 pairs has been observed to arise from *Œ. Lamarckiana* with seven; that triploid species have arisen either by dispermy or by a union of diploid and haploid gametes as in the *semigigas* group of *Œnothera* mutants; that hexaploid species have in turn arisen from the triploid ones; and that those forms with numbers not belonging to the regular multiple series have resulted from further irregularities in chromosome behavior. Among such irregularities are the failure of the two members of a homologous pair to separate at reduction (non-disjunction), and the segmentation of certain chromosomes at their points of constriction (Sakamura 1920; Kuwada 1919). Changes in chromosome number are thus looked upon as an important factor in the origin of new species.

The above conclusion is supported further by the observations of Sakamura (1918) and Kihara (1919) on wheat. Sakamura finds that the one-grained wheats (*Triticum monococcum*) have 14 chromosomes (diploid), the emmer wheats (*T. dicoccum, T. polonicum, T. durum,* and *T. turgidum*) 28 (tetraploid), and the spelt wheats (*T. spelta, T. vulgare,* and *T. compactum*) 42 (hexaploid). He concludes that the one-grained wheats are the ancestral forms from which the emmer and spelt wheats have arisen through changes in chromosome number. This is precisely the conclusion which Schulz (1913) and Zade (1914, 1918) had reached on other grounds. Kihara (1919) found further that by crossing emmer and spelt wheats fertile hybrids with 35 chromosomes could be obtained, and that these in future generations produced forms with varying chromosome numbers because of the irregular manner in which the chromosomes are distributed at the time of reduction. (See p. 253.) As a general rule hybrids produced by crossing forms with different chromosome numbers are sterile, but when they are fertile and their chromosome number is odd there is usually an irregularity in genetic behavior for several generations until the number again becomes settled. In some cases the number thus settled upon is that of the original ancestor with the lower number (the *Œnothera* mutants of deVries and Stomps cited above), whereas in other cases (*Triticum*) it is that of the ancestor with the higher number. The manner in which many such changes in number occur is not yet known.

A study of the chromosomes in the genus *Crepis* has been made from this point of view by Rosenberg (1918, 1920). He finds four species, including *C. virens*, with three pairs of chromosomes, eight species with four, four species with five, one species with eight, one species with nine, and three species with 21. In *Crepis* the chromosomes differ markedly in size, and Rosenberg concludes that the species with three, four, and five pairs have arisen through such irregular distribution of the smaller chromosomes as has actually been observed in the maturation divisions, together with recombinations occurring at fertilization. The segmentation of the larger chromosomes of the complement is not thought to occur.

In an extensive investigation of the chromosomes of *Zea Mays* Kuwada (1919) has found cytological confirmation of the conclusion of Collins (1912) that this species, which for some years has played a conspicuous rôle in genetical investigations, is in all probability a hybrid between *Euchlæna mexicana* (teosinte) and some other unknown form belonging to the nearly related Andropogoneæ. Owing to their inequality in size Kuwada is able to distinguish what he considers to be the chromosomes of the two supposed ancestral derivations in the cells of certain races of maize. Thus gemini with components of unequal size are frequently observed in the microsporocytes.

Animals.—The most complete description of the chromosomes in a large number of closely related animal species is that given by Metz (1914, 1916*b*) for the Drosophilidæ. In about 30 species Metz has identified no less than 12 main types of chromosome groups, all but one of them being found in the genus *Drosophila*. In Fig. 137 are shown diagrammatically the 12

principal types as they appear with their characteristic arrangements at
the time of cell-division. Type *A* is that found in *Drosophila melanogaster* (formerly known as *D. ampelophila*), upon which the greater part
of the genetic work in these flies has been done; it is also characteristic
of several other species. . Here there are two pairs of large bent "euchro-
mosomes," one pair of sex chromosomes, and one pair of very small
"m-chromosomes." In some of the other species it is seen that the
position of one or both of the large pairs is occupied by two pairs but

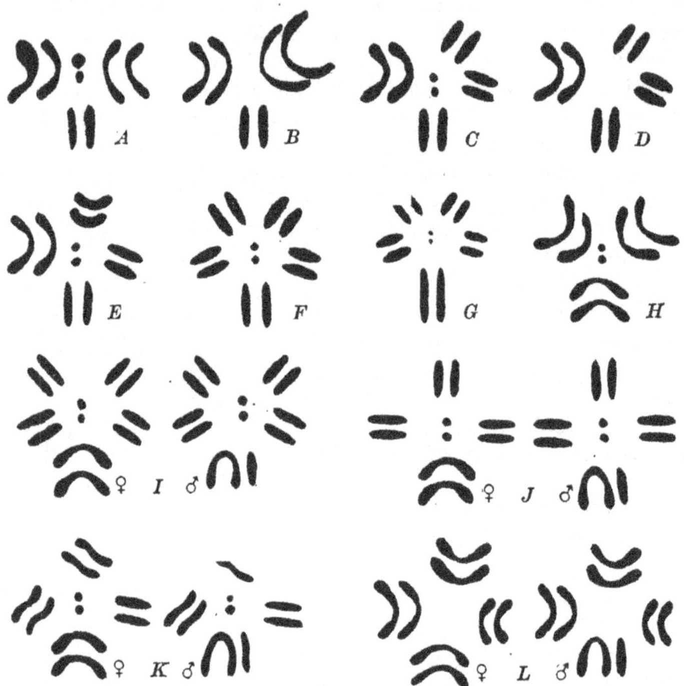

Fig. 137.—The 12 principal types of chromosome groups found in the cells of the Drosophilidæ. (*After Metz*, 1916.)

half as large, and there is much evidence to show that these have arisen
by a segmentation of the large chromosomes. Furthermore, the m-
chromosomes do not appear in some species, but it is not yet certain
whether they are actually lost through irregular mitoses or are fused
with some of the larger chromosomes of the group. Since irregular
mitoses resulting in abnormal distributions of chromosomes are actually
observed in *Drosophila*, and are known to be accompanied by changes
in the hereditary constitution, there can be little doubt that by this and
other means all the types of chromosome groups have been derived from

one or two original types, and that the specific differences exhibited by the organisms are related to these differences in their chromosome complements.

This conclusion is supported by the observations of McClung, Robertson, and others on the chromosomes of other insect families. In the grasshoppers, for example, Robertson (1916) finds that the various chromosomes of the complement form a regular graded series and can be identified in all the species and genera studied. The nearer the relationship the more nearly similar are the chromosome complements. Robertson states that the degree of relationship is as clearly expressed in the nucleus as in the externally visible characters, and that the evidence indicates that descent by variation from a common ancestral series of chromosomes is paralleled by degrees of variation in somatic structures.

Mutations Accompanied by No Change in Chromosome Number.— In most of the examples of this class it has been found that the mutant behaves in a strictly Mendelian manner, usually being recessive to the type from which it sprang. This observation falls into line with the fact that the number and behavior of the chromosomes remain the same; the operation of the Mendelian mechanism is not disturbed. Consequently if the origin of mutations of this class is dependent on the chromosomes it must be due to a change of some kind occurring within the chromosome and affecting the character of its factors, or genes. That such *factor mutations* do take place is the hypothesis upon which a large school of geneticists is attempting to account for many of the observed phenomena of inheritance. Such mutations may involve either a single gene only ("point mutation") or a group of genes occupying a given region of a chromosome ("regional mutation"). Furthermore, they may apparently occur either in the germ cells or in the somatic cells, but seem to be most frequent in the former at the time of maturation, for the reason that only one or two gametes (or spores followed later by gametes in most plants) among the large number produced reveal the presence of the altered gene in their effect upon the offspring. The mutation in the gene must here take place after the multiplication of the germ cells has been nearly or quite completed; otherwise the effect would be manifested by a larger number of gametes (or spores). If, as is true of the majority of cases, the mutation is such as to result in a recessive character, this character does not manifest itself until it meets a similarly mutated gene in the homozygous individual. Thus such an alteration may remain latent for many generations, or may never come to expression at all.

Factor mutations occurring in somatic (meristematic) cells result in what are known as "vegetative mutations." These are of two principal kinds: bud sports and chimeras. In the case of the bud sport, which is believed to be due to a factor mutation in the very young bud, the entire product of the bud—branch, flower, or flower cluster—has a new genotypic constitution and exhibits an appearance often strikingly different from that of the other branches or flowers of the plant. Such a factor mutation occurring in a partially developed shoot or organ results in a chimera, in which a distinct portion of the mature structure, commonly a sharply defined sector, differs genotypically and in appearance from the other portions. By some geneticists vegetative mutations are thought to be due to a somatic segregation of allelomorphic factors in a heterozygous individual and their consequent independent activity in different portions of the body.

The nature of the change which may thus occur in the gene is unknown, since the nature of the gene itself is entirely a matter of conjecture. Although it has been suggested that the gene, because of its relative stability, may be simply a molecule, it is more probably a more complex colloidal aggregate, possibly enzymatic in nature, which is capable of growth and division. As such it could not be expected to be absolutely stable, as some geneticists have thought, but changes would in all likelihood take place occasionally by addition, loss, or rearrangement of the constituent atoms of the molecule. The probable rate of change of the genes in *Drosophila* has been calculated by Muller and Altenburg (1919). What the agencies are which cause such changes is also unknown. Some evidence has been brought forward to show that genes may be modified by external influences, but by many it is regarded as of very doubtful value.. The stimuli to which the genes respond by undergoing some constitutional change are probably for the most part internal ones. Although the number of ways in which a gene may change is limited by its own organization, the possible changes are nevertheless numerous, so that it is very probable that many variations which constitute initial steps in evolution originate in this manner.[1]

Conclusion.—The following paragraphs are quoted from East and Jones (1919, pp. 76 ff.):

"The relation between fact and theory in the Mendelian conception of inheritance is this: Various kinds of animals and of plants were crossed and the results recorded. With the repetition of experiments under comparatively constant environments these results recurred with sufficient regularity to justify the use of a notation in which theoretical factors or genes located in the germ cells replaced the actual somatic characters found by experiment. Later, the observed behavior of the chromosomes justified localizing these factors as more or less definite physical entities residing in them. Now the data from the breeding pen or the pedigree culture plot and the observations on the behavior of the chromosomes during gametogenesis and fertilization are facts. The factors are part of conceptual notation invented for simplifying the description of the breed-

---

[1] See the discussion of these points by Conklin (1919–1920).

ing facts in order to utilize them for the purposes of prediction, just as the chemical atom is a conception invented for the purpose of simplifying and making useful observed chemical phenomena. As used mathematically, both the genetical factor and the chemical atom are concepts, but biological data lead us to believe that the term factor represents a biological reality of whose nature we are ignorant, just as a molecular formula represents a physical reality of a nature yet but partly known.

"With this distinction in mind, one may treat the factor—or the atom—from two points of view, either as a mathematical concept or a physical reality. As a mathematical concept it is the unit of heredity, and a unit in any notation must be stable. If one describes a hypothetical unit by which to describe phenomena and this unit varies, there is really no basis for description. He is forced to hypothecate a second fixed unit to aid in describing the first.

"The point at issue in this connection may be explained as follows: Characters do vary from generation to generation, and the question to be decided is, how much of this variation is due to the recombination of factors (considered now as physical entities) and how much is due to change in the constitution of the factors themselves . . .

". . . We believe there should be no hesitation in identifying the hypothetical factor unit with the physical unit factor of the germ cells. Occasional changes in the constitution of these factors, changes which may have great or small effects on the characters of the organism, do occur; but their frequency is not such as to make necessary any change in our theory of the factor as a permanent entity. In this conception biology is on a par with chemistry, for the practical usefulness of the conception of stability in the atom is not affected by the knowledge that the atoms of at least one element, radium, are breaking down rapidly enough to make measurement of the process possible."

Bibliography at end of Chapter XVIII.

CHAPTER XVI

SEX

In the present chapter attention will be devoted to the cytological aspects of the inheritance and determination of sex.[1] Much that is not cytological in nature will enter into consideration, but it is far from irrelevant: it has a direct bearing on the main cytological problem and must be included in order that the latter may be placed in its proper setting, and that the larger problem of sex may not be misrepresented by being considered only from the cytological point of view.

From early times few essentially biological matters have been of more interest to man than that of the determination of sex. Until recent years this interest has been prompted largely by practical motives: the ability to control sex in man and his domesticated animals is something which has long been desired. Of the many early ideas entertained on the subject the majority were the outcome of defective generalization and superstitious conjecture, and may be encountered in thinly veiled form at the present day, but a review of them all would be out of place here. The modern scientific interest in the problem of sex is far from being a purely practical one. The great bulk of recent research has been done not merely for the sake of the practical benefits which knowledge in this field might confer, but mainly in the hope that it may lead to an understanding of the origin, nature, and biological significance of sex itself, and to a solution of some of the problems of heredity. For this reason studies have not been confined to man and his economically important animals; any animal or plant, no matter how obscure, that will yield evidence is exhaustively investigated, and there can be no doubt that knowledge gained from such studies will, if sound, be directly applicable to practical ends.

Experimental Evidence for Sex-determination.—During the closing decades of the nineteenth century many researches were carried out in the hope of identifying the controlling agency in sex-determination with one or more of the environmental factors. The effects of light, temperature, moisture, and nutrition were examined, and although a number of workers believed their methods to be in a certain measure successful, the results were on the whole inconclusive. Among all the ideas put forward the most suggestive, in view of what has more recently been ascertained,

---

[1] See Correns (1907), Correns and Goldschmidt (1913), Morgan (1913), Doncaster (1914), and the works cited at the beginning of the preceding chapter.

was that of Geddes and Thomson (1889), namely, that the two sexes
differ primarily in the character of their metabolism, the female sex
being characterized by the preponderance of anabolic processes and the
male sex by those essentially katabolic in nature. This conception is
important not only in connection with the question of sex control, but
chiefly with respect to the more fundamental problem of the nature of
sex itself.

The long early period during which sex was looked upon as a character
more or less under the control of the environmental factors was succeeded
by one in which it came to be regarded as something automatically
regulated by some mechanism or condition within the cell, and as relatively unalterable by external agencies. This conclusion, definitely
reached by Cuénot (1899) for animals and by Strasburger (1900) for
plants, received the support of a number of experimental researches on
animals and diœcious plants. Some of the latter will first be mentioned.

It was found by Blakeslee (1906) that in certain strains of a mold,
Phycomyces, there are produced in the germ sporangium two kinds of
asexual spores, which give rise to "plus" and "minus" mycelia respectively. The "plus" (male?) mycelium later produces spores which
develop only into "plus" mycelia and so on indefinitely, while the
"minus" (female?) strain perpetuates only the "minus" condition: in
both cases the sex seems to be fixed by some mechanism functioning at
the time of spore formation.

In certain diœcious mosses (Él. and Ém. Marchal 1906, 1907) two
kinds of spores are produced in equal numbers in the capsule. Those of
one kind develop into male gametophytes (bearing antheridia only) and
those of the other kind produce female gametophytes (with archegonia
only). In no way were the Marchals able to alter the sexes of these
plants. Furthermore, new gametophytes formed by regeneration from
the old ones were just as rigidly fixed as to sex. Protonemata regenerated
from the tissue of the sporophyte, however, gave rise to leafy branches
bearing both antheridia and archegonia. Both sex potentialities were
therefore present in the sporophytic tissue and in the diploid gametophytes regenerated from it, whereas the normal haploid gametophytes
produced from spores were either purely male or purely female.[1] The
gametophytes of *Marchantia* (Noll; Blakeslee 1906) are similarly fixed
as to sex: if propogated repeatedly from gemmæ the sex in any given line
remains the same in spite of alterations in the environmental conditions.
In *Sphærocarpos* Douin (1909) and Strasburger (1909) were able to show
that two spores of a single tetrad produce male gametophytes while the
other two produce females.

The obvious conclusion to be drawn from the above cases is that in
such forms a separation of the sexes takes place during sporogenesis.

[1] Diagrams of these experiments are given by Morgan (1919*a*, pp. 152–3).

Both sexes are represented in the sporophyte (diploid) generation, as shown particularly by the mosses, but the spores, though morphologically similar, are of two distinct kinds: male-producing and female-producing. Since it is precisely at sporogenesis that reduction occurs, the natural inference is that a separation of qualitatively different sex-factors of some kind occurs in the heterotypic division, the sexes of the future gametophytes thus being automatically determined.

That a somewhat similar qualitative difference may exist in the microspores of many angiosperms, resulting in the frequent diœcious condition of the sporophyte, was concluded by Correns (1907) from his researches on *Bryonia* hybrids. He was best able to interpret the phenomena observed on the hypothesis that the eggs are all similar in having the female sex "tendency;" that there are two kinds of microspores and hence two kinds of male gametes, with male and female "tendencies" respectively; and that in the sporophyte the male tendency dominates the female. Darling (1909) was inclined toward a similar conclusion for *Acer Negundo*.

Strasburger (1910) in a general discussion of the subject growing out of his researches on *Elodea, Mercurialis,* and other plants, summarized the situation in plants as follows. In monœcious mosses the separation of the sexes occurs in the somatic divisions at the time the sex organs are formed. The separation has been secondarily joined with reduction, so that in the derived diœcious mosses it occurs at sporogenesis. In the homosporous pteridophytes it takes place at some stage in the gametophyte before the formation of the sex cells, as in monœcious mosses, though in some cases (*Equisetum, Onoclea,* and others) a marked physiological diœcism is present. In heterosporous pteridophytes and all seed plants the gametophytes are diœcious but the sporophytes may be either monœcious (hermaphroditic) or diœcious. In monœcious forms the sexes are separated at some stage prior to the development of megaspores and microspores, whereas in diœcious forms it must take place at some other point in the life cycle, since the two kinds of sporophytes (megaspore-bearing and microspore-bearing) are distinct from their initial stages. There is some evidence to show that such diœcism is due to a differentiation among the pollen grains and hence among the male gametes: some grains have a strong male tendency which dominates the female tendency of the egg, male progeny resulting; while other grains have a weak male tendency dominated by the female tendency of the egg, female offspring being produced. In brief, as sex separation became joined with reduction in forms with monœcious gametophytes, the diœcism of the gametophyte and heterospory (first physiological and then morphological) followed; and this in turn led to the diœcism of the sporophyte also. Finally, in such advanced forms there appears to be a differentiation among the spores of one sex, the microspores, giving male gametes of two types. The sex of the resulting offspring therefore depends here upon the type of male gamete functioning, as is known to be the case in so many animals. It has been suggested by Allen (1919) that the separation of the sex-factors in diœcious seed plants may possibly occur in the division which differentiates the two male nuclei in the pollen tube, rather than at the divisions producing the microspores. That this interpretation cannot be applied to Mendelian factors in general is evidenced by the fact that in maize hybrids the embryo, with very rare exceptions, has been found to be like the endosperm with respect to factors introduced by the pollen parent. So far as these factors are concerned, therefore, the two male nuclei must be qualitatively similar.

Strasburger's conclusion regarding monœcious mosses is confirmed by the recent experiments of Collins (1919) on *Funaria hygrometrica.* In this species the gametophytes arising from spores are bisexual (monœcious), but if gametophytes are produced by regeneration from the antheridia or perigonial leaves of a single "male flower," they all bear antheridia only. Collins thinks it possible that diœcism may have arisen as a result of vegetative multiplication following such a somatic segregation in the tissue of the monœcious gametophyte.

In animals also there is much evidence, aside from that afforded by the chromosomes to be discussed below, in favor of the view that sex is internally controlled. The following illustrative cases may be cited. The egg of the bee may develop either parthenogenetically or after fertilization by a spermatozoön: in the former case a male (drone) results and in the latter a female (queen or worker, depending on the nature of the food). In *Phylloxera* (Morgan 1906, 1908, 1909, 1910) there are two sizes of eggs produced by the females of the second parthenogenetic generation: both may develop parthenogenetically after forming one polar body, the larger ones into females and the smaller into males. Fertilized eggs always develop into females. In *Hydatina* (Whitney 1914, 1916, 1917) the female-producing eggs form one polar body while the male-producing eggs form two. Two kinds of eggs are also produced in *Dinophilus* (Malsen 1906; Nachtsheim 1919), but in this form both are regularly fertilized. In the nine-banded armadillo (Newman and Patterson 1909, 1910) one fertilized egg commonly gives rise to four new individuals, and the four are invariably all male or all female. Analogous instances of polyembryony are also known in insects. Human twins, if "identical" (produced by the same egg), are invariably of the same sex; if "fraternal" (produced by different eggs) they may or may not be of the same sex. It would therefore seem that sex in such cases as these must be determined either in the egg before fertilization or at the moment fertilization occurs.[1]

---

[1] The determination of sex in the egg before fertilization, as in *Phylloxera* and *Dinophilus*, is termed by Haecker "progamic" sex-differentiation; if determination occurs at the moment of fertilization, as in the bee, it is "syngamic" sex-differentiation; and if it occurs after fertilization, as may possibly be the case in some forms, it is "epigamic" sex-differentiation.

Sex-Chromosomes.—The theory of the automatic determination of sex and its relative, if not absolute, fixity has had one of its strongest supports in the results of certain researches on the spermatogenesis of animals. In 1891 Henking noticed in certain insects that half of the spermatozoa contain an extra body, which he thought might be a nucleolus. It was subsequently shown by Paulmier (1899), Montgomery (1901), and de Sinéty (1901) that this body is not a nucleolus but an extra or "accessory" chromosome. Henking's misinterpretation had apparently been due to the fact that the accessory chromosome often does not transform into a portion of the reticulum along with the other chromosomes ("autosomes"), but remains condensed and closely resembles a nucleolus. Half of the spermatozoa in these animals therefore have one more chromosome than the others: hence the male is said to be "heterogametic," or "digametic." It was at once suggested by McClung (1902) that the accessory chromosome in some way determines sex—that eggs fertilized by one kind of spermatozoön develop into females, while those fertilized by the other kind become males. This represents the first attempt to connect a given character with a particular chromosome. An extensive series of researches was now undertaken by Wilson, Miss Stevens, McClung, and a number of other cytologists, who discovered among insects many striking instances of the phenomenon. Accessory chromosomes (also referred to as sex-chromosomes, heterochromosomes, idiochromosomes, x-chromosomes, x-elements, and supernumerary chromosomes) of a number of different types were found, not only among insects, where they are best displayed, but also in certain echinoderms, nematodes, mollusks, and vertebrates, including birds and man. A number of representative cases will now be described.

Male Heterogametic.—In the threadworm, *Ascaris* (Boveri) (Fig. 138)[1] there is in each body cell and primary spermatocyte of the male a single heterochromosome, which seems to be attached to, or to constitute a portion of, one of the four autosomes. At the time of reduction this passes undivided to one daughter cell at the first division and divides at the second, so that half of the sperms only receive it. In the female there are two such heterochromosomes, every egg receiving one. If, now, an egg is fertilized by a sperm without a heterochromosome the resulting individual has only one (that from the egg) and develops into a male.

---

[1] For the sake of brevity and clearness these diagrams are drawn as if only one maturation mitosis occurred in spermatogenesis and oögenesis. It will be understood that there are two divisions, resulting in four sperms instead of the two shown, and in an egg and three polar bodies instead of the two eggs shown. The diagrams merely indicate that two sorts of sperms and one kind of egg are produced, and how this is brought about. In the cases of *Lygæus* and *Prionidus* the number of autosomes shown (4) is not the actual number present. See the review of the subject of sex chromosomes by Wilson (1911).

If, on the other hand, an egg is fertilized by a sperm with a heterochromosome, the resulting individual receives two, one from each gamete, and this individual develops into a female.

In *Ancyracanthus* (Mulsow 1912) (Fig. 138) the male has a single heterochromosome which, since it has no homologue with which to pair, passes to half the sperms, while in the female there are two such elements, every egg receiving one. The two types of union result in individuals of the two sexes, as in *Ascaris*. In *Ancyracanthus* Mulsow states that the five and six chromosomes can actually be counted in the living spermatozoa.

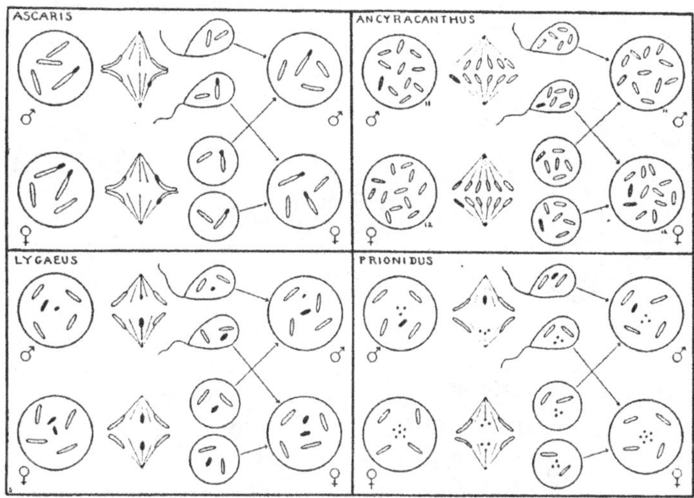

FIG. 138.—The behavior of the sex-chromosomes in *Ascaris* (Boveri), *Ancyracanthus* (Mulsow, 1912), *Lygæus* (Wilson, 1905), and *Prionidus* (Payne, 1909).

In *Lygæus* (Wilson 1905) (Figs. 138; 139, *A*) there are in the male two heterochromosomes, one small and one large (an "XY" pair); in the female there are two large ones ("XX"). Half the sperms receive the X and half the Y, and every egg has an X. Fertilization by an X sperm results in a female (XX), and by a Y sperm in a male (XY).

In *Prionidus* (Payne 1909[1]) (Figs. 138; 139, *B*) the male has three small heterochromosomes and also a much larger one. At the time of reduction the three small ones behave as a unit and pair with the large one: half of the sperms therefore carry the former and half the latter. In the female there are six small heterochromosomes, and the eggs are all alike in having three each. Fertilization now results in females with six small elements and males with three small and one large.

[1] In this paper Payne gives diagrams of several other types of heterochromosomes.

In the fruit fly, *Drosophila melanogaster* (Stevens 1907; Morgan 1911; Metz 1914) (Fig. 140), there are four pairs of chromosomes, including in the male an *XY* pair and in the female an *XX* pair. Reduction in

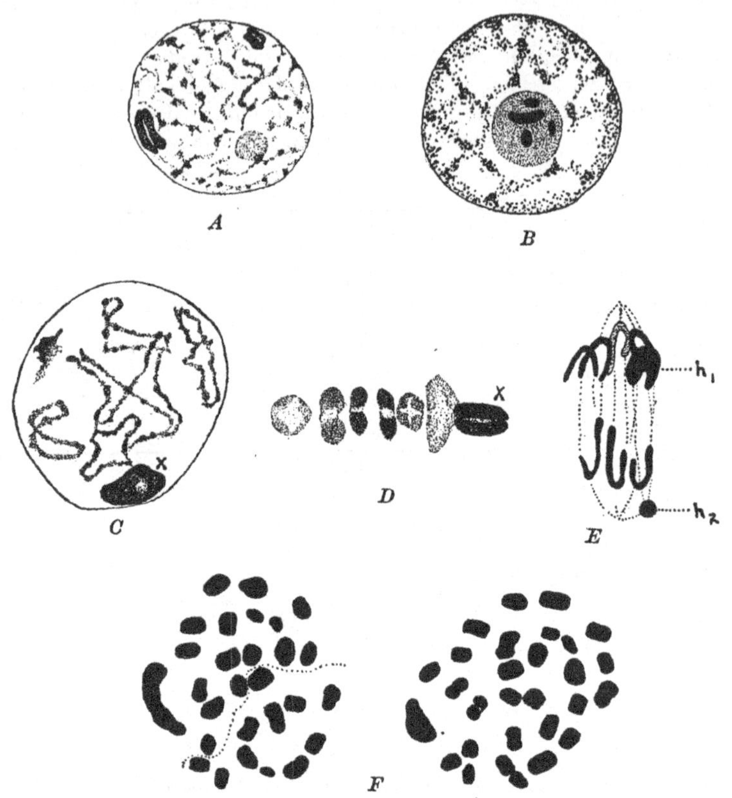

FIG. 139.—Sex-chromosomes in various insects.

A, spermatocyte of *Lygæus*, showing the *X*-chromosome at left and *Y*-chromosome above, both split. × 2250. (*After Wilson.*) *B*, prophase in spermatocyte of *Prionidus*, showing sex-chromosomes enclosed in plasmosome. × 2294. (*After Payne.*) *C*, prophase in spermatocyte of *Protenor*. × 2250. (*After Wilson.*) *D*, metaphase of heterotypic mitosis in spermatocyte of *Protenor*. × 2250. (*After Wilson.*) *E*, anaphase of heterotypic mitosis in spermatocyte of *Musca domestica; h*, heterochromosomes. × 1500. (*After Stevens.*) *F*, the two daughter chromosome groups in the anaphase of the heterotypic mitosis in the oöcyte of *Phragmatobia fuliginosa*, showing 28–29 distribution. × 4080. (*After Seiler.*)

spermatogenesis gives sperms of two sorts: all contain four chromosomes, but in half of them one of the four is the *X*, and in the other half it is the *Y*. Since every egg contains an *X*, two kinds of union are possible at fertilization: an *X* with a *Y*, giving a male fly, and an *X* with an *X*, giving a female. In this case, a typical example of the *XY* form of sex inheritance, extensive researches have shown that the *Y*-chromosome carries no factors for sex; the presence of one *X*-chromosome is associated with maleness and that of two *X*-chromosomes with femaleness. Morgan (1914, 1919*b*) and Morgan and Bridges (1919) find that gynandromorphism frequently appears in *Drosophila* females as the result of the elimination of one sex-chromosome in abnormal mitosis.

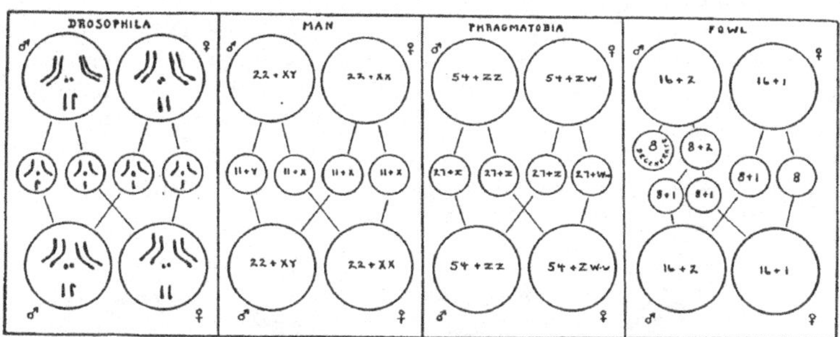

Fig. 140.—The behavior of the sex-chromosomes in *Drosophila* (Stevens, Morgan, Metz), Man (Wieman), *Phragmatobia* (Seiler), and the fowl (Guyer).

Malone (1918) reports that there are present in the spermatocyte of the dog 10 pairs of autosomes and one large unpaired *X*-chromosome. The *X* passes undivided to one pole in the first mitosis and divides in the second, so that half of the spermatids, and hence spermatozoa, receive an *X* while half do not. When measurements of these spermatozoa are plotted a bimodal curve results, showing that the chromosome difference is correlated with a size dimorphism. The same condition, except for the number of autosome pairs, is reported for the spermatozoa of horses, pigs, and cattle (Wodsedalek 1913, 1914, 1920).

In the case of man also the evidence at hand indicates a digametic condition on the part of the male, but certain striking discrepancies in the findings of various investigators have afforded a puzzle which up to the present time has not been satisfactorily solved. Flemming (1898) counted 24 chromosomes in the cells of the cornea, and Duesberg (1906) found 12 in the spermatocytes. The same numbers were found by Montgomery (1912). In 1910 Guyer reported that the spermatogonia of the negro contain 22 chromosomes; these in the spermatocyte form 10 bivalents and two distinguishable accessories. At the first maturation mitosis both of the latter go to one pole and at the second mitosis both divide, so that half of the sperms have 10 chromosomes and half have 12; the two accessories in the latter case are visible as "chromatin nucleoli" in the resting stage. This difference in the gametes Guyer regarded as probably associated with sex-determination. Gutherz (1912) failed to confirm Guyer's report of a dimorphism among the sperms, though he also observed the accessories. Guyer in 1914 reasserted his conclusions of 1912, but added that he was finding a much larger chromosome number in the cells of the white man.

Von Winiwarter (1912), also working upon the white man, found in the spermatogonia and spermatocytes 47 chromosomes, including one accessory. Since this accessory passes undivided to one pole in the first mitosis and divides in the second, half of the sperms receive 23 and half 24 chromosomes. In the cells of the female there are 48. From these data von Winiwarter logically concluded that the egg has 24 chromosomes, and that it develops into a male when fertilized by a sperm with 23 chromosomes, and into a female when fertilized by one with 24. Such a large number is found in the white man by Evans[1] also, but he finds 48 chromosomes rather than 47 in the spermatogonia, indicating the presence of an XY pair as in *Drosophila*.

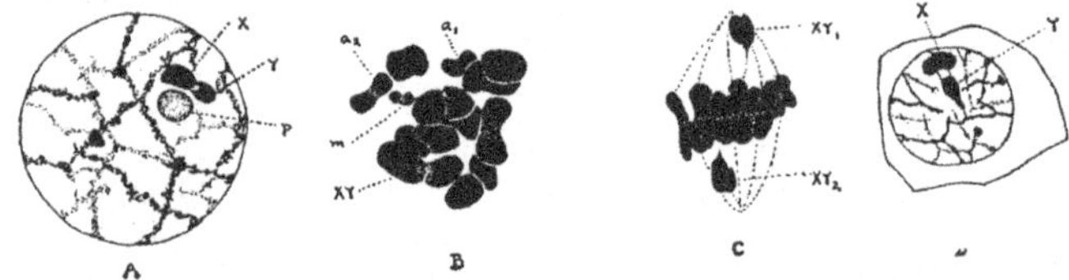

Fig. 141.—Sex-chromosomes in man. *A*, primary spermatocyte, negro. *B*, same, white. *C*, metaphase of heterotypic mitosis, negro. *D*, interkinesis, white. *XY*, the sex-chromosomes; *P*, plasmosome. (*After Wieman*, 1917.)

Wieman (1917), using both negro and white material, finds 24 chromosomes in the spermatogonia, two of them being distinguishable as an unequal XY pair which remains condensed while the autosomes form the reticulum (Figs. 140, 141). In the spermatocyte 12 pairs are evident, including the XY pair. At the first maturation mitosis the 11 autosome pairs separate into univalents as usual, but the X and Y divide longitudinally; thus each daughter cell (secondary spermatocyte) receives 11 autosomes and an XY pair. At the second mitosis the 11 autosomes divide longitudinally in the normal fashion and the X and Y separate. As a result all of the sperms receive 12 chromosomes: in half of them one of the 12 is the X and in the other half it is the Y. Although for a time it appeared that the white man had twice as many chromosomes as the negro, a difference ordinarily great enough to mark them as distinct species, Wieman shows clearly that in his material the two have the same number, and is inclined to regard von Winiwarter's material as in some way abnormal. Sex inheritance in man is evidently of the XY type, as Wieman's researches and genetic data indicate with considerable clearness; but why some material should plainly show twice as many chromosomes

---

[1] Unpublished work cited by Babcock and Clausen, p. 538.

as other material is a question which only future investigation can answer. It is not improbable, however, that a segmentation of the chromosomes at points of constriction may be mainly responsible for this condition.

In all of the cases reviewed above the male produces two sorts of spermatozoa differing visibly in chromatin content: in the language of Mendelism, the male is "heterozygous for sex." The female produces but one kind of egg; she is "homozygous for sex." The sex of the offspring is clearly dependent on the kind of spermatozoön which functions, and is therefore definitely correlated with the chromosome mechanism.

Female Heterogametic.—There are also on record a number of cases, chiefly among moths and birds, in which the female produces two kinds of eggs differing in chromosome content, while the male produces but one kind of spermatozoön: the female is heterozygous for sex, and therefore heterogametic, while the male is homozygous. Certain cases of this type will now be reviewed.

In the moth, *Phragmatobia*, Seiler (1913) has described the following condition. In the male the somatic number of chromosomes is 56, including 54 autosomes and two Z-chromosomes[1] (Figs. 139 *F*; 140). Each sperm receives 28 (27 + Z). In the female the somatic number is likewise 56, but includes a *ZW* pair instead of the *ZZ* pair of the male. Half the eggs receive 27 + Z and the other half 27 + *Ww* (the *W*-chromosome breaks temporarily into two parts during maturation). An egg with 28 (27 + Z) chromosomes fertilized by a sperm with 28 (27 + Z) develops into a male moth with 56 (54 + ZZ). An egg with 29 (27 + *Ww*) chromosomes fertilized by a sperm with 28 (27 + Z) develops into a female moth with 57 (54 + ZWw). The *W* and *w* subsequently reunite to form a single *W*, both sexes then having the same number, 56. Since some embryos show more than the normal number of chromosomes Seiler thinks it probable that the Z-chromosome is compound and may under certain conditions subdivide into smaller elements. The same investigator has recently (1919) reported a digametic condition in the female in two other moths, *Talæporia tubulosa* and *Fumea casta*. In the former the eggs have 29 and 30 chromosomes, and in the latter 30 and 31.

In the moth, *Abraxas grossulariata* (Doncaster 1914), sex inheritance is apparently of the *WZ* type, though there is often an aberrant behavior on the part of the chromosomes which has not been entirely explained.

In the common fowls Guyer (1909, 1916) has made observations which he interprets as follows (Fig. 140). In the male there are 18 chromosomes: 16 autosomes and two accessories. Both of the latter go to one pole in the first maturation mitosis, and in the second mitosis they sepa-

---

[1] It is customary to refer to the sex-chromosomes in species with sexually heterozygous females as *W* and *Z*, instead of *Y* and *X* as in the more common sexually heterozygous males.

rate. Half the spermatids, and therefore sperms, receive nine chromosomes (8 + 1 accessory), while the other spermatids failing to receive an accessory apparently degenerate. In the female there are 17 chromosomes: 16 autosomes and one accessory. Since the accessory passes undivided to one pole at the first maturation mitosis and divides at the second, half of the eggs receive nine chromosomes (8 + 1 accessory) and half receive eight. An egg with nine fertilized by a sperm with nine develops into a male with 18 (16 + 2 accessories). An egg with eight fertilized by a sperm with 9 develops into a female with 17 (16 + 1 accessory).

Cases of Parthenogenesis.—The cytological phenomena in those animals reproducing in part by parthenogenesis (see p. 357) are of much interest in this connection. In the honey bee the male, which develops from an unfertilized egg, has the haploid number of chromosomes in his cells, whereas the female, arising from a fertilized egg, is diploid. Similar in some respects is the case of the gall-fly (*Neuroterus*), in which eggs that have undergone reduction develop into haploid males, while other eggs are formed without reduction and develop into diploid females. In the male-producing eggs of *Phylloxera* there are two sex-chromosomes, two others being lost in the polar body; in the female-producing egg all four are present. Two kinds of sperms are produced, half of them with a sex-chromosome and half of them without it. The latter kind degenerate, leaving only the former functional. All eggs, if fertilized, develop into females. In *Hydatina senta* those eggs producing one polar body and developing into females are diploid, whereas those giving off two polar bodies and developing into males are haploid. In all of these cases maleness accompanies the haploid, and femaleness the diploid condition[1].

Plants.—Up to the present time a visible chromosome difference between the two sexes in plants has been established only in *Sphærocarpos*, the genus of diœcious liverworts in which Douin (1909) and Strasburger (1909) found two of the spores of a single tetrad to be male and the other two female. In *Sphærocarpos Donnellii* (Allen 1917, 1919) (Figs. 142, 143) there are in the cells of the female gametophyte seven autosomes which differ somewhat in length, and one very large *X*-chromosome. In the cells of the male gametophyte there are seven autosomes and a very small *Y*-chromosome. The sporophyte therefore has eight pairs: seven autosome pairs and the *XY* pair. Although all the stages in the divisions at sporogenesis have not been seen, the evidence is sufficient to show that the *X* and *Y* separate in the heterotypic mitosis and divide longitudinally in the homœotypic. Two spores of a tetrad therefore receive an *X*-chromosome in addition to the seven autosomes; these spores develop into female gametophytes. The other two spores of the tetrad receive

---

[1] Compare the case of the frogs developing by artificial parthenogenesis, page 319.

the Y in place of the X and develop into male gametophytes. The same condition has been reported for *Sphærocarpos texanus* (Miss Schacke 1919).

Although the situation in *Sphærocarpos* suggests the XY type of sex inheritance in *Drosophila* and other forms, it differs in several respects,

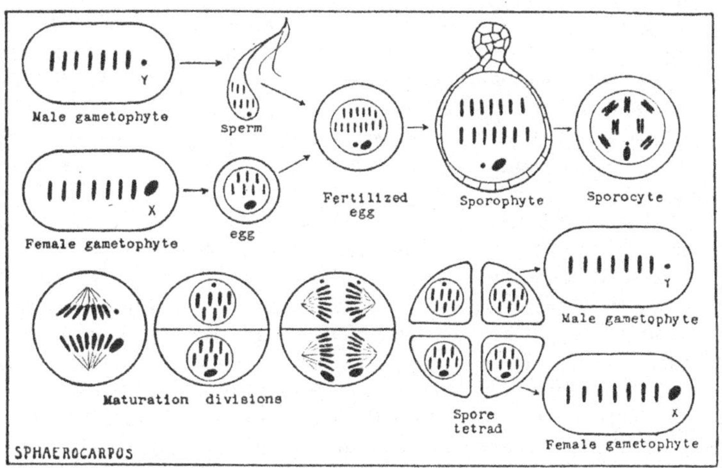

Fig. 142.—The history of the chromosomes in the life cycle of *Sphærocarpos*. (*After data of C. E. Allen, 1917, 1919.*)

as Allen (1919) points out. In *Sphærocarpos* the separation of the XY pair results in the production of two kinds of spores which develop directly into haploid organisms (gametophytes) of two sexes, whereas in *Drosophila* the corresponding separation results in two sorts of male gametes which determine the sexes of the diploid organisms developing from the eggs they fertilize. Furthermore, in animals with sex-chromosomes some forms show the presence of XX to be correlated with femaleness and X or XY with maleness (male heterozygous for sex), while in other forms XX is correlated with maleness and X or XY with femaleness (female heterozygous for sex[1]). There is evidence to show that the Y-chromosome carries no sex-factors in these cases,[2] though its absence may result in sterility (Bridges). (See Chapter XVII.) In *Sphærocarpos*, on the

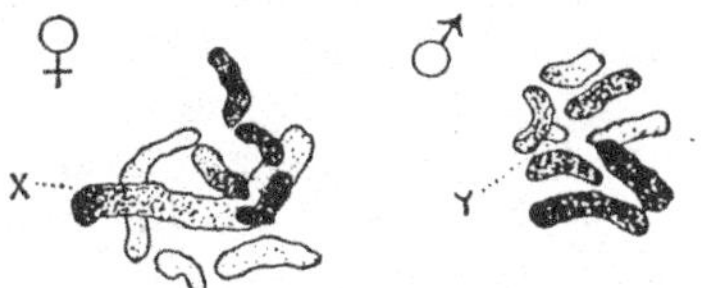

Fig. 143.—Chromosome groups from male and female gametophyte cells of *Sphærocarpos Donnellii*. (*After C. E. Allen, 1919.*)

[1] Sex-chromosomes referred to in this case as Z and W rather than X and Y.

[2] See in this connection Castle 1921.

other hand, the presence of X is correlated with femaleness, Y with maleness, and XY with the non-sexual condition of the sporophyte. Here the Y is apparently as important as the X in the transmission of sex-factors. Allen suggests that secondary sexual differences of the gametophytes, such as size, may be connected with the relative amounts of chromatin in the nuclei: the female gametophyte, having the large X-chromosome and therefore a distinctly greater mass of chromatin, develops more rapidly and becomes much larger than the male gametophyte with its small Y-chromosome. The primary sexual differences he regards as due to other factors.

Although *Sphærocarpos* affords the only known example of heterochromosomes in plants it is not improbable that other cases will be discovered.

Conclusion.—In all of the organisms included in the foregoing review the individuals of the two sexes differ visibly in their chromosome complements. Moreover, in most of them the sexual differences are definitely correlated with special distinguishable chromosomes, which are accordingly known as sex-chromosomes. The distribution of these bodies at the time of reduction results in the production of two kinds of male gametes or two kinds of female gametes, and in at least one case two kinds of spores. In all of these the chromosome differentiation in the cells correponds to the sexual differentiation of the organisms into which they develop. The conclusion appears unavoidable that *the differentiation of the sexes is here determined by a cell mechanism,* and that *the heterochromosomes have a definite causal relationship with sex.* How close this relationship may be, and to what degree it is a fixed one, are as yet by no means clear, but it is beyond question that the heterochromosomes are not the sole determining cause of sex, as some workers have hastily concluded. To this question we shall subsequently return.

Sex-chromosomes and Mendelism.—The heterochromosome phenomena are intimately bound up with the whole matter of Mendelian inheritance. According to the Mendelian interpretation the sexes are due, like other heritable characters, to factors carried by the chromosomes—by the heterochromosomes where these are present. The approximate $1:1$ ratio of the sexes in most organisms is accounted for in the following manner. Referring to our typical Mendelian pair of characters in the pea, tall and dwarf, it is found that when a plant heterozygous for tallness (Tt) is crossed with a pure recessive (tt) the resulting offspring are half tall (Tt) like one parent and half dwarf (tt) like the other, a $1:1$ ratio. If it is assumed in a similar manner that there is a pair of factors for sex, one sex (the male, Correns; the female, Bateson) being heterozygous and the other a homozygous recessive, a $1:1$ ratio of the sexes results.

Largely because of the observed behavior of sex-chromosomes the

Mendelian interpretation has been restated as follows. There is a single factor for sex. In some organisms the presence of two of these factors is correlated with femaleness and one with maleness (Fig. 144, *A*): the male, having only one sex-factor (*S*), is heterozygous and produces gametes of two kinds, with and without the factor; the female, having two sex-factors, is homozygous and produces eggs of one kind, with one sex-factor each. Two types of union are here possible, giving males and females with one and two sex-factors respectively. This interpretation is directly applicable to those cases in which the male has one sex-chromosome and the female two (*Ancyracanthus*, *Ascaris*), and also to those having an *XY* pair in the male and an *XX* pair in the female (*Drosophila*, *Lygæus*). Each sex-factor is thus thought to be located in an *X*-

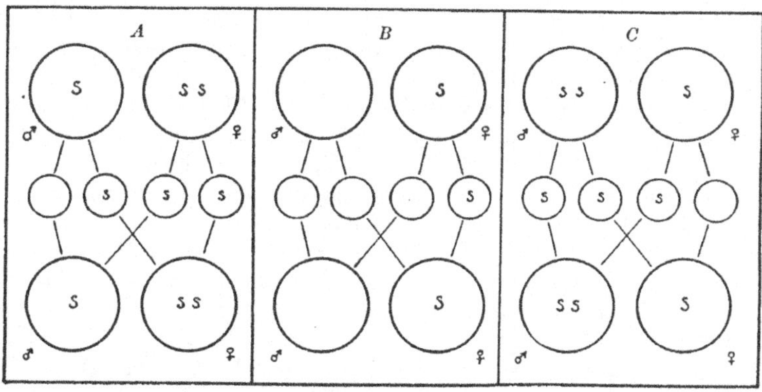

Fig. 144.—Mendelian interpretations of sex-inheritance. Explanation in text.

chromosome. In other organisms (moths and birds) these conditions are reversed (Fig. 144, *C*), the presence of two sex-factors being correlated with maleness and one with femaleness. The female is thus heterozygous and produces eggs of two kinds, with and without the factor. In the next generation the male receives two sex-factors (in the *Z*-chromosomes of the egg and sperm) and the female one (in the *Z*-chromosome of the sperm; the *W*-chromosome of the egg carries no sex-factors). In view of these two contrasted conditions as regards the quantitative relationship between factors and sex, it is probable that the sex-factors carried by the *X*-chromosomes are in some manner different from those in the *Z*-chromosomes. It has been suggested that in some cases (Fig. 144, *B*) the male may have no sex-factors at all, the heterozygous female thus having one more sex-factor than the male, as in the homozygous females of Fig. 144, *A*.

Experimental Alteration of the Sex Ratio.—Although most organisms approximate closely the 1 : 1 sex ratio called for on the basis of the

Mendelian theory of sex, constant deviations from this ratio are frequently found. Still more significant is the fact that in many cases the ratio can be markedly altered by changing the environmental conditions. Thus R. Hertwig (1906, 1912) and Kuschakewitsch (1910) found that if the eggs of the frog are allowed to become over-ripe before fertilization, in which case they take up an abnormal amount of water, the resulting individuals show an unusually high percentage (even 100 per cent) of males. Conversely, Miss King (1907–1912) lowered the water content of toad eggs, and with a mortality of only a little over 6 per cent obtained 80 per cent females.·

In *Dinophilus*, as already noted, there are two kinds of eggs laid: large ones developing into females and small ones developing into males. Malsen (1906) found that by altering the temperature the relative proportion of the two sexes could be changed, but this effect was brought about through an influence on the laying of the eggs: both kinds were produced as usual, but the laying of one kind was hindered.

The rotifer, *Hydatina senta* (Whitney 1914, 1916), if scantily fed on *Polytoma*, continues to produce generations of parthenogenetic females, but when copiously fed on *Euglena* females appear which lay male-producing eggs, and sexual reproduction then occurs. According to Shull and Ladoff (1916) the percentage of males is here correlated with the supply of oxygen which counteracts certain agencies (accumulated substances in the water) tending to decrease male production. Whitney (1919), however, contends that oxygen is not a factor affecting sex in *Hydatina*.

In interpreting such results as these considerable care should be exercised in distinguishing an actual determination of the sex of an individual from a number of other phenomena which, though they may appear like sex-determination, are not to be regarded as such in the strict sense. In many cases in which changed environmental factors have been shown to have an influence on the sex ratio it is clear that the results are not due to an actual reversal or determination of the sex in any individual, but rather to the fact that the new conditions imposed have caused a greater mortality among the eggs or embryos of one sex, so that those of the other sex preponderate. Although the ratio of the sexes may here be subject to an experimental control, the sex of no given individual is actually determined or altered.

Indirect control of another type is seen in organisms whose sex is dependent upon the form of reproduction (zygogenetic or parthenogenetic). Environmental conditions may in such cases influence the form of reproduction resorted to, and therefore the sex of the animals resulting; but the sex of no individual, once started, is altered. Morgan points out that the change of diet in *Hydatina*, instead of altering sex directly, induces the formation of a new type of female which may either function sexually or produce eggs which develop parthenogenetically into males.

Another case in point is that of *Dinophilus*, in which abnormal temperture conditions prevent the laying and development of eggs from which the individuals of one sex normally arise. As Thomson (1913, p. 502) remarks, ". . . if nutritive and other environmental influences are operative, it is, in the main, by affecting the production and survival of sexually-predestined germ cells."

A clear case of sex-determination in the strict sense would be one in which a given parthenogenetic egg, fertilized egg, spore, or young individual, in a unisexual organism, could be made to develop at will into either sex; or in which such an organism, already showing the essential characters of one sex, could be made to develop into the other sex (sex-reversal). Evidence that in certain cases such a determination or reversal can actually be accomplished has had much to do with the development of recent metabolic theories of sex.

Metabolic Theories of Sex—*Animals.*—According to the metabolic theories of sex, the difference between the two sexes is primarily one of metabolism, the difference being not necessarily in the kind of metabolic processes, as Geddes and Thomson thought, but more probably in their rate or level. One of the most prominent exponents of this type of theory is Riddle (1912, etc.), who has continued the important researches on pigeons begun by Whitman many years ago. At each laying these birds produce a pair, or clutch, of eggs. Under normal conditions the first egg laid develops into a male and the second into a female. By a careful analysis of a large number of eggs Riddle has been able to show that the yolk of the first egg is the smaller, and has the smaller amount of storage material (fats and phosphorus-bearing compounds), the higher water content, and the higher oxidizing capacity, or higher metabolism. The second or female-producing egg therefore differs in having a larger yolk, more storage material, less water, and a lower metabolism. The chromosomes of the pigeon are not accurately known, but it is probable, as indicated by the behavior of the sex-linked factors (see next chapter), that the *WZ* type of sex-chromosome differentiation found in other birds is present here. Whatever their cytological differences may be, *the two sexes are characterized in the egg stage by different levels of metabolism, the male having the higher level and the female the lower;* and this difference has been shown (as indicated by the metabolism of the blood) to persist into the adult stage.

The seasonal change in the amount of storage, and consequently in the level of metabolism, is definitely correlated with the percentage of a given sex and the degree of its manifestation. As the season advances both eggs of the clutch store more energy-containing materials, and the birds developing from the second (female-producing) egg, while often

24

quite "masculine" in their secondary sex characters early in the season, become increasingly "feminine" toward the end of the season.   Moreover, the total percentage of females increases.   By influencing storage, water content, and the general metabolic condition of these eggs Riddle has been able to induce experimentally an actual reversal of their natural sex tendencies.   Hence he concludes that sex is a quantitative, modifiable, and "fluid" character; the two sexes do not represent two qualitatively distinct, mutually exclusive properties, but are rather two conditions of one general property—two levels in a continuous series of metabolic states passing gradually one into another.   Accordingly, if the two levels normally maintained can be sufficiently altered a series of intergrades between the two sexes and an alteration of the sex itself should be possible, and this Riddle has apparently accomplished.

On the basis of this metabolic theory of sex Riddle interprets the results obtained by Miss King with toad eggs, by Hertwig with those of the frog, by Whitney and Shull with rotifers, and by other investigators to be mentioned below.   The correlation of high and low water content with maleness and femaleness respectively in toads and frogs, and that of change of food and increased oxygen supply with maleness in rotifers, are held to be in harmony with similar correlations which have been shown to exist in pigeons.

The theory that sex is a quantitative, reversible state closely associated with metabolic conditions is strongly supported by the researches of Goldschmidt (1916*ab*, 1917), Banta (1916), and Lillie (1917).   In the gypsy moth, *Lymantria dispar*, which is cytologically heterogametic, Goldschmidt has been able to induce experimentally a large series of "sex intergrades" between the male and female conditions.   It appears that an individual of either sex, after beginning its development, may be made to develop the characters of the other sex partially or completely, the degree of alteration depending upon the time at which the change sets in.   Goldschmidt concludes that normal individuals of both sexes must have both sex capabilities, the sex manifested depending upon the relative strength with which they are caused to act by certain conditions.   He thinks it probable that the hereditary characters have their material basis in enzymes or substances of a similar nature, those associated with femaleness and maleness being termed "gynase" and "andrase" respectively.   Although for a time he interpreted the behavior of the sexes in *Lymantria* on the basis of Mendelian factors, he is now inclined to view this form of explanation as inadequate.   The chromosome behavior in these moths is not well known, but breeding experiments with an intersexual female functioning as a male show the results which would be expected on the assumption that she had the *WZ*-chromosome constitution.   This would mean that the moth in question had had its sex reversed without any visible change in its chromosome complement, but this intrepretation awaits cytological confirmation.

In the phyllopod crustacean, *Simocephalus vetulus*, Banta (1916) finds that under certain experimental conditions ". . . the same individual, even the same sex gland, may develop eggs and sperms at the same time or sperms at one time and eggs at another time." Banta looks upon sex as not a fixed or definite state but rather "a purely relative thing" dependent upon the general physiological state of the organism which in turn is under the influence of environmental factors. He also reports sex intergrades in *Daphnia longispina* (1918).

Lillie (1917*abc*), as the result of a study of twins in cattle, concludes that "sex-*determination* in mammals is zygotic, but it does not imply an irreversible tendency to the indicated sex-*differentiation*. Intensification of the male factors of the female zygote from the time of onset of sexual differentiation by action of sex hormones may bring about very extensive reversal of the indicated sex-differentiation" (1917*c*).

Intersexes have recently been described in *Drosophila simulans* by Sturtevant (1920). All the intersexual flies are modified females and show the same grade of intersexuality. Sturtevant believes the condition to be due to the action of a mutant gene. If this interpretation is applied generally to the phenomenon of intersexuality it is evident that certain genes at least are modifiable by environmental conditions, for the reason that such conditions so often evoke the intersexual state.

A very striking instance of the control exercised by environmental factors over the sex of a given individual is found in *Bonellia* (Baltzer 1914). In this gephyrean worm the male is very small and degenerate, and lives parasitically upon the long proboscis and in the nephridium of the much larger female. If young individuals are kept in an aquarium by themselves they develop into females, but if they are placed with mature females they settle upon the proboscees of the latter and develop into males. By allowing them to remain on the proboscees for varying lengths of time Baltzer was able to obtain many intergrades between the typical male and female conditions. It is plain that the proboscis exercises an influence over the sex of the young animal, but whether this is through a secretion or some other agency is not known.

A somewhat similar case is that of *Crepidula plana*, described by Gould (1917). This mollusk is hermaphroditic but completely protandric, *i.e.*, the male and female phases are entirely separate in time, the former developing first. The development of the male phase is dependent upon the presence of a larger individual (not necessarily a female) of the same species. If a male is removed from the neighborhood of a larger individual the male organs degenerate, and after a period of sexual inactivity the animal becomes a female. The nature of the stimulus exerted by the larger individual has not been ascertained.

Plants.—Experiments of a corresponding nature on sex modification in plants are on the whole less conclusive than those on animals, for the reason that many of them have been carried out with angiosperms, in which hermaphroditism is so common and which are regarded by some botanists as all potentially bisexual. As a case illustrating the latter point may be cited the hemp plant, *Cannabis sativa*. This species is normally unisexual, but if the flowers are removed it will produce flowers of the other sex (Pritchard 1916).

Intersexes of many grades between the normal plants in diœcious angiosperms are frequently found, as in *Myrica Gale* (Davey and Gibson 1917), *Cannabis sativa*, *Salix amygdaloides*, and *Morus alba* (Schaffner 1919*ab*), whereas the relative proportions of maleness and femaleness in hermaphroditic forms are apparently very easily influenced by the environment. In *Plantago lanceolata* (Bartlett 1911; Correns 1908; Stout 1919) the stamens and pollen are developed in various degrees in different flowers, or even in the same flower; in some cases they are only rudimentary, leaving the flowers functionally female. Stout inclines to the view that diœcism results from the suppression of femaleness or maleness in organisms originally monœcious, and points out that the many "degrees of maleness" in *Plantago* fill the gap completely in this case. He accordingly concludes, in agreement with Riddle, Banta, and Goldschmidt, that sex is a labile, reversible character; that maleness and femaleness, both present in the somatic cells of all sporophytic individuals, are relative and not absolute conditions; and that "sex-determination, at least in hermaphrodites, is fundamentally a phe-nomenou of somatic differentiation that is ultimately associated with processes of growth, development, and interaction of tissues, and subject to modification or even complete determination by them." The sex-chromosome theory is held to be inadequate in the case of hermaphro-dites: sex is an "epigenetic," not a "preformed" character.

Yampolsky (1919, 1920) thinks it probable that male and female gametes with graded potencies are produced in *Mercurialis annua*, and that the sexes cannot be due to a segregation of sex-factors at reduction.

The sex of the haploid phase (gametophyte) of the plant life cycle, when this phase is unisexual (diœcious), behaves as a much more nearly irreversible character, as shown by the experiments on mosses, *Marchantia*, and *Phycomyces* cited early in the present chapter. How widely this holds true in thallophytes and bryophytes is not known. Allen regards the quantitative theory of sex as quite inapplicable to the case of *Sphærocarpos*. In the homosporous pteridophytes the gametophytes are commonly monœcious, and the fact that here, as in many liverworts, the male organs appear early and the female organs later suggests a physiological basis of sex-determination. Some fern gametophytes, on the contrary, are diœcious under all ordinary conditions. In *Onoclea* (Wuist 1913), such a normally diœcious form, the female gametophytes under certain experimental conditions produced antheridia in addition to their archegonia, but the male gametophytes could not be made to develop archegonia. In *Osmunda regalis japonica* and *Asplenium nidus*, which are monœcious, Nagai (1915) found that the concentration of the Knop's nutrient solution used has a controlling influence over the kind of sex organ appearing, the number of antheridia in general decreasing with the concentration.

These results, taken together with the fact that many gametophytes, especially those of homosporous pteridophytes, are monœcious, show that the capabilities of both sexes can be present in the haploid nucleus, as Strasburger thought, although in many forms the visibly developed sexes may be automatically determined by some cell mechanism.

General Discussion.—We have now reviewed some of the evidences which have led to two general theories of sex and its determination. One theory represents an attempt to account for the phenomena in question on the basis of a morphological cell mechanism whereby Mendelian or other factors are distributed in a definite and fixed manner, whereas on the other theory it is held that they are the results of a physiological differentiation manifesting itself chiefly in alterable levels of metabolism. Although these two conceptions may appear to be mutually exclusive if expressed in too uncompromising a form, both must contain elements of truth. It is beyond question that the two manifestations of sexual differentiation, the physiological and the morphological, are both of importance and cannot be ultimately irreconcilable: our task is to determine their relative significance and to discover the nature and degree of their mutual interdependence. It seems clear that the digametic condition when present in diœcious forms does regulate the ratio of the sexes: under all ordinary circumstances the sex of the individual is here dependent upon the kind of gamete which gives rise to it (or the kind of spore in the case of certain gametophytes). But it is to be emphasized that the dimorphism shown by such gametes or spores is not entirely a morphological one, or even mainly so: in many cases no morphological difference can be detected although the two are clearly different in physiological behavior, as shown by the spores of *Phycomyces* and certain bryophytes. It is generally inferred that here a structural difference, although invisible, is nevertheless present.

In this connection it may be recalled that the differences between the male and female gametes, irrespective of any differentiation which may be present among those of either kind, are both physiological and morphological. The primary characters of sex are those possessed by the gametes themselves, and the principal distinction between the male and female gametes seems to be a physiological one which is manifested in their mutual attraction and fusion. Any visible morphological differentiations that they may possess are to be regarded as secondary adaptations to unlike functions, because of the fact that in many of the lower organisms there is no discernible structural difference between the male and female gametes, and further because structural differences may be annulled in certain instances (some gregarines). Any material differences present in visibly similar gametes are more probably chemical in nature, the structural difference being of a molecular order. Indeed, there is probably no physiological difference without a structural difference of this sort. If the term structure be extended to include molecular constitution the discussion over the relative priority of structural and physiological differentiation becomes futile, for at this level the two are aspects of one and the same change. It is only when we restrict the term structure to the grosser, visible features that we can speak of physiological differentiation as preceding alteration in structure. Ultimately structural and functional changes are indistinguishable. Just as in the gametes of the two sexes, so also in the unisexual individuals which the gametes produce there may be striking differences of both morphological and physiological natures; but if we use the term morphological only with reference to visible features the primary distinction between the sexes in organisms of all grades is apparently one of physiological state, this distinction and its result (sexual reproduction) being of the greatest biological importance.

Taking into consideration all organisms, low and high, it seems probable that any dimorphism among the gametes of one sex or the other has in some way been developed in connection with the maintenance of the above mentioned difference in physiological state in organisms of a certain level of advancement. Different organisms show all degrees in the differentiation among the gametes of one sex: some are marked by an absence of any visible difference either in the gametes or in the hermaphroditic individuals produced; in others the gametes (of one sex) are visibly similar but result in male and female individuals in regular ratio; and finally there are those in which the gametes are of two kinds both physiologically and morphologically, the two kinds controlling the production of individuals of the two sexes. It is in organisms of the last type especially that the question of sex-determination finds the adherents of the chromosome-Mendelian theory and those of the metabolic theory in disagreement. Is the sex of the individual inevitably dependent upon the type of gamete functioning (usually the kind of sperm fertilizing the egg), or is it possible to overcome the effect which the chromosome mechanism may have by influencing sufficiently the metabolism of the organism? If the sex of an individual is so changed, does the chromosome mechanism undergo a corresponding alteration?

Those who have developed the chromosome-Mendelian theory have perhaps too often held that the two sexual states, maleness and femaleness, are in their ultimate analysis mutually exclusive—that they are two fundamental and qualitatively different alternative characters depending unalterably upon unit factors. The hereditary factors or genes are usually regarded by geneticists as unmodifiable except by sudden mutations. Failure to distinguish between the modification of genes and the modification of the interaction of genes during ontogenesis has led many to the view that the sex of the individual must be rigidly fixed at fertilization in digametic and dioecious forms, or at reduction in the case of dioecious gametophytes like those of *Sphærocarpos*. It is very difficult to reconcile any inflexible theory of this nature with the great diversity of situations known without resorting to hypotheses of somatic segregation of factors, alterations in dominance, and other assumptions not well supported by observational evidence. Such difficulties are encountered in the common hermaphroditic condition of gametophytes, which are haploid; in the possibility of causing the development of the second sex in gametophytes normally unisexual (*Onoclea*); and especially in the numerous cases of sex intergrades, in which it is possible not only to control the relative amounts of maleness and femaleness in hermaphroditic forms, but also to produce all intermediate grades between male and female individuals, and furthermore to reverse the sex in unisexual forms, even in certain species with sex-chromosomes (moths and probably pigeons).

If, in accordance with the ideas of many biologists, sex is held to be a quantitative, "fluid" character associated with a continuous series of physiological states which may pass into one another, the way is open for the explanation on a common basis of all cases of hermaphroditism in both haploid and diploid individuals, of sex intergrades, and of the experimental modification of sex. At the same time the influence of the sex-chromosomes or even of smaller factors within them may be allowed. As Riddle (1917) states, organisms have had the problem of producing germs of two metabolic levels, and in some cases this has led to the establishment in the two sexes of two amounts of chromatin or even of two different chromosome complements. The sex-chromosomes, or units contained within them, act with others in the maintenance of two diverse levels of metabolism in the gametes and in the offspring, and with these levels are correlated the two conditions which we distinguish as male and female. Even if the sexes in such cases do not differ in the quality of their chromatin, they at least differ quantitatively in this respect, in agreement with the theory that sex is a quantitative character. The chromosome difference being only one factor in a complex system producing the two sexual states, and no single element in this system being the sole determiner of sex, it is not impossible that the effect of this one factor should be annulled by sufficiently altering the other factors and thus modifying the action of the factorial system as a whole. The same is to be said of other characters also. What an organism inherits is not simply this or that character or sex, but rather a tendency to develop a definite group of characters, including a particular sex, under a given set of environmental conditions.

In those organisms possessing heterochromosomes the sex of the individual under all ordinary circumstances is dependent upon the kind of sperm (or, in some cases, the kind of egg) functioning at fertilization, and does not change thereafter. Furthermore, it apparently cannot be changed by many methods commonly supposed to be efficacious in this respect. So far the chromosome theory is valid; but it does not follow from this that the sex which is characterized by a certain physiological state and is correlated with a certain type of chromosome complement and a variety of secondary sexual characters, is so firmly fixed that it cannot be altered by any extraordinary means. The metabolic state, even though its regulation may be accomplished in part through a visible mechanism, is the resultant of a complex series of reactions which may be interfered with at many points. In some cases this metabolic state has been artificially altered to a degree sufficient to bring about an actual reversal of the sex. It is admitted that other heritable characters definitely associated with constant genes are greatly modified in the manner and degree of their expression by environmental influences, and the evidence now at hand indicates that no exception to this rule can be made in the case of sex.

In criticizing the results and interpretations of Riddle, Morgan (1919a) points out that the behavior of a certain sex-linked character worked out by Strong (1912) indicates that the females which were "changed into males" have the male chromosome complement, and that sex is as much a matter of chromosomes here as elsewhere. He declares further that there is as yet no known case in which the sex determined by a chromosome mechanism has been changed by other agencies in spite of the chromosome arrangement. The evidence here points to the conclusion that when an alteration of the sex is induced, this does not occur without a corresponding alteration in the chromosome mechanism. In *Drosophila*, however, Sturtevant (1920) finds that the intersexes observed by him are modified females with the usual two X-chromosomes. Here, therefore, certain male characters at least are present in an organism with the female chromosome complement.

The number of instances of change of sex in forms normally controlled by a chromosome mechanism will probably increase as the nature of sexual differentiation becomes more fully known and as experimental technique improves; but it is also probable that in animals the sex of which we are most desirous of controlling, practical difficulties may prevent the attainment of satisfactory results. Slight differences in the responses of the two kinds of male gametes (or female gametes) might conceivably make possible a control over the kind functioning, but it seems more probable that the sex of the individual will be found to be more easily influenced, if at all, after fertilization. But it is quite unknown whether or not an individual which had undergone a reversal of sex would be as successful biologically as a normal one. Unusual features may be expected in a sexually functional organism with the chromosome complement normally accompanying the other sex, if such an organism should be found; but such interesting questions can only be answered by the facts yielded by further research.

All questions of sex control are secondary to the main problem of the ultimate nature of sex, a problem which reveals itself with increasing clearness as primarily a physiological one. The question of the origin and significance of sex is one which lies outside the scope of this work.

Bibliography at end of Chapter XVIII.

CHAPTER XVII

LINKAGE

In Chapter XV attention was directed to the remarkable parallelism which exists between the distribution of the Mendelian characters and that of the chromosomes. A vast number of breeding experiments with both plants and animals have shown that new combinations of characters are formed at the time of fertilization, when two parental sets of chromosomes are brought together, and that a segregation of characters occurs at the time of reduction, when the chromosomes are sorted out into two groups. Moreover, the distribution of a single allelomorphic pair of Mendelian characters parallels precisely that of a single homologous pair of chromosomes. These facts indicate clearly that chromosomes and characters are in some manner causally related. This conclusion is strongly supported by the cytological aspects of sex inheritance, maleness and femaleness in a large number of reported cases being definitely correlated with the activity of certain distinguishable chromosomes.

It has also been pointed out that the hypothesis upon which these phenomena are generally interpreted is that the characters are represented in the chromosomes by material factors, or genes, which in some way control the development of characters in the individual. Since, now, an organism usually has many more Mendelian character pairs than it has chromosome pairs, one pair of chromosomes must as a rule carry genes for more than one pair of characters. Furthermore, the different pairs of chromosomes are entirely independent of one another in distribution. It would therefore follow that if two allelomorphic character pairs have their genes located in different chromosome pairs, they will be quite independent of each other in their inheritance through a series of generations; whereas, if their genes are located in the same pair of chromosomes, they will be inherited together. The latter condition—the persistent association of characters belonging to different allelomorphic pairs through a series of generations—actually exists and is known as *linkage.*

The phenomenon of linkage was discovered in 1906 by Bateson and Punnett in the sweet pea. They found flower color to be linked with the shape of the pollen grain: purple flowers nearly always had long grains, while red flowers had round grains. The possible relationship between linkage and the chromosome hypothesis was pointed out by Lock in the same year. Linkage relations have since been worked out in a considerable number of plants and animals, and are especially well known in the case of the fruit fly, *Drosophila melanogaster*, owing to the exhaustive researches of Morgan and his coworkers.

A Typical Case of Linkage.—As a typical example may be taken the following case of linkage in *Drosophila* (Fig. 145). A male fly with white eyes and yellow body is mated to a female with red eyes and gray body.

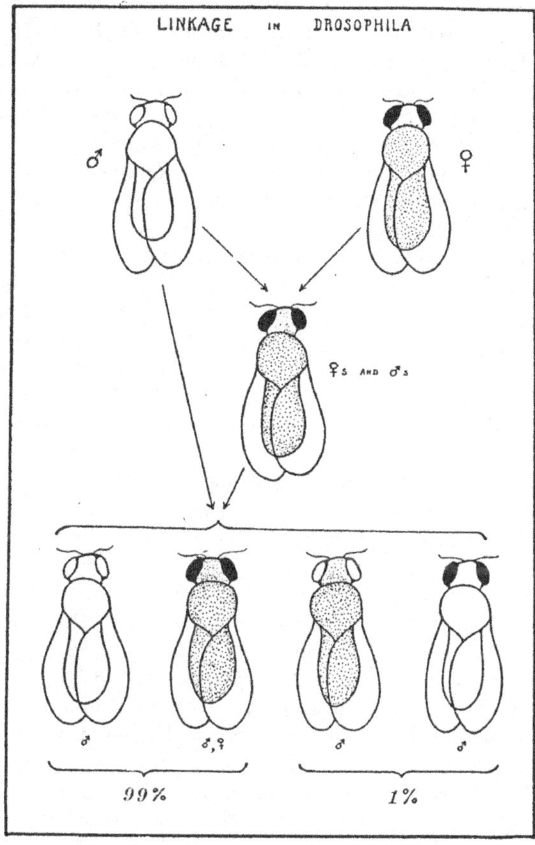

Fig. 145.—Linkage in *Drosophila*. Red eyes in black; gray bodies stippled; white eyes and yellow bodies unshaded.

The flies of the F_1 generation all have red eyes and gray bodies, since red is dominant over its allelomorph white, and gray is dominant over its allelomorph yellow. If, now, the females of this generation are "back-crossed" to males with both of the recessive characters,[1] white and yellow,

---

[1] This back-crossing to the pure recessive is the common method of testing the genotypic constitution of hybrids.

flies of four types appear in the next generation, as shown in the diagram: white-yellow, white-gray, red-yellow, and red-gray. Since one parent in this last cross is a double recessive, these results show that the F_1 red-gray female must produce eggs of these four genotypic constitutions.

All four types, however, are not produced in equal numbers. Those flies with the same combinations as were shown by their grandparents, white-yellow and red-gray, together comprise 99 per cent of the total number; only 1 per cent show the other possible combinations, white-gray and red-yellow. It thus appears that if the two characters, red eyes and gray body, enter a hybrid together (*i.e.*, are contributed by the same parent) they come out together in the next generation in nearly all cases: they are definitely linked, and this is explained on the hypothesis that their genes are located in the same chromosome. The same is obviously true of their respective allelomorphs, white eyes and yellow body: their genes are carried in the other chromosome of the homologous pair. Were the two allelomorphic pairs of genes carried in different pairs of chromosomes no such linkage would occur: the two characters red and gray, and similarly the two characters white and yellow, would then be present together in 50 per cent of the flies, the chance frequency, rather than 99 per cent. How it is that the linkage is broken in some cases, giving 1 per cent with exceptional combinations, is a question to which we shall return later in the chapter.

Sex-linkage.—A very interesting special case of linkage is seen in the phenomenon of sex-linkage, which may be illustrated by the following example (Fig. 146) (Morgan 1910). If a red-eyed wild male is mated to a white-eyed female (a member of a race descended from white-eyed mutants) the F_1 individuals are white-eyed males and red-eyed females—each eye color has been transferred from one sex to the other, a case of "criss-cross" inheritance. If these F_1 flies are bred together, the F_2 generation comprises four types: red-eyed males and females, and white-eyed males and females. Turning now to the chromosomes, if the distribution of the sex-chromosomes is followed through these generations this striking fact is revealed: red eye color appears in every fly, male or female, which possesses the X-chromosome of the original red-eyed male; and white eyes appear in every male which receives one of the X-chromosomes of the original white-eyed female, and in every female receiving two of them. This is taken by Morgan to mean that the original male's X-chromosome carries the dominant factor for red eyes, while each of the X-chromosomes of the original white-eyed female carries a recessive factor for white eyes. In all the flies it is seen that the presence of one X-chromosome is correlated with maleness, and that of two X-chromosomes with femaleness (compare Fig. 140); and that the two types of eye color under consideration follow the distribution of these chromosomes—they are *sex-linked characters*.[1] So far as is known, the *Y*-chromosome of the male carries no sex- or sex-linked factors. This general interpretation is directly applicable to the reciprocal cross (white-eyed male × red-eyed female), in which, however, the relative proportions of red-eyed and white-eyed flies in F_1 and F_2 are different: in F_1 all flies of both sexes have red eyes, while in F_2 all the females and one-half of the males have red eyes, white eyes appearing only in one-half of the males. (See Morgan *et al.* 1915, pp. 16–20; Babcock and Clausen 1918, pp. 74–77.)

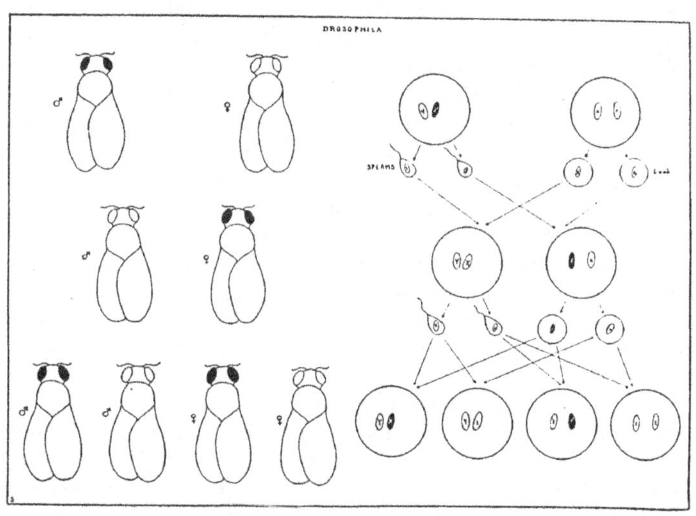

FIG. 146.—Sex-linkage in *Drosophila*. Three successive generations at left; red eyes shown in black. The history of the sex-chromosomes through these generations shown at right; X-chromosome of original male shown in black. (*Adapted from Morgan.*)

In such cases as the above it is evident that characters other than sex may be referred to certain chromosomes of the complement: it is possible not only to tell which chromosomes have to do with sex, but also to identify the ones concerned in the production of red and white eye colors. A large number of such sex-linked characters have been identified in *Drosophila*, and several have been found in other animals. Human colorblindness is a character which is inherited in a manner analogous to that of sex-linked characters in *Drosophila*, and its mechanism is apparently the same. The presence of this defect more commonly in men than in women, and its appearance in so few individuals in affected lines, are due to the fact that it is both a recessive and a

---

[1] Sex-linked characters are not to be confused with sex-limited characters. The latter are those found exclusively in one sex, and are now referred to as secondary sexual characters.

sex-linked character, precisely like white eyes in *Drosophila*. It occurs in a woman only if both of her X-chromosomes bear factors for it, which means that such a factor must have been received from each parent; whereas one such factor is sufficient to produce colorblindness in a man, because his Y-chromosome carries no factors which might dominate it. Furthermore, since the X-chromosome of the male is always derived from the mother, a man can inherit colorblindness from his mother but not from his father. From these facts it follows that a colorblind woman transmits the defect to all of her sons and to half of her grandsons and granddaughters; whereas a colorblind man transmits the defect to none of his children and only to one-half of his grandsons.[1]

The first sex-linked character known in plants was that of narrow rosette leaves in the red campion, *Lychnis dioica* (Shull 1910, 1911), a plant which appears to have the XY type of sex inheritance, but in which no distinguishable sex-chromosomes have been identified.

Non-disjunction.—The chromosome interpretation of sex and of sex-linkage has received an interesting confirmation in a phenomenon discovered by Bridges (1913). In a certain strain of *Drosophila* the white-eyed females were observed to give rise to a certain proportion of "unexpected" forms. Most of their offspring (92 per cent) were red-eyed females and white-eyed males, as expected in such an experiment, but some of them (8 per cent) were white-eyed females and red-eyed males. A long series of crosses showed that these results could be accounted for if it were assumed that in the original white-eyed female both of the X-chromosomes passed together to one pole in the reduction division instead of separating as they should. This was termed *non-disjunction* (Fig. 147). As a result the eggs, instead of having the normal single X-chromosome, would have either two or none, and the distribution of the sexes and the sex-linked characters would be altered in the manner observed. In a cytological examination of the flies in which these abnormal phenomena appear Bridges showed that this non-disjunction of the X-chromosomes does occur occasionally in the female (Fig. 148). The chromosome theory thus received confirmation. "An abnormal distribution of the sex-chromosomes goes hand in hand with an abnormal distribution of all sex-linked factors" (Morgan). Additional genetic and cytological data have since been furnished by Bridges (1916) and Safir (1920).

Linkage Groups.—An extensive series of studies on linkage relations in various plants and animals has brought out the fact that the Mendelian characters of a given organism fall into a certain number of "linkage groups," the members of each group being linked to one another in various degrees but showing no linkage with the members of other groups. It appears further, when the relationships of enough characters have been worked out, that the number of linkage groups is the same as that

---

[1] This case is fully explained by Babcock and Clausen (1918, p. 197).

of the chromosome pairs. *Drosophila melanogaster*, in which linkage relations have been most fully analysed, has four pairs of chromosomes (Fig. 148): two large "euchromosome" pairs, one pair of sex-chromosomes,

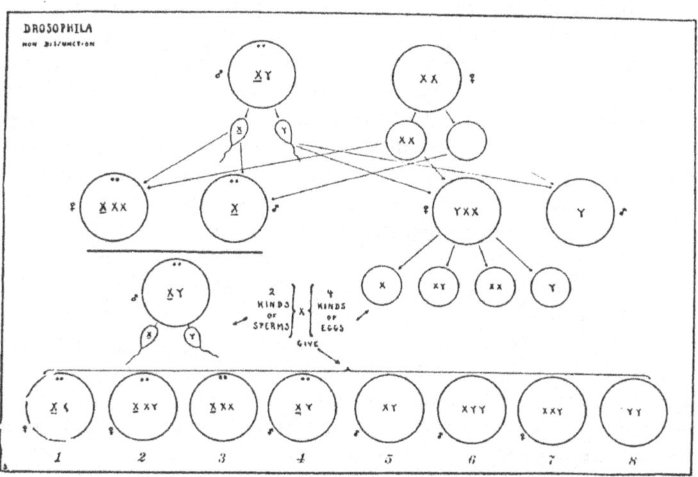

FIG. 147.—Non-disjunction and its results in *Drosophila*. The two large circles in first row represent male and female flies producing sperms and eggs respectively. Non-disjunction in the female gives 2 kinds of eggs, with XX and with no sex-chromosomes, instead of the normal single kind with one X. At fertilization there are possible 4 combinations rather than 2, as shown in the large circles of second row. Owing to the several ways in which her 3 sex-chromosomes may be distributed at maturation, the female represented by the third circle produces 4 kinds of eggs. When mated to a normal male (below horizontal line) with his 2 kinds of sperms, 8 combinations are possible (last row). Nos. 1, 4, and 5 are normal flies and give the usual types of progeny. Nos. 2, 6, and 7, owing to the presence of 3 sex-chromosomes, give exceptional results when bred. Types No. 3 and No. 8 do not appear in the cultures, probably because they die very early. The original male has red eyes and the original female white eyes. Red eyes (represented by dots) appear in every fly bearing the X-chromosome of the original male, as in Fig. 146. Compare Morgan 1919a, Figs. 93 and 94. (*Diagram based on data of Bridges and Morgan.*)

FIG. 148.—The chromosomes of *Drosophila melanogaster* as they appear during mitosis in a female, a male, and a non-disjunctional female. (*After Morgan.*)

and one pair of very small "m-chromosomes." The Mendelian characters in *Drosophila* fall into four linkage groups, and it is noteworthy that one of these groups contains only two known characters. Each chromosome pair therefore seems to be responsible for a certain group of characters. It has been shown above that one of these groups, the sex- and sex-linked characters, can be definitely assigned to the pair of sex-chromosomes; and Morgan further believes that the factors for the two characters of the small linkage group are located in the m-chromosomes. The two remaining linkage groups, which contain many characters each, are assigned to the large euchromosome pairs. Each chromosome is accordingly regarded as a body containing the factors or genes for a considerable number of characters; and on the basis of the evidence to be presented below it is concluded that these genes, differing thus in their hereditary potencies, are arranged in the chromosome in a linear series as suggested by Roux many years ago.

In plants the best known cases of linkage are in *Zea Mays*, in which Emerson and his students at Cornell have identified six linkage groups, and *Pisum*, which has so far shown four linkage groups (White 1917). Since maize has 10 pairs of chromosomes, four more groups may be expected, while in *Pisum*, which has seven pairs, three more groups will probably be established; in fact seven independently inherited characters are known. It is an interesting fact that Mendel, in his famous researches on *Pisum*, happened to select for study seven pairs of characters belonging to the seven different groups, and so did not detect the phenomenon of linkage.

From the foregoing considerations there arises an interesting and very important question. If two homologous chromosomes, each carrying factors for a certain group of characters (those of one group allelomorphic to those of the other group), separate into different gametes (or spores) at the time of reduction, how does it happen that occasionally there appears an individual with some of the characters of each group? And if a single chromosome carries a series of factors for a certain group of characters, how shall we account for the occasional individual with some of these characters but not the rest? To state the problem in the terms of linkage, if each group of linked characters is represented by a series of genes in a given chromosome, how is the linkage broken in a certain percentage of cases, with the resulting formation of new linkage groups, as shown by the exceptional red-yellow and white-gray flies in the experiment described at page 379? A solution to this problem has been offered in the Chiasmatype Theory.

The Chiasmatype Theory.—In our discussion of chromosome conjugation it was pointed out (p. 257) that various opinions have been entertained regarding the nature of the association between the members of the synaptic pair. Some workers have held that the chromosomes fuse completely and lose their identity, and that the two chromosomes appearing on the first maturation spindle are not to be looked upon as identical with those which entered into conjugation. On the contrary, there are those who deny any fusion at all between the members of the pair, holding rather that their identity is not impaired in any way during

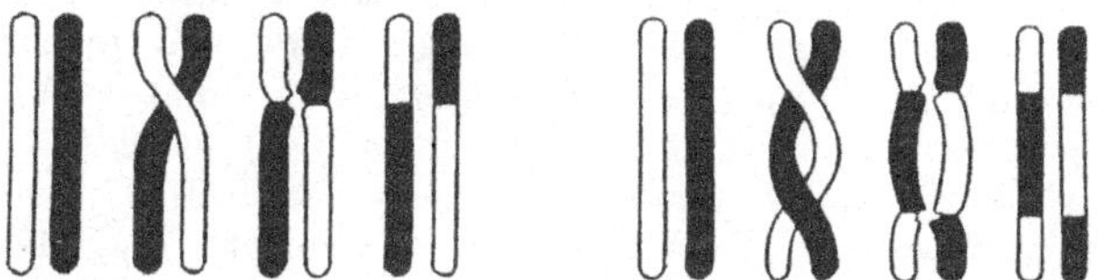

Fig. 149.—The behavior of the conjugating homologous chromosomes according to the Chiasmatype Theory of Janssens. Single crossing over at left; double crossing over at right. (*Adapted from Babcock and Clausen.*)

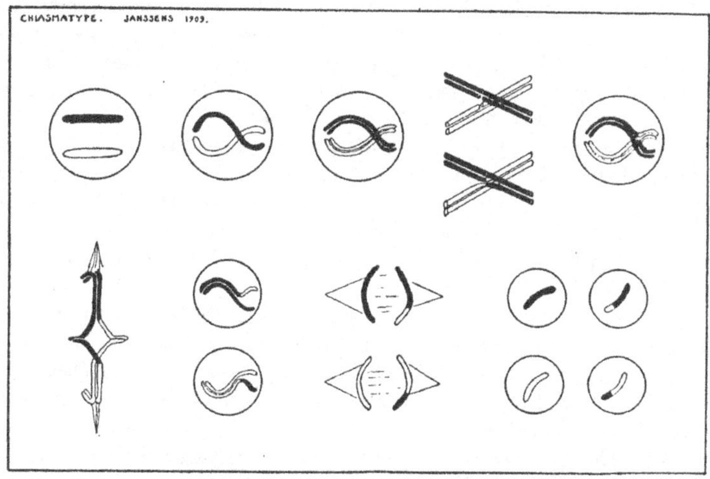

Fig. 150.—Crossing over between 2 of the 4 chromatids of the chromosome tetrad, giving 2 crossover and 2 non-crossover gametes. (*Adapted from Janssens, 1909.*)

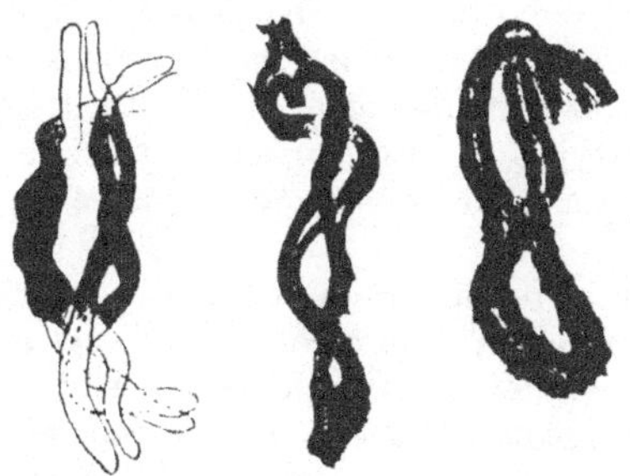

Fig. 151.—Chromosomes of *Batracoseps attenuatus*, showing chiasmas. (*After Janssens.*)

their intimate association: the two chromosomes appearing on the first maturation spindle are exactly the same as those which conjugated.

Between these two extremes lie other views, the most suggestive of them being that proposed by Janssens (1909).[1]

The theory of Janssens in its simplest possible form may be stated as follows. The members of the conjugating pair twist about each other and come into very intimate association at certain points. When they again separate a break occurs at the point or points of closest contact, but along a new plane, so that each of the two separating chromosomes is made up of portions of both conjugating members (Fig. 149). This process is known as *chiasmatypy* or *crossing over*. Such a behavior might occur at various stages in the heterotypic prophase: most probably it takes place at an early stage, when the conjugating chromosomes are in the form of simple thin threads (Fig. 83). In other cases it may take place at a later stage, when, in the case of animals, each of the chromosomes has split preparatory to the second mitosis, forming a tetrad of chromatids. Here the crossing over may occur between only two of the four chromatids (Fig. 150, Janssens's typical case; see also Fig. 151), or between all four. If only two of the four chromatids are concerned, only two of the resulting gametes (or spores) will be "crossover gametes" (or spores), as in Fig. 150; whereas, if the crossing over takes place between all four of the chromatids, or between the two yet unsplit threads in the earlier prophase, all four of the gametes (or spores) will be "crossover gametes" (or spores).

Application of the Chiasmatype Theory to the Problems of Linkage.— It is the above interpretation of the nature of chromosome conjugation that lies at the basis of the work of Morgan and his students on *Drosophila*. As already pointed out, these workers have found good evidence for the conclusion that each chromosome is responsible for a certain group of characters, the members of the group showing a strong tendency to remain associated because their genes are borne by a common carrier. They further believe that the evolution of new character groupings has been brought about not only through the crossing of different hereditary strains, but also through the evolution of chromosomes with new constitutions by the process of crossing over. On the basis of the frequencies in which the new types of grouping occur the relative positions (loci) of the genes for the different characters have been plotted in the chromosomes.

The above points are illustrated in Fig. 152, which summarizes what is supposed to have occurred in a certain series of crosses between flies with yellow body, white eyes, and miniature wings, and flies with gray body, white eyes, and long wings. In the cells of the hybrid there is

---

[1] Janssens has recently (1919*ab*) published an outline of his views of the maturation phenomena in Orthoptera in which he again makes use of the chiasmatype interpretation. His results are discussed in some detail from the cytological and genetic points of view by Wilson and Morgan (1920).

a chromosome from one parent bearing the genes for the three linked characters, *Y*, *W*, and *M*, and a chromosome from the other parent with the genes for the three linked characters, *G*, *R*, and *L*. The first three characters are allelomorphic to the last three respectively, and the two chromosomes are homologous.[1] Upon breeding from the females of these hybrids it is found that in the majority of cases (first column) the F_2 flies show the same genetic consitution as do the grandparents (with respect to these particular characters). This is taken to mean that the two homologous chromosomes forming the synaptic pair and separating at reduction have maintained their substance intact—there has been

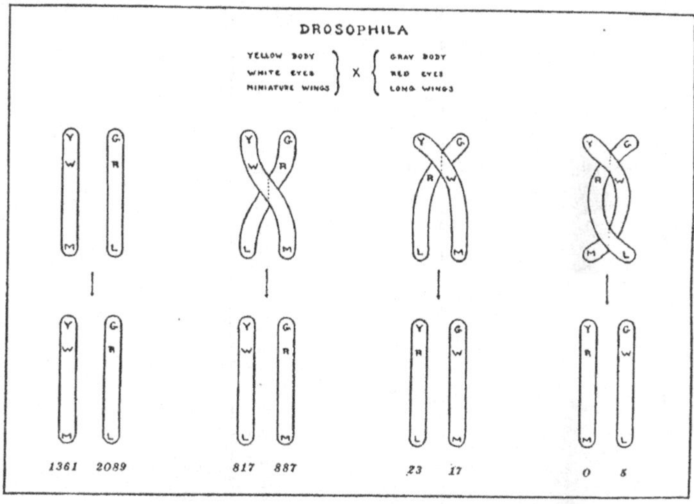

Fig. 152.—Diagram illustrating the evolution of new linkage groups through crossing over. Explanation in text. (*Adapted from Morgan.*)

no crossing over. In a certain number of cases (second column) new groupings of the characters in question are observed in the F_2 flies: some have *Y*, *W*, and *L*, while others show *G*, *R*, and *M*. This is interpreted on the assumption that a break has occurred along a new plane (dotted line) at a point of contact, so that at reduction some of the gametes, and hence the F_2 flies to which they give rise, receive a chromosome with *Y*, *W*, and *L*, while others receive *G*, *R*, and *M*. In a smaller number of cases (third column) the new combinations *YRL* and *GWM* are formed in a similar manner, the break and reunion occurring at another point. In a very small number of cases (fourth column) the combinations *YRM* and *GWL* appear, which can be explained on the basis of a double

[1] The hybrids have gray bodies, red eyes, and long wings, because *G*, *R*, and *L* are dominant over *Y*, *W*, and *M*. (Ordinarily the genes have other designations.)

CHROMOSOME I

Position	Gene
0.0	yellow, scute 0.0±
0.3	lethal 7
0.6	broad
0.8	prune
1.5	white, etc.
3.0	notch, facet
3.1	abnormal
5.5	echinus
7.3	bifid
7.5	ruby
13.7	crossveinless
16.7	club
20.0	cut
21.0	singed
27.5	tan
33.0	vermilion
36.0	tiny-bristles
36.1	miniature
37.5	dusky
38.0	furrowed
43.0	sable
44.4	garnet
53.5	small-wing
54.5	rudimentary
56.5	forked
57.0	bar
58.5	small-eye
59.5	fused
65.0	cleft

CHROMOSOME II

Position	Gene
-2.0	telegraph
0.0	star
1.0	aristaless
4.0	expanded
9.0	truncate
11.0	gull
13.0	pink-wing
14.0	streak
22.5	cream-b
28.0	flipper
29.0	dachs
33.0	ski
35.5	squat
44.0	minute-6
46.5	black
46.7	jaunty
48.5	apterous
52.5	purple
58.0	safranin
61.0	trefoil
65.0	vestigial
66.5	telescope
67.0	dash
70.0	lobe
71.0	minute-5II
73.5	curved
75.0	dachsous
77.0	roof
85.0	minute-2
88.0	humpty
95.0	purpleoid
97.5	aro
98.5	plexus
100.0	lethal-IIa
101.0	brown
103.0	blistered
104.5	morula
105.0	speck
105.5	balloon

CHROMOSOME III

Position	Gene
0.0	roughoid
25.3	sepia
25.8	hairy
32.0	divergent
33.5	cream-III
34.0	dwarfoid
35.0	scarlet
38.3	tilt
38.5	dichaete
40.5	ascute
41.5	deformed
43.0	maroon
44.0	curled
45.0	dwarf
45.5	pink
53.8	two-bristle
54.0	spineless
55.0	bithorax
56.0	bithoraxoid
59.0	glass
60.0	kidney
60.5	giant
61.0	spread
63.5	delta
65.5	hairless
67.5	ebony
68.0	band
70.0	CIII
72.0	white-ocelli
86.5	rough
89.0	beaded
95.4	claret
95.7	minute

CHROMOSOME IV

Position	Gene
0.0	bent
	eyeless 0.0±

crossing over as shown in the diagram. Most of the F_2 flies show the same combinations as their grandparents, which indicates that no crossing over has occurred in the majority of the gamete-forming cells. Since there are many more F_2 individuals with YWL or GRM (second column) than with YRL or GWM (third column), it is believed that the distance between the genes W and M (and between R and L) in the original chromosome must be greater than that between Y and W (and between G and R), so that there is more chance for crossing over to occur in the lower part than in the upper. Thus if two linked characters are often separated (*i.e.*, have their linkage broken) their genes are thought to lie relatively far apart in the chromosome, whereas linked characters separated only rarely are supposed to have their genes located very near each other. Recombinations involving a double or even more complex crossing over (fourth column) would be expected very rarely. In this way the genes for the various characters have been assigned to their loci in the chromosomes on the basis of the frequencies in which the various new combinations in F_2 appear.

In Fig. 153 is shown the "map" of the chromosomes of *Drosophila* as determined by Morgan and his associates, each factor being placed a certain number of units of distance from the end. Many other known genes are not shown in the diagram. As the unit of distance is taken that space separating two factors whose linkage is broken (*i.e.*, between which crossing over occurs) in 1 per cent of the cases. Thus crossing over occurs between "yellow" and "bifid" in 7.3 per cent of the observed cases; the factors are therefore placed 7.3 units apart (first chromosome). Furthermore, if some linkage relations are known, it is possible to calculate certain other linkages in advance. For example, if it were known that crossing over occurred between "sepia" and "pink" (third chromosome) in 20.2 per cent of the flies, and also that "sepia" and "kidney" showed a 34.7 per cent crossing over, the prediction that "pink" and "kidney" would show a 14.5 per cent crossing over (not including the modifying effects of double crossing over, should this occur) would be borne out by experimental results. Such an agreement of the results of new crosses with predictions made on the basis of known linkages has occurred over and over again in the experiments of Morgan and his students. The chiasmatype hypothesis as thus elaborated obviously fits the observed facts remarkably well.

Interference.—Another piece of evidence brought forward in support of the hypothesis that the factors have a linear arrangement within the chromosome is the phenomenon of interference, which has been elucidated by Sturtevant (1913), Weinstein (1918), and particularly by Muller. If the factors or genes are arranged in a series as supposed on the above hypothesis, it would be expected that when crossing over occurs at a given point in a pair of chromosomes, the regions immediately on either side of this point would be prevented from crossing over at the same time, for the reason that the twisting of the chromosomes about each other is not close enough to allow two crossovers so near each other. Thus in Fig. 154, if crossing over occurred between the two factor pairs *Dd* and *Ee*, breaking the linkages between *D* and *E* and between *d* and *e*, there would be at the same time no such break between *Cc* and *Dd* or between *Ee* and *Ff*, since for mechanical reasons a second crossing over could not take place at either of these points simultaneously with that between *Dd* and *Ee*. Crossing over at one point would thus interfere with crossing over which

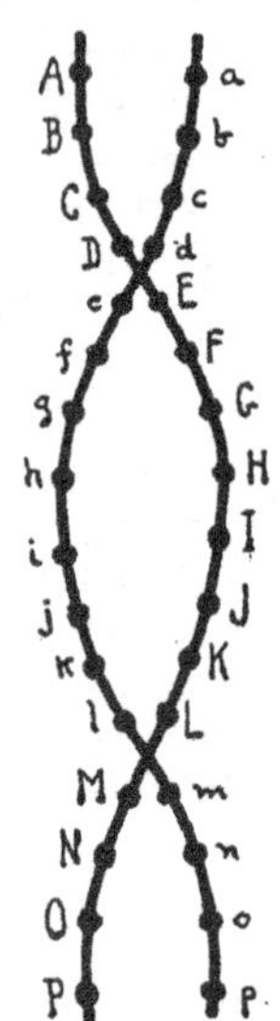

FIG. 154.—Diagram illustrating interference. Explanation in text.

might otherwise occur at nearby points. The amount of this interference would progressively decrease at points farther and farther from the first crossover point, until at a certain distance (measured by the length of the loops usually formed by the twisting chromosomes), as at *LM*, it would vanish entirely; here crossing over would occur with its normal frequency irrespective of any crossover at *DE*.

Muller has found that the characters behave according to these expectations. If two characters have their linkage broken in a certain percentage of cases, this percentage is noticeably lowered if breaks in linkage occur between two other characters having normally a fairly close linkage with the first two. In other words, one linkage break interferes with other linkage breaks within the same linkage group; and the degree of this interference varies from a high value in the case of a closely linked series of characters to zero in the case of characters very loosely linked. This, it is pointed out, is just what should occur among characters represented by a linear series of genes in chromosomes which undergo crossing over, but which cannot twist about one another with more than a certain degree of closeness.

The phenomenon of interference thus indicates another point in which the chiasmatype hypothesis as developed by Morgan fits the experimental facts.

A further point is of interest in this connection. In *Drosophila* it is only in the females that crossing over takes place; it is in the eggs, and not in the spermatozoa, that new factor combinations appear as the result of this process. The absence of crossing over in the male may be associated with the fact that the *Y*-chromosome carries no known factors;[1] the male is heterozygous for sex. In the fowl, in which the female rather than the male is heterozygous for sex, it has been shown that crossing over occurs in the male but not in the female. Crossing over for

[1] See, however, Castle (1921).

some reason is limited in these cases and some others to the sex which is heterozygous for the sex-factors. On the contrary, in the grasshopper, *Apotettix*, Nabours (1919) has shown that some crossing over occurs in both sexes, and the same appears to be true in *Primula* (Gregory; Altenburg 1916), the rat (Castle and Wright 1916), and *Zea* (Emerson). In *Paratettix*, in which no crossing over has been demonstrated, Miss Harman (1920) reports that the homologous chromosomes do not conjugate until the end of the prophase, and suggests that their independence during the early stages may account for the absence of crossing over.

General Discussion.—In the foregoing pages a brief account has been given of the main points in the theory developed by those who have made the most thoroughgoing attempt to relate the phenomena of heredity to a visible cell mechanism. To follow out the details of its application does not lie within the scope of this chapter: it is here intended only to furnish a starting point for cytological studies in this field by indicating the common ground upon which cytology and genetics meet. It is important, however, to differentiate between evidence which is genetical and that which is cytological in nature; and further to remind ourselves to what extent observed fact and hypothesis respectively have been woven into the theory. Caution is particularly necessary in this latter regard, since the general nature of many of our ideas of inheritance is traceable in part to the speculative theories of Weismann. Weismann's theories of heredity and development, which are summarized in the next chapter, were primarily "corpuscular" or "particulate" theories: the phenomena of heredity and development were referred to distinct material units which in some way were able to bring about the development of the heritable characters in the individual and their transmission from one generation to the next. Bearing in mind the phenomena of inheritance reviewed in the preceding chapters, especially the behavior of the Mendelian characters, it is difficult to escape the conclusion that differential factors of some sort, which in an unknown manner initiate the series of reactions resulting in the several characters, are carried in the nucleus. To determine the nature of these factors and to discover the real relation existing between them and the developed characters are among our greatest problems. That the factors or genes are discrete units is a hypothesis which is not only plausible, but has also proved itself to be most useful. If such factors exist, the chromosomes afford a means of precisely the kind required to account for the observed distribution of characters throughout a series of generations. Hence from Roux and Weismann onward the factors have been lodged in the chromosomes.

But it is when these factors are directly sought with the aid of the microscope that disappointment is met. The frequently observed granules or chromomeres in the chromatin thread or chromosome are accepted by some geneticists as the desired material genes; but, as pointed out in the chapters dealing with somatic mitosis and reduction, many cytologists are very uncertain as to the morphological status and significance of these bodies, which seem to them to be far too inconstant in number and behavior to represent the units in question. Although it is tempting to look upon the chromatic granules as the units which current theories of heredity seem to require, it must be admitted that the observational evidence is insufficient to warrant the categorical statements frequently made to the effect that the chromosome is composed of a definite number of more elementary visible chromatic units, which have definite space relations and are the significant units in the cellular mechanism of heredity. On the other hand, the careful observations of Wenrich (1916) have shown that in the grasshopper, *Phrynotettix* (Fig. 155), the chromatic granules are relatively constant in size and position

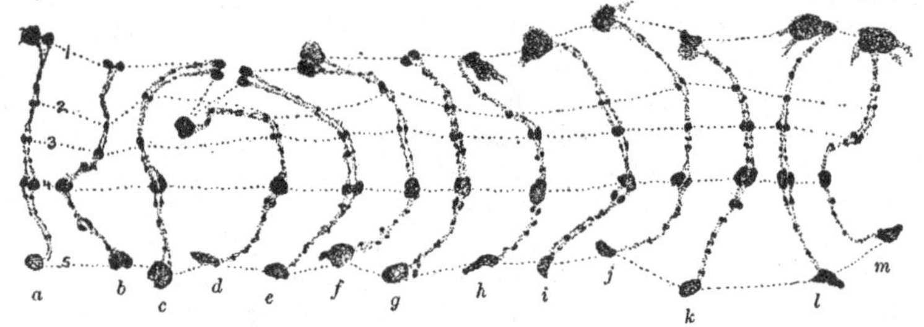

Fig. 155.—Chromosome pair "B" in conjugation from the spermatocytes of 13 different individuals of *Phrynotettix magnus*, showing constancy in size and arrangement of the principal chromomeres. The same constancy is shown in the different cells of a single individual. × 1500. (*After Wenrich*, 1916.)

in a given member of the chromosome complement, even in different individuals; and they furthermore show a close correspondence in the two homologous chromosomes as they pair at synapsis. This is one of the most striking pieces of direct cytological evidence yet brought forward in support of the theory that the chromosome is a "chain of factorial beads" (Harper), and heightens the probability that the postulated units of inheritance will turn out to be more than purely conceptual ones.

Whatever may be the value of the chromatic granules, one can hardly fail to recognize the highly suggestive nature of the arrangement of the chromatin in a thin thread, its frequently beaded appearance, and its accurate longitudinal fission into two equal parts at the time of cell-division. In the absence of direct and convincing cytological evidence for the presence of various "qualities" arranged in a series along the thread, we may still look hopefully for the support which it would seem that the theory of Roux must sooner or later have. It must be admitted that at present the evidence for the existence of genes is in the main genetical rather than cytological.

Similarly unsatisfactory is the cytological evidence for the breaking and reunion of the chromatin threads required by the crossing over hypothesis. Since the hypothesis was put forward by Janssens (1909) adequate and convincing descriptions of this process have been singularly wanting.

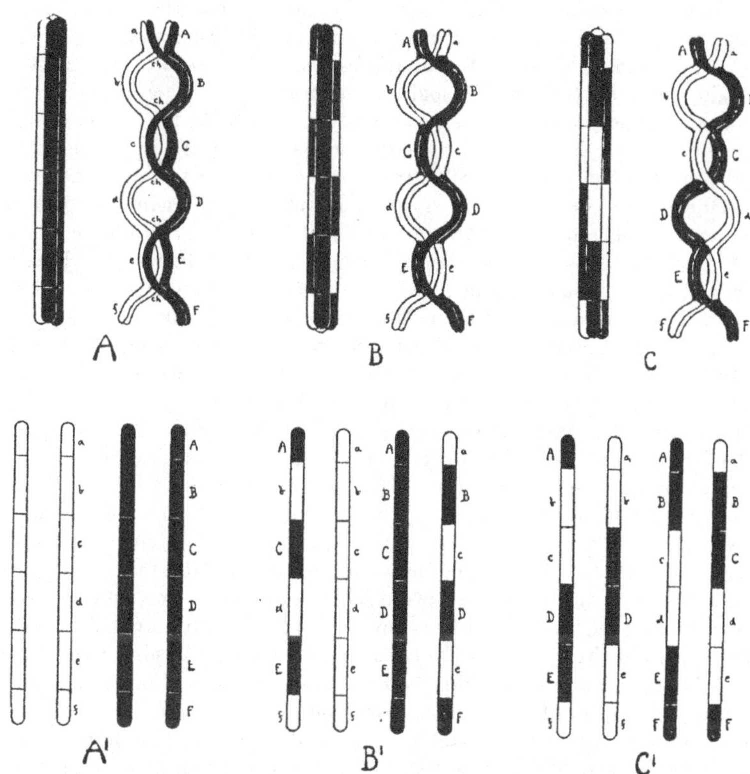

Fig. 156.—Diagrams illustrating various possibilities concerning the compound ring tetrads in Orthopteran spermatocytes, following the outlines of Janssens's figures, but showing also the relations of the chromatids. At the left in each of the upper figures is the longitudinal tetrad-rod from which the ring-series arises, showing results of assumed early cross-overs in B^1 and C^1. *A*, the compound ring as conceived by McClung, Robertson, etc., with the four resulting chromatids at A^1 (no cross-overs). *B*, a compound ring, such as might follow a two-strand cross-over at each node, giving the results shown in B^1. *C*, a compound ring giving the results shown in Janssens's diagrams, resulting from a two-strand cross-over between two pairs of threads, in regular alternation at successive nodes. The result (C^1) is four classes of chromatids, as shown in C^1. (*Figure and legend from Wilson and Morgan*, 1920.)

particularly in those cases in which experimental results would make its establishment most desirable. Wilson (1912*b*), Robertson (1916), and Wenrich (1916, 1917) point out that the figures formed by the chromosome tetrads in the spermatogenesis of certain insects may be interpreted without recourse to the hypothesis of chiasmatypy, and that the observations and figures of Janssens do not prove the existence of that phenomenon. Robertson, for example, holds that the "chiasma" figures are more simply explained as the result of a tendency on the part of the four chromatids to open out partly along the conjugation plane and partly along the plane of splitting, without any actual breaking and recombination (Fig. 156). Wilson (Wilson and Morgan 1920), however, thinks it "highly probable that the cytological mechanism of crossing-over must be sought in some process of torsion and recombination in the earlier stages of meiosis—perhaps during the synaptic phase or slightly later—and that this process may leave no visible trace in the resulting spireme-threads."

During the time when the homologous chromosomes in the form of slender threads are twisted about each other in the early prophase of the heterotypic mitosis there is abundant opportunity for the required breaking and union to occur, and appearances often lend themselves well to such an interpretation. If the side-by-side position is not assumed until later in the prophase (Scheme *B*, Chapter XI) the time during which such interchanges might occur is much shorter. But it is a matter of extreme practical difficulty in either case to determine whether or not the plane of separation is the same as the plane of union at a given crossing point. The forces controlling the breaks and recombinations as well as the frequencies with which they occur in the different chromosome pairs are even more difficult to imagine. The difficulty of accounting for these phenomena, however, does not weigh heavily as an argument against their occurrence. Whether or not chiasmatypy actually takes place is a question which must be settled primarily by direct evidence, and the need for careful search for such evidence cannot be too strongly emphasized. In the opinion of the cytologist the behavior of each chromosome as a whole must be much more thoroughly known before cytological interpretations of the phenomena of inheritance involving any smaller units which the chromosome may contain can be regarded as more than hypotheses, valuable as these hypotheses may be in the correlation of genetic data.

At this point it may be well to recall (see Chapter XI) that all cytologists do not agree that the synaptic mates maintain such an independence (except at crossover points) as is presupposed by the advocates of the chiasmatype theory. A number of observers have reported an actual fusion of the conjugating threads, the resulting pachytène thread subsequently splitting, probably along the line of this fusion. It may accordingly be suggested, in line with the hypotheses discussed by Allen (1905), that if the threads actually carry or consist of discrete units or genes, and if these units do not themselves fuse during the process, such a resplitting of the pachytène thread, if not wholly along the fusion plane because of a twist of the thread or of the plane itself, would bring about a redistribution of the genes as effectively as would the chiasmatype process as originally proposed.[1] The splitting of chromatic threads is, moreover, a process already known to occur in all other mitoses. But interpretations involving the actual fusion of the conjugating threads have been adversely criticized with much effect by Grégoire (1910), and only further research on the cell can lead us to an adequate evaluation of the above outlined suggestion.

Other Theories of Linkage.—The chiasmatype hypothesis is not the only one which has been advanced to account for the phenomenon of linkage. Most prominent among other attempts to solve this problem is that of the English geneticists, especially Bateson, Punnett, and Trow, who have advanced what is known as the *Reduplication Hypothesis.* Instead of accounting for the new factor combinations manifested by a certain percentage of the gametes on the basis of an interchange of factors in the chromosomes at the time of reduction, these investigators seek to explain them by postulating a series of differential divisions in the earlier cells of the germ cell lineage, whereby Mendelian factors, not carried by chromosomes, are segregated in such a manner that the observed types of gametes are produced. Furthermore, the various factors are supposed to be segregated successively, and at such stages in the cell lineage that the proliferation or reduplication of the cells with new combinations of factors shall account for the ratios in which the new types appear. Although the differentiation of the somatic and early germ cells is accompanied by visible differences in the constitution of their cytoplasm (see p. 406), there is at hand no cytological evidence for such a segregation of hereditary units as is thought to occur by the proponents of the Reduplication Hypothesis. So long as this is the case, discussion of the hypothesis, together with the subhypotheses formulated to meet certain serious objections, hardly belongs to cytology. One fact, however, pointed out by Morgan (1919a) as very significant in this connection, is found in the results of some experiments by Plough (1917). Plough investigated the effects of temperature on the frequency of linkage breaking (crossing over) in *Drosophila.* Not only did he find that temperature does affect the amount of crossing over, but the effect was clearly produced at the time of maturation and not earlier. This evidence is directly opposed to the view that the new factor combinations are formed during cell-divisions some time prior to reduction.

Another effort to account for the results of crossing over without resorting to the chiasmatype process is represented in a suggestion (Goldschmidt 1917b) to the effect that the two factors of an allelomorphic pair are held to their places in the two homologous chromosomes by a pair of variable forces, which allow them to exchange places in a certain

---

[1] A form of this interpretation of synapsis has suggested itself also to Dr. C. W. Metz and Dr. E. G. Anderson, who inform the author that such a hypothesis will in all probability conform satisfactorily to the data of genetics.

proportion of cases without involving a rupture of the chromosomes. These forces, however, are purely conjectural. It is pointed out by Jennings (1918), moreover, that the gametic ratios theoretically resulting from such a process do not agree with the actual ratios observed in *Drosophila*.

Value of the Chromosome Theory of Heredity.—Whatever judgment may ultimately be rendered on the chromosome theory of heredity as outlined in these chapters, it must be agreed that the value of this theory in the present state of our knowledge can hardly be overestimated. Through its use a huge number of the observed facts of inheritance are being reduced to order: the painstaking investigation of the interrelationships of all the known heritable characters of even a single organism such as *Drosophila* cannot fail to be a great service to biological science. Its appeal to the cytologist, as Wilson states, is largely through the manner in which it seeks to make use of known cell mechanisms rather than entirely hypothetical processes. Those portions of the theory which are as yet unsupported by the results of direct cytological observation, though not contradicted thereby, at least have the virtue of affording a useful and graphic representation of the mutual behavior of hereditary characters. Notwithstanding the statement that "the graphic representation of the location of the factors is a type of representation common to every set of phenomena which can be expressed as percentages" (Trow 1916), these hypotheses are of great value, for by aiding in the correlation of the facts of inheritance they serve to increase the number of observed phenomena statable in terms of order; and the reduction of experience to order and the statement of this order in simple formulæ, together with the search for new truth, constitute the principal tasks of science. If this work of correlation has been well done the whole body of facts can readily be placed under another theoretical interpretation and described in a new set of terms should occasion require.

Although it may be that the chiasmatype hypothesis of linkage is in certain points inadequate, mathematically (Trow 1916) or otherwise, it is nevertheless true, as we have already seen, that it fits the case very well. At the same time we may remind ourselves that the fact that a hypothesis works well is no guarantee of its ultimate truth. But even if the chiasmatype interpretation should have to be ever so greatly modified as new facts accumulate, it is scarcely to be doubted that the chromosome theory of heredity in some form will turn out to be in accord with the truth. With respect to this general theory Wilson (1909) writes as follows:

"I stand with those who have followed Oscar Hertwig and Strasburger in assigning a special significance to the nucleus in heredity, and who have recognized in the chromatin a substance that may in a certain sense be regarded as the idioplasm. This view is based upon no single or demonstrative proof. It rests upon circumstantial and cumulative evidence, derived from many sources. The irresistible appeal which it makes to the mind results from the manner in which it brings together under one point of view a multitude of facts that otherwise remain disconnected and unintelligible. What arrests the attention when the facts are broadly viewed is the unmistakable parallel between the course of heredity and the history of the chromatin-substance in the whole cycle of its transformation. In respect to some of the most important phenomena of heredity it is only in the chromatin that such a parallel can be accurately traced. It is this substance, in the form of chromosomes, that shows the association of exactly equivalent maternal and paternal elements in the fertilization of the egg. In it alone do we clearly see the equal distribution of these elements to every part of the body of the offspring. In the perverted forms of development that result from double fertilization of the egg and the like it is only in the abnormal distribution of the chromatin-substance by multipolar division that we see a physical counterpart of the derangement of development. Only in the chromatin-substance, again, do we see in the course of the maturation of the germ cells a redistribution of elements that shows a parallel to the astonishing disjunction and redistribution of the factors of heredity that are displayed in the Mendelian phenomenon."

With more particular reference to the chiasmatype hypothesis Wilson (1913) says:

"This, admittedly, is a bold venture into a highly hypothetical region. Its justification is the pragmatic one that it 'works.' The hypothesis gives us the only intelligible explanation that has yet been offered for a series of undoubted facts; and it is certainly worthy of the most attentive further examination . . . . We have much to gain and nothing to lose by the use of explanatory hypotheses that are naturally suggested by the facts and help us to formulate them for analysis, so long as such hypotheses are not allowed to degenerate into dogmas accepted as an act of faith, but are only used as instruments for further observation and experiment."

Bibliography at end of Chapter XVIII.

CHAPTER XVIII

WEISMANNISM AND OTHER THEORIES

The theory of heredity described in the foregoing chapters, though resting on its own foundation of observational and experimental evidence, shows in some of its features the influence of certain earlier speculative hypotheses, particularly those set forth by Weismann.[1] Some of the conceptions embodied in these hypotheses are consequently involved in cytological and genetical discussions of the present day, and for this reason we shall here outline their main points, briefly indicating wherein our modern theory has advanced beyond them.

Although conceptions of other types arose very early, many of the hypotheses in question were based on the assumption that the phenomena of heredity and development are the result of the activity of ultimate living particles of ultramicroscopic size. Thus Herbert Spencer (1864) built up a theory of considerable proportions about his 'physiological units,' and these formed the prototype of the units postulated in many later theories. Of these theories the most prominent were those of Darwin, Nägeli, de Vries, and Weismann.

Darwin's Hypothesis of Pangenesis.—In his *Variations of Plants and Animals under Domestication* (1868) Darwin included a chapter on his "Provisional Hypothesis of Pangenesis," which, though offered only as a suggestion, excited great interest in the field of biology, especially after the advances made in cytology a few years later. In several points it closely resembled a theory propounded by Buffon more than a century earlier (1749). Darwin clearly saw in the cytological aspects of heredity one of the great biological problems of the future, but his only speculations on the subject were embodied in the pangenesis hypothesis, which may be stated as follows:

All the cells of the organism at all stages of development give off small particles, or *gemmules*, which multiply by fission and circulate throughout all parts of the body. These gemmules pass to the germ cells, carrying with them the power to reproduce cells like those from which they came. In this way units representing all the kinds of cells composing the organism are collected in the gametes (or spores or buds) and are thus passed on to the next generation. During the embryogeny of the new individual the gemmules are so distributed that at the proper times and places they

[1] See Kellogg (1907), Delage and Goldsmith (1913), Thomson (1899, 1913), and Conklin (1915).

develop into cells like those from which they originally migrated, in this manner building up a new individual like the parent. Some of the gemmules do not function until a comparatively late stage in the ontogeny, and others may remain latent through several generations: on these two assumptions it is possible to account for the late appearance of certain characters and for the fact that others may "skip" one or more generations. It is further supposed that some gemmules remain latent in the individuals of one sex: thus, for instance, the characters normally present only in the male may be transmitted through the female.

In the many criticisms of this hypothesis the tendency has been to judge and condemn it solely on the ground of the supposed transportation of the gemmules from the body cells to the germ cells, for which no direct evidence has ever been discovered. It should not be forgotten, however, that Darwin suggested an explanation for the phenomena of heredity on the basis of representative material units in the cells, a conception which was of the greatest importance in that it constituted the starting point for later fruitful investigations and theories. The migration of the gemmules was postulated largely to account for the phenomena of regeneration and the inheritance of acquired somatic modifications. Since regeneration may be explained as well on other grounds, and since the evidence for the inheritance of acquired somatic modifications is for the most part of such extremely doubtful value, such a migration of representative units, first denied by Galton (1875), has come to be regarded as unnecessary. A theory postulating representative particles but no such migration was that of de Vries.

De Vries's Theory of Intracellular Pangenesis.—According to de Vries (1889) the particles of hereditary substance, or *pangens*, do not represent different kinds of cells as Darwin thought, but stand rather for different elementary characters or qualities out of which the many visible characters of the organism are built up. Furthermore, these living elements or pangens do not pass from cell to cell, but merely circulate between the nucleus, where a complete outfit of them is conserved, and the other parts of the cell—hence the term "intracellular pangenesis." In this way the characters brought into the new individual through the nucleus are delivered to the cell as a whole. Contrary to the idea of Darwin and especially to that of Weismann (see below), all the cells (or nuclei) of the body contain pangens for all the hereditary characters: they are not sorted out as development proceeds.

Nägeli's Idioplasm Theory.—A highly speculative theory of a somewhat different type was that formulated by Nägeli (1884) five years before that of de Vries. Protoplasm was thought by Nägeli to be made up of a vast number of fundamental living units; these he called *micellæ*. As a result of the ways in which these molecular complexes or micellæ may be arranged, there are in the cell two kinds of protoplasm: in nutritive protoplasm, or *trophoplasm*, the micellæ have no regular orientation, whereas in *idioplasm* they are oriented in a particular manner. According to Nägeli the phenomena of heredity are due to the constitution and transmission of this idioplasm; idioplasm is the physical basis of inheritance. It is not confined to the nucleus, but forms an elaborately constituted network extending throughout all the cells of the organism. By arranging themselves in various groupings within this network the micellæ are able to bring about the development of the many specific characters. The further details of this highly "fragile" hypothesis are summarized in convenient form by Delage and Goldsmith (1913), who point out that in spite of its many unsupported assumptions it did involve two fertile ideas: first, that there are two kinds of protoplasm, one of which carries the characters of the organism; and second, that there are a limited number of elementary characters which combine in various ways to produce the many visible characters.

Weismann's Theory.—The most highly developed and influential of all such speculative theories was that of Weismann. On the basis of the conception of pangens Weismann built up the highly involved system of hypotheses set forth in his *Das Keimplasma* of 1892 and in more elaborated form in his *Evolution Theory* of 1902. Certain modifications were later made.

As Delage and Goldsmith have noted, Weismann incorporated in his theory several of the stronger points of earlier theories, such as Darwin's conception of representative particles, Nägeli's elementary characters, and de Vries's intracellular migration of particles. With Nägeli he distinguished between nutritive *morphoplasm* and hereditary *idioplasm* or *germ-plasm*, but unlike Nägeli he identified the idioplasm with the chromatin of the nucleus. His conception of the constitution of the idioplasm was essentially as follows:

The ultimate unit in all living matter is the *biophore*, which is a minute living particle capable of growth and reproduction—a vital unit. The many kinds of biophores in a given cell represent the many characters of that individual cell: they are not bearers of the characters as such (though Weismann often spoke of them in this fashion), but are rather factors upon whose presence the development of the characters depends. The biophores are grouped to form vital units of a higher order, known as *determinants*. The determinant, since it is composed of the many kinds of biophores in the cell, has the power of determining the development of a certain type of cell or group of cells. In general, therefore, there are as many sorts of determinants in the organism as there are types of cells, or "independently variable parts," to be developed. The determinants are in turn grouped into *ids*. A single id contains all the kinds of determinants, and so stands for the sum of all the characters of the organism: it contains the "complete architecture" of the germ-plasm. The ids in a given species differ only slightly among themselves, the differences corresponding to the variations observed within the species: they are the "ancestral germ-plasms" which have been contributed by past generations. The ids are identified with the visible chromatin granules in the nuclear reticulum or in the chromatin thread during mitosis. In most cases the ids are grouped to form *idants*, or chromosomes. In some forms which have a large number of granular chromosomes it is possible that each is composed of but one id. The id therefore, rather than the chromosome, is the unit of primary importance. In case there are several ids in a chromosome (idant) they are arranged in a linear series. The idea that the chromosomes are all alike since they carry closely similar ids was later (1913) modified by Weismann, largely as the result of the demonstration that very minute characters are segregated in Mendelian fashion.

With the aid of this elaborate mechanism Weismann explained ontogenetic development in the following manner. In the fertilized egg from which the individual is to develop all the kinds of determinants are present: thoes of the female parent are contained in the egg nucleus and those of the male parent are brought in by the nucleus of the spermatozoön. During the long series of cell-divisions beginning with the fertilized egg and ending with the completion of the mature organism, the many kinds of determinants are sorted out through a progressive disintegration of the ids, and are distributed in a definite and orderly manner to the different parts of the body. Many somatic mitoses are therefore regarded not as equational (*erbgleich*), but in reality qualitative (*erbungleich*). When a given determinant finally reaches the proper cell. *i.e.*, when that cell is finally formed, the determinant splits up into its constituent biophores; and these, through their action upon the cell elements, give to the cell its specific characters. The general character of a cell is accordingly due to the type or types of determinant which it receives. For Weismann, therefore, development (ontogenesis) was definitely bound up with the evolution or unfolding of a complex structure contained in the fertilized egg. Although he did not hold that the units in the egg have the same spatial relations as their corresponding characters or structures in the adult, it has been said with some degree of truth that he transferred preformationism to the nucleus.

Such being Weismann's conception of development, how did he account for heredity? If the various kinds of body cells in an individual are characterized by different types of determinants, how is it that the germ cells, or gametes and fertilized egg through which this individual is to give rise to the next generation, possess a complete outfit of determinants? According to Darwin's hypothesis, outlined in the foregoing pages, representative particles or gemmules are contributed by all the body cells at all stages to the germ cells, by which they are transmitted

26

to the next generation. (See Fig. 157, *A*.) Such a contribution of elements from the body cells to the germ cells was denied completely by Weismann. He held rather that a certain portion of the complete germ-plasm (idioplasm; chromatin) of the fertilized egg is carried along un-

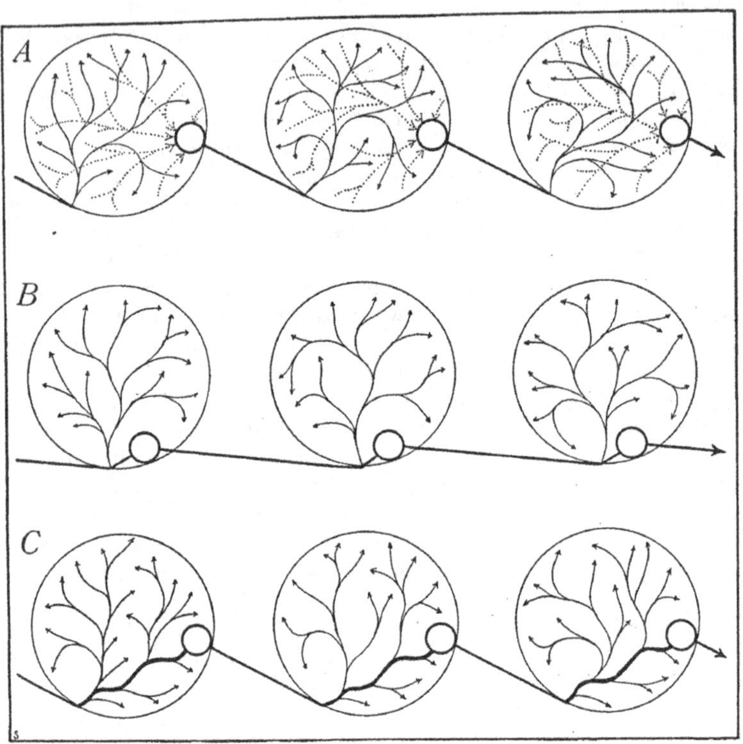

Fig. 157.—Diagram illustrating the hypotheses of Darwin and Weismann. The large circles represent successive generations of individuals, and the small circles their germ cells. For the sake of simplicity inheritance is shown as uniparental rather than biparental. *A*, Darwin's Hypothesis of Pangenesis. The branching solid lines ending in arrows represent the sorting out of the hereditary units (gemmules) during ontogenesis; the dotted arrows show the migration of gemmules from the body cells to the germ cells, by which they are carried into the next generation. *B, C*, Weismann's theory of the continuity of the germ-plasm, with no contribution of hereditary units from the body cells to the germ cells. In one case (*B*) the germ cells are set aside at the beginning of ontogenesis, and in the other (*C*) much later. In both cases the "complete germ-plasm" is delivered to the germ cells through a shorter or longer series of equational divisions (heavy lines).

changed and delivered intact to the germ cells. It had been shown (Haeckel 1874; Rauber 1879; Jaeger 1878; Nussbaum 1880; Galton) that in certain animals the primitive germ cells are set aside at once when development begins, and Weismann pointed out that they are therefore differentiated before any sorting out of the hereditary units can have taken place.   Hence the germ cells are really produced by the germ cells of the previous generation and not by the individual's own soma (body) at all: they are present from the beginning of development with the full hereditary outfit, and by a few equational divisions they give rise to the gametes.   This is represented in Fig. 157, *B*.   In the more usual case of those animals and plants in which the germ cells appear later in the onto-geny Weismann held that, although a sorting out of the units occurs in the majority of the cells during ontogenesis, those meristematic cells which constitute the chain connecting the fertilized egg with the germ cells— the *germ track (Keimbahn)*—maintain the undiminished germ-plasm (Fig. 157, *C*).   Thus in this case as in the other there is a continuity of the germ-plasm, if not a continuity of the germ cells (unless meristematic cells also be regarded as germ cells).   Since the germ-plasm of any generation is derived directly from that of the preceding one, it is continu-ous through an unlimited number of generations; and the successive somas (bodies) are, so to speak, side branches given off at intervals from the main stream of the germ-plasm.

In elaborating the above views Weismann (1885, 1892) insisted strongly upon the independence of the "potentially immortal" germ-plasm and the transient and mortal soma[1].   He argued that since there is no contribution of hereditary elements from the soma to the germ cells, somatic changes being in no way impressed upon the germ cells from which the next generation is to arise, there can be no inheritance of acquired somatic modifications.   In multicellular animals the only inherited variations are those originating in the germ-plasm of the germ cells or germ track as responses to internal (nutritive etc.) or external environmental stimuli, or as the result of recombinations of hereditary units at the time of fertilization (amphimixis).   Weismann admitted that the germ-plasm, though remarkably stable, might be altered directly by the environment or even by modifications in the surrounding soma; but he denied that in the latter case the alteration would be of such a nature as would cause the reappearance of the same somatic modifica-tion in the next generation.   With Weismann, as with Mendel, the main problem of heredity was not to discover how the characters of the organ-ism get into the germ cells which it produces, but rather how the char-acters of an organism are represented in the germ cell from which it is pro-duced (Darbishire 1911, Chapter 12).   He attempted to show how it is that the stream of germ-plasm on the one hand maintains a stability suffi-cient to account for the resemblance between the successive bodies spring-ing from it at intervals, and on the other hand undergoes orderly changes responsible for the evolutionary advance shown in a long series of generations.   In the words of Agar (Bower, Kerr, and Agar, 1919), "According to Darwin, parents truly *transmit* their characteristics to

[1] See discussion of senescence in Chapter VII.

their offspring (by means of the gemmules). According to the modern view [Mendel; Galton; Weismann], however, children resemble their parents not, strictly speaking, because the latter have passed something on to them, but because both have been produced from the same germ-plasm" (p. 91). "The parent is rather the trustee of the germ-plasm than the producer of. the child" (Thomson 1913).

Weismann attempted further to account for the variations effective in evolution on the basis of his theory of *Germinal Selection*. He supposed that the determinants, while multiplying in the germ cells, are subject to selection like all other organic units. Some determinants, being better placed with respect to the nutritive conditions, are favored thereby and grow stronger and more influential, while others undergo changes in the opposite direction. The cells or parts of the organism receiving the determinants which have had the advantage in the struggle become better developed than those receiving the weaker determinants. As this process continues from generation to generation the new variation gradually increases until it becomes pronounced enough to be laid hold of by natural selection. In this manner Weismann accounted for the preservation of small variations not yet of selective value, and for continued variation along definite lines (orthogenesis) in both plus and minus directions. Thus for him selection was the cardinal principle which ruled not only over organisms, but also over cells, ids, determinants, and biophores. As he himself stated it, "This extension of the principle of selection to all grades of vital units is the characteristic feature of my theories."

Some Modern Aspects of Weismannism.—Although the distinction between soma cells and germ cells is not now drawn so sharply as in the days of Weismann, it is nevertheless of interest to note certain facts adduced in support of his contention that the germ-*plasm* is continuous.

In *Ascaris megalocephala* (Boveri 1887c, 1889, 1891, 1892, 1904; Zacharias 1913) it is observed that at the second cleavage mitosis the chromsomes in one blastomere remain entire, while in the other blastomere they become broken up into smaller pieces, some of which are lost in the cytoplasm and are not included in the daughter nuclei (Fig. 158, A). This process is called "chromatin diminution." At the third and fourth cleavage mitoses a similar diminution occurs in all the blastomeres but one; in this one the chromosomes remain entire. At the fifth division it is seen that in the one undiminished cell no further diminution occurs as it divides, and its descendants become the germ cells. The primary germ cell is therefore set apart at the fourth mitosis; and, whereas the other embryonic cells giving rise to somatic structures have undergone a diminution, the entire chromatin outfit is delivered to the germ cells through the undiminished cells of the germ track. A similar condition is present in *Miastor* (Kahle 1908; Hegner 1912, 1914). In *Ascaris canis* (Walton 1918) the germ cells are similarly set aside at the seventh cleavage mitosis.

In criticizing this supposed evidence for the independence and continuity of the germ-plasm Child (1915) points out that, since undiminished cells may give rise to other cells as well as germ cells in the early divisions, the process observed may represent merely a segregation of different organs rather than a separation of the germ-plasm from the soma; and that the non-diminution of the chromatin in the germ track may be the result of the differentiation of the germ cells rather than its cause, the differentiation at this stage being primarily a physiological

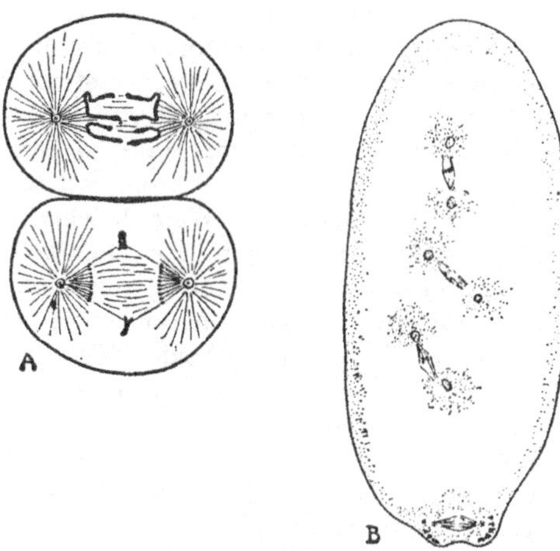

Fig. 158.

A, chromatin diminution in *Ascaris megalocephala*. The second cleavage mitosis is in progress: all the chromatin is retained in the upper blastomere, from which the germ cells are to arise, whereas chromosome diminution occurs in the lower blastomere, which is to give rise to the somatic cells. (*After Boveri.*) *B*, third nuclear division in the yet unsegmented egg of *Chironomus confinis* showing the early setting aside of the primitive germ cells at the lower end. (*After Hasper.*)

(metabolic) one. He refers to certain later researches of Boveri (1910), which apparently show that "the occurrence or non-occurrence of chromatin diminution in a nucleus depends, not upon its qualitative constitution, but upon its cytoplasmic environment." From this it is concluded that "the 'germ path' is a feature of the cytoplasm, and the cytoplasm is not, properly speaking, a part of the germ-plasm at all, but represents the soma of the cell" (p. 327).

In support of this conclusion we may cite, as does Child, those cases among insects (Hasper on *Chironomus*, 1911 ; Hegner on *Miastor*, 1912, 1914) and copepods (Haecker 1897, 1902; Amma 1911) in which the substance of the future germ cells may be distinguished very early in the embryogeny, even in the undivided egg, either as a visibly differentiated region of the cytoplasm (*Keimbahn-plasma*) (Fig. 158, *B*), or by the presence of certain cytoplasmic granules or inclusions (*Keimbahn determinants*). These latter are ultimately delivered to the definitive germ cells, the nuclei at the same time showing no differences in the germ and soma cells. Although such cases seem to show that "the factors determining what shall become germ cells and what somatic structures apparently exist in the cytoplasm and not in the nuclei" (Child 1915, p. 329), it is nevertheless very significant for the chromosome theory of heredity that only in the germ cells, whatever the cause of their differentiation from the other cells of the body, should the chromatin be retained in the complete state in the cases of *Ascaris* and *Miastor*. Whatever may be the relation of the chromatin to differentiation, and whatever may be the degree of its independence of the soma-plasm, it is noteworthy that here it is precisely in the germ cells and in the cells of the germ track—the cells especially important in heredity—that the chromatin shows an unbroken continuity from cell to cell and consequently from generation to generation. Were the chromosome mechanism disturbed in these cells as it is in the somatic cells, or should "diminished" cells regenerate a completely normal organism, a serious obstacle would be in the path of the chromosome interpretation of heredity as now formulated. The actual behavior of the chromatin in the germ track of *Ascaris* argues for rather than against the chromosome theory, at least as regards hereditary transmission.

A much used argument against Weismann's theory of development (ontogenetic differentiation) is found in the phenomenon of regeneration. It is well known that in certain animals and especially in plants a portion of the body consisting solely of differentiated cells may under certain conditions give rise to a complete individual with functional germ cells. Weismann accounted for such regeneration on the basis of an additional hypothesis which stated that during the sorting out of the hereditary units in the process of cell differentiation certain "supplementary determinants" are carried along unaltered, and that later, if occasion arises, these cause the development of the differentiated cells into an organism with all the usual characters. Since in certain cases (*Begonia*) almost any cell of the body may undergo regeneration into a complete plant, it is evident that all of the body cells must have a "complete" germ-plasm. Hence the distinciton between a germ-plasm limited to cells capable of producing an entire individual, and a soma-plasm present only in somatic cells without such power, becomes of no value. Every cell capable of regeneration—germ cell, meristem cell, or differentiated somatic cell—contains the complete germ-plasm, which appears to be simply the chromatin possessed by all the cells alike. Lack of power to regenerate is not due to a lack of complete germ-plasm but to other conditions associated with the degree of differentiation shown by the cells. In thus using the terms germ-plasm and soma-plasm (somato-plasm) synonymously with chromatin and cytoplasm respectively, Weismann's conception of the chromatin as the substance especially important in heredity remains, although his theory of the dependence of ontogenetic differentiation upon a sorting out of qualitatively different units of this substance during development is no longer held.

This use of the term germ-plasm is general among geneticists, who are concerned with the problems of heredity, and may be distinguished from that of certain students of the physiology of development, by whom germ-plasm is regarded as "any protoplasm capable, under the proper conditions, of undergoing regression, rejuvenescence, and reconstitution into a new individual, organism, or part" (Child 1915, p. 462). From this latter point of view the germ-plasm would be regarded as either the complete protoplast capable as acting as so described, or, as Child is inclined to believe, only an abstract idea—merely a term standing for heredity.

Weismann's theory of the sorting out of hereditary units during onto-genesis was abandoned not only because of its inapplicability to the results of certain experiments, but also because no support for it could be found in a direct study of the cell mechanism. Strasburger and other investigators insisted strongly that so far as can be ascertained the division of the chromatin at each somatic mitosis is exactly equational, there being not the slightest indication of such a difference in the chro-matin of the two daughter cells as might be expected were the divisions qualitative (*erbungleich*). To this Weismann had only to reply that since the differentiation is a matter not of ids or of idants but of determi-nants, the two nuclei would be visibly alike in spite of their qualitative difference. Although certain cases have been described in which growth is not equal in all of the chromosomes during the early stages of develop-ment, and although the two daughter nuclei may become differentiated through unequal nutrition after their formation, as Strasburger suggested, most biologists have adopted the view that all of the somatic nuclei are qualitatively alike in their chromatin content so far as its hereditary powers are concerned. They have thus followed de Vries (1889) in holding that factors for all of the hereditary characters are present in all of the somatic cells, a conclusion strongly supported by the facts of regeneration. The ontogenetic differentiation of the cells which mani-fests itself largely in cytoplasmic changes, as well as the relative regen-erative powers which these cells possess, are attributed for the most part to physiological causes, the latter in large measure determining what hereditary capabilities of the various cells shall come to expression. The distinction between the view of Weismann and that of more recent investigators is made clear in the two diagrams of Fig. 159, which have been copied from Conklin (1919–1920).

Notwithstanding the fact that many changes have been made in its details, Weismann's theory of heredity proved to be of much greater value than his theory of development. Morgan (Morgan *et al.* 1915, pp. 223–227) points out that Weismann made three contributions to the study of genetics, which may be stated in three propositions: (1) The germ-plasm

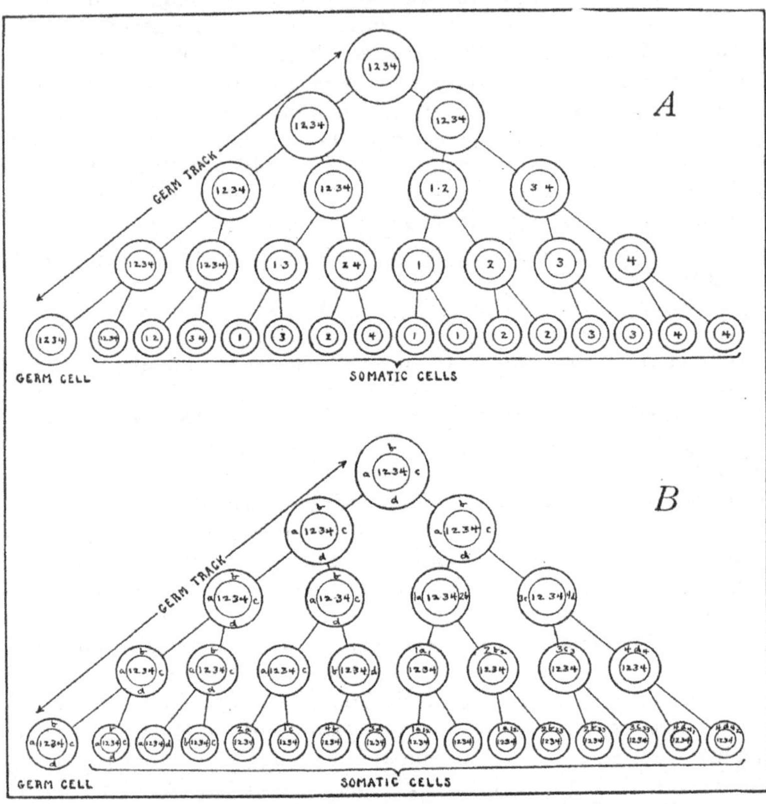

FIG. 159.—The behavior of the hereditary units in ontogenesis according to Weismann (*A*) and the current interpretation (*B*). In *A* the determinants in the nucleus (1, 2' 3, 4) are supposed to be distributed differentially to the various somatic cells. In *B* the genes (1, 2, 3, 4) are distributed equally to every cell, but the cytoplasm is distributed differentially. The same genes working upon different cytoplasms produce different results in various somatic cells. (*Diagrams and legend from Conklin, 1919–1920.*)

contains independent elements which may be substituted one for another without undergoing change; (2) a segregation of maternal and paternal factors, pair by pair, occurs at one period in the history of the germ cells; (3) the behavior of the chromosomes is specifically applicable to the problems of heredity. In these principles are found "the basis of our present attempt to explain heredity in terms of the cell," for upon them is founded the Factorial Hypothesis, now supported by a large mass of experimental evidence.

In our conception of the nature of the heredity units or factors we have departed widely from Weismann. For him each of the ids arranged in a series in the chromosome represented the sum of the characters of a complete organism; the smaller parts were represented by the smaller units (determinants) composing the id, and these units in turn were made up of biophores, which were ultimate and independent living particles. According to our modern hypothesis each of the serially arranged factors or genes exerts an influence on the development of one or more characters, but does not stand for a complete organism as did the id, or for a part of it as did the determinant. Moreover, it is generally regarded as a mass of some complex chemical substance whose activities are due to its definite though imperfectly known physico-chemical properties, rather than to forces exerted by hypothetical vital units.

In justice to Weismann it should be pointed out that the frequently made criticism that his theory was a vitalistic one is warranted only to a limited extent. Although his ultimate hereditary units, the biophores, were regarded as actually living particles, Weismann stated that "they are not composed in their turn of living particles, but only of molecules, whose chemical constitution, combination, and arrangement are such as to give rise to the phenomena of life." He was careful to point out that in spite of the fact that it cannot be proved that no peculiar vitalistic principle exists, we should hold fast to a purely physico-chemical basis of life "until it is shown that it is not sufficient to explain the facts, thus following the fundamental rule that natural science must not assume unknown forces until the known ones are *proved* insufficient . . . We can quite well believe that an organic substance of exactly proportioned composition exists, in which the fundamental phenomena of all life— combustion with simultaneous renewal—must take place under certain conditions by virtue of its composition" (1902, lecture 36).

The manner in which hereditary factors are segregated at gametogenesis has been found to be different from that conjectured by Weismann. As indicated in the chapter on reduction, he supposed it to occur through a transverse division of the chromosome, whereas it is now known that it is accomplished by the disjunction of pairs of entire chromosomes, the separating members of each pair being qualitatively different. The "reduction" predicted by Weismann was found to occur, but not in the manner he supposed. As shown above, his idea of a further qualitative segregation of units of a lower order in the somatic divisions has not been substantiated. Notwithstanding the abandonment of his theory of development and the changes made in his theory of heredity, Weismann's influence on both cytology and genetics was enormous, largely because of his emphasis upon the need for careful studies of the cell mechanism at the critical stages of the life history, and upon the idea that this mechanism is in some way bound up with the phenomena of heredity. "It has been Weismann's great service to place the keystone between the work of the evolutionists and that of the cytologists, and thus bring the cell-theory and the evolution-theory into organic connection" (Wilson 1900, p. 13).

We may further point out, with Morgan (1915), that the factorial hypothesis assumes only three things about the factors: they are constant, they are usually in duplicate in each body cell and immature germ cell, and they usually segregate in the maturing germ cells. The hypothesis, and the Mendelian theory in general, therefore have to do only with heredity: they do not attempt to explain the causes of development. They seek rather to account for the initial resemblances or differences in hereditary potentiality which are observed to exist between the germ cells from which successive generations arise. Between the materials composing the initial factors and the fully expressed characters of the organism "lies the whole world of embryonic development," to which the application of the theories under consideration has not yet been extended in any systematic or satisfactory manner. Nevertheless many investigators, though realizing the failure of Weismann's attempt to explain development in terms of representative particles, are strongly inclined to the view that since the Mendelian characters appearing toward maturity behave as though associated with discrete units in the germ, the course of ontogenetic development in its earlier stages must also be due in large part to the activity of factors carried by the nucleus. Development is thus held to be predetermined or controlled by an internal mechanism: external agencies act only by affecting the operation of this mechanism. The factors control the character and behavior of the cells, and upon these in turn the organism, which is a cell aggregate, is alone dependent for its characters and activities. In place of the early hypothesis on which it was supposed that the development of characters is controlled by the migration of determiners or pangens from the nucleus into the cytoplasm at precisely the right times and places, we now have the theory that the factors in the nucleus probably produce their effects by initiating series of chemical reactions which involve all parts of the cell. As Morgan (1920) states, "Granting that differences may exist in the nuclei of different species, different end products are expected. The evidence that such differences may be related to specific substances in the nucleus is no longer a speculation but rests on the analytical evidence from Mendelian heredity. In what way and at what times the nuclear materials take part in the determination of characters we do not know. The essential point is that we are in no way committed to any interpretation. Stated negatively we might add that there is nothing known at present to preclude the possibility that the influence is a purely chemical process."

Non-factorial Theories.—The above theory of the dependence of the course of development upon the operation of an internal factorial mechanism is essentially an "elementalistic" conception: the attempt is made to explain the organism in terms of its constituent parts, namely, the cells and smaller elements contained by them. As noted in our historical sketch, a number of botanists and zoölogists many years ago called attention to the fact that limits must be set to the conception of the cell as the unit of structure and function; and they have been followed by a school, made up largely of experimental embryologists, which holds that organization is not the result of cell formation, but rather precedes and regulates the latter. From this "organismal" standpoint the organism as a whole, and not one or another of its elementary parts, is regarded as the primary individual. This individual is something more than the cell aggregate pictured by Schleiden and Schwann: it dominates the activity of its constituent members from the beginning of the life cycle onward, and behaves as a unit irrespective of the manner and degree of its subdivision into special centers of action, the cells. The condition present in cœnocytic plants is especially noteworthy in this connection, as are also those cases among animals in which a derangement of the early embryonic cells does not prevent the eventual attainment of the normal form. As urged with much force by Ritter (1919), "the organism in its totality is as essential to an explanation of its elements as its elements are to an explanation of the organism."

·The factorial theory may also be said to represent preformationism in a very modern form. "We are sailing nearer the preformation coast than at any time since the modern study of development began under von Baer" (Conklin 1913). Directly opposed to corpuscular and factorial theories of development are those which seek to explain the course of ontogenesis not by an internal mechanism but rather as the result of the influence of external agencies and the physiological responses shown by protoplasm in the form of cells to such influence: development is held to be truly epigenetic. The control exercised by environmental factors during the organism's early developmental stages, and the effects of various tropisms and tactisms between the component cells upon the type of organization resulting, have been especially emphasized by t). Hertwig, Hartog, Roux, Herbst, Driesch, and others. The most suggestive recent work of this nature in plants is that of Harper (1908, 1918ab) on colonial algæ. In *Hydrodictyon* and *Pediastrum* a number of free-swimming cells come together and build up colonies of very definite forms, and a series of experiments has shown that the position in the colony of any given cell is in no way predetermined. As already pointed out in Chapter XIV, Harper contends that the type of multicellular organization thus built up in successive life cycles is to be explained as the result of physico-chemical interactions between independent cells organized as swarm spores, and not as the product of the activity of a system of spatially arranged factors in a special germ-plasm.

In this connection the name of Driesch (1907–8, 1914) has become particularly prominent, not only because of his great experimental ingenuity, but also because of his decision that the facts of ontogenetic development cannot be accounted for on the basis of any mechanical theory, either now or in the future. As a result he takes the unscientific step of assuming the existence of a non-mechanical, non-spatial, non-psychic, non-energetic "entelechy," which presides over and controls development. Such non-experiential agencies, manufactured for the purpose of solving difficult problems, lead to experimental indeterminism and tend only to obscure the points at issue: they may furnish convenient names for great gaps in our knowledge, but they never give more than pseudo-explanations. Nevertheless, in spite of his tendencies to mysticism, as Harper (1919) remarks, Driesch has shown the impossibility of an exact parallelism in spatial configuration between the germ-plasm and the multicellular organism as a whole: there can be no strict preformation in development. On the other hand, the work of the Mendelians shows clearly that development cannot be completely epigenetic: nothing seems clearer than that development is at least in part dependent upon the orderly operation of an internal organization or mechanism. Wilson (1909, pp. 106 ff.), in discussing the relation of the chromatin to heredity and development, writes as follows:

"But do we really need to employ the pangen symbolism in the consideration of this question? It seems a sufficient basis for our present attack on the problem to assume that the control of the cell-activities is at bottom a chemical one and is effected by soluble substances that may pass from nucleus to protoplasm and from protoplasm to nucleus. Certainly it is to such a view that very many of the chemical and physiological studies in this field are now unmistakably pointing. The opinion is gaining ground that the control of development is fundamentally analogous, perhaps closely similar, to the control of specific forms of physiological action by soluble ferments or enzymes . . . We are thus led to something more than a suspicion that the factors of determination, and therefore of heredity, are at bottom of chemical nature . . . The conclusion thus becomes highly probable that the characteristic differences of metabolism between different species, including those involved in development, are traceable to initial chemical differences in the germ cells. In so far as the chromatin theory expresses the truth, the primary basis of these differences may be sought in the nuclear substance."

A Chemical Theory of Heredity.—Among the theories based on the conception of the idioplasm as a substance with a special chemical constitution, rather than as a system of determinants, may be mentioned that of Adami (1908, 1918). As indicated in Chapter III, Adami attributes the phenomena of life to the activities of a protein-like "biophoric molecule," which is made up of a chain or ring of amino-acid radicles to which side-chains of various kinds may become attached. With regard to individual development it is supposed that "in the ovum there is one common idioplasm of simple type, to which, when distributed in the various cells derived from that ovum, different side-chains become attached, according to the relationships assumed by those cells, so that the cells of different orders are controlled and formed around proto-plasmic or idioplasmic molecules composed of those central rings plus varying series of side-chains" (p. 145). With Driesch it is held that "the structure of the cells in a multicellular organism is a function of their position," since "the position of the cell determines the modification under-gone by its idioplasm." Furthermore, "the greater the change impressed upon the idioplasm of these cells, and the longer that idioplasm is sub-jected to the conditions inducing this change, the more permanently will the daughter cells exhibit the peculiar alteration in the idioplasm, with con-sequent modified structure wherever they find themselves in the economy. We have, in short, to recognize that two orders of forces determine the structure of every cell in the body: (1) the previous influences acting upon its idioplasm and causing it to be of a particular chemical constitution; and (2) the position in which the cell finds itself, and the forces acting momentarily and immediately upon its idioplasm. Or, briefly, these two series of forces are inheritance and environment, and inheritance and environment determine the constitution of the idioplasm and the struc-ture of the cells" (p. 151).

"In terms of this theory, therefore, inheritance essentially depends upon the chemical constitution of the idioplasm or the life-bearing or biophoric protoplasm of the germ cells, not upon the number of the sepa-rate ids or biophores or ancestral plasms or pangens contained in the idio-plasm; and variation, whether slight and individual, or extensive and leading to the production of new species, is ultimately the expression of modification in the constitution of that idioplasm brought about by envi-ronment. Whereas Weismann's theory lays stress upon relative fixity in the constitution of the idioplasm, this theory admits freely the capacity for change in structure of the same. So long as the surrounding condi-tions are unaltered the idioplasm is unchanged; alter these conditions and the idioplasm is liable to variation in constitution" (pp. 152–3).

Adami cites certain calculations of the probable size of inorganic and organic molecules to show that the existence of a system of determinants or other representative particles of the Weismannian type is a physical impossibility. He also points out that since the idioplasm must increase enormously in bulk by the addition of new material and become repeat-edly subdivided as cells and individuals multiply, there can be no actual continuity of the germ-plasm through countless generations: what is eternal is rather a potential continuity of molecular arrangement and constitution, *i.e.*, the physical and chemical properties of the germ-plasm rather than the substance itself.

Conclusion.—In the foregoing pages we have touched upon some of the most important biological problems toward the solution of which cytology must make her further contributions. With regard to individual development it must be determined on the one hand to what extent the course of ontogenesis is dependent upon the operation of an internal cell mechanism and how this mechanism brings about its results, and on the other hand how far it is controlled by external environmental agencies: a way must be found between the "Scylla of preformation and the Charybdis of epigenesis" (Conklin 1913). Furthermore, the manner and the causes of the progressive modification of the hereditary mechanism must be better known in order that evolutionary advance may be accounted for. With respect to both development and heredity the rôles of the two individualities, the cell and the organism as a whole, must be more fully ascertained and correlated.

It is obvious that no adequate solution of any of these problems can be reached until the physico-chemical constitution of protoplasm, especially that of the idioplasm or inheritance material, is more clearly disclosed. Only further research can show whether we shall continue to regard the idioplasm or chromatin as a heterogeneous system of discrete molecules or molecular complexes (factors or genes) with a definite spatial arrangement, as is supposed on our current Mendelian theories, or shall come to look upon it as a single enormously complex chemical substance in which varying side-chains or other portions of the molecule are responsible for the variety of results observed. It is at any rate a striking fact that "in the Mendelian phenomenon we see a synthesis, splitting apart, and recombination of determinative factors that is singularly like that of chemical elements or radicles" (Wilson 1909, p. 108); and nothing appears more clearly evident than the truth of Wilson's assertion that ". . . . in the union of cytology and biochemistry lies our greatest hope of future advance."

Bibliography 14

Heredity; Sex

Adami, J. G.　1918.　Medical Contributions to the Study of Evolution.　London.

Allen, C. E.　1917.　A chromosome difference correlated with sex differences in *Sphærocarpos*.　Science **46**: 466–467.

　1919.　The basis of sex inheritance in *Sphærocarpos*.　Proc. Am. Phil. Soc. **58**: 289–316.　figs. 28.

Altenburg, E.　1916.　Linkage in *Primula sinensis*.　Genetics **1**: 354–366.

Amma, K.　1911.　Ueber Differenzierung der Keimbahnzellen bei den Kopepoden.　Arch. Zellf. **6**: 497–576.　figs. 25.　pls. 27–30.

Babcock, E. B. and Clausen, R. E.　1918.　Genetics in Relation to Agriculture.　New York and London.　(Bibliography.)

BALTZER, F. 1910. Ueber die Beziehung zwischen dem Chromatin und der Entwicklung und Vererbungsrichtung bei Echinodermbastarden. Arch. Zellf. 5: 497–621. pls. 25–29. figs. 19.

1914. Die Bestimmung des Geschlechts nebst einer Analyse des Geschleschsdimorphismus bei *Bonellia*. Mitt. Zool. Sta. Neapel.

BANTA, A. M. 1916. Sex intergrades in a species of Crustacea. Proc. Nat. Acad. Sci. 2: 578–583.

1918. Sex and sex intergrades in Cladocera. Ibid. 4: 373–379.

1919. The extent of the occurrence of sex intergrades in Cladocera. Anat. Record 15: 355–356.

BARTLETT, H. H. 1911. On gynodiœcism in *Plantago lanceolata*. Rhodora 13: 199–206. figs. 3.

1915. The experimental study of genetic relationships. Am. Jour. Bot. 2: 132–155.

BATESON, W. 1913. Mendel's Principles of Heredity. Cambridge and N. Y.

BATESON, W. and PUNNETT, R. C. 1906. Report to Evol. Comm., Roy. Soc.

1911a. On the interrelations of genetic factors. Proc. Roy. Soc. B 84: 3–8.

1911b. On gametic series involving reduplication of certain terms. Jour. Genetics 1: 293–302. pl. 40. 1 fig.

BATESON, W. and SUTTON, IDA. 1919. Double flowers and sex-linkage in *Begonia*. Jour. Genetics 8: 199–207

BAUR, E. 1909. Das Wesen und die Erblichkeitsverhältnisse der "Varietates albomarginatæ hort." von *Pelargonium zonale*. Zeit. Ind. Abst. Vererb. 1: 330–351. figs. 20.

1918. Mutationen von *Antirrhinum majus*. Ibid. 19: 177–193. figs. 10.

VAN BENEDEN, E. 1883. Recherches sur la maturation de l'oeuf, la fécondation et la division cellulaire. Arch. de Biol. 4.

BLACKMAN, V. H. 1911. The nucleus and heredity. New Phytol. 10: 90–99.

BLAKESLEE, A. F. 1906. Differentiation of sex in thallus gametophyte and sporophyte. Bot. Gaz. 42: 161–178. pl. 6. 3 figs.

BOVERI, TH. 1887. Ueber Differenzierung der Zellkerne während der Furchung des Eies von *Ascaris megalocephala*. Anat. Anz. 2: 688–693.

1891. Befruchtung. Ergeb. d. Anat. u. Entw. 1: 386–485. figs. 15

1892. Die Entstehung des Gegensatzes zwischen den Geschlechtszellen und den somatischen Zellen bei *Ascaris megalocephala*. Sitz. Ges. Morph. Phys. München 8.

1895. Ueber die Befruchtungs- und Entwicklungsfähigkeit kernloser Seeigeleier und über die Möglichkeit ihrer Bastardierung. Arch. Entw. 2: 394–443. pls. 24, 25.

1899. Die Entwicklung von *Ascaris megalocephala* mit besonderer Rücksicht auf die Kernverhältnisse. Festschr. f. C. von Kupffer, 383–430. pls. 40–45. figs. 6.

1902. Ueber mehrpolige Mitosen als Mittel zur Analyse des Zellkerns. Verh. Phys.-Med. Ges. Würzburg 35.

1903. Ueber den Einfluss der Samenzelle auf die Larvencharaktere der Echiniden. Arch. Entw. 16: 340–363. pl. 15. figs. 3.

1903. Ueber die Konstitution der chromatischen Kernsubstanz. Vortrag Deutsch. Zool. Ges.

1904a. Ueber die Entwicklung dispermer Ascariseier. Zool. Anz. 27: 406–417.

1904b. Ergebnisse über die Konstitution der chromatischen Kernsubstanz. Jena.

1907. Zellen-Studien. VI. Die Entwicklung dispermer Seeigel-Eier. pp. 292. pls. 10. figs. 73. Jena.

1909a.	Ueber Beziehungen des Chromatins zur Geschlechtsbestimmung. Sitzber. Phys.-Med. Ges. Würzburg 1908-9.

1909b.	Ueber "Geschlechtschromosomen" bei Nematoden. Arch. Zellf. **4**: 132-141. 2 figs.

1909c.	Die Blastomerenkerne von *Ascaris megalocephala* und die Theorie der Chromosomenindividualität. Ibid. **3**: 181-268. pls. 7-11. figs. 7.

1910a.	Die Potenzen der *Ascaris*-Blastomeren bei abgeänderter Furchung. Festschr. f. R. Hertwig III, 131-214. pls. 11-16. figs. 24.

1910b.	Ueber die Teilung centrifugierter Eier von *Ascaris megalocephala*. Arch. Entw. **30**: II: 101-125. 32 figs.

1911.	Ueber das Verhalten der Geschlechtschromosomen bei Hermaphroditismus. Verb. Phys.-Med. Ges. Würzburg **41**.

1914.	Ueber die Charaktere von Echinidenbastardlarven bei verschiedenen Mengenverhältniss mütterlicher und väterlicher Substanzen. Ibid. **43**.

1918 (posthumous).	Zwei Fehlerquellen bei Merogonieversuchen und die Entwicklungsfähigkeit merogonischer und partiellmerogonischer Seeigelbastarde. Arch. Entw. **44**: 417-471. 3 pls.

BOWER, F. O., KERR, J. G. and AGAR, W. E. 1919. Lectures on Sex and Heredity London.

BRIDGES, C. B. 1913. Non-disjunction of the sex-chromosomes of *Drosophila*. Jour. Exp. Zool. **15**: 587-606.

1916.	Non-disjunction as proof of the chromosome theory of heredity. Genetics **1**: 1-52, 107-163.

BRIDGES, C. B. and MORGAN, T. H. 1919. The second group of mutant characters in *Drosophila*. Carnegie Inst. Wash. Publ. 278, Memoir II.

BUFFON, G. L. 1749. Histoire Naturelle: Histoire des Animaux, Chapter 3.

CASTLE, W. E. 1909. A Mendelian view of sex heredity. Science **29**: 395-400.

1916, 1920.	Genetics and Eugenics. Cambridge, Mass.

1921.	A new type of inheritance. Science **53**: 339-342.

CASTLE, W. E., COULTER, J. M., DAVENPORT, C. B., EAST, E. M., and TOWER, W. L. 1912. Heredity and Eugenics. Chicago.

CASTLE, W. E. and WRIGHT, S. 1916. Studies of inheritance in guinea-pigs and rats. Carnegie Publ. 241. pp. 192. pls. 7.

CHILD, C. M. 1915. Senescence and Rejuvenescence. Chicago.

COLLINS, E. J. 1919. Sex segregation in the Bryophyta. Jour. Genetics **8**: 139-146. pl. 6.

COLLINS, G. N. 1911. Inheritance of waxy endosperm in hybrids of Chinese maize. Proc. 4th Internat. Conf. Genetique, Paris.

1912.	The origin of maize. Jour. Wash. Acad. Sci. **2**.

CONKLIN, E. G. 1913. Heredity and Responsibility. Science **37**: 46-54.

1915.	Heredity and Environment. Princeton.

1919-1920.	The mechanism of evolution in the light of heredity and development. Sci. Monthly **9**: 481-505; **10**: 52-62, 170-182, 269-291, 388-403, 498-515. figs. 23.

CORRENS, C. 1906. Die Vererbung der Geschlechtsformen bei den gynodiöcischen Pflanzen. Ber. Deu. Bot. Ges. **24**: 459-474.

1907a.	Die Bestimmung und Vererbung des Geschlechtes nach neuen Versuchen mit höheren Pflanzen. Berlin.

1907b.	Zur Kenntniss der Geschlechtsformen polygamer Blütenpflanzen und ihre Beeinflussbarkeit. Jahrb. Wiss. Bot. **44**: 124-176. figs. 4.

1908.	Die Rolle der männlichen Keimzellen bei den Geschlechtsbestimmung der gynodiöischen Pflanzen. Ber. Deu. Bot. Ges. **26a**: 686-701.

1909a.	Vererbungsversuche mit blass (gelb) grünen und buntblätterigen Sippen bei *Mirabilis jalapa*, *Urtica pilulifera* und *Lunaria annua*. Zeit. Ind. Abst. Vererb. **1**: 291-329. figs. 2.

1909b. Zur Kenntniss der Rolle von Kern und Plasma bei der Vererbung. Ibid. **2**: 331–340.

CORRENS, C. und GOLDSCHMIDT, R. 1913. Die Vererbung und Bestimmung des Geschlechtes. Berlin.

COULTER, J. M. and COULTER, M. C. 1918. Plant Genetics. Chicago.

CRAMPTON, H. E. 1896. Experimental studies on Gasteropod development Arch. Entw. **3**: 1–18. pls. 1–4.

CUÉNOT, L. 1899. Sur la determination du sexe chez les animaux. Bull. Sci. France et Belg. **32**. (For later papers see Morgan, 1919a.)

DARBISHIRE, A. D. 1911. Breeding and the Mendelain Discovery. London.

DARLING, C. A. 1909. Sex in diœcious plants. Bull. Torr. Bot. Club **36**: 177–199. pls. 12–14.

DARWIN, C. 1868. Animals and Plants under Domestication.

DAVEY, A. J. and GIBSON, C. M. 1917. Note on the distribution of sexes in *Myrica Gale*. New Phytol. **16**: 147–151.

DAVIS, B. M. 1909. Cytological studies on *Œnothera*. I. Pollen development in *Œnothera grandiflora*. Ann. Bot. **23**: 551–572. pls. 41, 42.

    1910. Cytological studies in *Œnothera*. II. The reduction divisions of *Œnothera biennis*. Ibid. **24**: 631–652. pls. 52, 53.

    1911. Cytological studies on *Œnothera*. III. A comparison of the reduction divisions of *Œnothera Lamarckiana* and *O. gigas*. Ibid. **25**: 941–974. pls. 71–73.

DELAGE, Y. and GOLDSMITH, M. 1913. The Theories of Evolution. N. Y.

DIGBY, L. 1919. On the archesporial and meiotic mitoses of *Osmunda*. Ann. Bot. **33**: 135–172. pls. 8–12. 1 fig.

DONCASTER, L. 1914a. Chromosomes, Heredity and Sex. Quar. Jour. Micr. Sci. **59**: 487–522. figs. 4.

    1914b. The Determination of Sex. Cambridge.

DOUIN, C. 1909. Nouvelles observations sur *Sphœrocarpos*. Rev. Bryol. **36**: 37–41. 10 figs.

DRIESCH, H. 1892a. Entwicklungsmechanisches. Anat. Anz. **7**: 584–586.

    1892b. Entwicklungsmechanische Studien I II. Zeit. Wiss. Zool. **53**: 160–184. pl. 7. 2 figs.

    1893. Entwicklungsmechanische Studien III–VI. Ibid. **55**: 1–62. pls. 1–3.

    1895. Zur Analysis der Potenzen embryonaler Organzellen. Arch. Entw. **2**: 169–203. pl. 15.

    1907. Die Entwicklungsphysiologie 1905–1908. Ergeh. Anat. u. Entw. **17**: 1–157.

DRIESCH, H. and MORGAN, T. H. 1895. Zur Analysis der ersten Entwicklungsstadien des Ctenophoreneies. I II. Arch. Entw. **2**: 204–224. pls. 16, 17.

DUESBERG, J. 1906. Sur le nombre des chromosomes chez l'homme. Anat. Anz. **28**: 475–479. figs. 3.

EAST, E. M. and HAYES, H. K. 1911. Inheritance in maize. Conn. Agr. Exp. Sta. Bull. **167**: 106–107.

    1915. Further experiments in inheritance in maize. Ibid. **188**.

EAST, E. M. and JONES, D. F. 1919. Inbreeding and Outbreeding. Philadelphia and London. (See especially Chapter 4.)

EMERSON, R. A. 1918. A fifth pair of factors, *Aa*, for aleurone color in maize, and its relation to the *Cc* and *Rr* pairs. Cornell Univ. Agr. Exp. Sta. Memoir 16.

EVANS, H. M. (See Babcock and Clausen.)

FICK, R. 1907. Vererbungsfragen, Reduktions- und Chromosomenhypothesen, Bastardregeln. Ergeb. Anat. u. Entw. **16**: 1–140. (Review.)

FLEMMING, W. 1898. Ueber die Chromosomenzahl beim Menschen. Anat. Anz. **14**: 171–174. 1 fig.

27

GALTON, F. 1875. A theory of heredity. Jour. Anthrop. Inst.

GATES, R. R. 1909a. The behavior of the chromosomes in *Œnothera lata* × *O. gigas*. Bot. Gaz. **48**: 179–199. pls. 12–14.

1909b. The stature and chromosomes of *Œnothera gigas*, De Vries. Arch. Zellf. **3**: 525–552.

1911. Pollen formation in *Œnothera gigas*. Ann. Bot. **25**: 909–940. pls. 67–70.

1913. Tetraploid mutants and chromosome mechanisms. Biol. Centr. **33**: 93–99, 113–150. figs. 7.

GATES, R. R. and THOMAS, N. 1914. A cytological study of *Œnothera* mut. *lata* and *Œ.* mut. *semilata* in relation to mutation. Quar. Jour. Micr. Sci. **59**: 523–571. pls. 35–37. figs. 4.

GEDDES, P. and THOMSON, J. A. 1889. The Evolution of Sex. London and N. Y.

GEERTS, J. M. 1911. Cytologische Untersuchungen einiger Bastarde von *Œnothera gigas*. Ber. Deu. Bot. Ges. **29**: 160–166. 1 pl.

GODLEWSKI, E. 1906. Untersuchungen über die Bastardierung der Echiniden und Crinoidenfamilie. Arch. Entw. **20**: 579–643. pls. 22, 23. figs. 7.

1909. Das Vererbungsproblem im Licht der Entwicklungsmechanik betrachtet. Vortr. aufs Entw. Org. **9**.

GOLDSCHMIDT, R. 1911. Einführung in die Vererbungswissenschaft. Leipzig.

1912. Bemerkungen zur Vererbung des Geschlechtspolymorphismus. Zeit. Ind. Abst. Vererb. **8**: 79–88.

1916a. Experimental intersexuality and the sex problem. Am. Nat. **50**: 705–718.

1916b. A preliminary report on further experiments in inheritance and determination of sex. Proc. Nat. Acad. Sci. **2**: 53–58.

1917a. A further contribution to the theory of sex. Jour. Exp. Zool. **22**: 593–611. 53 figs.

1917b. Crossing over ohne chiasmatypie? Genetics **2**: 82–95.

GOULD, H. N. 1917. Studies on sex in the hermaphrodite mollusk *Crepidula plana*. II. Jour. Exp. Zool. **23**: 225–250.

GRÉGOIRE, V. 1911. Les recherches de Mendel et des mendelistes sur l'hérédité. Rev. des Questions Sci., Oct., 1911, April, 1912.

GREGORY, R. P. 1912. The chromosomes of a giant form of *Primula sinensis*. (Prelim. Note.) Proc. Camb. Phil. Soc. **16**: 560.

1914. On the genetics of tetraploid plants in *Primula sinensis*. Proc. Roy. Soc. London B **87**: 484–492.

GROSS, J. 1912. Heterochromosomen und Geschlechtsbestimmung bei Insekten. Zool. Jahrb., Allg. Teil, **32**: 99–170.

GUTHERZ, L. 1912. Ueber bemerkenswertes Strukturelement (Heterochromosom) in der Spermiogenese des Menchen. Arch. Mikr. Anat. **79**: 79–95. pl. 6. figs. 2.

GUYER, M. F. 1909. The spermatogenesis of the domestic chicken (*Gallus gallus dom.*) Anat. Anz. **34**: 573–580. 2 pls.

1910. Accessory chromosomes in man. Biol. Bull. **19**: 219–234. pl. 1.

1914. A note on the accessory chromosomes of man. Science **39**: 721–722.

1916. Studies on the chromosomes of the common fowl as seen in testes and embryos. Biol. Bull. **31**: 221–268. 7 pls.

HAECKEL, E. 1866. Generelle Morphologie. Jena.

HAECKER, V. 1897. Die Keimbahn von *Cyclops*. Arch. Mikr. Anat. **49**: 35–91. pls. 4, 5. 5 figs.

1902. Ueber das Schicksal der elterlichen und grosselterlichen Kernanteile. Jen. Zeitschr. **37**: 297–400. pls. 17–20. 16 figs.

1907. Die Chromosomen als angenommen Vererbungsträger. Ergeb. u. Fortschr. Zool. **1**.

1912. Allgemeine Vererbungslehre. 2 Aufl.

HANCE, R. T. 1918. Variations in the number of somatic chromosomes in *Œnothera scintillans* de Vries. Genetics **3**: 225–275. 7 pls.

HARMAN, M. T. 1920. Chromosome studies in Tettigidæ. II. Biol. Bull. **38**: 213–230.

HARPER, R. A. 1918a. Organization, reproduction and inheritance in *Pediastrum*. Proc. Am. Phil. Soc. **57**: 375–439. 2 pls. 35 figs.

1918b. The evolution of cell types and contact and pressure responses in *Pediastrum*. Mem. Torr. Bot. Club **17**: 210–240. figs. 27.

1919. The structure of protoplasm. Am. Jour. Bot. **6**: 273–300.

HASPER, M. 1911. Zur Entwicklung der Geschlechtsorgane von *Chironomus*. Zool. Jahrb., Anat. Abt., **31**: 543–612. pls. 28–30. 14 figs.

HEGNER, R. W. 1912. The history of the germ cells in the pædogenetic larvæ of *Miastor*. Science **36**: 124–126. 1 fig.

1914a. Studies on germ cells. I. The history of the germ cells in insects with special reference to the keimbahn determinants. II. The origin and significance of the keimbahn determinants in animals. Jour. Morph. **25**: 375–509. figs. 74.

1914b. The Germ-cell Cycle in Animals. New York.

VON HENKING, H. 1891. Untersuchungen über die ersten Entwicklungsvorgänge in den Eieren der Insekten. Zeit. Wiss. Zool. **51**: 685–736. pls. 35–37.

HERBST, C. 1909. Vererbungsstudien. VI. Arch. Entw. **27**: 266–308. pls. 7–10.

HERTWIG, O. 1875. Beiträge zur Kenntniss der Bildung, Befruchtung, und Theilung des tierischen Eies. I. Jen. Zeitschr. **18**.

1909. Der Kampf um Kernfragen der Entwicklungs- und Vererbungslehre. Jena.

HERTWIG, R. 1906. Ueber Knospung und Geschlechtsentwicklung von *Hydra fusca*. Festschr. f. J. Rosenthal.

1907. Untersuchungen über das Sexualitätsproblem. III. Verb. Deu. Zool. Ges.

1912. Ueber den derzeitigen Stand des Sexualitätsproblem nebst eigenen Untersuchungen. Biol. Centr. **32**: 65–111, 129–146.

HIS, W. 1874. Unsere Körperform und das physiologische Problem ihrer Entstehung. Leipzig.

IKENO, S. 1917. Studies on the hybrids of *Capsicum annuum*. II. On some variegated races. Jour. Genetics **6**: 201–229. pl. 8. figs. 2.

JANSSENS, F. A. 1905. Spermatogénèse dans les Batrachiens. III. Evolution des auxocytes males du *Batracoseps attenuatus*. La Cellule **22**.

1909. La theorie de la chiasmatypie. La Cellule **25**: 389–411. pls. 2.

1919. (a) A propos de la chiasmatypie et la théorie de Morgan. (b) Une formule simple exprimant ce qui se passe en réalité lors de la "chiasmatypie" dans les deux cinèses de maturation. Soc. Biol. Belg. 917–920, 930–934.

JENNINGS, H. S. 1917. Observed changes in hereditary characters in relation to evolution. Jour. Wash. Acad. Sci. **7**: 281–301.

1918. Disproof of a certain type of theory of crossing over between chromosomes. Am. Nat. **52**: 247–261.

KAHLE, W. 1908. Die Pædogenese der Cecidomyiden. Zoologica **21**.

KELLOGG, V. L. 1907. Darwinism Today. New York.

KIHARA, H. 1919a. Ueber cytologische Studien bei einigen Getreidearten. I. Bot. Mag. Tokyo **32**: 17–38. figs. 21.

1919b. II. Ibid. **33**: 95–98.

1921. III. Ibid. **35**: 19–44. pl 1.

KING, H. D. 1907. Food as a factor in the determination of sex in amphibians. Biol. Bull. **13**: 41–56. figs. 2.

1909. Studies on sex determination in amphibians. II. Ibid. **16**: 27–43.

1910. Temperature as a factor in the determination of sex in amphibians. Ibid. **18**: 131–137.

1911. Studies on sex determination in amphibians. IV. Ibid. **20**: 205–236.

1912. Studies on sex determination in amphibians. V. Jour. Exp. Zool. **12**: 319–336.

KUSCHAKEWITSCH, S. 1910. Die Entwicklungsgeschichte der Keimdrüsen von *Rana esculenta*. Festschr. f. R. Hertwig. II.

KUWADA, Y. 1919. Die Chromosomenzahl von *Zea Mays* L. Ein Beitrag zur Hypothese der Individualität der Chromosomen und zur Frage über die Herkunft von *Zea Mays* L. Jour. Coll. Sci. Tokyo **39**: 1–148. pls. 2. figs. 4.

LILLIE, F. R. 1917*a*. Sex determination and sex differentiation in mammals. Proc. Nat. Acad. Sci. **3**: 464–470.

1917*b*. The free martin: a study of the action of sex hormones in the fœtal life of cattle. Jour. Exp. Zool. **23**: 371–452.

1917*c*. Same title. Science **45**: 354.

LINDSTROM, E. W. 1918. Chlorophyll inheritance in maize. Cornell Agr. Exp. Sta. Memoir 13

LOCK, R. H. 1906. Recent Progress in the Study of Variation, Heredity, and. Evolution. London and New York.

1907. On the inheritance of certain invisible characters in peas. Proc. Roy. Soc. London B **79**: 28–34.

1908. The present state of knowledge of heredity in *Pisum*. Ann. Roy. Bot. Garden **4**.

LOCY, W. A. 1915. Biology and its Makers. New York.

LOEB, J. 1918. Further experiments on the sex of parthenogenetic frogs. Proc. Nat. Acad. Sci. **4**: 60-62.

LUTZ, A. M. 1912. Triploid mutants in *Œnothera*. Biol. Centralbl. **32**: 385–435. figs. 7.

1916. *Œnothera* mutants with diminutive chromosomes. Am. Jour. Bot. **3**: 502–526. pl. 24. figs. 7.

1917. Fifteen- and sixteen-chromosome *Œnothera* mutants. Am. Jour. Bot. **4**: 53–111. figs. 9.

MALONE, J. Y. 1918. Spermatogenesis of the dog. Trans. Am. Micr. Soc. **37**: 97–110. pls. 9, 10.

VON MALSEN, H. 1906. Geschlechtsbestimmende Einflüsse und Eibildung des *Dinophilus apatris*. Arch. Mikr. Anat. **69**: 63–99. pl. 2.

MARCHAL, ÉL. et ÉM. 1906. Recherches experimentales sur la sexualité des spores chez les mousses dioiques. Mem. Couronnés, Cl. de Sci., Dec. 15, 1905. pp. 50.

McCLUNG, C. E. 1902*a*. The spermatocyte divisions in the Locustidæ. Sci. Bull. Univ. Kansas **11**.

1902*b*. The accessory chromosome—sex determinant? Biol. Bull. **3**: 43–84.

1905. The chromosome complex of Orthopteran spermatocytes. Biol. Bull. **9**: 304-340. figs. 21

1908. Cytology and Taxonomy. Kansas Univ. Bull. **4**.

1914. A comparative study of the chromosomes in Orthopteran spermatogenesis. Jour. Morph. **25**: 651–749. pls. 1–10.

MENDEL, G. 1865 (1866). Versuche über Pflanzenhybriden. Reprinted in Flora **89**: 364–403. 1901.

METZ, C. W. 1914. Chromosome studies in the Diptera. 1. A preliminary survey of five different types of chromosome groups in the genus *Drosophila*. Jour. Exp. Zool. **17**: 45–56. 1 pl.

1916. Chromosome studies in the Diptera. III. Additional types of chromosome groups in the Drosophilidæ. Am. Nat. **50**: 587–599. pl. 1. 1 fig.

MEVES, F. 1908. Die Chondriosomen als Träger erblicher Anlagen, usw. Arch. Mikr. Anat. **72**: 816–867. pls. 39–42.

MONTGOMERY, T. H. 1901*a*. A study of the germ cells of Metazoa. Trans. Am. Phil. Soc. **20**: 154–236. pls. 4–8.

1901*b*. Further studies on the chromosome groups of the Hemiptera Heteroptera. Proc. Acad. Nat. Sci. Phila. **53**: 261–271. pl. 10.

1905. The spermatogenesis of *Syrbula* and *Lycosa* with general considerations upon chromosome reduction and the heterochromosomes. Proc. Acad. Nat. Sci. Phila. **57**: 162–205. pls. 9, 10.

1906*a*. The terminology of aberrant chromosomes and their behavior in certain Hemiptera. Science **23**.

1906*b*. Chromosomes in the spermatogenesis of the Hemiptera Heteroptera. Trans. Am. Phil. Soc. **21**: 97–174. pls. 9–13.

1910. Are particular chromosomes sex determinants? Biol. Bull. **19**: 1–17.

1911. The spermatogenesis of an Hemipteran, *Euschistus*. Jour. Morph. **22**: 731–718. pls. 1–5.

MORGAN, T. H. 1895. Half-embryos and whole-embryos from one of the first two blastomeres of the frog's egg. Anat. Anz. **10**: 623–628.

1906. The male and female eggs of phylloxerans of the hickories. Biol. Bull. **10**: 201–206. figs. 4.

1908. The production of two kinds of spermatozoa in phylloxerans. Proc. Soc. Exp. Biol. and Med. **5**.

1909*a*. Sex determination and parthenogenesis in phylloxerans and aphids. Science **29**.

1909*b*. A biological and cytological study of sex determination in phylloxerans and aphids. Jour. Exp. Zool. **7**: 239–352. 1 pl. 23 figs.

1909*c*. The effects produced by centrifuging eggs before and during development. Anat. Rec. **3**: 157–161. 9 figs.

1910*a*. The chromosomes in the parthenogenetic and sexual eggs of phylloxerans and aphids. Proc. Soc. Exp. Biol. and Med. **7**.

1910*b*. The method of inheritance of two sex-limited characters in the same animal. Ibid. **8**.

1910*c*. Sex-limited inheritance in *Drosophila*. Science **32**: 120–122.

1910*d*. Chromosomes and heredity. Am. Nat. **44**: 449–496.

1910*e*. Cytological studies on centrifuged eggs. Jour. Exp. Zool. **9**: 593–655. pls. 8.

1911. An attempt to analyse the constitution of the chromosomes on the basis of sex-limited inheritance in *Drosophila*. Ibid. **11**: 365–413. pl. 1.

1912. The elimination of the sex-chromosomes from the male producing eggs of pylloxerans. Ibid. **12**: 479–498. figs. 29.

1913. Heredity and Sex. New York. (Bibliography.)

1914. Sex-limited and sex-linked inheritance. Am. Nat. **48**: 577–583.

1919*a*. The Physical Basis of Heredity. Philadelphia. (Bibliography.)

1919*b*. Several ways in which gynandromorphism in insects may arise. Anat. Record **15**: 357.

1920. Whitman's work on the evolution of the group of pigeons. Science **51**: 73–80.

MORGAN, T. H. and BRIDGES, C. B. 1916. Sex-linked inheritance in *Drosophila*. Carnegie Inst. Wash. Publ. 237.

1919. Contributions to the genetics of *Drosophila melanogaster*. I. The origin of gynandromorphs. Carnegie Inst. Wash. Publ. 278.

MORGAN, T. H., STURTEVANT, A. H., MULLER, H. J., and BRIDGES, C. B. 1915. The Mechanism of Mendelian Heredity. 1st ed. New York.

MULLER, H. J. 1916. The mechanism of crossing-over. Am. Nat. **50**: 193–221, 284–305, 350–366, 421–434. figs. 13.

MULLER, H. J. and ALTENBURG, E. 1919. The rate of change of hereditary factors in *Drosophila*. Proc. Soc. Exp. Biol. and Med. **17**: 10–14.

MULSOW, K. 1912. Der Chromosomenzyklus bei *Ancyracanthus cystidicola* Rud. Arch. Zellf. **9**: 63–72. pls. 5, 6. figs. 5.

NABOURS, R. K. 1919. Parthenogenesis and crossing over in the grouse locust, *Apotettix*. Am. Nat. **53**: 131–142.

NACHTSHEIM, H. 1912. Parthenogenese, Eireifung und Geschlechtsbestimmung bei der Honigbiene. Sitzber. Ges. Morph. Physiol., München.

1913. Cytologische Studien über die Geschlechtsbestimmung bei der Honigbiene (*Apis mellifica*). Arch. Zellf. **11**: 169–241. pls. 7–10.

1919. Zytologische und experimentelle Untersuchungen über die Geschlechtsbestimmung bei *Dinophilus apatris* Korch. Arch. Mikr. Anat. **93**: II, 17–140. pls. 2–5. figs. 5.

NAGAI, I. 1915. On the influence of nutrition upon the development of sexual organs in the fern prothallia. Jour. Coll. Agr. Univ. Tokyo **6**: 121–164. pl. 10. figs. 7.

VON NÄGELI, C. 1884. A Mechanico-physiological Theory of Organic Evolution. (Engl. transl.) (See Wilson 1900, p. 401.)

NAWASCHIN, S. 1910. Näheres über die Bildung der Spermakerne bei *Lilium Martagon*. Ann. Jard. Bot. Buit. **2**: Suppl. 3, 871–904. pls. 33, 34.

NEWMAN, H. H. and PATTERSON, J. T. 1909. A case of normal identical quadruplets in the nine-banded armadillo, and its bearing on the problems of identical twins and sex determination. Biol. Bull. **17**: 181–187. figs. 3.

1910. The development of the nine-banded armadillo from the primitive streak stage to birth; with special reference to the question of specific polyembryony. Jour. Morph. **21**: 359–424. pls. 9. figs. 15.

OSBORN, H. F. 1917. The Origin and Evolution of Life. N. Y. (Chapter V.)

PARMENTER, C. L. 1920. The chromosomes of parthenogenetic frogs. Jour. Gen. Physiol. **2**: 205–206.

PAULMIER, F. C. 1898. Chromatin reduction in the Hemiptera. Anat. Anz. **14**: 514–520. 19 figs.

1899. The spermatogenesis of *Anasa tristis*. Jour. Morph. **15**: Suppl. 223–272. pls. 13, 14.

PAYNE, F. 1909. Some new types of chromosome distribution and their relation to sex. Biol. Bull. **16**: 119–166. pl. 1. figs. 12.

1912. The chromosomes of *Gryllotalpa borealis* Burm. Arch. Zellf. **9**: 141–148. figs. 2.

1916. A study of the germ cells of *Gryllotalpa borealis* and *Gryllotalpa vulgaris*. Jour. Morph. **28**: 287–327. pls. 4. figs. 5.

PEARL, R. and PARSHLEY, H. M. 1914. Data on sex determination in cattle. Biol. Bull. **24**: 205–225.

PFLÜGER, E. 1884. Ueber die Einwirkung der Schwerkraft und anderer Bedingungen auf die Richtung der Zelltheilung. Arch. f. ges. Physiol. **34**: 607–616.

PLOUGH, H. H. 1917. The effect of temperature on crossing over. Jour. Exp. Zool. **24**: 147–210. figs. 9.

PRITCHARD, F. J. 1916. Change of sex in hemp. Jour. Hered. **7**: 325–329.

PUNNETT, R. C. 1913, 1917. Reduplication series in sweet peas. I, II. Jour. Genetics **3**: 77–104; **6**: 185–194.

1919. Mendelism. 5th ed. London.

RIDDLE, O. 1914. The determination of sex and its experimental control. Bull. Am. Acad. Med. **15**.

1916. Sex control and known correlations in pigeons. Am. Nat. **50**: 385–410. See also Jour. Hered. **7**: 158–164.

1917a. The control of the sex ratio. Jour. Wash. Acad. Sci. **7**: 319–356.

1917b. The theory of sex as stated in terms of results of studies on pigeons. Science **46**: 19–24.

RITTER, W. E. 1919. The Unity of the Organism. Boston.

ROBERTSON, W. R. B. 1915. Chromosome studies. III. Inequalities and defi_ ciencies in homologous chromosomes: their bearing upon synapsis and the loss of unit characters. Jour. Morph. **26**: 109–141. pls. 1–3.

1916. Chromosome Studies. I. Jour. Morph. **27**: 179–332. pls. 1–26.

ROSENBERG, O. 1918. Chromosomenzahlen und Chromosomendimensionen in der Gattung *Crepis*. Ark. f. Bot. **15**: 1–16. figs. 6.

1920. Weitere Untersuchungen über die Chromosomenverhältnisse in *Crepis*. Svensk Bot. Tidskr. **14**: 320–326. figs. 5.

ROUX, W. 1888. Beiträge zur Entwicklungsmechanik des Embryo. Arch. Path. Anat. **114**: 113–153, 246–290. pls. 2, 3.

SAFIR, S. R. 1920. Genetic and cytological examination of the phenomena of primary non-disjunction in *Drosophila melanogaster*. Genetics **5**: 459–487. pl. 1.

SAKAMURA, T. 1918. Kurze Mitteilung über die Chromosomenzahlen und die Verwandtschaftsverhältnisse der *Triticum*-Arten. Bot. Mag. Tokyo **32**: No. 379.

SCHACKE, M. A. 1919. A chromosome difference between the sexes of *Sphærocarpos texanus*. Science **44**: 218–219.

SCHAFFNER, J. H. 1919a. Complete reversal of sex in hemp. Science **50**: 311–312.

1919b. The nature of the diœcious condition in *Morus alba* and *Salix amygdaloides*. Ohio Jour. Sci. **19**: 409–416.

SCHLEIF, W. 1912. Geschlechtsbestimmende Ursachen im Tierreich. Ergeb. u. Fortschr. Zool. **3**: 165–328. figs. 22. (Review.)

SCHULTZE, O. 1903. Zur Frage von den Geschlechtsbildenden Ursachen. Arch. Mikr. Anat. **63**: 197–257.

SCHULZ, A. 1913. Die Geschichte der kultivierten Getreide. Neberts Verlag. Halle.

SEILER, J. 1913, 1914. Das Verhalten der Geschlechtschromosomen bei Lepidopteren. Zool. Anz. **41**: 246–251. 4 figs. Arch. Zellf. **13**: 159–269. pls. 5–7.

1919. Researches on the sex-chromosomes of Psychidæ (Lepidoptera). Biol. Bull. **36**: 399–404. pl. 1. fig. 1.

SHULL, A. F. 1910. Studies in the life cycle of *Hydatina senta*. 1. Jour. Exp. Zool. **8**: 311–354. 1 fig.

1915. Inheritance in *Hydatina senta*. II. Ibid. **18**: 145–186.

SHULL, A. F. and LADOFF, S. 1916. Factors affecting male-production in *Hydatina*. Ibid. **21**: 127–162. 1 fig.

SHULL, G. H. 1910. Inheritance of sex in *Lychnis*. Bot. Gaz. **49**: 110–125. figs. 2.

1911. Reversible sex mutants in *Lychnis dioica*. Ibid. **52**: 329–368. figs. 15.

1914. Sex-limited inheritance in *Lychnis dioica* L. Zeitschr. Ind. Abst. Vererb. **12**: 265–302. pls. 1, 2.

DE SINÉTY, R. 1901. Recherches sur la biologie et l'anatomie des Phasmes. La Cellule **19**: 119–278. pls. 5.

SMITH, G. 1912. Studies in the experimental analysis of sex. Pt. 9. Quar. Jour. Micr. Sci. **58**: 159–170. pl. 8.

SPENCER, H. 1864. Principles of Biology.

STEVENS, N. M. 1905. Studies in spermatogenesis with especial reference to the "accessory chromosome." Carnegie Inst. Publ. **36**.

1907.   The chromosomes of *Drosophila ampelophila*.   Proc. 7th Internat. Zool. Congress, Boston.

1908*a*.   The chromosomes in *Diabrotica vittata*, *Diabrotica soror*, and *Diabrotica* 12-*punctata*.   Jour. Exp. Zool. **5**: 453–470.   3 pls.

1908*b*.   A study of the germ cells of certain Diptera, with reference to the heterochromosomes and the phenomena of synapsis.   Ibid. **5**: 359–374.   pls. 4.

1911*a*.   Preliminary note on heterochromosomes in the guinea pig.   Biol. Bull. **20**: 121–122.   figs. 5.

1911*b*.   Heterochromosomes in the guinea pig.   Ibid. **21**: 155–167.   figs. 35.

1911*c*.   Further studies on heterochromosomes in mosquitoes.   Ibid. **20**: 109–120. figs. 38.

1912.   Supernumerary chromosomes and synapsis in *Ceuthophilus* (sp?).   Ibid. **22**: 219–231.   figs. 35.

STOMPS, T. J.   1912.   Mutation in *Œnothera biennis* L.   Biol. Centr. **32**: 521–535. 1 pl.

1912.   Die Entstehung von *Œnothera gigas*.   Ber. Deu. Bot. Ges. **30**: 406–416.

1919.   Gigas-mutation mit und ohne Verdoppelung der Chromosomenzahl.   Zeit. Ind. Abst. Vererb. **21**: 65–90.   pls. 3.   figs. 4.

STOUT, A. B.   1919.   Intersexes in *Plantago lanceolata*.   Bot. Gaz. **68**: 109–133.   pls. 12, 13.

STRASBURGER, E.   1877.   Ueber Befruchtung und Zelltheilung.   Jen. Zeitschr. **11**.

1884.   Neue Untersuchungen über die Befruchtungsvorgang bei den Phanerogamen, als Grundlage für eine Theorie der Zeugung.   Jena.

1900.   Versuche mit diöcischen Pflanzen an Rücksicht auf Geschlechtsverteilung. Biol. Centr. **20**: 657–665, 689–698, 721–731, 753–785.

1908.   Chromosomenzahlen, Plasmastrukturen, Vererbungsträger und Reduktionsteilung.   Jahrb. Wiss. Bot. **45**: 479–568.   pls. 1–3.

1909.   Zeitpunkt der Bestimmung des Geschlechts, Apogamie, Parthenogenesis und Reduktionsteilung.   Hist. Beitr. **7**: 1–224.   3 pls.

1910*a*.   Ueber Geschlechtsbestimmende Ursachen.   Jahrb. Wiss. Bot. **48**: 427–520. pls. 9, 10.

1910*b*.   Chromosomenzahl.   Flora **100**: 398–446.   pl. 6.

STRONG, R. M.   1912.   Results of hybridizing ring-doves, including sex-linked inheritance.   Biol. Bull. **23**: 293–320.

STURTEVANT, A. H.   1913.   The linear arrangement of six sex-linked factors in *Drosophila*, as shown by their mode of association.   Jour. Exp. Zool. **14**: 43–59.

1915*a*.   The behavior of the chromosomes as studied through linkage. Zeitschr. Ind. Abst. Vererb. **13**: 234–266.

1915*b*.   No crossing over in the female of the silkworm moth.   Am. Nat. **49**: 42–44.

1917.   Crossing over without chiasmatype?   Genetics **2**: 301–304.

1920.   Intersexes in *Drosophila simulans*.   Science **51**: 325–327.

STURTEVANT, A. H., BRIDGES, C. B., and MORGAN, T. H.   1919.   The spatial relation of genes.   Proc. Nat. Acad. Sci. **5**: 168–173.

SUTTON, W. S.   1902*a*.   On the morphology of the chromosome group in *Brachystola magna*.   Biol. Bull. **4**: 24–39.   figs. 11.

1902*b*.   The chromosomes in heredity.   Biol. Bull. **4**: 231–251.

SWINGLE, W. W.   1917.   The accessory chromosome in a frog possessing marked hermaphroditic tendencies.   Biol. Bull. **33**: 70–90.   pls. 1–6.

TÄCKHOLM, G.   1920.   On the cytology of the genus *Rosa*.   A preliminary note. Svensk. Bot. Tidskr. **14**: 300–311.   figs. 3.

THOMPSON, W. P.   1912.   Artificial production of aleurone grains.   Bot. Gaz. **54**: 336–338.

THOMSON, J. A.   1899.   The Science of Life.   London.

1913.   Heredity.   2d ed.   London and New York.

TISCHLER, G. 1918. Untersuchungen über den Reisenwuchs von *Phragmites communis* var. *pseudodonax*. Ber. Deu. Bot. Ges. **36**: 549–558. pl. 17.

1920. Ueber die sogenanannten "Erbsubstanzen" und ihre Lokalisation in der Pflanzenzelle. Biol. Zentralbl. **40**: 15–28.

TROW, A. H. 1913. Forms of reduplication—primary and secondary. Jour. Genetics **2**: 313–324. figs. 6.

1916. A criticism of the hypothesis of linkage and crossing over. Jour. Genetics **5**: 281–297.

TUPPER, W. W. and BARTLETT, H. H. 1916. A comparison of the wood structure of *Œnothera stenomeres* and its tetraploid mutation *gigas*. Science **43**: 292.

DE VRIES, H. 1889. Intracelluläre Pangenesis. Jena. (For other papers of de Vries see Morgan 1919a).

WALKER, C. E. 1911. On variations in chromosomes. Arch. Zellf. **6**: 491–496. 1 fig.

WALTER, H. E. 1914. Genetics. New York.

WALTON, A. C. 1918. The oögenesis and early embryology of *Ascaris canis* Werner. Jour. Morph. **30**: 527–604. pls. 9. 1 fig.

WEINSTEIN, A. 1918. Coincidence of crossing over in *Drosophila melanogaster* (*ampelophila*). Genetics **3**: 135–173.

WEISMANN, A. 1885. Die Kontinuität des Keimplasmas als Grundlage einer Theorie der Vererbung. Jena.

1891–1892. Essays upon Heredity. Oxford.

1892. Das Keimplasma. (Engl. transl., "The Germ Plasm," 1893.)

1902. The Evolution Theory.

1913. Vorträge über Deszendenztheorie. Jena.

WELSFORD, E. J. 1914. The genesis of the male nuclei in *Lilium*. Ann. Bot. **28**: 265–270. pls. 16, 17.

WENRICH, D. H. 1916. The spermatogenesis of *Phrynotettix magnus* with special reference to synapsis and the individuality of the chromosomes. Bull. Mus. Comp. Zool. Harvard Coll. **60**: 55–136. 10 pls.

1917. Synapsis and chromosome organization in *Chorthippus* (*Stenobothrus*) *curtipennis* and *Trimerotropis suffusa* (Orthoptera). Jour. Morph. **29**: 471–516. pls. 1–3.

WESTER, P. J. 1914. The determination of sex. Jour. Heredity **5**: 207–208. (Account of work of CIESIELSKY on *Cannabis* and *Spinacia*.)

WHITE, O. E. 1917a. Inheritance studies in *Pisum*. II. The present state of knowledge of heredity and variation in peas. Proc. Am. Phil. Soc. **56**.

1917b. Inheritance studies in *Pisum*. IV. Interrelation of the genetic factors of *Pisum*. Jour. Agr. Res. **11**: 167–190.

WHITNEY, D. D. 1914. The influence of food in controlling sex in *Hydatina senta*. Jour. Exp. Zool. **17**: 545–558.

1916. The control of sex by food in five species of rotifers. Ibid. **20**: 263–296. figs. 6.

1917. The relative influence of food and oxygen in controlling sex in rotifers. Ibid. **24**: 101–138. figs. 4. 4 diagrams.

1919. The ineffectiveness of oxygen as a factor in causing male production in *Hydatina senta*. Ibid. **28**: 469–492.

WIEMAN, H. L. 1917. The chromosomes of human spermatocytes. Am. Jour. Anat. **21**: 1–22. pls. 4.

WILSON, E. B. 1892. The cell-lineage of *Nereis*. Jour. Morph. **6**: 361–480. pls. 13–20.

1900. The Cell in Development and Inheritance. 2d ed.

1905a. Studies on chromosomes. I. The behavior of the idiochromosomes in Hemiptera. Jour. Exp. Zool. **2**: 371–406. figs. 7.

1905*b*. Studies on chromosomes. II. The paired microchromosomes, idiochromosomes, and heterotypic chromosomes in Hemiptera. Ibid. **2**: 507–546.

1905*c*. The chromosomes in relation to the determination of sex. Science **20**: 564.

1906. Studies on chromosomes. III. The sexual differences of the chromosome groups in Hemiptera, with some considerations on the determination and heredity of sex. Jour. Exp. Zool. **3**: 1–40. figs. 6.

1909*a*. Studies on chromosomes. IV. The "accessory' chromosome in *Syromastes* and *Pyrrochoris* with a comparative review of the types of sexual difference of the chromosome groups. Ibid. **6**. 69–100. pls. 2. figs 2.

1909*b*. Studies on chromosomes. V. The chromosomes of *Metapodius*. A contribution to the hypothesis of the genetic continuity of chromosomes. Ibid. **6**: 147–206. 1 pl. figs. 13.

1909*c*. The cell in relation to heredity and evolution. In "Fifty Years of Darwinism." New York.

1910. The chromosomes in relation to the determination of sex. Science Progress **16**: 570–592. figs. 3.

1911*a*. The sex chromosomes. Arch. Mikr. Anat. **77**: II 249–271. 5 figs..

1911*b*. Studies on chromosomes. VII. A review of the chromosomes of *Nezara*, etc. Jour. Morph. **22**: 71–110. 1 pl.

1912*a*. Some aspects of cytology in relation to the study of genetics. Am. Nat. **46**: 57–67.

1912*b*. Studies on chromosomes. VIII. Observations on the maturation-phenomena in certain Hemiptera and other forms, with considerations on synapsis and reduction. Jour. Exp. Zool. **13**: 345–449. pls. 9.

Wilson, E. B. and Morgan, T. H. 1920. Chiasmatype and crossing-over. Am. Nat. **54**: 193–219. figs. 8.

von Winiwarter, H. 1912. Études sur la spermatogénèse humaine. Arch. de Biol. **27**: 91–189. pls. 6, 7.

Wodsedalek, J. E. 1920. Studies on the cells of cattle with special reference to spermatogenesis, oögonia, and sex-determination. Biol. Bull. **38**: 290–316. pls. 5.

Wuist, E. D. 1913. Sex and development of the gametophyte of *Onoclea struthiopteris*. Physiol. Res., Prelim. Abstr. **1**: 93–94.

Yampolsky, C. 1919. Inheritance of sex in *Mercurialis annua*. Am. Jour. Bot. **6**: 410–442. pls. 37–40.

1920*a*. The occurrence and inheritance of sex intergradation in plants. Ibid. **7**: 21–38.

1920*b*. Sex intergradation in the flowers of *Mercurialis annua*. Ibid. **7**: 95–100. pl. 5.

Zacharias, E. 1913. Die Chromatin-Diminution in den Furchungszellen von *Ascaris megalocephala*. Anat. Anz. **43**: 33–53. figs. 2.

Zade, A. 1914. Serologische Studien an Leguminosen und Gramineen. Zeitschr. f. Pflanzenzüch. **2**: 101–151.

1918. Der Hafer. Eine Monographie auf wissenschaftlicher und praktischer Grundlage. pp. 355. figs. 32. Jena.

Zoja, R. 1895. Sullo sviluppo dei blastomeri isolati uova di alcune meduse (e di altri organismi). Arch. Entw. **1**: 578–595. pls. 21–23; **2**: 1–37, pls. 1–4.

INDEX

Bold-face numbers indicate pages bearing illustrations

Gromia (protozoan), **45**
Ground substance, 33
Growth period in oöcyte, 158, 221, 233, **234**, 274, **275**
Gruber, 61, 69, 70
Gryllotalpa (mole cricket), **118**, 119
Guignard, blepharoplast, 90, 94
    eyespot, 111
    fertilization, 14, 298, 300, 325
    reduction, 220
Guilliermond, bacteria, 67
    centrosome, 80
    chondriosomes, 115–122
    reduction, 224
    yeasts, 211, 292
Guinea pigs, **274, 341,** 342
Gurwitsch, 213
Gutherz, 361
Gutta percha, 135
Guyer, 285, 361, 363
Gymnogramme (fern), 90
Gyrodactylus (flatworm), 160

H

Haberlandt, 46, 61, 103, 104, 105, 193
Haeckel, 16, 48, 324, 402
Haecker, fertilization, 279
    hereditary substance, 210
    individuality of chromosome, 157, 159
    Keimbahn-plasma, 405
    nucleolus, 66
    pseudoamitosis, 211, 213
    reduction, 227, 244, 245
    sex-determination, 357
Hæmatochrom, 112
Haller, 6
Hamm, 14
Hammarsten, 40
Hance, 165, 346
Hannig, 196
Hanstein, 12, 24, 33, 46, 133
Hardy, 62
Hargitt, 213
Harman, 160, 213, 391
Harper, centrosomes in ascus, 80, **81**
    chondriosomes, 123
    cleavage furrow, 187
    development in colonial algæ, 334, 411
    evolution of cell structure, 207
    fertilization in ascomycetes, 290

Harper, inheritance in colonial algæ, 331, 334
    mitosis in myxomycetes, 208
    plastids, 103, 105, 107, 115
    polarity, 138, 140
    protoplasm, 38, 49, 51
    reduction in ascomycetes, 223
Hartmann and Nöller, 84
Hartog, 104, 185
Harvey, E. E., 162
Harvey, E. N., 284
Harvey, Wm., 4
Hasper, 405
Hatschek, 36, 49, 138
Hautschicht, 42
Heartwood, 194
Hegler, 108, 202, 203
Hegner, 62, 404, 405
Heidenhain, 64, 70, 138, 144, 183
Heilbrunn, 189, 286, 287
Helix (snail), 96, 256
Helleborus (angiosperm), 159
Helvella (fungus), 136, 223, 291, 313
Hemiptera, 239, 255
Hemp, 372
Henking, 227, 244, 251, 256, 358
Henle, 10
Henneguy, 70, 95, 96
Herbst, 213, 325, 326
Heredity, cell organs in, 323 *ff.*
    chemical theory of, 412
    in colonial algæ, 331, 334
    early theories of, 398 *ff.*
    Mendelian, 336 *ff.*
    non-factorial theories of, 411
    Weismann's theory of, 401
Herla, 157, 164
Herlant, 285, 286
Hermann, 144, 180, 183
Herrick, 274
Hersilia (copepod), 237
Hertwig, O., basal granules, 96
    cell-division, 143
    fertilization, 15, 273, **276**
    nucleus in heredity, 16, 324
    spindle, 180
Hertwig, R., amitosis, 211, 213
    chromidia, 133
    division in protozoa, **207**
    heterotypic prophase an abortive mitosis, 259
    nucleoplasmic ratio, 63, 70
    sex-determination, 368, 370

T

Made in the USA
Monee, IL
07 July 2026